Encyclopedia of Quality Assurance and Management

Encyclopedia of Quality Assurance and Management

Edited by **Stacy Escobar**

Clanrye International

New Jersey

Published by Clanrye International,
55 Van Reypen Street,
Jersey City, NJ 07306, USA
www.clanryeinternational.com

Encyclopedia of Quality Assurance and Management
Edited by Stacy Escobar

International Standard Book Number: 978-1-63240-198-4 (Hardback)

Printed in China.

Contents

Preface

It is often said that books are a boon to mankind. They document every progress and pass on the knowledge from one generation to the other. They play a crucial role in our lives. Thus I was both excited and nervous while editing this book. I was pleased by the thought of being able to make a mark but I was also nervous to do it right because the future of students depends upon it. Hence, I took a few months to research further into the discipline, revise my knowledge and also explore some more aspects. Post this process, I begun with the editing of this book.

This book encompasses encyclopedic information regarding quality, its assurance as well as management. The objective of this book is to discuss novel concepts, state-of-the-art technology and developments in researches related to quality. New ideas and latest advancements in the area of quality assurance and related issues are showcased in this book, which is organized according to the application fields. The book discusses the fundamental ideas and historical perceptions regarding quality, and also, quality assurance procedures in education, healthcare, medicine, software advances, service sector and other technical fields. This book primarily targets students, researchers, quality engineers, production and process managers, and professionals who are interested in quality assurance and related fields. The book is a significant addition to the literature in the area of quality assurance and management.

I thank my publisher with all my heart for considering me worthy of this unparalleled opportunity and for showing unwavering faith in my skills. I would also like to thank the editorial team who worked closely with me at every step and contributed immensely towards the successful completion of this book. Last but not the least, I wish to thank my friends and colleagues for their support.

Editor

1

The Use of Quality Function Deployment in the Implementation of the Quality Management System

Elena Condrea, Anca Cristina Stanciu
and Kamer Ainur Aivaz
"Ovidius" University of Constanta
Romania

1. Introduction

Nowadays a strong accent focus on quality, this period of time being considered like one of the "quality years".

Firms are forced to reduce losses from the sales more and lower and also to extend their dispatch markets concomitantly with new customers gain. The competition more and more intense brought in the first plan the idea that quality is not something added, not something good to be owned but a condition to survive.

Being a complex item, quality couldn't be directly and easily measured and expressed, except for some technical characteristics and the majority of the economic one.

Quality depends also on corresponding materials, the equipments' performances, the technology, the technical control accuracy, the employee's suitable qualification etc. On the market, at a certain moment, several products could be found although they accomplish the same functions and have quite different performances and prices. To each set of performances correspond certain advantages, which could be quantified, but for each user, the advantages curve is different, in respect of the destination of the products.

Things look quite different if we try a quantitative expression of the quality or of a research project, an innovation initiative or of a business proposal. We could not talk about a quality "standard", or in any case, about a "universal" one. That's why it exists a need to find a very general method, with universal appliance, to permit each time the use some specific instruments.

As well, the conditions in which the products quality is determined are extremely complex, due to both objective and subjective elements.

Starting from here, B. Boehm, J.A. McCall, P. Richard and G. Walters structured a number of principles to allow a quantitative and objective measurement, in conformity with the scheme:

Subject quality

↓

Characteristics (externals)

↓

Elements (internals)

↓

Measurements (quantitative of the internal elements)

The main idea at the foundation of their studies is that the measure to express quality must result in an amount of numerous measurements, each having in view a certain characteristic. In conformity with the scheme, each product presents several characteristics to be appreciated. Each characteristic depends on an amount of internal elements of the subject to be measured, elements which could be quantitatively expressed [values of elements are to be found into numerical values of the characteristics and these, in their turn, will conduct to a quantitative indicator of the quality].

Meanwhile, with the reconsideration of the quality notion there appear also methods to allow finding solutions to better satisfy a certain buyer's segment. Such a method, known in the literature as *QFD* (*Quality Function Deployment* - extension of the quality function) or under the more familiar name of *House of Quality*, name given by the shape of one of the diagrams used which seemed to be a house, like having been designed by a child.

2. Quality function deployment. Theoretical aspects

Named also the "voice of customer", QFD is a systematic method to develop products/ services based on expectations and desires of customers, the position on the market of these products and services and their efficiency.

The basic principle of the method is represented by the customer's requirements in each step of their trajectory.

Specific for this method is the fact that all the development and renewal of products and services activities are perceived from the customer's perspective.

QFD is a team method, being applied by a team of 6-8 people, who must be involved in all the firm's departments.

QFD represents in fact a planning process, made to help the design, production and marketing of some products and services taking into account the customer's opinion.

The method issued in 1966, initiated by the Japanese Yoji Akao, being used in fact for the first time in 1972 by Mitsubishi and starting with the 80s, the method gained a large applicability both in USA and in Europe (1988), in order to design the development of different products, processes or projects.

Yoji Ajao defines QFD as being *a method which transforms the consumers requirements in quality characteristics and designs the quality of the finished product/service, through the systematic development of the relationships between demand and characteristics, starting with the products/services functions, followed by its characteristics, its components characteristics and ending with the stages and characteristics of the processes from which it results.*

The objectives of the method could be structured as follows:

- Structuration of the design process;
- Reduction of the design cycles;
- Transformation of the customer "voice" in technical requirements and quality plans;
- Increasement of the quality level of the final products.

The advantages of using QFD are to be:

- The customer needs are restored more accurate into the specifications of the products/process design;
- Shorter design and development cycles;
- Lower costs, high productivity;
- Documentary orientation;
- The team involvement;
- Experience and information are structured into a concise format, easy to be assimilated.

The method allows the elaboration of a project concerning the clients' requirements. First of all it must take place an inquest to establish which are the functions of the products expected by the customers and also their importance. Then the characteristics are settled and after that they are correlated with functions. In the meantime, some comparisons with other firms' performances are made and also the characteristics are analyzed in their relationship, in order to observe which ones are correlated and which ones are opposite.

Then, similarly, from the characteristics, the methodology follows with technical measurable performances and then with materials technology.

2.1 Short history of QFD development and application

QFD method was developed in Japan at the end of the 60s by the professors Shigeru Miyuno and Yoji Akao. At that moment, the statistic control of the quality, introduced after the 2nd world wide war, had already roots in Japan. New quality methods were introduced, with the contribution of the quality control involvement in the business management, process known afterwards as TQC or TQM.

Professors Mizuno and Akao intended to develop a method to guarantee the quality, a way to make a product suitable for clients, before its issue on the market. Till that moment, the quality control methods were focused on the settlement of difficulties issued during the production or after.

The first important scale application was presented in 1966 by Kiyotaka Oshiumi from Bridgestone Tire in Japan. He used a "fish bone" diagram to discover the customer's expectations (outputs), and those characteristics and factors of the process (causes) which influenced the respective result.

This method was first used for the growth of the Naval Shipyards' performance, belonging to Mitsubishi. Initially, it was used to improve the quality of the company's products; in time, however, they realized that, using the same analysis technique, the method could be used for the improvement of the quality belonging to every activity within an economic unit which produces or carries out services.

In time, certain famous Japanese companies had confirmed the method's efficiency, Toyota being among them, who introduced it in 1977 and which, in a 7 years application period, lowered the fabrication costs of an automobile by 40%, while significantly raising the quality and lowering the fabrication cycle.

In 1986, Ford and Xerox, in the United States, adopted the method.

QFD means a *mot-a-mot* translation of the Japanese words "*hinshitsu kino tenkai*", but first was translated like evolution of the quality function; name suggested by dr. L.T. Fan in 1978.

At the first workshop (seminar) about QFD in USA, the sponsor Masaaki Imai seemed that "*evolution*" doesn't reflect the sense of "*change*" and consequently, "*hinshitsu tenkai*" could be better translated like "*the quality development*". In that manner appears the name of QFD (*Development of the Quality Function*).

At the foundation of QFD method there is the *House of Quality*, a set of matrix used to link the voice of customer with the technical needs of a product, the control plans of the process and the production operations.

In the scheme (fig.1) we can observe the structure of the *House of Quality* and explanation of each component:

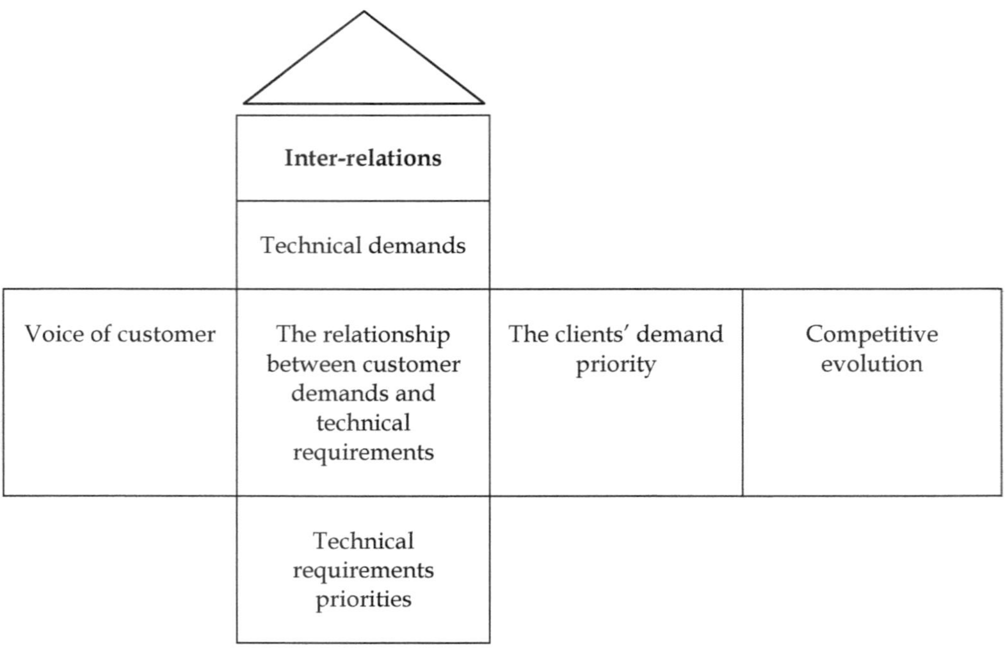

Fig. 1. The structure of the House of Quality

2.2 QFD methodology

The central point of the diagram looks like a table with two entrances. On the rows there are the customers requirements and on the columns is underlined the correspondance between the customer's expectations and the quality characteristics of the respective product or service. This matrix or table is named *the matrix of relations*.

The QFD is also called the "House of Quality" because the solving solutions of the discussed problem are found in a series of matrix arranged as a house consisting of foundation, first floor, attic and roof, as you can see in fig. 2.

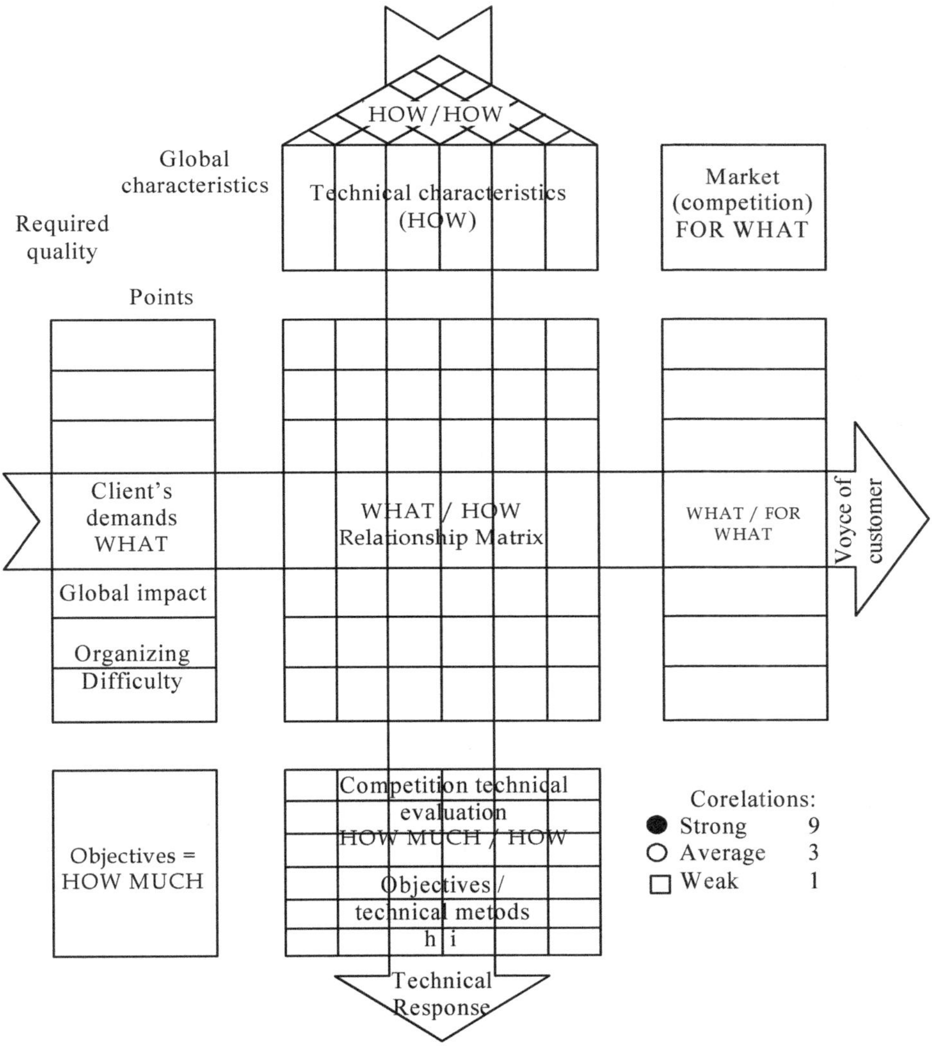

Fig. 2. QFD example

Thus:

The first floor is build out of "the client's voice" and here we find:

- The problem's/client's demands;
- The correlation matrix of the problem's demands with the characteristics required by their solving;
- The matrix which represents the firm's position in comparison with the main competitors, for each demand of the analyzed problem;

The house's foundation consists of:

- The technical evaluation matrix of the main competitors;
- The objectives and technical measures needed in order to solve the analyzed problem;

The house's attic consists of the characteristics and/or methods needed to solve the problem's demands.

The house's roof represents the correlation matrix of the technical characteristics.

In addition to these building elements of the *House of Quality*, the analysis also requires to be made the correlation between the elements that form the *House of Quality*.

Generally, there are three types of correlations:

- Strong;
- Average;
- Weak.

It comes by itself the fact that there is the possibility of not having a correlation between some elements of a matrix.

Each correlation category will be given a certain score, as well. A codified representation of each correlation category is recommended, in the purpose of x having a more suggestive matrix representation.

The steps of developing a QFD analysis are as follows:

1. The selection and preparation of the problem;
2. Establishing the demands (WHAT), which means:
    - Identifying the problem's demands or the client's wishes (the client's voice);
    - Their importance (giving points, percentages or hierarchizing the demands).
3. The perception and judging of the competition (FOR WHAT), which presumes the following:
    - Gathering information about the market;
    - Commercial information;
    - Comparing with the main competitors , focusing on each of the clients' demands (WHAT/FOR WHAT);
4. Establishing the technical characteristics and/or methods that could compete to solve the analyzed problem (HOW);
5. Completion of the correlations in the WHAT/HOW and HOW/HOW matrix;
6. Technical competition evaluation, which means:

   - Comparing each technical characteristic with the competition (HOW MUCH/HOW);
   - Calculating the importance of the technical characteristics and their hierarchizing;
   - Filling out the "Global impact" and "Organizing difficulty" rows.
7. Exploitation, binocular vision, which means analyzing the WHAT/HOW matrix's coherence;
8. Carrying out the quality improvement process, which presumes:
   - The process of improving the product's/service's quality;
   - Extending the analysis.

The QFD analysis can be pictured as a branching graph, meaning that from each of the resulting problems' diagrams, new QFD analyses can be made.

With the QFD method, regarding the quality control and assurance, there can be put into practice other methods, techniques or instruments taken from the quality theory or management, such as:

- National, regional, international standards;
- Market studies;
- The statistical control of the products/services;
- The Pareto Diagram, difference cause-effect Diagrams, Brainstorming Method etc.;
- The Quality Circles etc.

QFD does not limit itself only to technological problems; it can also be applied to aspects regarding the reliability and costs, allowing for the definition of prioritary action directions (opportunities of technological nature or a difference nature, risks' analysis etc.).

It can be said that, an important advantage of the QFD method is given by the *possibility of identifying the clients' latent demands* (demands which have not manifested themselves, yet).

In the application of the method some steps are to be followed:

1. Determination of the customer's requirements and the importance for them.
2. Identification of the quality characteristics by the work team. The degree of requirements coverment through characteristics is evidenced by a score system (graph).
3. Determination of the quality characteristics which are to be performed for the new product and evaluation of the difficulty graph to be obtained. Concomitantly is processed also the sense of variation preferred for this values (increasement, decreasement, indifference).
   The evaluation of interaction and correlation between characteristics is emphasised in the correlation matrix (the superior zone of the diagram which forms the roof of the house).
4. The compared analysis of product/service planned with the product/service of competitors from 2 points of view:
   a. From the clients point of view;
   b. From the technical point of view.
   Comparing the quality characteristics of products/services with those of the competitors.

5. Establishment of the final values of quality characteristics for the new product.

The building of a *House of Quality* requires 6 basic steps:

1. Identification of the customer needs

The voice of customer remains at the base of the QFD process. Here below there are some essential approaches regarding gathering information from the clients:

c. Official polls
d. Focus groups
e. Direct contacts with clients
f. Claims analysis
g. Online monitorization.

2. Identification of technical needs

Technical needs are characteristics that describe the customer needs in the designer language. There must be measurable because the result is controlled (checked) and compared with the target objectives. The roof of the house shows the relationship being converted, using a series of symbols. A typical scheme uses the "●" symbol for very strong relationships and a "☐ " for weak relationships. For example, two technical requirements of certain superior services are the capacity, staff and equipment of a clinic. The relationship between them is strong and, in order to increase the capacity, more staff and equipment are needed.

3. The link between the customer needs and the technical needs

The customer needs must be written in the left column and the technical ones on the top. Inside the matrix the symbols indicate the type or relation in a similar way with those used at the roof of the House of Quality. The purpose of this matrix is to show if the final technical needs cover the customer needs. This kind of evaluation is usually based on the experts' experience, customer's reaction or controlled experiments.

The lack of a solid link between the customer needs and the technical ones show both that the needs are not covered and that the final product will hardly accomplish them.

And if a technical need doesn't affect a customer need it could be useless, or the designers might forget an important need of the customer.

4. Addition of the competitive evaluation and of the sale key points

In this stage the importance of every customer need is evaluated and the competitor's products and services are the ones which cover these needs and are also researched. These evaluations are very important and reflect the customer's expectations. Competitive evaluation underlines the strengths and weaknesses of the competition. Due to this stage, designers could discover methods to improve products and QFD method and the strategic vision of the company shows the priorities of the important customer are not satisfied by the competitor's products (such as family activities), thus, the company could obtain advantages by focusing on these aspects. The respective needs become sale key points and lie at the foundation of the marketing strategies.

5. Evaluation of technical needs of the competitive products and services and establishment of targets

This stage happens usually on the base of information added or products tested. These evaluations are compared with the competitive evaluation of the customer's needs in order to determine the disparities between the customer's needs and the technical ones.

If it is proved that a competitor product satisfies the customer's needs but the evaluation of the technical needs shows something else, then even the measurement was wrong, even if there is a difference of image (be it positive for competitor, or negative for the company product) which affects the consumer perception. For example, customers say they give a great importance to family and in the meantime, the competitive evaluation shows that these aspects are not accomplished. The establishment of a target regarding this need will satisfy the consumer need and will offer an advantage against the competitor's products.

6. Selection of the technical needs to be modified in the process

In this stage there are identified the technical requirements which have a strong link with the customer needs, that are considered sale key points. In the rest of the process, the customer voice will be taken into account. Features not being considered critical don't need a greater attention, for example, the key factors in a fitness Centre are: the program, the equipment, the fee and the access to the Internet.

The six stages are just the beginning of the QFD process. There are used three houses of quality to develop the main parts of the customer needs, the process plan and the quality control.

The second house is very similar with the first, but it refers to the subsystem and components.

The technical needs of the first house of quality are described in detail (fig.3.).

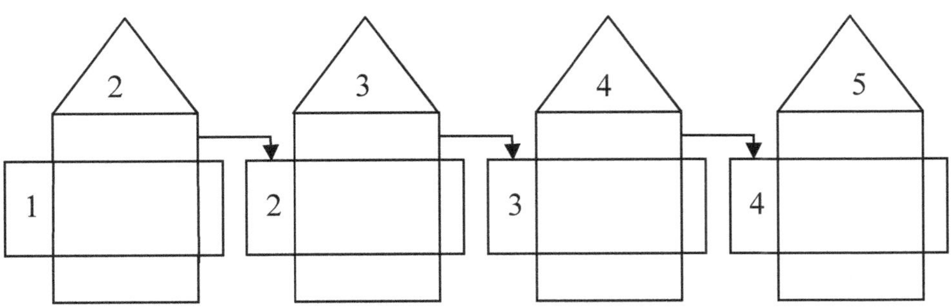

Fig. 3. The four quality houses – fill them in

Legend:
1. Client's needs;
2. Technical needs;
3. Components' characteristics;
4. Process operations;
5. Quality control plan.

At this moment, the target values, function and aspect are to be settled. For example, the program of a fitness Centre could be partaged into: program for children, for family, etc. each with their specific needs and therefore, each with its own house of quality.

In the field of production, the majority of the QFD activities are represented by the first two houses which are displayed by developing the product and the engineering function. The next stage refers to needs surveyors and line operators.

In the third house, the process plan makes the link between the characteristics of the components and the key operations. That makes possible the passage from the plan to the application.

In some cases, there are used more simple houses of quality, which exclude the competitive analysis. For example in the health national organizations, competition doesn't interest anybody.

3. Optimizing the activity of S.C. ELDA MEC S.R.L. through the application of the Quality Function Deployment method

QFD was applied within ELDA MEC SRL Constanta in order to establish a product, which optimally corresponds with the dairy products consumers.

3.1 Presentation of S.C. ELDA MEC S.R.L. Constanta

S.C. ELDA MEC S.R.L. is a company with limited responsibility, with a completely private capital, established in 1996; activity field: *production and commerce of dairy products.*

The company's headquarters is on Dumbrava Rosie Str. no 5, Constanta.

Work points - the company's headquarters and, respectively, in Topraisar, Constanta County, Romania (starting from July 2007).

In the purpose of carrying out this activity, the company owns fabrication licenses for the following products:

- Milk for consumption;
- Acidophil products;
- Fresh cow cheese, creams;
- Sour cream for consumption and whipped cream;
- Hard paste cheeses.

The products made and commercialized by Elda Mec are presented in table 1.

Currently, the unit is in conformity with European sanitary-veterinary norms regarding the production on a national level (L 41) and is carrying out the program for preparing the prime materials' suppliers for export quality.

Presently, the company has 15 employees, who assure a good functioning for the current production capacity.

Nr. crt.	Product name	Pakage	Quantity
1.	Fat yoghurt	PET bottle	900 g
2.	Fat yoghurt	PET bottle	2 kg
3.	Sana	PET bottle	900 g
4.	Yoghurt cream	PET bottle	900 g
5.	Diet yoghurt	PET bottle	900 g
6.	Fresh cow cheese	Plastic casserole	500 g
7.	Fresh cow cheese	Plastic bucket	5 kg
8.	Făgăraş cheese	Plastic casserole	250 g
9.	Sour cream 25% fat	Plastic casserole	450 g
10.	Sour cream 20% fat	Plastic casserole	450 g
11.	Sour cream 20% fat	Plastic bucket	5 kg
12.	Cow telemea	Plastic box	15 kg
13.	Delicatesa Elda	Plastic bucket	8 kg

Table 1. The products made and commercialized by Elda Mec SRL Constanta, Romania

3.1.1 The analysis of the marketing environment

- The consumption market of these products is in continuous expansion due to the curing qualities of the dairy products. An important factor in this growth is represented by *The Alliance for Educational Milk Advertisement,* a nation-wide program for informing the people about the benefits of consuming milk and industrially-processed dairy products. This program informs people about the benefits of milk and industrially-processed dairy products rather than the unprocessed ones, about the increase of hygiene-sanitary quality and nutritional value through industrial processing.
- This type of products is available to any consumer, milk being a food rich in calcium and phosphorous which decisively contributes to the growth and upkeep of the bone system and a good functioning of the muscles and nervous impulses transmission. Moreover, milk contains the vitamin complex B (B1, B2, B6, B12), which has an important role in the prevention of fatigue and nervous states.
- Given the fact that, on the Constanta market, there are not that many companies to cover the demand for fresh dairy products, S.C. ELDA MEC S.R.L. has the opportunity to impose itself among the other competitors, because it offers higher quality products to the consumers.
- For distribution, S.C. ELDA MEC S.R.L. uses its own transportation, equipped with storage installations for optimal product storage.
- S.C. ELDA MEC S.R.L. distributes its products especially through the retail network, together with the wholesale, having sealed contracts with the main local trade chains and with some large public alimentation unit chains, restaurants with commercial vocation on the whole Romanian seashore. Moreover, the company intends to build its own trade network, in time.
- *The consumers' demand regarding the market's offer:*
  - The production of milk and dairy products has risen by 95% in 2000-2009, while the annual average consumption of milk and dairy products has grown with 19% in the same period.

- It is important the fact that, in the last years, the consumers are orientating themselves more and more towards high quality products, which offer consumption safety and the natural characteristics of milk. Thus, the yoghurts and cheeses obtained through pasteurizing technological processes, followed by implanting of carefully selected bacteria and molds. The products obtained by the company respects the consumers' norms regarding quality and taste. Furthermore, after the implementation of the current project, the product diversity and quality is to be improved.
- In conformity with the Nielsen press statements and those of the Romanian Milk Industry Patronal Association, the yoghurt and sour cream segment is the only one which has grown by 3-5% even in the crisis periods.
- In conformity with Nielsen, the dairy product buyers choose, mostly, hypermarkets (over 40%) when buying these products, followed by supermarkets (20%), discounters (15%) and local shops (10%).
- The yoghurt segment remains, for the Romanian consumer, an item in the daily basket = an item present in day to day consumption of each Romanian household.
- What regards the milk and dairy products demand, we can say that, while under a certain price level, it is inflexible. However, it is sensible to the growth of the consumers' incomes, proof being the consumption evolution in the last years.

3.1.2 SWOT Analysis S.C. ELDA MEC S.R.L. Constanta

Strengths • High quality products; • The implementation of the HACCP Plan for most of the manufactured products; • Varied product types equipment, orientation towards traditional products; • Intermediate price situation, between economy and middle range • Safe and secure, own distribution • Location (are with high economic and touristic potential)	**Weaknesses** • Fluctuating personnel • Low number of distributors • Poor product advertisement • The lacking of a performant complete laboratory analysis;
Opportunities • The opening of new commercial units chain and units for public alimentation, especially restaurants with commercial vocation; • Multiple possibilities for assortment diversification • Accessing nonrefundable funds for development	**Threats** • Direct competition • EU norms imposed to Romania in January 2007 • The fluctuation of the company's employees, especially after joining the EU

3.1.3 Activities proposed for completion within the firm after the SWOT Analysis

- Widening the product equipment, as to fulfill the demands of as many consumers as possible
- Fabrication of new products based on traditional Romanian recipes
- Intensification of the advertisement actions regarding the products made by ***ELDA***, of the Elda Mec company in general
- Increasing of the distributors' number , and that of the distribution channels
- Expansion of the market coverage; selling the products to as many commercial chains as possible (national and international)
- Commercialization of some products (Yoghurt, Sana) under other distributor's brand
- Accessing nonrefundable funds so that the whole production-selling process responds to the EU norms' requirements
- Increasing the investments towards establishing a performant laboratory analysis

3.1.4 S.C. ELDA MEC S.R.L. Constanta objectives

- To increase production capacity, at the same time following the reaching and maintaining of a high quality level;
- To certify the Quality assurance system in conformity with ISO 9001 and the HACCP Plan (Hazard Analysis. Critical Control Points) in the new production unit.
- To implement the ISO 14000 environment standards.
- To satisfy the demands (needs) of the clients through the assurance of products which are as diverse as possible and of a high quality.
- To inform and educate the consumers in order to differentiate natural products, respectively the dietetic ones from the other products from the same array.
- To accentuate within the advertisement campaigns the quality difference and the therapeutically qualities of the natural dietetic dairy products in comparison with other dairy products available on the market.
- To increase the market share in the Constanta area - currently we have a share of over 40% in the city of Constanta, being found in numerous supermarkets, Selgros and Mega Image.
- To increase, on a yearly basis, the sales with at least 10% in order to, in up to 3 years, make the ELDA firm known through its quality not only on the Dobrogea market, but on the whole Romanian one and even in some European Union markets.

3.2 Applying the QFD method within ELDA MEC S.R.L. Constanta

The application of the QFD method at the firm Elda Mec S.R.L. Constanta, Romania aims to establish the optimal ratio between the functions and the quality characteristics of products in order to optimal correspond to the customer's requirements.

The analyse focus on the 3 products presented as follows:

- Fat Yoghurt;
- Low fat yoghurt;
- Cream of yoghurt.

The quality characteristics of the mentioned products are found in the product papers below:

Fat Yoghurt product paper

Made from: cow milk;
Types: fat;
Organoleptic and physic-chemical characteristics:

Characteristics	Fat
Aspect and consistency:	Thinner consistence, without zer-eliminating gas bubbles
Color:	White, milk color or a slightly yellow tint
Smell and taste	Yoghurt specific, pleasant, a bit bitter, without foreign taste or smell (sour, moldy).
Fat, %, minimum	2.8
Dry substance, %, minimum	9.0
Acidity, dgr. T	75 - 140
Proteic substances, %, minimum	3.2
Delivery tempterature, dgr.C, maximum	8
Whey, %, maximum	5

Quality control, marking, storage, transport:
Lot establishment: lot = max 5000 kg, yoghurt of the same type, package, presented at the same time at verifying.
Packaging: 900g PET bottle, 500g PET bottle, 2 kg PET bottle and 5 kg bucket.
Marking: marking or labeling with: factory brand, product name, type or fat content, net weight, expiration date, fabrication standard.
Storage: clean refrigeration spaces clean, at 2-8 dgr. Celsius.
Transport: clean, dry covered vehicles, at 8-12 dgr. Celsius.
Consumption indications: it can be consumed by all consumer categories, which do not have allergies or medical contraindications regarding the product's components.

Low fat yoghurt product paper

Made from: cow milk;
Types: light;
Organoleptic and physic-chemical characteristics:

Characteristics	Easy
Aspect and consistency:	Thinner consistence, without zer-eliminating gas bubbles
Color:	White, milk color or a slightly yellow tint
Smell and taste	Yoghurt specific, pleasant, a bit bitter, without foreign taste or smell (sour, moldy).
Fat, %, minimum	0.1
Dry substance, %, minimum	8.5
Acidity, dgr. T	75 - 140
Proteic substances, %, minimum	
Delivery temperature, dgr.C, maximum	8
Whey, %, maximum	5

Quality control, marking, storage, transport:
Lot establishment: lot = max 5000 kg, yoghurt of the same type, package, presented at the same time at verifying.
Packaging: 900g PET bottle.
Marking: marking or labeling with: factory brand, product name, type or fat content, net weight, expiration date, fabrication standard.
Storage: clean refrigeration spaces clean, at 2-8 dgr. Celsius.
Transport: clean, dry covered vehicles, at 8-12 dgr. Celsius.
Consumption indications: it can be consumed by all consumer categories, which do not have allergies or medical contraindications regarding the product's components.

Cream of yoghurt product paper

Made from: cow milk;
Types: fat;
Organoleptic and physic-chemical characteristics:

Characteristics	Easy
Aspect and consistency:	Thinner consistence, without zer-eliminating gas bubbles
Color:	White, milk color or a slightly yellow tint
Smell and taste	Yoghurt specific, pleasant, bitter-sweet, without foreign taste or smell (sour, moldy).
Fat, %, minimum	2.8
Dry substance, %, minimum	8.5
Acidity, dgr. T	75 - 140
Proteic substances, %, minimum	7
Delivery temperature, dgr.C, maximum	8
Whey, %, maximum	0.5

Quality control, marking, storage, transport:
Lot establishment: lot = max 5000 kg, yoghurt of the same type, package, presented at the same time at verifying.
Packaging: 900g PET bottle.
Marking: marking or labeling with: factory brand, product name, type or fat content, net weight, expiration date, fabrication standard.
Storage: clean refrigeration spaces clean, at 2-8 dgr. Celsius.
Transport: clean, dry covered vehicles, at 8-12 dgr. Celsius
Consumption indications: it can be consumed by all consumer categories, which do not have allergies or medical contraindications regarding the product's components.

QFD method was selected as an instrument for planning and development of the quality functions in conformity with de quality characteristics expected by customers and to permit the achievement of this project.

Having in view to develop the methodology, there was formed a multidisciplinary team, whose members are involved in the departments of production-quality, acquisition and marketing/commercial.

In a first stage, an inquest was developed in order to establish the functions of the three products from the "yoghurt" family of products, expected by the customers and their importance, based on a questionnaire, containing the following questions:

1. Do you know the products offered by S.C. ELDA MEC S.R.L.?
2. If the answer is" Yes", name three products.
3. Which is/are the products that you frequently buy?
4. Which are the characteristics you appreciate for the three products named above?
5. Do you buy/use the yoghurt from the S.C. ELDA MEC S.R.L.?
6. If "Yes", which kind of yoghurt and which are the characteristics of item selected?

The objective of this team is to discover which are the quality characteristics of the products from the "yoghurt" category which respond better to the customer's expectations, in order to be sustained and developed in the benefit of the clients.

Taking into account the consumers of the firm's products opinion, it was observed the following correlation between the function and characteristics of products:

- The "nutritive" function was associated with the characteristics: *content of fat* and *content of proteins;*
- The "sensorial" function – with the characteristics: *appearance, firmness, smell* and *taste;*
- The "hygienically-sanitary" function: *temperature at distribution* and *the presence of foreign items;*
- The "aesthetic" function – associated with the *packaging system;*
- The "commercial" function – *trade mark and labelling system.*

Using the questionnaire it was identified the degree of importance of the requirements, granted by customers, with points from *1 – low important* to *5 – very important*, as follows:

Crt. No.	Needs (expectations) of the customers	Defined like	Points granted
1	Nutritional contribution	1. Content of fat %	5
		2. Content of proteins %	3
2	Sensorial needs satisfaction	3. Appearance and firmness	4
		4. Smell and taste	5
3	Safety in consumption	5. Temperature at distribution	3
		6. Presence of foreign items	4
4	Comfort and commodity in use	7. Packaging system	3
		8. Tightness	4
5	Information	9. Trade mark	3
		10. Labelling	3

Table 2. Hierarchical values of the customers appreciations

The degree of covering the requirements through characteristics was evidenced with the relation matrix, with the following significance:

* - Very low cover (possible)
o - Low
- Very strong

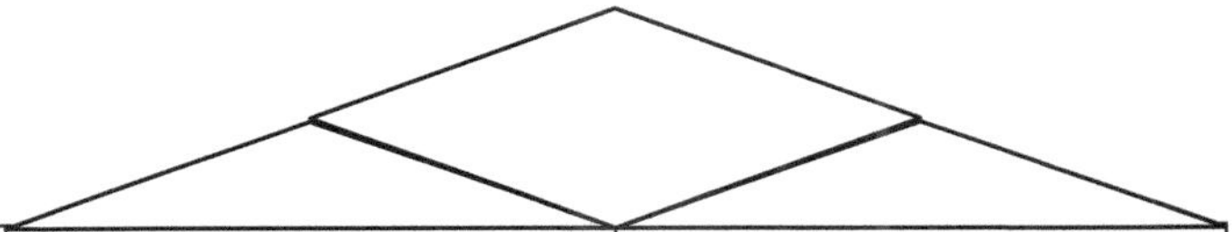

Nutritional contribution	Content of fat % #	Content of proteins % *
Sensorial needs satisfaction	Appearance and firmness #	Smell and taste #
Safety in consumption	Temperature at distribution o	Presence of foreign items #
Comfort and commodity in use	Packaging system o	Tightness #
Information	Trade mark o	Labelling o

Fig. 4. The requirements cover matrix

Consequently to the definition and measurement of the consumers needs, the QFD team establishes that the respondents grant a major importance to the "fat content", "appearance and firmness", "tightness", "presence of foreign items " and "smell and taste". That fact proves the necessity to sustain and develop further the product "Fat yoghurt", otherwise, the product the best positioned on the market. The study demonstrated also the necessity to pay a greater attention to the trade mark and labelling systems.

Further on, the degree of correlation between the selected characteristics was evidenced in the correlation matrix.

The degree of correlation between the selected characteristics demonstrated by the QFD team that the majority of characteristics are sustained each other but it is necessary to improve the characteristics referring to the commodity and comfort in use, respectively the shape of the package, temperature at distribution, trade mark, labelling and the possibility to correlate the ratio price/quantity.

Taking into account the result after the consumer's investigation, the QFD team proposed the realization of a new package with improved characteristics that consequently assured a significant growth of sales for the analysed products, especially for the article "fat yoghurt".

4. Conclusions

Firms from all domains face the difficulty due to the modification more and more rapid of the requirements and expectations of the customers which vary significantly also between the different market shares.

The technical progress, the more and more important complexity of the production and the greater and greater pressure of innovation represent only several of the requirements in increasement that firms face.

QFD must contribute to the effective and efficient transformation of the customer's requirements in the capabilities specific for a firm.

QFD method could be used everywhere the customers requirements (internal or external customers) must be transformed in specific capabilities for the firm, like in development/engineering, production or logistics.

In order to understand the QFD philosophy is supposed the basic knowledge of the quality management. In this way are enabling those who intend to become initiated in taking decisions: if QFD must be introduced and where precisely and which would be the successful procedure.

The mixed presentation of the concepts, principles and examples stimulate the understanding of the method and clarify the fact that QFD could be used in the consumption of goods area, in investments and even in the services one. To ensure the success of application is recommended a pilot-project with the assistance of an experienced moderator. The quality development must be sustained from the beginning till the use (products or services), intermediated by the clients. That means the definition, development, building, distribution, installation and if necessary some added services provided in order not only to satisfy but delight the customer also.

The chance to survive in competition is conditioned by the acceptance of the transformation. Nowadays firms must apply methods and procedures in order to develop and plan in conformity with the customers requirements. As a quality development method, QFD is oriented to the customer and involve the management in the orientation towards the consumer's process.

Orientation towards consumers is at the moment crucial for all the firms. But for all firms that the customer's requirements are continuously changing, faster and faster and the importance for different groups vary quite important.

For SME's an efficient and certified quality system in conformity with the international standards represents a real chance to survive in competition. Thus, the little business could be enabled to:

- Satisfy a necessity or accomplish a well defined objective (customer satisfaction);
- Produce in conformity with standards and specifications;
- Comply with the environment exigencies;
- Obtain competitive prices;
- Obtain benefit.
- With these purposes, the firm must:
- Understand, know and evaluate all the requirements, expectations and necessities of the potential customer;
- Design the quality of products and services;
- Achieve the quality of all processes, products and services in conformity with the initial project;
- Evaluate permanently the satisfaction degree for all the real customer;
- Assure a post-sale assistance, suitable for all customers.

As in all these stages could appear gaps, generating unconformities - non quality - it must be prevented or eliminated by correction.

In this manner could be contradicted the prejudice very speed that the quality costs (control, evaluation, correction). In fact, costs of the non quality, a well done job from the beginning and prevention permit the obtainance of quality with minimal costs.

5. References

Akao, Y., Kogure, M., Yasushi, F. (1983). Seminar on Company - Wide Quality Control and Quality Deployment. October 31 - November 3, 1983, the Arlington Park Hilton, Chicago. Sponsored by the Cambridge Corporation and c-sponsored by ASQC

Akao, Y. (1997). QFD: Past, Present, and Future; *International Symposium on QFD 97 - Linkoping*, http://stat.haifa.ac.il/~quality-study/4306/ReadingMaterial/QFD_History.pdf

Baloiu, L.M., Frasineanu, I. (2001). Gestiunea inovatiei, Economica Editure, ISBN*: 973-590-542-6, Bucuresti, Romania*

Condrea E., Stanciu A. (May 2006). QFD Method, in *Tribuna Economica* Magazine nr.5/2006, ISSN 1018-0451, pp.56-57, 2 pg., Bucuresti, Romania

Condrea, E., Stanciu, A.C. (2008). Quality Management, Lumina Collection, Libertatea Panciova Editure - Novi Sad, Serbia, ISBN 86-7001-166-2, Printed by SC INFCON SA Constanta, Romania

Condrea E., Aivaz K.A., Muhcină S., Nicodim L. (2008). New Evaluation Instruments on the Quality Management System, Conferința internațională *TEHNONAV 2008. The 6th edition of the International Conference on Naval and Mechanical Engineering - New technologies for future clean mobility and environment*, organized by the Faculty of Mechanical Engineering, „Ovidius" University Constanța, 22-24 mai 2008. The paper was publicly presented at *Materials technologies and production management* and published in the volume *TEHNONAV 2008. The 6th edition of the International Conference on Naval and Mechanical Engineering - New technologies for future clean mobility and environment*, Ovidius University Press 2008, ISBN 978-973-614-447-9, pp.579-583, 5 pg., Constanta, Romania

Evans, J.R. and Lindsay, W.M. (2005). The Management and Control of Quality, Sixth Edition, ISBN-13: 978-032-420-223-6, Ohio: South-Western

Olaru, M., Isaic-Maniu, Al., Lefter, V., Pop, A.N., Popescu, S., Dragulanescu, N., Roncea, L., Roncea, C. (2000). Techniques and instruments used in quality management, Economica Editure, ISBN 973-590-256-7, Bucuresti, Romania

ReVelle, J.B., Moran, J.W., Cox, C.A. (1998). The QFD Handbook, John Willey & Sons, ISBN 047-117-381-9, New York, USA

Shigeru, M. and Akao, Y. (1978). Quality Function Deployment: Company Wide Quality Approach (in Japanese), JUSE Press

Shillito, M.L. (1994). Advanced QFD: Linking Technology to Market and Company Needs, Wiley-Interscience, ISBN 978-047-103-377-6, New York, USA

Shillito, M.L. (2000). Acquiring, Processing, and Deplying: Voice of the Customer, CRC Press, ISBN 978-157-444-290-8

Terninko, J. (1997). Step-by-step QFD (Quality Function Deployment): Customer Driven Product Design, CRC Press, ISBN978-157-444-110-9

Vachette, J.E. (1990). Amélioration continue de la Qualité, Editeur d'Or, ISBN 270-811-151-5, Paris, France

2

Five Essential Skills for 21st Century Quality Professionals in Health and Human Service Organisations

Cathy Balding
Qualityworks P/L and La Trobe University
Australia

1. Introduction

Society's demand for quality in all spheres has never been higher. In health and human services industries in particular, consumers and funding bodies demand both technical excellence and outstanding customer service. Industries such as health, aged care and community services are struggling to meet these challenges, as the numbers of consumers grow, technology adds new a layer of complexity that solves some problems and creates others, and staff are expected to provide excellent customer service as well as technically effective services. The role of the quality improvement professional in these organizations is expanding in line with these growing expectations and has never been more important. Traditional quality systems focused on compliance and monitoring are no longer sufficient to create an excellent consumer experience, and quality managers need to add to their skills base to effectively support their organizations in this rapidly evolving environment. This chapter proposes five essential skills for quality professionals in the new millennium that build on, and go beyond, those associated with traditional monitoring and improvement, and are essential for taking organizations beyond compliance to transformation of the consumer experience. The five essential skills for 21st century quality managers discussed in this chapter are:

1. Support robust quality governance
2. Work effectively in complex systems
3. Develop a balance of rule based and proactive approaches to quality
4. Develop strategic quality plans
5. Create impact and improve outcomes through sustained systems change

The content is derived from the literature and from the author's 20 years experience working as a quality manager and with quality managers in health and aged care.

2. Support robust quality governance

Transforming the consumer experience cannot be achieved without effective governance for quality. We now need quality governance and systems that address the impact we have on our consumers – not just the outcomes we achieve. People across the organisation, from the boardroom to the customer interface, need to be clear on their individual responsibility for the quality of the services they provide and supported to enact it. Quality managers must be able to work with governing bodies and executives to design and develop systems that support staff to fulfil their responsibilities. This section discusses the governance systems required to enable and empower personnel across the organisation to enact their role in creating high quality services every day.

2.1 Understanding and implementing quality governance

The concept of quality governance is a relatively recent phenomenon. When the author started working as a quality manager in the 1980s, we thought that if we were accredited, doing some auditing and clinical review and engaging staff in quality projects then we were doing well. We knew that leadership was important, but we didn't know how important it was or indeed how best to lead. It took various studies and inquiries into suboptimal care and adverse events in healthcare to demonstrate that safe and high-quality care in a complex environment requires more than good staff trying hard. Clinical governance largely emerged from the findings of public inquiries into poor care that found that the majority of these organisations were not the victims of deliberately negligent practitioners. What they lacked were systems: for including consumers in their care, for supporting staff to provide quality care, for clarify accountabilities and for measurement and improvement. Nor did they exhibit consumer and safety-oriented cultures, with 'blame and shame' the common response to adverse events and passive response to data indicating suboptimal results. (Hindle et al., 2006)

Of course, quality care can't be achieved without good staff doing their best. But to create great care consistently, healthcare staff also need sturdy organisational supports behind them. Staff are 'front of house' - out there working with the customers. Governance is 'back of house' - the behind-the-scenes systems that support staff and enable them to provide a great consumer experience. To make the components of great care happen for every consumer, every day you'll need to ask:

- What do we currently have in place that supports great care as we've defined it?
- What do we need to enhance/change to achieve our quality goals?
- What new processes/supports do we need that we don't currently have?

Providing safe, quality care and guarding against organisational weaknesses that allow poor care requires commitment and accountability to be embedded in the organisational structures and culture, but also requires a targeted plan. Setting goals and targets for the quality of care your organisation wants to deliver, and implementing strategies to achieve them is part of the governance of any health or aged care organisation. The emergence of clinical governance over the past decade has been healthcare's approach to providing this accountability, planning and support. In aged and primary care, this can be reframed using more appropriate terms such as 'quality governance' or 'care governance'. The key components of governance can be organised into four generic cornerstones:

- strategic leadership, planning and culture
- consumer participation
- effective and accountable workforce
- quality and risk systems.

The importance of a quality governance system cannot be overstated; it provides the foundation for the myriad pieces of a quality system and gives people a role in that system, which in turn makes the implementation of the various governance systems easier.

2.1.1 Clarifying accountabilities for creating safe, quality care

The concept of governance arose from the need to ensure greater and clearer accountability for the quality and safety of care experienced by the consumer. This is still a work in progress in healthcare. There are many health service organisations in which individuals are not aware of the clear, specific, personal responsibility they have for the quality of care and services they provide. This makes it difficult for staff to carry out their responsibilities, and even harder to create a consistently safe, quality experience for consumers. Governance is where the governing body, executives and managers play their critical role in creating safe, quality care. The executive must translate the strategic quality goals into operational plans and strategies to facilitate their implementation as part of organisational business. Those on the frontline of care create the consumer experience, but the organisational supports for this must come from the top, as staff require leadership, policy, systems and an investment of time and resources to implement the strategies. And, of course, the quality manager provides technical support across the organisation to enable staff to fulfil their responsibilities. An example of generic governance roles for quality care is described in Table 1.

2.1.2 Developing dynamic quality committees

Another aspect of accountability is the way in which committees support the quality system. Driving the achievement of the quality plan through line management will generally occur in partnership with working groups or committees, particularly where implementation requires cooperation across staff groups or services. When committees are action focused they are invaluable in tracking and driving progress with the quality goals. When committees are just information recipients, staff will have difficulty understanding their purpose - and may try to avoid them. Quality managers need to be alert to directionless committees - and get them on track before they erode the credibility of the quality system. Committees should take an active role in quality goal monitoring and action at the local department/service level (where they might take responsibility for driving one component of a goal) right through to board committee level (which monitors progress with achieving the quality goals). Committees that have an explicit responsibility for achieving a quality goal are more likely to be proactive decision makers and less likely to be passive recipients of information.

To be useful, committees need a clear purpose and something that they are responsible for so they can make decisions and take action. Giving a quality committee responsibility for driving and monitoring a quality goal, objective, strategy or governance support will add some life and energy to proceedings. A clear purpose also helps determine a committee's agenda and membership. Quality committee agendas can be structured according to the quality goals and their objectives and components, which makes it easier to see how data monitoring and improvement activities link to the achievement of great care. All reporting should help a committee determine if progress is being made towards implementing governance cornerstones or achieving the relevant quality goals. Committee membership is always tricky to get right. Members can be invited on the basis of who has to be on this committee - there will always be political and relationship imperatives in a complex system - and who you need on the committee to fulfil its purpose. Some members may need to be there because they are decision makers and have formal power. Depending on the committee's role, you may also want people with informal power - the influencers. If the committee is responsible for addressing improvement in a particular area of the organisation, you will need some who have a deep understanding of the relevant systems, relationships and mental maps. Everyone on a quality-related committee should understand its purpose and exactly what each of their roles is - be it sharing their knowledge, experience or influence - and be invited to contribute to discussions and decisions on that basis.

Organisational level	Quality Governance Responsibilities
Governing Body Accountable for the quality of care, services and consumer experience	• Make the achievement of great care a priority • Set strategic direction and the line in the sand for the quality of care and services to be achieved • Lead a just, proactive culture • Ensure management provides the necessary system supports and staff development to provide great care for each consumer, and monitors progress towards achieving the strategic quality goals
Chief Executive and Executives Accountable for and lead great care and services	• Make the achievement of great care a priority • Set strategic goals for great care and operationalise them through effective governance, resources, data, plans, systems, support, tools, policy and people development • Monitor and drive progress towards the strategic quality goals • Develop a thinking organisation and a just culture, wherein staff are supported to take a proactive approach to achieving safe, quality care and services
Directors and Managers Responsible for the quality of care in each service	• Make the achievement of great care a priority and take a proactive approach to achieving it • Operationalise the strategic quality goals by translating them into local initiatives • Understand the key organisational safety and quality issues and the broader quality agenda • Monitor and drive progress by implementing the drivers of great care within their services • Develop staff and systems to create quality care and services for each consumer • Make the right thing easy for staff to do
Clinicians and Staff Responsible for quality of care at point of care	• Make evaluation and improvement a routine part of care • Develop, implement and evaluate initiatives to contribute to the organisational quality goals • Support and enable all staff to create great care • Create a great experience for each consumer through positive behaviours and attitudes and a proactive approach

Table 1. Examples of governance roles in creating quality care (Australian Commission on Safety and Quality in Healthcare [ACSQHC], 2010; Victorian Quality Council [VQC], 2003)

2.2 Work effectively in complex systems

Organizations providing human services are complex systems. They have a large number of inputs and processes, and are continually exposed to outside pressures and influences. It is imperative that quality managers working in these environments understand how these systems work to be successful. This section explains what complex systems are, how they work and, most importantly, why these things are important for quality managers, because of the way they directly impact on the pursuit of high quality services in an organisation. Working in a complex system, but treating it as if it is a simple or complicated system, makes it difficult to achieve consistently high quality services. Change and improvement in complex systems require a particular approach, tailored to the unique characteristics of the complex environment.

2.2.1 An overview of some key complex systems characteristics

Complex systems operate according to distinctive and often counter-intuitive rules. It is important that quality managers understand these rules and, in particular, their implications for creating change and improving safety and quality. Traditional, production line approaches to quality are only half the story in a complex environment such as a health or aged care service.

All complex systems have a goal, which may be as simple as survival, or maintaining the current situation. Be prepared for push back from the system if you interfere with it achieving its goal. Systems enjoy their status quo and strive to maintain it. If you change one part of the system, this will result in resistance from the other parts of the system it is linked to because it means they will have to change as well. The more parts of the system there are and the more possible connections between them, the harder it is to change and the easier it is to create chaos (Meadows, 2008). So whenever you take action within a complex system, there will be side effects. These may be positive or negative, depending on your perspective. In our health services, we usually expect that effect will follow cause. This is production line thinking. We recognise these as false conclusions when we can't then replicate the same result in another part of the organisation. The result may have been due to the natural variation inherent in every system. Or it may have been due to your intervention - but this intervention won't work the same way in another part of the system. Generally speaking, real change in complex systems requires a lot of different parts of the system to be working towards the same change.

A complex system acts like a web of elastic bands so that when you pull one piece out of position it will stay there only for as long as you exert force on it. When you let go, you may be surprised and annoyed that it springs back to where it was before. In addition, a complex system may or may not be stable. Stable complex systems that have not been subject to a lot of change become more resistant to change as time goes on. All of us have experienced this in organisations, where one service or department has somehow escaped the force of change experienced by other parts of the organisation. When their turn comes, they find change very difficult. In an unstable system, however, pressure to make changes can cause the system to burst like a balloon. If the system is under a lot of pressure routinely, this may only take a small trigger, just as a small crack in a dam can lead to its collapse because of the constant pressure of water behind it. So if you put an unstable system under enough pressure for long enough, it can suddenly disintegrate.

Despite these characteristics, complex systems work because people make them work. But to do this, processes in the system are often changed as the system evolves, and then the relationships between the processes have to change to keep the system working. The relationship between different parts of the system determines how the system overall works, so each process change, however minor, can affect the behaviour of the whole. This is an important point! All processes in a system are interdependent and they all interact. The key to change is not to just focus on one process in isolation, but to look at how it relates to the other processes in the system. Systems can also become self-organising and can generate their own hierarchies of power and influence. These hierarchies may not be the same as those seen on your organisational chart. Each person, wherever they sit in the system, has the power to affect the way the system behaves. Relationships within each subsystem are denser and stronger than relationships between subsystems. For example, there are likely to be more interdependencies and networks up and down a silo in a health service than across and between silos. Interaction within the silos occurs mainly between members of the same professional group: nurses interacting with nurses, and doctors interacting with doctors. These tribes give the people within them an important sense of belonging but it can be hard to break down the walls and build bridges between them (Braithwaite, 2010).

Complex systems do not necessarily operate according to the policies of the organisation. On the contrary, complex systems can be exceedingly policy resistant. This resistance particularly arises when an introduced change threatens the goal of the system or when policies are implemented that are not based on the reality and unwritten rules of those having to implement them. We've all experienced policies developed on the run, or even painstakingly over a long period, that have only been partially adhered to by those they were designed for. If there is too great a mismatch between the policy requirements and the way that things really get done or the goals of the system, the policy will generally fail. At worst, people will disregard it; at best, they will work around it to meet their goals of getting their work done in the most effective, efficient and easiest way - a way that has probably been crafted over time and is protected by and embedded in the way the system operates and the unwritten beliefs of those who work within it. The way in which policy is implemented can also influence the degree to which it is enacted as intended. Poor implementation opens up a policy to all sorts of change and interpretation by those using it. This may drive policy enactment to drift away from the original intention.

The importance of quality professionals being able to adjust to and deal with these characteristics cannot be underestimated. It can mean the difference between the creation of consistently safe and quality services, and implementing monitoring and improvement with few gains. The implications of these complex systems characteristics are discussed throughout the remainder of this chapter.

2.3 Develop a balance of rule based and proactive approaches to quality

Human services have traditionally relied on rules to enforce standards and ways of working. But, as we can see from the characteristics of complex systems, more than traditional approaches are required to create consistently safe and high quality health and human services. Of course some rules and standardization are important, but too many rules can do as much damage as too few. Staff work around rules that are not a good fit for their environment and all systems and procedures gradually erode in complex systems, where they are open to a myriad of influences and changing circumstances. What is required is a balance of rules, systems and thinking, proactive staff.

Improving reliability through systems that force and guide safe decisions, provide backups, remind staff of preferred behaviour and catch fallible humans when they make a mistake, are key aspects of creating safety. In fact, their use is in its infancy in healthcare - compared to other high-risk industries - and there would probably be significant benefit in fast-tracking the implementation of proven safety systems. Rule-based decision making, such as the use of protocols and checklists is also extremely useful in many situations; for example, by inexperienced practitioners who are learning standard procedures for frequent high-risk situations. Standard procedures can be useful for experts as well - particularly if they find themselves in a situation that they do not often experience (Flin et al, 2008). Not all aspects of standardisation and reliability are foolproof, however, and there is danger in thinking that they are a set and forget solution to safety. There are many reasons for this in a complex system. Remember the 'policy resistant' aspect of complex systems? Complex systems - and the people working within them - do not always respond well to overly restrictive rules, and they may react in unexpected ways. Creating a standardized approach, unless based on a forcing function, does not guarantee that it will be followed. And forcing functions, while useful in creating safety, can give rise to complacency and a lack of staff alertness. So standardisation is *one* answer to improving safety and quality, but not the *only* answer.

Why is this? We often find that there is such a strong emphasis on procedures, checklists and protocols that organisations attempt to write one for every eventuality. But it is almost impossible for a procedure to be written for every situation in a complex system, and unlikely that staff will refer to all procedures if there are too many of them (Amalberti et al, 2006). Reliability in high reliability organisations is accomplished by standardisation and simplification of as many processes as possible. But your health service is a dynamic organism with a high level of variability, production pressure, professional autonomy and rapid creation of new knowledge. Not everything can be fixed and standardised so when trying to reduce variability and improve reliability, it is better to focus on the variation that is creating real problems, rather than variation more broadly. All safety policies have a natural lifespan as the context around them is constantly changing. The challenge of creating and maintaining safety within this context requires a mix of standardisation and proactive, flexible, thinking solutions.

Over reliance on rule-based decision making is another flaw in mechanistic approaches to safety and quality in health services. It may cause a degree of skill decay; if an unexpected and unfamiliar situation arises and no rule exists, will the person making the decisions be able to formulate an effective course of action? (Flin et al, 2008). Protocols too may reduce or discourage the ability of people to be proactive, practice situational awareness, identify deviations from normal situations - in short, to think for themselves (Dekker, 2005). Bad decisions can also occur in rule based situations if the wrong rule or protocol is selected. It is human nature to prefer a familiar rule, whether or not it is the right one to match the situation in which the decision maker finds themselves. A mechanistic rule-based approach to safety is based on the premise that safety is the result of people following procedures, but staff work around rules and procedures that do not meet their needs for efficiency and streamlining. Developing checklists and protocols in response to risks may provide a sense of action having been taken, but can send the message that reliable, safe care requires nothing more than insisting upon routine standardised procedures. Nothing threatens safety like the belief that the problem is solved (Bosk et al, 2009).

2.3.1 Moving beyond standardisation to create safety and quality

When developing safety policies and protocols, it is better to give staff fewer rules that can be reliably followed around the clock than to write 'perfect' protocols based on ideal conditions that require workarounds to fit the situation at 11pm on a Saturday night. Try to resist the pressure to develop a new rule in response to every adverse event or root cause analysis finding because you'll end up with a mix of 'should follow' and 'must follow' rules that will muddy the safety waters. 'Should follow' rules that have little credibility or apparent consequence are unlikely to be followed in a messy, high-risk, high-stress environment, so why bother? Erosion of compliance with 'should follow' rules can, in turn, negatively influence compliance with the more important 'must follow' rules. When people are violating a protocol, find out why! It may be for a good reason and may give you an insight into what's going on in practice - and what's required to improve. Use observation and discussion to work out what's really happening. And when introducing a new protocol to reduce a risk, do the troubleshooting around whether or not it's likely to be followed, before people's lives depend on it. Quality managers who understand and can explain the value of not constraining the system any more than necessary, and who encourage challenging a new protocol with 'why won't it work?' and 'how are people likely to work around it?' are more likely to effect positive change in their organisation's approach to safety and quality than those obsessed with rules and compliance.

Another strategy for creating safety and quality in complex organizations is to develop the resilience of the staff. Resilience engineering is a concept derived from human factors engineering - the discipline that studies the interface between machines and systems and human beings, and improves design so that humans can operate safely and effectively. From a human factors perspective, resilience refers to the ability, within complex and high-risk organisations, to understand how failure is avoided and how to design for success. It describes how people learn and adapt to create safety in settings that are fraught with gaps, hazards, tradeoffs and multiple goals. Resilience can be described as a property of both individuals and teams within their workplace (Jeffcott et al, 2009). It fits well with James Reason's observation that his 'Swiss Cheese Model of Accident Causation' (Reason, 2008). requires another slice of cheese - cheddar, not Swiss - at the end of the line. This slice represents humans as the final barrier and defence against unsafe situations turning into harm, when all other systems fail. Practising resilience requires organizations to investigate how individuals, teams and organisations monitor, adapt and act effectively to cope with system failures in high-risk situations, and to apply and develop these lessons.

In the end, rules don't create safety - people do. Quality care and services are created by systems and standardisation, and also by proactive staff working in partnership with consumers to create the organisation's vision for great care. Building resilience is a component of this approach that combines elements of creating safety, human factors, high performing teams, job satisfaction and empowerment in a way that may assist with winning the hearts and minds of the staff at point of care. These are the staff we ultimately depend on to create and deliver the safety and quality of care we want our consumers to experience every day.

We cannot expect to eliminate human error and systems failure, but we can develop organisations that are more resistant to their adverse effects. Achieving this balance within a high-risk and ever changing environment is a critical challenge for healthcare managers and staff. But this approach reflects more realistically the environment within which we work every day. An environment that cultivates both systems and people not only supports the creation of a safer environment, but improved quality of care and services more broadly.

2.4 Develop strategic quality plans

Health services have traditionally measured inputs and outputs, and to a lesser extent outcomes, as valid and reliable outcome data can be difficult to obtain. They have been less concerned with measuring and addressing their impact on the consumer experience. We often see quality systems focused on compliance and small scale improvement, resulting in task focused programs with little purpose or direction. Like a jigsaw puzzle without the picture, there are many pieces, but no one is quite sure how to put it together. Yet engaging staff in playing their part in quality requires an inspiring vision of the service quality the organisation is committed to provide for each consumer, and a clear pathway to get there. Creating consistently high quality consumer experiences in complex organisations requires a strategic approach. Quality professionals must be able to work with their executives and managers to create a blueprint wherein goals, strategies, leadership and governance converge on a specific target: great technical care and customer service. Strategic quality planning and implementation within complex healthcare environments is a key skill for quality managers in the 21st century.

So, what is goal-based quality planning - and why do we need it? Staff involved in health and aged care quality systems are often frustrated because they don't understand why they are being asked to collect data, develop new processes or go to meetings. Simply, they can't see how these efforts fit into the bigger picture. All they see are tasks that interfere with their capacity to do 'real' work. A goal based quality plan is the blueprint for how the quality system components work together to achieve a quality consumer experience. A clear, strategically focused quality plan can help quality professionals to clarify and fulfil their role and support managers and staff to better understand their part in achieving quality care. It also demonstrates that participation in the quality system is about a lot more than achieving accreditation, as the focus of the quality system becomes the impact of monitoring and improvement activities on consumers, rather than fulfilling accreditation requirements. And this is of much more interest to clinicians and staff.

There are three key aspects to a quality system in health and aged care:

- Maintenance - minimise risk, maintain processes and standards of care, detect problems, monitor compliance
- Improvement - identify and drive operational improvements in processes designed to solve problems and improve consumer experiences and outcomes
- Transformation - develop and pursue a strategic view of consistently 'great' care for every consumer (Balding, 2011).

Most quality systems address maintenance and improvement, but too few use their quality and governance structures and processes to pursue transformation. So how does goal-based quality planning address this? More importantly, how does it address this in a complex environment? Your quality plan and system are only as good as the extent to which they impact on the care the consumer receives - supporting it to be good today, and driving it to be great over the long term. Helping managers and staff understand this, and their role in it, is a key responsibility of the quality manager. And it's not just the managers and staff who need to understand it; a quality manager will often have to explain it to the organisation's executive and governing body. When it comes to quality, governing bodies needs something tangible to govern and leaders need something concrete to lead.

The strategic approach to quality planning and creating great care in this chapter is based on the characteristics of successful strategic planning processes used in healthcare and other industries, and is a good fit with complex systems characteristics. They include:

- the use of vision statements that inspire and stretch the organisation
- the development of revolutionary goals to achieve the vision
- a horizontal approach to the planning process where input and participation are equalised across the organisation
- using learning, information and rewards to increase the strategic view of the entire organisation
- encouragement and the cultivation of strategic thinking and culture change at all levels of the organisation
- having strategic decision making driven down to all levels of the organisation so that achieving the strategic direction becomes part of everyone's job. (Zuckerman, 2005).

Organisations using this dynamic approach develop their quality plan as the platform for achieving the organisational strategic vision for quality. The strategic planning process is managed centrally or corporately and the leaders, managers and staff who are closest to the consumer are the key implementers. A dynamic quality plan is a map and a vehicle for reaching a destination. That means that a strategic approach to maintaining, improving and transforming great care and services requires you to know the where (where are we now and where do we want to go?), the why and what (why are we doing this and what do we want to achieve?) and the how (how will we get there?).

2.4.1 Setting goals is key to success

One of the most valuable skills a quality manager can offer an organisation is the development of clear and measurable goals. Do you really know what your organisation is trying to achieve? What do you want to be known for in terms of the quality of care and services you provide? Where do you stand in terms of the key quality and safety issues in your industry?

The research points to the need for a shared purpose if real change is to be made. Engaging people's hearts and minds in a common purpose requires us to paint a rich, specific picture of what they will gain if they participate and what the end result will look like. This is a staple of effective strategic planning. But it is still rare to see health services with a specific vision for the quality of care and services they wish to provide for their consumers. The pressures of short-term budget cycles and political and corporate demands do not lend themselves to a comprehensive, longer-term approach. However, stretch goals can have a transformational effect on an organisation. A strategic approach should be designed to take your organisation somewhere better than it is now, and that requires a quality plan based on the vision of care that your organisation wants to move towards. It must also be based on current reality, achievable enough so that people can believe it can happen and enough of an improvement that it is worth pursuing. If you want people to lay the quality bricks, you have to engage them in developing a rich picture of what the finished house will look like.

It is important to define quality care from both the consumer and provider perspectives. One without the other is only half the story. It is not an easy undertaking to pull the threads of your organisation together to achieve a common vision for the quality of care your organisation wants to provide. And it is likely to be nearly impossible unless it is clearly defined, ruthlessly prioritised and pursued with laser-like focus. It also needs to fit with existing system goals. To achieve all of this, plans should not contain too many ingredients and focus on achieving the essentials of great care for every consumer, every time. This means that these essentials must be defined. Engaging people across the organisation, including consumers and the governing body, is a good way to ensure this picture of quality care is both aspirational and achievable. Frontline staff and 'frequent flyer' consumers are central to this process. No one understands the difference between great and unacceptable care like those engaged in the care and service delivery transaction. The conversation around developing the vision might go something like this:

- How would we like each of our consumers to experience our care and services in three years time?
- How would we like to describe our care and services?
- How would we like our consumers to feel about our services and describe their experience with us?
- What would we like the media to be saying about us - or not saying?

Consumers, staff, executives and the governing body can - and should - contribute to these conversations. But it is not always easy to take the next step and turn this rich picture of quality care into concrete, strategic goals. This is where many organisations falter. Without goals, your quality plan may look like a long to-do list with no specific purpose. The vision for the care you want to provide must be rich, and also translated into concrete goals to describe the way things could be. Goals must be attractive and describe real, desirable, achievable changes, as seen in Table 2.

Our strategic goals for the care and services each of our consumers will experience by the end of 20XX are: • Care and services are designed and delivered to create the best possible experience for each individual (person-centred). • Care and services are designed and delivered to minimise the risk of harm (safe). • Care is based around the consumer as an individual, and is designed to achieve optimal outcomes (effective and appropriate). • Consumers are provided with, and experience, care and services in a logical, clear and streamlined flow (continuous, accessible, efficient)

Table 2. Examples of strategic goals for an organisation's quality of care (Balding, 2011).

People are attracted to ideas they feel they are involved in generating. Involving the staff affected in developing the goals for change can help create both buy-in, and the goal clarity that people need before deciding if and how they will participate. Goal clarity appears to be another problem area in creating change. If you aim at nothing in particular - or something ambiguous - that's probably what you'll hit. And yet it is not uncommon to see changes and improvements implemented with only a vague idea of what they will achieve and no clear objectives against which to measure success. The goals for your change must be SMART: specific, measurable, achievable, realistic and time-bound. Goals are about turning your vision into something achievable. Goals are not tasks; goals describe the desired future achievement. A SMART goal will encompass: How well? By when? How will we know? These are then broken down into objectives and the key tasks or stepping stones that have to be traversed, depending on where you're starting from, to achieve the final goal.

2.4.2 Select priorities carefully

A traditional problem with quality plans is that they are over ambitious. But it's far better to do fewer things and get them right. That's why any good plan has short, medium and long-term goals. Developing an annual Quality Action Plan, derived from the strategic quality plan, is a good way to keep the strategic quality plan current and dynamic. The annual plan contains the priorities to be achieved over the coming 12 months. It ensures the strategic quality plan can evolve with changing external and internal circumstances, while maintaining the overall direction towards achieving the quality goals over the longer term.

So what should be done in the first year of the plan? The selection of your first year objectives will be based on the activities that:

- have the greatest impact in creating a positive experience for each consumer
- maximise safety
- address components of great care that are currently suboptimal - or non existent
- minimise and eliminate the things that shouldn't happen
- solve significant problems and manage key risks
- meet legislative, policy and accreditation requirements
- get something going that will take a long time to achieve
- cover a lot of the quality plan's intent, using the 80:20 principle.

The 'first among equals' priority for consumers is safety and this requires robust processes across all services to reduce risk in key areas. Priorities may also be selected based on safety and indicator data, consumer and staff feedback and identified problems in specific areas. Policy, funding issues and key risks must also be addressed as priorities - that's a reality. If compliance and safety issues are at the head of your quality priorities queue, try to also include some aspirational objectives for improving the consumer experience from other dimensions of quality on the Year One list, or you may lose the momentum and energy created by the planning process. Internally, you will already have many activities in place that will help you achieve your goals. You could start by conducting a gap analysis to ascertain where current quality activities are or are not addressing or supporting the key priorities. Other organisations can also supply ideas for achieving your quality goals. Above all, don't get caught up in the detail of planning to the extent that you lose sight of your purpose. Keep the care you want every consumer to experience at the centre of your activities.

2.5 Create impact and improve outcomes through sustained systems change

Once high quality care and services are achieved, they must be embedded in everyday work. This is one of the most challenging aspects of a quality system, particularly in complex, dynamic organizations, and effective change skills are pivotal to the quality role. Quality managers often underestimate the difficulties of achieving sustained change in this environment, resulting in re-work and waste as changes that don't 'take' are re-implemented. Lasting change to effect improvement requires both systems and people change.

2.5.1 Understand the current system before you try to change it

In a complex system you need to understand what drives current processes before you can achieve a sustained impact and improvement in outcomes. Observe the humans in their natural systems environment. This may be the most important of all the 'change basics' steps - and one of the least practised. With the goal of determining organisational fit and readiness for change, you can look for systems factors such as:

- the degree to which the system participants perceive the change as beneficial
- who and what drives the current system
- the key relationships between processes and people
- the degree of fit between the goals of the system and the goals of the change
- the timing and context of the change. What else is changing or happening in this system?
- the perception of the need for change
- personal attitudes towards change generally, and past experiences with change in the organisation
- the social and values anchors that are important to the change targets and that maintain the status quo. Which of these are non-negotiable?
- aspects of the current situation that the change targets don't like. Can these be eliminated or improved as part of the change?
- driving and restraining forces for change and the degree to which it looks like the drivers outweigh the restraints (NHS 2002, 2004).

This should help you build an informative picture of the current situation. What has to change to achieve your vision? Work policies and practices? Physical surrounds? Emotional ties? Cultural norms? Understanding and working with the current culture is critical to success - even if that culture is the very thing you want to change. Use your mud map of the current situation to assess, identify and build on what currently works. 'Appreciative inquiry' is a process of identifying something that works consistently well within a system and finding out how this happens (NHS, 2002). Have you ever performed a root cause analysis on something that works to find out why it works well? This makes a nice change from looking at things that don't work well, which is a more common approach in healthcare. Tools such as process mapping, direct observation and conversations with the various players are useful here to tease out the positive characteristics of the current system that will help anchor the changed system. Not only will this help inform your preplanning, but you will be laying a foundation for buy in.

2.5.2 Develop your strategies for change – And impact

Your strategies for change will be based on your mud map of the current situation, particularly the anchors keeping the current situation in place, and represent the flight plan for how to get to your goals from where you are. Where possible, learn from others who have introduced the same or similar changes, whilst adapting their strategies to your own environment. There is no guarantee that strategies that have been successful elsewhere will work as well in your organisation due to the many layers of interactions that make your system unique. Change, transformation and improvement cannot be delivered through the adoption of an imported recipe or formula without adapting it to the current environment. If you introduce a new procedure, software system, data collection or form on a Monday morning without investing in preparing and equipping the people who will use the innovation, it is unlikely to be automatically adopted. The process may have changed, but the people haven't - they are the same as they were on Friday afternoon. Process change is not the same as people change. Process change is transactional and concrete. People change is transitional and involves a psychological process to come to terms with a new situation and change behaviour to enable the new situation to occur. Unless this transition is well managed, change will not work and things can get stuck. Even with obviously positive changes, there are transitions that begin with having to let go of something and there will be push back because your change adds to the staff 'to do' list and new behaviours take longer, both of which result in lost time. At worst, staff are losing something they are strongly wedded to and may actively resist or get stuck in a neutral zone where they are aware of the change but not actively engaged - a sort of change no man's land (Bridges, 1997).

It is important to remember that all staff feel that they are doing their best for each patient. Change for improvement should always be presented as something that helps good practitioners achieve even more. They may maintain that their only desired benefit of change is improved patient outcomes and these, of course, are likely to take some time to become apparent after the initial change. So what are some of the short-term benefits of change you can use to get people's attention? This is where you have to talk about impact as well as outcome. Impact what we are trying to achieve by change, for both consumers and for staff. It's not only about trying to improve the results of care. It's about consumers feeling the impact of your change through a different, more positive experience. Does the change mean that staff are more active listeners - so consumers feel heard? Is it that the change can form part of an action research project and that you can assist staff to write it up for a journal or a conference paper? Will it help both consumers and staff feel more informed and in control of what's going on? Can a process be made more efficient and simpler as part of the change? Can you save them time and money? (Frankel et al, 2011).

Within this framework, as far as possible, give staff as much freedom as possible to devise their own ways of achieving the goals, based on their intimate knowledge of their own systems. But empowering people to create change is not just saying 'make it so' and then being disappointed when they don't achieve the desired result. Empowering people to change in complex systems is not straightforward. But there are some common actions that have been shown to be essential in assisting people to take ownership of a task or change: direction, knowledge, resources and support - the DKRS model of empowerment (Balding, 2011). For the DKRS model to succeed, each of these four components must be present to fully enable people to take ownership of the task or a change. We often see one or two of these employed in healthcare change but it is unusual to see an individual or team supplied with all four (Balding, 2009). Empowerment does not mean abandonment. Giving people permission to do something differently is not helpful if they are unable to do it. That permission just sets them up to fail. Setting the context for change means preparing the players, understanding what they know and don't know, working with them, watching their performance, giving them feedback and creating an ongoing dialogue with them (Meadows, 2008). It may be more effort at the front end of a change to work with staff to ensure they have all four components, but it will save you a lot of time and trouble at the back end of the change if they are able to embrace, own and run with the change in their local environment.

2.5.3 Test and implement the changes

Rapid cycle piloting of change using the Plan Do Study Act (PDSA) cycle is a useful approach to change in a complex system. PDSA fits the changeable and adaptable nature of complex systems and enables you to test ideas under a variety of circumstances (Reason, 2008). It's also a good way to pick up on the feedback and side effects of your change. This model also includes the possibility that the change being tested will not be successful, but because these tests are done on a small scale the risk of failure can be kept to a level that's manageable. PDSA also helps achieve quick wins, even if small, that are integral to gaining stakeholder acceptance of change. Success on a small scale builds confidence, which allows larger risks and changes. Pilot projects work best under the following circumstances:

- Pilots are limited to small samples and short cycles of change with the people who want to be involved
- They use solutions that have worked for others, but are adapted to fit the local situation
- The easiest change with the most leverage for the biggest impact is made
- An action learning process is used to frequently review progress and the change leader stops to ask: 'how did we go?', 'what did we learn?', 'what were the unintended consequences and side effects?' and 'how should we do it differently in the next cycle?'
- Participants are not afraid to stop a test change that's clearly not working. This is part of change in complex systems (Haines, 1998; Reason, 2008).

Staff involved in the pilot will be watching, judging and weighing up whether or not to hitch their wagon to the new way. It is imperative that your process has credibility. When you pilot a change, use a simple but rigorous project management approach and do exactly what you have promised. If you want to change people's beliefs about how things should be done, you must change what they see. A memo or an email about doing something differently will not make it happen. If you want people to believe that changing their behaviour will result in a certain positive outcome, that outcome must occur. If you commit the leadership group to behaving in a different way, they must behave in that way. This is where many change initiatives break down: we make the plan and say what will happen, but don't follow through.

Early wins are required to show that change is possible and can have positive outcomes. Action sends a strong message, more than any memo ever could. Don't be surprised by unexpected or negative outcomes, and don't expect a linear cause followed by effect chain with your change. Look for the unintended negative side effect of your change. For example, if you have streamlined the new consumer registration process, does this leave clients feeling that they have been hurried and not heard? Don't ignore or downplay these negative side effects - they are not failure, but the way of the world in complex systems.

2.5.4 Reinforce, embed and spread the change

Creating buy-in is one thing. 'Stay in' is something else altogether. Systems need a constant supply of new energy to survive and, until your new change starts to create its own energy, it requires yours! Sustainability is a process, not an ending (NHS, 2002). Many managers want to get everything up and running on auto pilot as soon as possible, but this is the antithesis of what actually sustains change.

In complex systems, sustainability and spread are dynamic processes that need focus and attention. So, define sustainability. What do you mean by it? What do you want to still be happening in one/three/six months from now? People need to be reminded of the goal and the vision, and the way in which these are achieved requires monitoring and course correction in a shifting complex environment. Involve people in developing solutions to overcome the unexpected problems that arise, ensure they are equipped for their role in the change and reinforce where their contribution to the change makes things better for patients. Use the sceptics to help you identify the problems and the roadblocks and show you value their input. Arguing with them will not change their mind and you may lose valuable information (Haines, 1998).

If you've done a good job of your change process by giving the participants a positive experience, ensuring the change is an improvement for patients and staff and finding those quick wins, the initiative should have its ownership and should just about spread itself. This is the 'tipping point' concept, which provides a useful summary of spread (Gladwell, 2002). The 'law of the few' and the 'stickiness factor' are tipping point concepts, which provide us with direction on how to go about reaching the point where the change takes on a life of its own. The law of the few means that a few influential, popular people can effectively spread a message, so use the people who have influence - the 'players' in your complex system - and also the people who just get around and talk a lot. Stickiness means that a message has impact: you can't get it out of your head, it sticks in your memory. Are your messages 'sticky' or dull and forgettable? (Gladwell, 2002). Are they presented in the language of the people - or in complex bureaucratese?

Once you've got the change right, embed it in job descriptions, policies and procedures, competencies and performance reviews. Reinforce it. Remove the old way - if you don't, people will cling to it because it's familiar, and it will make the new way seem like an extra, rather than a replacement. Keep the change on meeting agendas as a specific review item for at least six to twelve months, depending on the size of the change. Appoint a 'keeper' of the change - someone influential whose job it is to keep an eye on the new way of doing things and the people involved, and to identify regression and unintended side effects. Ensure it continues to be linked to broader organisational initiatives.

3. Conclusion

As the pressure on our health and aged care services grows, so too do the demands on the quality professional. Continuing to increase the efficiency and quality of healthcare will require new knowledge and savvier ways of working. To meet these challenges, quality professionals will need to expand their role beyond traditional compliance, measurement and improvement skills and tasks. They will be required to understand their workplaces as complex systems and be experts in supporting their complex organisations to create high quality care. To do this they will support and lead their organisations to develop robust governance, to create safety through a mix of effective systems and resilient people and to achieve sustainable change that positively impacts the consumer experience as well as improving outcomes. These are the new skills for 21st century quality managers.

4. Acknowledgment

With grateful thanks to my family for indulging my preoccupation with all things quality.

5. References

ACSQHC. (2010). *Australian Safety and Quality Framework for Healthcare: Putting the Framework into Action: Getting Started,* Australian Commission on Safety and Quality in Healthcare, Australia

Amalberti, R.; Vincent, C.; Auroy, Y.; & de Saint Maurice, G. (2006). Violations and Migrations in Health Care: A Framework for Understanding and Management. *Quality and Safety in Health Care,* December 15 (Suppl 1): i66–i71

Balding, C. (2005). Strengthening Clinical Governance through Cultivating the Line Management Role., *Australian Health Review,* vol. 29, no. 3

Balding C. (2011). *The Strategic Quality Manager,* Arcade Custom, Australia.

Bosk, CL.; Dixon-Woods, M.; Goeshel, CA.; & Pronovost, PJ. (2009). The Art of Medicine: Reality Check for Checklists, *The Lancet,* vol. 374, August 8

Braithwaite, J., (2010). Between-group Behaviour in Health Care: Gaps, Edges, Boundaries, Disconnections, Weak Ties, Spaces and Holes. A Systematic Review, *BMC Health Services Research,* vol. 10, no. 330, *http://www.biomedcentral.com/1472-6963/10/330* (Accessed February 2011)

Bridges, W. (1997). *Managing Transitions,* Addison Wesley Publishing Company, USA

Dekker, S. (2005). *Ten Questions About Human Error,* Lawrence Earlbaum Associates Inc, USA

Flin, R.; O'Connor, P.; & Crichton M. (2008). *Safety at the Sharp End – A Guide to Non-Technical Skills,* Ashgate Publishing, UK

Frankel, A.; Leonard, M.; Simmonds, T.; Haraden, C.; & Vega, K. (eds) (2009) *The Essential Guide for Patient Safety Officers.* Joint Commission on Accreditation of Healthcare Organisations and Institute for Healthcare Improvement, USA

Gladwell, M. (2002). *The Tipping Point,* Backbay Books, USA

Haines, SG. (1998). *Systems Thinking and Learning,* HRD Press, USA

Hindle, D.; Braithwaite, J.; Travaglia, J.; & Idema, R. (2006) *Patient Safety: A Comparative Analysis of 18 Enquiries in 6 Countries.* Centre for Clinical Governance Research, UNSW, Australia

Jeffcott, SA.; Ibrahim JE.; & Cameron, PA. (2009). Resilience in Healthcare and Clinical Handover, *Quality and Safety in Healthcare,* vol. 18, pp. 256–260

Meadows, DH., (2008). *Thinking in Systems – A Primer,* Sustainability Institute, USA

NHS. (2002). *The Improvement Leaders Guide to Managing the Human Dimensions of Change: Working with Individuals,* NHS Modernisation Agency, UK

NHS. (2004). *Engaging Individual Staff in Service Improvement,* NHS Modernisation Agency, UK

Reason, J. (2008). The Human Contribution, Ashgate Publishing Company, UK

Victorian Quality Council. (2003). *Better Quality, Better Healthcare: A Safety and Quality Framework for Victorian Healthcare,* Department of Human Services, Victorian Government, Australia

Zuckerman, A. (2005). *Healthcare Strategic Planning*. Health Administration Press, USA

3

ISO-GUM and Supplements are Utilized for QA of BCA Data

Yasuo Iwaki
Chaos Applied Research Office
Japan

1. Introduction

International Organization of Standard- Guide to the expression of Uncertainty in Measurement (ISO-GUM)[1] was published as guidance for making a measurement result into an assurance performance by ISO in 1993 and in 1995 corrected. ISO has inquired by adding new thinking to the conventional method. To apply the GUM approached two main assumptions must hold. One is used expression of uncertainty from error so that it might be suitable for an assurance in evaluation of ambiguity measurement data. Another one used Exploratory Data Analysis (EDA) for ambiguity data from classical statistical analysis. It is now useful in order that the Blood Chemical Analysis (BCA) may also grade up the reliability of an analysis result with assurance performed.

ISO was additional new issue that is published many supplements in order to fully utilize ISO-GUM. Particularly, it is come to be for Markov Chain Monte Carlo (MCMC) in Bayesian inference and. the come to be used Multi-viable Analysis (MA) that is useful the multi regression analysis by multi-nonlinear least squares method. One of them is a production procedure for obtaining an assurance process. ISO-GUM is respectively as one by one step, it is progressing [2]. The result is also required by improvement of in measurement accuracy and Quality Assurance (QA).

2. Background of ISO-GUM

In planning measurement system is very important for Measurement Systems Analysis (MSA) [3]. It is a specially designed experiment that seeks to identify the uncertainty components in the measurand. MSA is used to evaluate the quantitative analysis in medical test system which is entire process of obtaining data. The inspector has to understand well the process of measurement system used in order to ensure the obtained Data Quality Object (DQO) and has strived for quality analysis in good assessment. Many mistakes in the total testing process are called "laboratory errors" [4], although these may be due to poor communication or poorly designed, all of which are beyond the laboratory error control. The uncertainty element in total measurement system is recognized to be an error of measurement management poor. At the management all the processes need to be managed from the preceding paragraph of laboratory analysis to the processing an after and an end. It has to remove the fault element and risk element of clinical healthcare by the uncertainty data. In order to improve supply service of data from laboratory, it require to QA of the data offered.

The role of global and regional metrological organizations is also to be discussed to get a mutual confidence between these test laboratories. The main targets of above activities may be summarized by the Mutual Recognition Agreement (MRA) between participating economies.

ISO-GUM published for an international consensus based on this concept, it is emerging that analysis values are expressed in combination with uncertainty of their measurement to indicate their reliability. In the field of measurement science and clinical chemical analysis, there is exists world wide requirements for the reliable and competitive evidence to confirm the measurement process and measurement results in many stages. A goal of object is improvement in the reliability as Good Laboratory Practice (GLP)

The purpose is in construction of MSA for without hesitating diagnosis since some inspect data is ambiguous. It is selecting a point of healthcare for always suitable diagnosis, and losing the futility of a health resource. For this reason, it aimed at starting with improvement in the accuracy in the field of BCA as a trial to MSA and assuring the quality of result completely by a laboratory implement guide. The goal of good practice guidance research has continued in the post to support the development of Quality Assurance (QA) and Quality Control (QC) in Quality Engineering (QE) completely by high order accuracy. ISO-GUM was legislated an accuracy assurance for dealing of measurement data.[5]. It is ISO15193:2009 that is defined as in vitro diagnostic medical devices.

ISO/IEC Guide 98-1:2009 provides a brief introduction to be GUM in order to indicate the relevance of that new fundamental guide and promote its use. It also outlines documents related to the GUM that are intended to extend the application of that guide to broader categories and fields of practical problems. It also considers various concepts used in measurement science [6] that is included a science (thinking) and an engineering (thinking). In particular, it covers the need to characterize the quality of a measurement though appropriate statements of measurement uncertainty. ISO edited international vocabulary in metrology (VIM) simultaneously.

2.1 QA and QC

Measurement data condition are roughly divided into within-laboratory and between-laboratory for QC. Under within-laboratory measuring condition, the uncertainty due to within-day variations are estimated from repeatedly measured values of the test sample within the same laboratory. As for research target in clinical examination, the accuracy of Internal Quality Control (IQC) in house must be keep always more than two sigma. The under between-laboratory measuring condition, the uncertainty due to compared between another laboratory variation data. The uncertainty due to between-laboratory variation are estimated from simultaneously measured values of the test sample obtained at more than one laboratory, and an accuracy of External Quality Control (EQC) secures more than three sigma levels, it is an international level. In both case, the individual component are composed to obtain standard uncertainty due to measurement conditions. The lack of certainty, a state of having limited knowledge where it is impossible to exactly describe existing state or future outcome, more than one possible outcome. Necessity of more than the three sigma accuracy was carried out to IT medical system for world wide base medical healthcare. Especially an External Quality Assurance Scheme (EQAS) is an importance it can use also for common view of medical cognitive diagnosis technology. If the final report value of BCA became a commercial transaction article, it requires to follow the QA by ISO standard. Quality Engineer (QE) is a means also to prevent a misdiagnosis effectively. Furthermore, the Statistical Quality Control (SQC) of a patient individual's data is also important for prevent from a clinical misdiagnosis. The result of a statistical analysis is working not only get to know condition of disease, but it is utilized for exacted judge decisions to support a point of care program. The accomplishing to this requirement, some international regulations and guides has been edited as the results of joint works among several international organizations for data quality.[7][8]

QA is doing its best also in the field of a clinical examination to be able to respond to a patient or a donor effectively. ISO-GUM is edited in series to 98-1 from 98-5 as a guide of ISO/IEC. QA of clinical test by ISO-GUM is made utilization in 2006.

2.2 ISO and QE

2.2.1 Measurement error

The conventional statistical technology was researching for error of random data by the subjective statistical work by "law of large number", it is come out complicated random error and systematic error. Systematic error can be removed as bias since it can be made a fixed numerical value. Random error was difficult work in order to remove. Therefore, the result in which reading out and clear not able to achieved. Further, the first type error (false positive) and the second type error (false negative) are achieved among the errors. So it had become a cause of the misdiagnosis. For prevent a misdiagnosis, it is required to correct all of the errors factor. Work of an improvement of these errors has been also studied by fault state through many years in the Quality Engineering (QE). The central pole theorem became important recently. Many technologies were useful commonly in QE and in ISO-GUM [8].

2.2.2 QE

QE assumes how often a fault state generative, it is starting analysis from hierarchical gradient-based motion estimation of fault factors though "Failure Mode and Effects Analysis (FMEA)". QE determines the root cause of the fault element though "Root Cause Analysis (RCA)" and "Fault Tree Analysis (FTA)". These are developed based on tracer technology. In professional Test (PT), the procedure of decision followed one by one, in order to discern the importance level of fault factors.

FTA is versatile methods for dealing with probabilistic risk, reliability and availability. Although FTA was developed in the 1690s for hardware system, it is an adaptable logic-based technique that has been applied to combined hardware and software systems. This research was led it to make the result of the BCA The relation between QE and ISO-GUM is shown in table 1.

Term	ISO-GUM	QE
process	key comparison. and law of propagation of uncertainty.	FMEA. FRACAS FTA and RCA
Estimation	uncertainty.	Random error
Algorithm	MCMC model Multi-variable analysis.	Quadratic equation.
expression	illustration.	formula
result	confidence interval	tolerance
critical	reference	standard
unit	SD	%
system	legislation	Scientific thinking Technical thinking
Focus	normal and abnormal distribution	normal distribution
test	effective free degree	parametric nonparametric
Estimation	Uncertainty.	error
Algorithm	MCMC modeling. Multi-variable analysis.	Quadratic equation.
expression	illustration.	formula
result	confidence interval	tolerance
critical	reference	standard
Focus	normal distribution and abnormal distribution	normal distribution only
test	Nonparametric & nonlinear	Parametric & linear

Table 1. Comparison ISO-GUM and QE

2.2.3 DL/AMD

Present are procedures based on modern Bayesian statistics which are used calculate characteristic limit i.e. the decision threshold, detection limit and confidence limit in BCA. Indicated are also key elements of this statistics which can be used for measurement of Decision Level/Amount Minimum Detectable (DL/AMD), DL applied to the activity result. AMD was the detection criterion that was insensitive to sample specific variables such as chemical yield and detector efficiency. The example of instrumental enzyme activation analysis provides an illustration of the issues discussed.

DL/AMD was able to profit by the operation. QE has developed for the QC of industry as 6 sigma level. However, these have been processed by the regression analysis by making subjective frequency probability of a statistic value into normalized distribution. Regression analysis is applied the least squares method. An occurrence probability of the statistics value of a fault element which is accompanied element by a natural variance, it becomes an abnormal distribution in many cases.

In the result of research, "Decided level/Minimum Detectable Concentration (DL/MDC)" are other different taxonomy of uncertainties and decisions that include a more broad sense of uncertainty and how it should be approached from an ethics perspective. Vagueness and ambiguity are some times described as "second order uncertainty", there is uncertainty even about the definitions of uncertain states or outcomes. The difference result is that this uncertainty is about human definitions and concepts not an objective fact of nature. It has been avoidable while uncertainty (first order).

2.2.4 Probability Density Function (PDF)

PDF expresses distribution of uncertainty (see Fig1). Assignment of a PDF to a quality analysis is using the Principle of Maximum Entropy (PME). There are existence two great traditions. One is probability theory what think is likely trueness of population mean, and second is confidence interval analysis what we know to be total error. Total error is system error plus random error. Probability bounds analysis gives the same answer as confidence interval analysis does when only range information is available. It being careful is that arises between total error and a confidence interval in many cases that example is abnormal distribution, which better as for evaluation uncertainty for QA. A normal distribution is comfortable in estimating the width of variation; it is fundamental statistical quantity by analysis of variance (ANOVA). If also gives the same answers as Monte Carlo analysis does when information is abundant. If it is an abnormal distribution, a setup will became difficult about a trueness value, and the reference value by ISO-GUM is then calculated for QA [10]

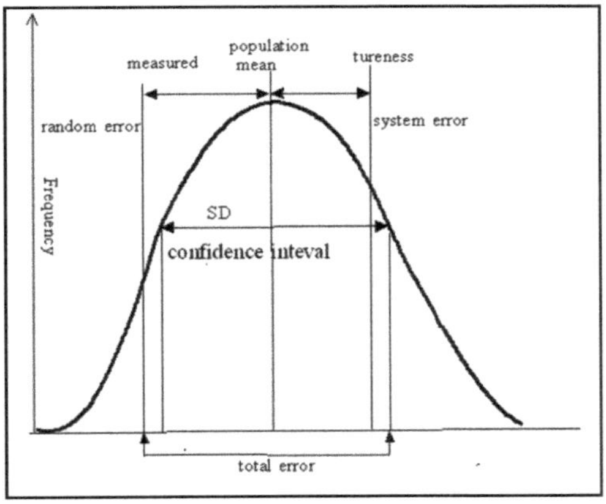

Fig. 1. Express of error and confidence interval

2.2.5 ANOVA

In measurement model, input quantities are measuring data of random variable that is interest as many components, $x=(x_1,x_2,\ldots,xi)$, then it can quote a functional relationship between measurement result Y and input quantities $f(x_i)$, as (1)

$$Y=f(x_1,x_2,\ldots,x_i). \tag{1}$$

Where, Y stands for the output quantity, that is the measurand in VIM, whereas x_i for multi input quantity. This is a model with one output which is adopted in the current ISO-GUM. A normal distribution is comfortable in estimating the width of variation of fundamental statistical quantity; namely, Root Sum Square (RSS), arithmetical average (mean), Standard Deviation (SD), Coefficient of Variance (CV) and etc. [18]

One of most import indicator of random error is time. In ANOVA, QA of measurement data must be considered error theory and effective ecologically so that basically, when null hypothesis is set up in kinetic state. It quotes a time series function. This rule is shown that the dependence is trueness function f(x,t) and it is include error factors g(x,t), as (2)

$$dx/dt=f(x,t)+g(x,t) \tag{2}$$

These are assumed as independent function. When an error factor exists by plurality, Burger's formula is adapted to Multi-variable analysis (MA) that has one or more g(x.t). ISO-GUM evaluates by reference value, even when a trueness value is unknown, and error serves as uncertainty.

Bayesian theorem grows from the simple principle that two random variables factor t and x remain in the following dependence: as (3). Vertical arrow | indicate conditional distribution.

$$P(xt)=P(x \mid t)*P(t)=P(t \mid x)*P(x) \tag{3}$$

$$P(x \mid t)= P(x \mid t)*P(x)/ P(t)$$

2.2.6 Key comparison and reference value

If the trueness value was unknown, ISO-GUM is changed into reference value from trueness value, it is gating in the posterior data base that is set up by the EDA as frequency hypothesis and it does check with a Key Comparison (KC) method. KC is comparing between null hypothesis and frequency hypothesis.

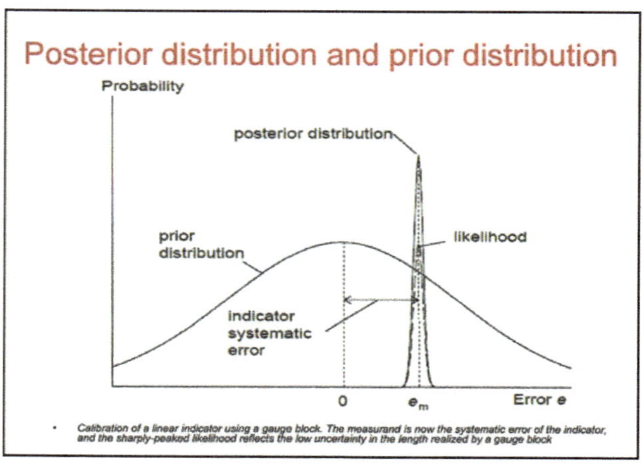

Fig. 2. Posterior distribution and prior distribution

Null hypothesis is making guesstimate uncertainty from prior distribution with an experiment data. Frequency hypothesis is made posterior distribution on data base that is standard posterior distribution. Reference value estimated with in likelihood position in Fig.2. [15]

2.2.7 EDA

Exploratory Data Analysis (EDA) in ISO-GUM calculates one by one until an assurance execution that is according to "Law of the Propagation of Uncertainty (LPU)". It is same as RCA and FTA [9] in QE. Work progresses aiming at goal of full implementation of an assurance. ANOVA is used "Law of the Propagation of Laboratory error (LPE)".

2.2.8 MCMC

Markov Chain Monte Carlo (MCMC)[4] determines numerically a PDF. A set of possible states or outcomes it where probabilities is assigned to abnormal distribution. This also includes the application of PDF to continuous variables.

MCMC is proposed for the calculation uncertainty which can be considered to the primary other method than statistical method to EDA in ISO-GUM. Result is illustrated to easy understand. MCMC follows from "Derivation" of Markov formula and the Monte Carlo method is based on the central limit theorem, it is include Gibbs sampling and Metropolis - Hastings (M-H) algorithm.

In probability and statistics, the t-distribution or Student's t-distribution is probability distribution that arises in the problem of estimating the mean of a normally distributed population when the sample size is small. It is the basis of the popular student's t-test for the statistical significance of the difference between two sample means, and for the difference between two population mean. The Student's t-distribution is a special case of the generalized hyperbolic distribution. Student's distribution arises when the population standard distribution is unknown and has to be estimated from the data. As it is in nearly all practical statistical work, problems treating the standard deviation as if it were known are of two kinds:

1. Those in which the sample size is so large that one may treat a data-based estimate of the variance as if it were certain.
2. Those that illustrate mathematical reasoning, in which the problem of estimating the SD is temporarily ignored, because that is not the point that the author or instructor is then explaining.

Example of MCMC is recommended in Fig 3 by NIST. There are making the three kind distributions from measured data of abnormal distribution. One is the rectangular (uniform) probability distribution, that are possible for setting six value of primary confidence interval easily and none is preferred against any other value, the probability for any value to be on top after throwing one dice is 1/6 that is derived by Markov formula. Second is t-distribution that is important in both theory and practice. And confidence intervals derived from Student's t-distribution with n-1 degree of freedom. The parameter is called the number of degree of freedom. It is the same as a 95% confidence interval. Third is normalized distribution that is converted from abnormal distribution by regress analysis. Three kinds distribution is analyzed in Fast Fourier Transform (FFT) series, and carry out is a combined Fourier synthesis that is making new confidence interval by ISO-14253-1 (see 3,5) in Gaussian distribution. In next step, Inverse Fast Fourier Transform (IFFT) is made the convolution of Gaussian distribution F(u) there, F(u) is made to the Cumulative Distribution Frequency (CDF) that is alternatively referred to in the literature as the distribution function. CDF is compared in comparator, it be able to find out the fault point on the curve, and it is required to exploratory the cause of generating fault state further. U is carried out a final standard uncertainty Us that is affinity (Binding ratio P*Q/Po= %) in this case. Fig.3 has a spare input port for special form distribution e.g. triangular distribution, U form distribution. A result same as a fuzzy member function instead of FFT is useful.

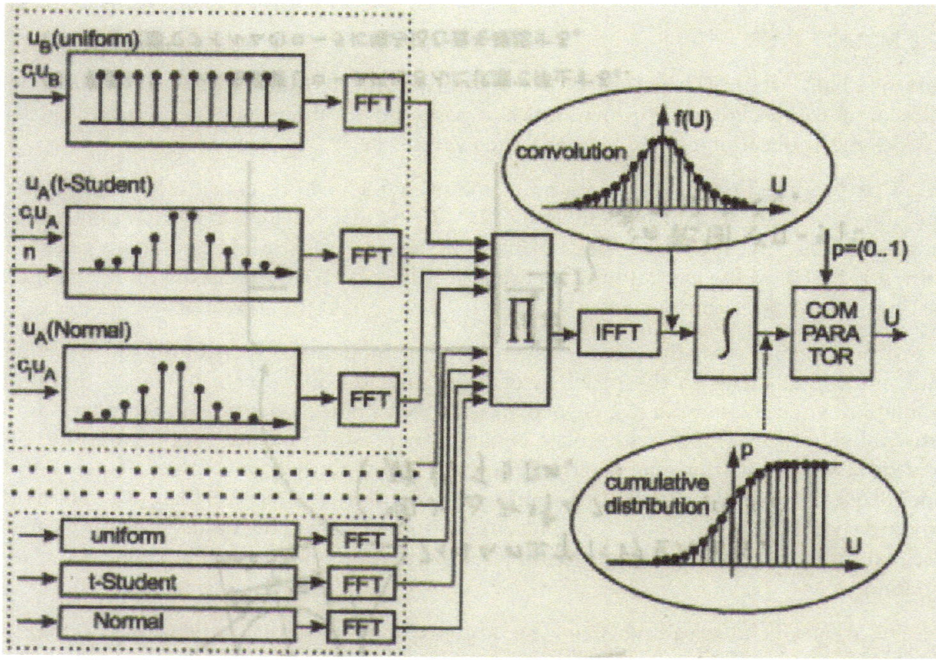

Fig. 3. MCMC of EDA in 2002 by rule of NIST

2.3 Uncertainty

2.3.1 Uncertainty theory

Uncertainty is defined "A parameter, associated with a result of a measurement, that characterizes the dispersion of the values that could reasonable be attributed to the measurand." in VIM.

The uncertainty data can usefully in 2 stage ways according to how they are estimated.

1. The first stage is utilizing the Bayesian inference of key comparison according to LPU.
2. The second stage analyzes is grouped into two categories the measurement result into type A and type B, according to determined the distribution condition. An importance distinction between both types of statistics lies in a quite different approach to the concept of probability.

The conventional concept of probability in statistics is associated with the relative frequency of random events. Such a statistics fails in case of systematic effect. Non-linear measurement models, uncertainty values measured close to detection limits. Gibbs sampling is used for collection of data by double samples.

Uncertainty associated with measuring operations within the same laboratory is estimated by applying the nested analysis of multi-variance method though experimental data with between-day and within-day variation and sample vial as relevant factors. Specifically, a calibration line is generated at a time of every experiment during same days of experiment period. The thus obtained measured values examined for outliers. If outlier is found, its cause is identified and the rarely value is removed. If a problematic finding is obtained in the measurement, a new measurement is performed, after investigation, two stage nested analysis of variance is applied to estimated individual variation

A state of uncertainty where some possible outcomes have an undesired effect or significant loss as a risk quantitative uses of the terms uncertainty and risk are fairly consistent from fields such as probability theory, actuarial science, and information theory. Some also create new terms without substantially changing the definition of uncertainty or risk.

2.3.2 Procedure of uncertainty evaluation

Evaluation of measurement uncertainty by ISO-GUM recommendations can be summarized in the following step

1. Estimation of the standard uncertainties of the main sources: Definition of the quantity being measured data by the sensitivity coefficient
2. Calibration of the components of uncertainty for each main sources::
3. Calibration of the effective degree of freedom of the standard combined uncertainty
4. Calibration of the expand uncertainty: However, the ISO-GUM approach exhibits some limitations:
5. Model linearization: The principle of error propagation applied to obtain the standard uncertainty truncates the Taylor's series expansion in first order terms. This is a linear approximation that in some cases could need terms of higher order.
6. Assumption of normality of measurand in common practice, the distribution of the result is taken as normal and consequently, expanded uncertainty Ue is calculated as the product of the coverage factor k and the combined uncertainty Uc.
7. Record the data evaluated for uncertainty in an open document

2.3.3 Standard uncertainty

It performs operation of assurance by ISO-GUM in order of standard uncertainty U, Uncertainty is expressed standard deviation. Standard uncertainty Us is obtained MCMC of process .in Fig.3. Us is defined "Uncertainty of the result x of a measurement expressed as a standard deviation" by ISO/BIPM guide 98.

2.3.4 Combined standard uncertainty

The combined standard uncertainty Uc is adding many standard uncertainties Us of fault elements. Uc is defined "Standard uncertainty of the result y of a measurement when the result is obtained from the values of a number of other quantities". by ISO/BIPM guide 98.

Existence of two or more fault elements will use the multivariate analysis of supplement 3 in ISO-GUM. Each element is sampling as uncertainty elements (s_1. s_2, s_n).

$$Uc=\sqrt{U^2s_1+U^2s_2+\cdots+U^2sn}\ , \tag{4}$$

2.3.5 Expand uncertainty

A expand standard uncertainty Ue for assurance performance is calculated as multiple factor by numerical coverage factor k to combined uncertainty as (5). Factor k is called coverage factor.[3] An assurance value is authorized by the final expand uncertainty Ue [11][13].

$$Ue=k*Uc \tag{5}$$

2.3.6 Coverage factor

The coverage factor.[3] is computed it by Welch-Satterwaite formula as (6) with Effective Free Degree (EFD) that is shown (V_{eff}). Free degree is important the number of samples based on maximum likelihood method. It is improved an (Akaike) information criterion (AIC). k is defined "Quantity defining and interval about the result of a measurement that may be expected to encompass a large fraction of the distribution of values that could reasonably be attribution to the measurand." By ISO/BIPM guide 98.

If the measurand distribution is approximated to a student's distribution, k is taken as the tabulated Student's t-value for given significance level. In the general cases, the analytical evaluation of the EDF is still an unsolved problem, in type B, generally contributing with infinite degree of freedom.

$$V_{eff} = \frac{u_c^4(y)}{\sum_{i=1}^{N} \frac{c_i^4 u^4(x_i)}{v_i}} \tag{6}$$

An example, the use of a value of *k* other than 2 is taking k, equal to a t-factor obtained from the *t* -distribution when Uc has low degrees of freedom in order to meet the dictated requirement of providing a value that defines an interval having a level of confidence close to 95 percent.

3. ISO-GUM and assurance

ISO-GUM was legislated in enforcement an QA based on Quality Engineering (QE).

The standardization of measurements is high priority in laboratory medicine, its purpose being to achieve closer comparability of results obtained routine measurement procedures. The GUM has been increasingly adopted as a de facto standard procedure by calibration laboratory,(see ISO17025). The GUM-GUM has been formally adopted as a US National Standard in the form of ANSI/NCLS Z540-2-1997, ISO is edited ISO-15189 to medical use.

3.1 To apply the ISO-GUM

GUM approach two main assumption must hold (12).

1. The system is modeled using a functional relationship between measured quantities X=f(x) and the measurements result y in the form y=f(x). The adequacy of the formula for uncertainty carried out u(y) which is derived by propagating uncertainties in a first-order approximation to the model of the measurement system. Production of analysis model is explained to the supplement 2.
2. The distribution of y is known, e.g. Gaussian or student distribution in order to obtain the value of coverage factor k for gaiting assurance interval with result.

3.2 Type A uncertainty estimation

Type A evaluation of standard uncertainty may be based on any valid statistical method for treating data. Example is calculating the standard deviation of mean of a series of independent observation, using the method of least squares to fit a curve to data in order to estimate the parameters of the curve and their standard deviations is calculated.

Type A is defined "Uncertainties are evaluated by the statistical analysis of a series of observation" by ISO guide. In type A, that performs only normal probability distribution of measurement data, and parametric test method is useful ANOVA, so called classical statistics. The normal distribution is also called the Gaussian or the bell shape curve, it is ubiquitous in nature and statistics useful the central limit theorem. Every variable element can be modeled as sum of many small variations which are approximated normal distribution. it is the only stable distribution having all of its moment finite.

3.3 Type B uncertainty estimation

Type B is upgrading evaluation of uncertainty than type A for abnormal distribution in order to obtain assurance, it needs employs other method than the statistical method, and based on non-parametric test method. The other method than statistical was proposed MCMC to type B. Type B is defined "Uncertainties evaluated by means other than the statistical analysis of a series of observations." in ISO/BIPM guide.

Assignment of PDF to a quantity analysis is using the Principle of Maximum Entropy (PME). Therefore the PME tells us that the Assignment of PDF is rectangular, t-distribution and normalize distribution. (see Fig.3). Type B evaluation of standard uncertainty is usually base on scientific judgment using all information available. Type A evaluation of uncertainty based on limited data are not necessarily more reliable than soundly based Type B evaluations.

3.4 Gibbs sampling by Bayesian theory

Gibbs sampling is an algorithm to generate a sequence of samples from the joint probability distribution of two or more random variable. The purpose of such a sequence is to approximate the joint probability distribution, or to compute an integral (such as an expected value). Gibbs sampling is a special case of the M-H algorithm, and thus an example of practical use is MCMC algorithm. The algorithm is named after the physicist J.W. Gibbs, in inference to an analogy between the sampling algorithm and statistical physics. The algorithm was devise by brothers Stuart and Donald German, some eight decades after the passing of Gibbs.

Gibbs sampling is applicable when the joint distribution is not known explicitly, but the conditional distribution of each variable is known. The Gibbs sampling algorithm generates an instance from the distribution of each variable in turn, conditional on current values of the other variable. It can be shown that the sequence of samples constitutes a Markov chain and the stationary distribution. Markov chain is just the sought after joint distribution. Gibbs sampling is particularly well adapted to sampling the posterior distribution of a Bayesian networks, since Bayesian networks are typically specified as a collection of conditional distribution. The point of Gibbs sampling is that given a multivariate distribution, it is simpler from a conditional distribution than to marginalize by integrating over a joint distribution.

It is now the definitive document supplement of ISO-GUM on evaluating.

3.5 Assurance proceed

Quality Assurance (QA) of a measurement result is substantial by proposal new almost every year, QA is doing its best also in the field of a clinical examination to be able to respond to a patient or a donor effectively. The assurance performance of ISO-GUM is come out by set up of the confidence interval [4] and is decided. Therefore, it is considered that the value acquired by regulation of ISO-GUM is an assurance performance. ISO-GUM is edited in series to 98-5 from 98-1 as a guide of ISO/IEC. It has published by Joint Committee guide Measurement (JCGM). JCGM document number is JCGM 100-107.And ISO14253 [10] was publish as ISO standard.

ISO14253 (See Fig.4) contains decision rules which require the tolerance zone to be reduced by the measuring uncertainty. The measurement data are made to prove conformance a specification and expand by the measuring uncertainty. And it is attempting to prove non-conformance to a specification. It has known the legal phrase "prove beyond a reasonable doubt".

Specification has two clear limits lines. Uncertainty makes the question of conformance more complex. In a drawing specification, it is usually clear what the limits tolerance are, it may be a maximum acceptable cover factors value of upper limit and a low limit is reject line.

The method of setting up a confidence interval (zone) is defined by ISO-14253-1 as same as conformance zone. And measurement uncertainty is increased.[14]

However, the ISO-GUM approach exhibits some limitations, like:

1. Model linearization: The principle of error propagation applied to obtain the standard uncertainty truncates the Taylor's series expansion in first order terms. This is a linear approximation that in some cases could need terms of higher order.

2. Assumption of normality of measurand (z): In common practice, the distribution of the result is taken as normal and consequently. Expand uncertainty U(e) is calculated as the product of the coverage factor k and the combined uncertainty U(c). Thus k=2 is very commonly declared value, which corresponds to a level of significance of the approximately 95% in 2 sigma zone.
3. In calibration of the effective degree of freedom. If the distribution of z is approximated to a student's distribution, the coverage factor k is taken as the tabulated Student's t-value for given significance level and effective degree of freedom calculation by the Welch-Satterthwaite equation (6). In the general cases, the analytical evaluation of the effective degree of freedom is still an unsolved problem, type B uncertainties, generally contributing with infinite degree of freedom. However, the ISO-GUM 95 approach exhibits some limitations.
4. Modern concept of evaluation of measurement result uncertainty is based on the model function. Model linearization, the principle of error propagation applied to obtain the standard uncertainty truncates the Taylor's series expansion in first order terms. This is a linear approximation that in some cases could need terms of higher order as $Y=f(x_1,x_2,\ldots,x_i)$ as Eq.(1).
5. Eq.(1) applied to the use base on a first-order Taylor series expansion is approximation. Uc is gotten by doing a geometric mean of the type A and B result. This is model with one output which is adopted in current ISO-GUM. Knowledge of input quantities, which is compete, comes from their PDF. While the PDF has good theoretical foundations, the process of measurement modeling does not yet have them. There are no clues about it in the ISO-GUM. This is depended on experience data and prior knowledge.

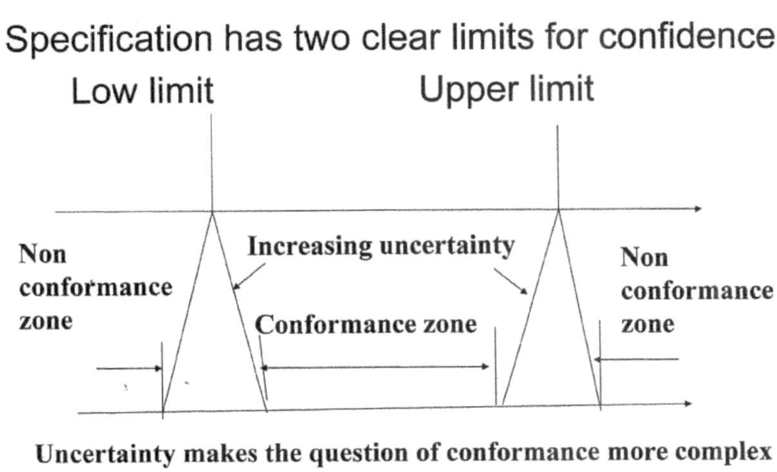

Fig. 4. Assurance interval by ISO-14253-1

If the measurand in expand uncertainty goes into a conformance (confidence interval or confidence zone), it value will be authorize the final report value as assurance performance.

4. Supplement of ISO-GUM

4.1 List of supplement

The list of supplements of ISO-GUM is shown below. (The mark * is under plans).

Introduction to the GUM related documents, there were published already three supplements.

1. Supplement1: Numerical methods for the Propagation of distributions using a MCMC method [3] by document JCGM101.
2. Supplement 2: Extension to any number of output quantities is useful models. JCGM102.
3. Supplement 3: Modeling for useful multivariable analysis. JCGM103.

An introduction to the GUM is under planning documents of supplement for ISO-GUM, there are four supplements (3).

4. *Supplement 4: An introduction to be GUM and related documents. Published 2009 JCGM104.
5. *Supplement 5: Concepts and basic principle of measurement uncertainty evaluation.
6. *Supplement 6:The role of measurement uncertainty in deciding conformance to specified requirements.
7. *Supplement 7: Application of the least squares method.

Data Quality Assurance Object (DQAO) development process consists of the following seven steps by supplement 4.

Step 1. State the Problem
Step 2. Identify the Decision
Step 3. Identify the Inputs to the Decision
Step 4. Define the Study Boundaries
Step 5. Develop a Decision
Step 6. Specify Acceptable Limits on Decision Errors
Step 7. Optimize the Design for Obtaining Data.

4.2 MA in supplement 3

Practical use of multi-variable analysis (MA) is required for BCA from two or more uncertainty elements being inherent. MA is defined in supplement 3. As for an algorithm, the theoretical formula (3) is usefully. The supplement 7 is considering using a least squares method as a base and using Burgers Equation and Crank-Nicolson method for a MA Propagation of uncertainty for several variable can be simple considerably if it is simple multiplicative of secondary variable.

QA of a measurement result is substantial by proposal new almost every year.

The practical use of MA is required for BCA from two or more uncertainty elements. It is expected that the analysis result can begin to find a new uncertainty factor.

5. Measurement principal

5.1 Radio-Immuno Assay (RIA)

The experiment should be use test reagents of RIA that is a kind of BCA. RIA is a scientific method used to test antigen (example, hormone levels in blood) without the need to use a bioassay. It involves mixing known quantifies of radioactive antigen with cold antibody to that antigen, then adding unlabeled (cold antigen) is measuring the amount of labeled antigen displaced. Initially, the radioactive antigen is bound to the antibody. When cold antigen is added, the two compete for antibody binding sites. The bound antigens are separated from unbound one. Radioactive isotope is used gamma emit of I-125. This is both extremely sensitive and specific, but it requires special precaution because radioactive substances are used, sophisticated apparatus, and is expensive. In this research, it is useful main data of Erastrse-1 regent which is a sort of pancreas hormone.[22]

Immunoassays are a form of macromolecular binding reaction; no covalent chemical bonding is involved. Antibodies interact with their antigen by weak hydrogen bonding and van der Waals force. Antigen-antibody reactions are dependent on complementary matching shapes being assumed by antibody variable regions of the immunoglobulin. Almost all polyclonal antibodies used in immunoassay reactions are of the immunoglobulin G class. The N-terminal 110 amino acid residues of both the heavy and light chains of the immunoglobulin molecules are variable in sequence and interact to form the antigen binding site. Nevertheless, this variability gives rise to a vast array of difference antibodies binding to different molecules. Attributes of an ideal immunoassay is shown to follow

1. The immunochemical reaction behavior should be identical and uniform for both the inference preparation and the analysis in the homogeneity sample.
2. The immunochemical reaction of the antibody reagent is uniform from batch to batch.
3. The immunochemical method is well standardized to ensure that the size of measurement signal is caused only by the antigen-antibody product.
4. For macromolecules the results are declared in arbitrary unit, i.e. international Unit (IU) conversion to mol/L(SI) unit is constant and is dependent on many factors.

The immuno-reaction of an antigen and a catalyst (an antibody activity) is led to measurement theory of RIA with reaction kinetics. Reaction kinetics is expressed with the chemical equivalent amount compound by unit time t. The chemical reaction is shown in Fig 5. The process of chemical change is divided into three phases that are a signs phase at growth phase and stagnation phase. A signs phase shows the resistance characteristic over a reaction in the preparation step of a reaction. A growth phase shows reaction capability in the stage where a reaction grows rapidly. A stagnation phase with asymptote which is a stage where the chemical reacting finally and reaches a chemical equilibrium. An uncertainty element can be expressed with the abnormal state of reaction time and reaction capability.

The strength of binding is determined by equilibrium binding constant for the antigen-antibody complex formations. The binding follows the basic thermodynamic principles with reversible reaction between two molecules. This relationship is described by the chemical reaction model as (7) which is useful a chemical kinetics reaction R_1 of immunity with a reversible reaction R_2 in a polynomial expression.

$$P+P^*+Q \overset{R1}{\underset{R_2}{<=>}} P^*Q+PQ+PI+P^*I \rightarrow P^*Q \quad (7)$$

Where are:

- P: antigen,
- Q: antibody,
- P*: Labeled antigen with detectable marker,
- PQ: Reaction compound,
- P*Q: reaction compound with P*,
- $R_{1;:}$ association constant
- R_2: disassociation constant
- R; affinity.(Binding ratio P*Q/Po= %)
- Po is total antigen Po=PQ+P*Q+PI+P*I
- PI: Abnormal reaction compound.

Po is invariably an usual constant of nature in the "law of mass action". In this case, it must be the measured P*Q/Po of the effective binding ratio. Affinity is the same as the kinetic reaction rate. An affinity increases until the deactivation of saturation that the reaction is based in the reaction process in time. Kinetic chemical reaction is accelerated by bio-materials as a quadratic differential polynomial equation with reaction time t. The secondary order differential equation is as follows according to three phases as (8)

$$Ad(P^*Q)^2/dt_2+Bd(P^*Q)/dt+C(P^*Q) \quad (8)$$

The extent of reaction is shown in Fig.5. Secondary order differentiation of the first item shows the rate of acceleration included a special resistance phase. It is shown to the portion of start of the reaction curve in Fig.5. A primary differentiation of the next item is shown reaction velocity. The last item shows the amount of chemical compound after reached chemical equilibrium as stoichiometry. The first item has role important for security of reaction and the analysis of an outlier.

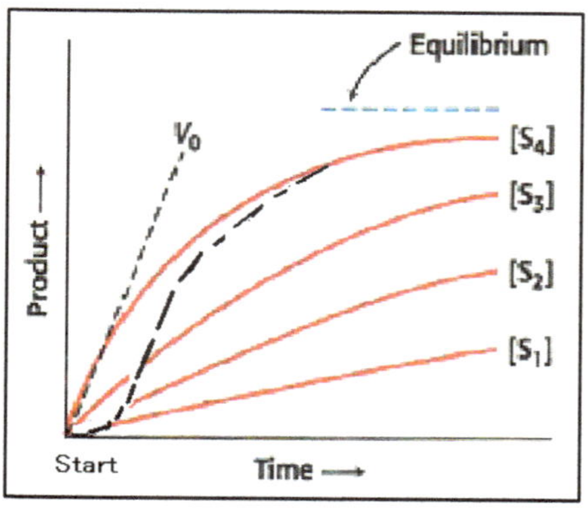

Fig. 5. Elastase-1 chemical reaction

Kinetic reaction ratio is estimated as follow

$$d[P^*Q]/dt=R1[P^*][Q]-R2[P^*Q]... \quad (9)$$

$$k1[P^*][Q]=(R1-R2)[P^*Q]............ \quad (10)$$

The final reaction a stagnation phase is reached on chemical equilibrium. It become to $d[P^*Q]/dt=0$ and change from (9) to (10). An affinity state is shown on the reaction curve at time.

5.2 Calibration curve

In this research, a purpose is improving the accuracy of the calibration curve used for quantitative analysis and making it level which can be assurance. The accuracy of calibration curve is important as intermediate accuracy of the whole measurement system. The measurement data was collected by equalization of double sample in order to make high order accuracy. The Data Quality Object (DQO) of the improvement accuracy targeted the calibration curve for a chemistry quantity analysis.

Even in the case of measurements requiring multipoint linear (non-linear) calibration with five or more different concentration reagents of reference homogeneity material. The uncertainty of routine test values can be quantified using basically the same procedure as "Estimating the uncertainty of routine test values" except that the uncertainty of the reference material is calculated as a combined standard uncertainty using the mean value obtained by averaging the standard uncertainty (see to Fig.6).

The reference material used should be an actual sample, the property of which is similar to the patient spaceman to be assayed. Even in the multi-point linear calibration with three or more different rations concentrations of reference material, the uncertainty of routine test values can be quantified using basically procedures. Test reagent is called calibrator.

All the product calculation curves are having the quality verified by formula of Mechaelis-Menten formula.

Key aspect is the Antigen-Antibody interaction.

1. Reaction is reversible and favors complex formation under physiological condition.
2. Binding depends on hydrogen bonds, van der vaals forces, ionic bonds, hydrophobic interaction.
3. Binding is very specific and requires the correct 3-D structure of an antigen.
4. The amount of complex formed depended on concentration of antibody and antigen. Both antigen and antibody (if large enough) have multiple sites for binding to occur. Therefore, extensive cross-linking can occur when both are present in solution. When antibody and antigen reach equivalence large immune complexes form which can precipitate out of solution.

6. Experiment method

6.1 Progress of work

In immunochemical analysis can be nonlinear analysis that is applicable of estimating to the uncertainty of assigned value of calibrators and QA control sample.

The purpose of such a sequence is to approximate the joint probability distribution, or to compute an integral (such as an expected value). The estimation of a possible discrepancy takes into account both random error and in the measurement process. The distinction to keep in mind and with regard to be able corrected or cannot be corrected theoretically least.

Gibbs sampling is particularly well adapted to sampling the posterior distribution of a Bayesian networks, since thus are typically specified as a collection of conditional distribution. The point of Gibbs sampling is that given a multivariate distribution, it is simpler to sampler from a conditional distribution than to marginalize by integrating over a joint distribution. The algorithms are useful Softcomputing method, fundamental statistical method, MCMC and MA, Softcomputing method is fuzzy function and chaos function.

6.2 Data quality object(DQO)

This experiment used the test reagent kits of the Elastase-1 that is one of test reagent of pancreas hormone mainly. The test method is RIA [17], RIA is a kind of BCA with detectable marker that is labeling the radioactive material of I-125. A measurement method is detection of the gamma ray which I-125 emits.

The calibration curve made of the test reagent of six sorts of concentration (dose) that is the arrangement harmonious in the shape of a stairs state. A calibration curve is created by applying regression analysis to a frequency density of measurement result. The composition of reagent concentrations are six sorts of 50, 150, 500, 1500 and 5000 dose. Dose is international catalyst unit (IU). Test reagent is called as calibrator. The number of sample size is total 320 sets of calibrators that is divided three groups (lots) that consists one group of 120 sets and two groups of 100 sets. The number of one group is recommended the least in ANOVA. It was made three groups here in order to investigate whether a difference exists between groups. Fig.6 shows the procedure which creates distribution of statistics from the population calibration curves to which regression analysis was applied the frequency density distribution is created by measurement data and analyzed. In an immunoassay the total amount of antigen for standard solution is known by each dose. Plotting affinity against total antigen produces the standard curve. The value for affinity is determined by radioactivity amounts remaining bound divided by the total amount of radioactivity added in the beginning. The radioactivity is proportional to the concentration of the labeled antigen. It is using this standard curve and the affinity for an unknown sample the antigen concentration can be calculated. The dose response curve shown is that a typical of RIA standard curve that is used calibration curve.

In an immunoassay the total amount of antigen for standard solution is known by each dose. Plotting affinity (binding ratio) against total antigen produces the standard curve. The value for affinity is determined by radioactivity remaining bound divided by the total amount of radioactivity added in the beginning - the radioactivity is proportional to the concentration of the labeled antigen. It is using this standard curve and the affinity for an unknown sample the antigen concentration can be calculated. The dose response curve shown is that a typical of RIA standard curve.

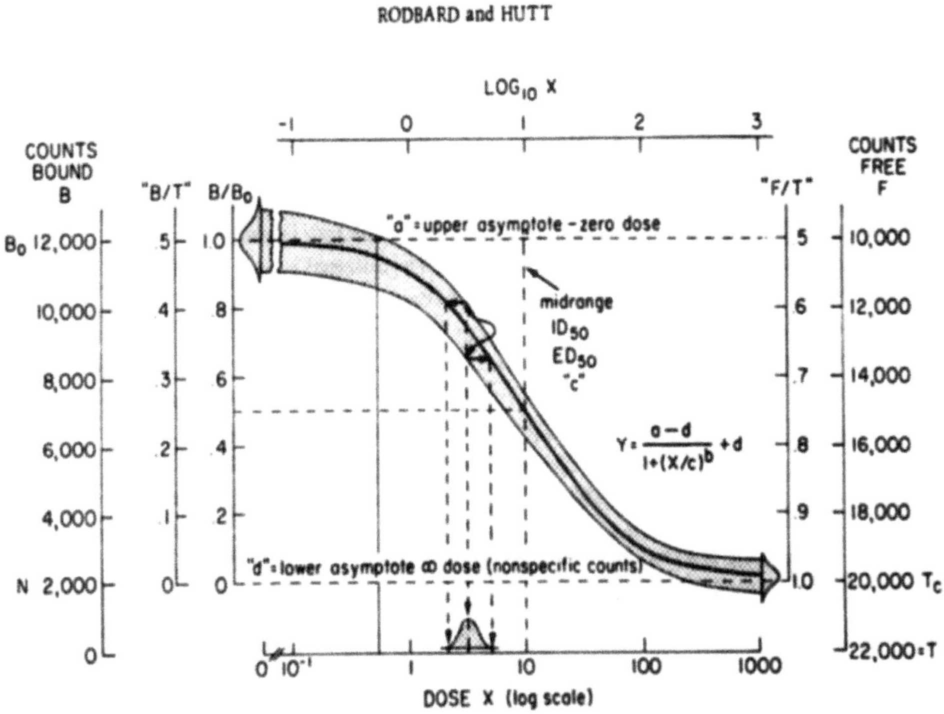

Fig. 6. Make a probability distribution of Calibration curve batch, for QC[16].

6.3 Additional reagent samples

In this experiment, Elastase-1 sample data is verified whether it would be added other test regents of two homeopathic test reagents of the same usage RIA and test chacking.. Furthermore, in order to investigate an information criterion number and likelihood test and it used multivariable analysis in the time series.

Two sorts of homeopathic reagents having chosen, there are Thyroxin, and Testosterone.

The reagent of Testosterone is one female hormone and the composition of reagents concentrations are 6 sorts dose which are 25, 50, 100, 250, 500 and 1000. The reagent of Thyroxin is one of thyroid hormone and the composition reagent concentrations are 5 sorts dose which are 0, 3.0, 6.0, 12, and 24.

7. Experiment result

7.1 PDF data of elasyase-1 reagents

Measured data is shown in Fig.7 and Fig 8. Both figure quotes the pareto graph that is a pile of the quasi normal distribution curve (solid curve B) on the measured PDF graph (bar graph A). Fig.7 shows 0 dose reagents data as the largest affinity in six sorts of calibrators of reagent, Fig.8 shows 5000 dose reagent data as the lowest affinity in the six dose sorts. A bar data serves as basic data from now on advancing analysis which is observation data as A in Fig.7 and Fig.8.

Quasi normal distribution is created by regression analysis from the measured frequency distribution as B in Figs.7 and Fig.8. The ordinate shows the account of frequency density of total 320 sample sets. The abscissa shows affinity (%) divided into 20 steps between maximum and minimum affinity.

PDF graph is carrying out random change in any range of sample groups. The PDF of random variation exists completely characterized in both figures.

Both figures are chosen as example of representation from six dose data. Six sorts all concentrations of PDF data were shown the abnormal distribution. All PDF data of measurement results of six sorts were not same as the form. Abnormal distributions are cannot computed both skewness and kurtosis.

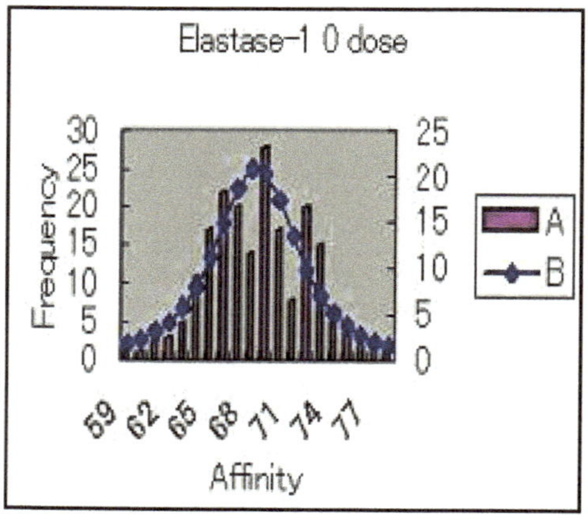

Fig. 7. PDF of 0 dose data of Elastase-1

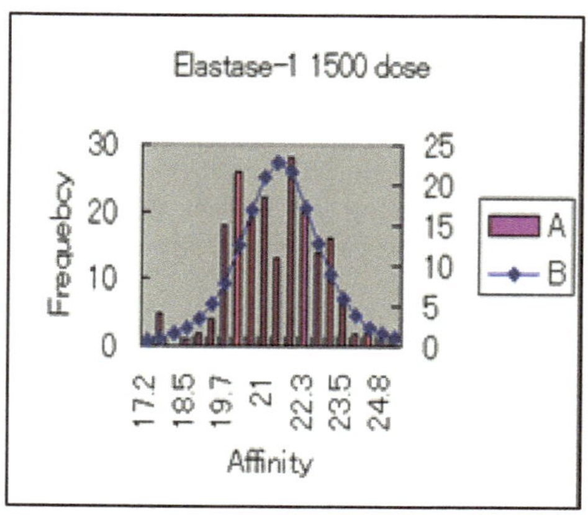

Fig. 8. PDF of 5000 dose data of Elastase-1

7.2 The superposition graph of six PDF data

Fig 9 is shown a superposition graph by six measured PDF forms on the variation of affinity (%) of maximum and minimum positions. All of PDF are shown not same form.

Fig.10 is shown a superposition graph the abnormal distribution of six sorts curve obtain by regression analysis, because all distributions are not take same as peak position.

The ordinate is the account of frequency of total 320 sample sets. The abscissa is shows affinity (%=P*Q/P_0). The abnormal PDF distribution should be required to create the CDF follow up MCMC.

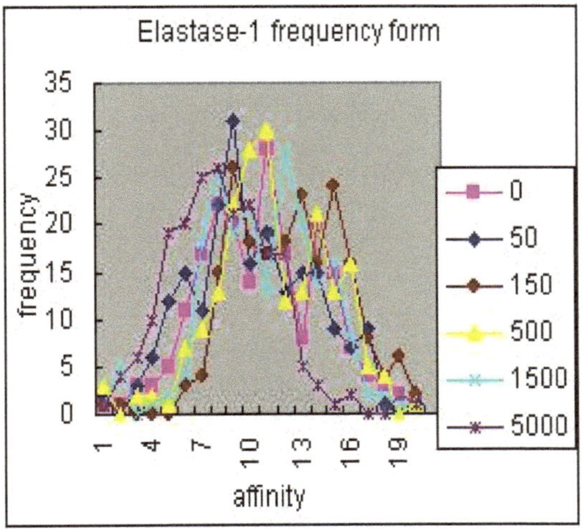

Fig. 9. A superposition graph of six measured PDF.

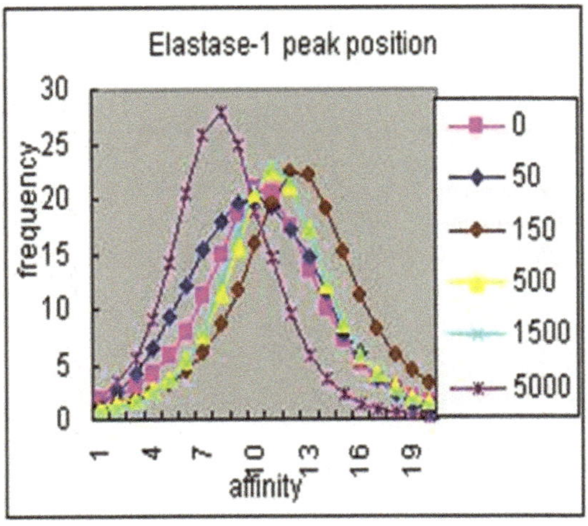

Fig. 10. A superposition graph of six the abnormal PDF.

7.3 fundamental statistics quantity

Table 2 shows what summarized the fundamental statistics quantity (AIC, mean, RSS, Medi, peak, Max, Min, F-test and t-test) and the data of EDA added the operation result by chaos theory and Fuzzy logic. Max is the maximum value in upper limit of confidence interval and Min is the minimum value in the lowest limit of confidence interval. Max and Min data use primary confidence interval. Fuzzy logic use also 20 steps member function in fuzzy logic, Chaos theory use difference equation in nonlinear.

Some value shows the same value based on many columns in table 2. Six columns (Fuzzy, AIC, mean, RSS medi and peak) are shown in red bold letter. The focus inside of the same value is the mean. The mean value can be referenced as a central value by data in this experiment. This value can be set as a reference value by Type A for internal quality control (IQC). The assurance interval by type A was calculated by 95% level and coverage factor k=2. It is 2 sigma of a routine test level and it can be assurance against 95% level which is given by the 0,025 and 0.975 fractions limited [7]. If measured data is over the limited line, it is unclear that dispersion is large, so that the affinity was a low value.

Dose	0	50	150	500	1500	5000
Chaos	71.84	65.94	54.93	36/.77	22.5.2	12.17
Fuzzy	**68.5**	**61.8**	**52.1**	**34.8**	**20.7**	**10.4**
AIC	**68.46**	**62.42**	**53.0**	**34.97**	**21.34**	**10.59**
mean	**68.4**	**61.8**	**52.2**	**34.7**	**20.8**	**10.5**
RSS	**68.3**	**61.9**	**52.1**	**34.7**	**20.7**	**10.4**
Medi	**68.3**	**61.7**	**53.1**	**34.8**	**20.6**	**10.4**
peak	**68.3**	**62.2**	**52.1**	**34.9**	**21.5**	**10.6**
Max	78.2	71.8	60.0	43.1	28.9	17.1
Min	46.7	41.1	34.4	21.7	12.4	6.9
F-test	2.17	2.16	2.17	2.17	2.17	2.17
t-test	2.03	2.02	2.03	2.03	2.03	2.02
EFD	0.18	0.17	0.19	0.17	0.17	0.18

Cipher cod of result
Chaos chaos theory: Difference equation in nonlinear
Fuzzy: 20 steps membership' function
AIC: Likelihood function (Akaike Information Criterion)
RSS: Root Sum Square
Medi: Central value
Peak: peak point.
Max: Maximum value in distribution
Min Minimum value in distribution
F-test F-distribution
t-test t-distribution
EFD: Effective Freedom Degree

Table 2. Fundamental statistics quantity by type A

7.4 CDF data

CDF data shows in Fig. 11 and Fig 12. These data quoted a pareto graph that is a pile of CDF distribution curve (solid curve B) on the PDF (bar graph A). Fig.11 shows the data on 0 dose reagents as the largest affinity in six sorts of reagent, and Fig.12 shows the data on 5000 dose reagent as the lowest affinity in six sorts of reagent.

This research analyzed of abnormal portion looked on CDF curve in Figs.11 and 12. By ISO-GUM of type B. it acquire further higher measurement accuracy and, it is required to analyze a fault element by FTA and EDA. Here, it has to process by type B for abnormal distributions. Fig 7 and Fig,8 data was useful basic data in this research. In both Figs 11 and 12, the reagent kit is divided into three groups (lots) and it shown influence to verify of difference between in reagent kits. The ordinate of is the account of frequency of total 320 sample sets. The abscissa is shows affinity (%=P*Q/P$_0$) which All graphs quote variation of affinity that divided the affinity into 20 ranks between maximum and minimum. The ordinate quotes generating frequency counts.

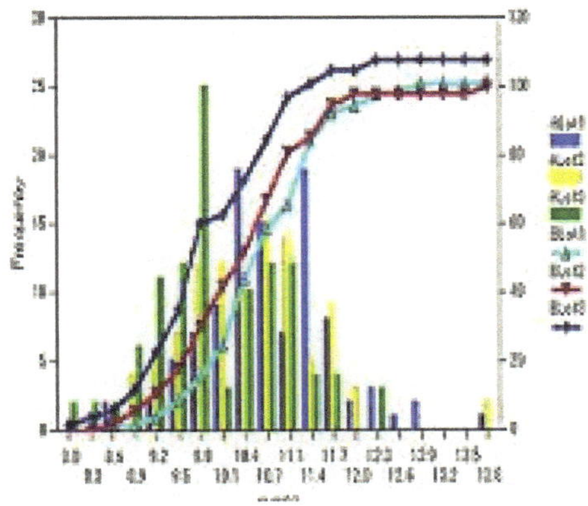

Fig. 11. Elastase-1 0 dose of 3 rots graph.

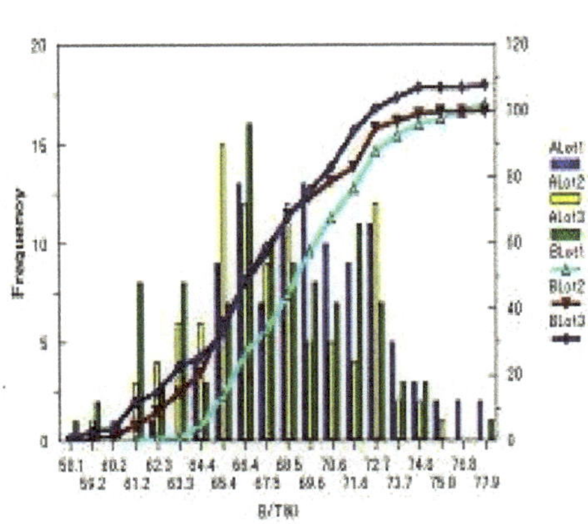

Fig. 12. Elastase-1 5000 dose of 3 rots graph

The CDF curve shown the form of sigmoid and bending has seen in some portion. The bend of a curve suggested that abnormalities existed in a chemical reaction.

All of CDF result (including the four sorts of other dose) showed a state of an abnormal distribution similarity and not same form. The abnormal distribution of PDF should be required to useful MA that is nonparametric test and nonlinear analysis.

7.5 Confidence zone (interval)

The protocol of QA was development to estimate the uncertainty of measurement of a chemical analysis by utilizing in house validation studies. The approach was to generate an estimate of the uncertainty across the analytical concentration range. [21]

Table 3 shows amount of calculation result by the Welch Satterthwaite as Eq (3) for coverage factor simulation. Value in table 3 is converted into an effective free degree from the Welch Satterthwaite factor by standard statistical table showing EFD in table 3. The experiment result needs to set coverage factor k to 2.2 by standard statistical table. Because coverage factor 2.0 is generally used as standard. It is necessary to make narrower than a general value the confidence interval which can be assured in this case.

Table 4 shows the confidence interval and related data for QA, a result is expressed numerically and it is. Assurance Interval (AI) data shows final assurance value. Reject zone is shown also in the t-distribution and the square distribution in MCMC of Fig.3.

The assurance interval (zone) determined based on type B of ISO-GUM, in this case is required to set the narrower interval than getting ANOVA value for external quality control (EQC) that is required 99.7% as more than three 3 sigma. Only the measured value which exists inside an assurance interval turns into assured performance.

The measurement vale whose assurance is attained only a mean value which is exist in the confidence zone (see Fig 2) inside assurance interval (AI in table 4) that is authorize.

In the case of a type A evaluation of uncertainty, repeated measurement indications are regarded as independently drawn from normalized frequency distribution and according to its suggestion the uncertainty evaluation method of supplement 1 is applied after having assigned scaled and shifted t-distribution to corresponding input quantities.

Next research investigated the cause of the abnormal part shown in curve of CDF. FTA and RCA performed the method.

Sample size	0	50	150	500	1500	5000
102	17.18	16.83	16.28	16.62	16.64	18.22
100	18.22	16.87	16.28	16.45	16.88	16.84
110	18.27	18.27	18.68	16.07	18.19	18.51
29	7.31	7.44	7.3	7.31	7.27	7.29
341	60.98	59.39	64.24	56.45	58.97	34.1

Table 3. Calculation result by the Welch Satterthwaite

	Dose	0	50	150	500	1500	5000
SD	Central	62,4	56.4	47.2	32.4	20.6	12.0
	0.95 UA	8.64	8.43	7.03	5.86	4.53	2.8
	CIA	53.8 71,0	48. 64.9	40.2 54.2	26.5 38.3	16.1 25.2	9.2- 14.8
N.D	Central	68.3	62.2	52.1	34.9	21.5	10.6
	0.95 UB	7.26	6.54	6.11	451	3.11	1.84
	CIB	61.- 75.6	55. 68.7	46- 58.21	30.3 49.6	18.4 24.6	8.76- 12.4
Us	UC	11.2	10.67	9.22	7.32	5.40	3.29
AI	CIC	**62.4** **73.9**	**56.8-** **67.7**	**4.61-** **56.7**	**31.3** **38.6**	**18.8** **24.2**	**9.0-** **12.2**

Cipher cod of result
SD: Standard Deviation
ND: Normalize Distribution
Us: Standard Uncertainty
AI: Assurance Interval
Central value
0.95UA: 0.95%x2 Uncertainty type A
0.95UB: 0.95%x2 Uncertainty type B
CIA: Confidence Interval type A
CIB: Confidence Interval type B
UC: Combine Uncertainty
CIC: Confidence Interval for assurance
UC: Combined uncertainty

Table 4. Accuracy interval and related data

7.6 MA date

7.6.1 Data of elastase-1

Uncertainty associated with measuring operation which are calibration dispersion, within-day and between days, within-laboratory and between- laboratory dispersion, and the like, including factor due to reagent preparation and instrument variation. In co-data which is inherent in a measurement result, two or more uncertainty factor analyze by MA. The time depended uncertainty analysis is day to day variance in this experiment.

In former research, it used multiple-regression for MA based on "Law of propagation of Uncertainty" in EDA and has found out a new uncertainty factor. Fig.13 and Fig.14 is importance data in which is new variable element as the influences for storage days. While the reaction capability is quoted fall down of a test reagent for storage days.

Fig.13 shows changing of the reaction capability by storage days of 0 dose reagent. Data of Elastese-1, it is data comes to deteriorates in right going down in proportion to the increase in days to day. To while the reaction capability of a test reagent downs for storage days.

Fig.14 shows changing of the reaction capability by the storage period days of 5000 dose reagent. Data shows a degradation of reaction capability that reaches a detectable limit and it is meaning the unstable state of large uncertainty.

In both graph, the ordinate is the affinity (%). The abscissa is shown storage days.

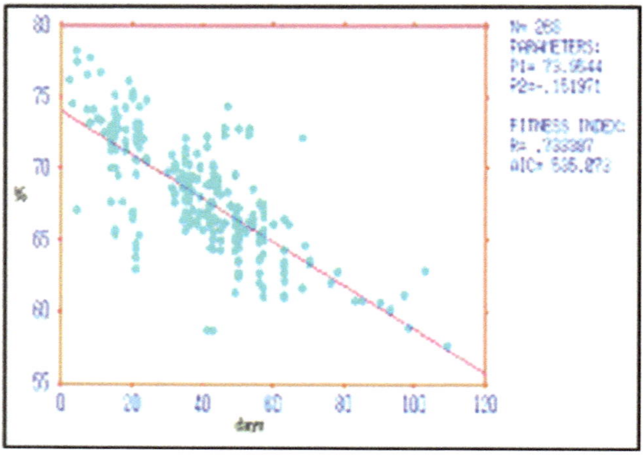

Fig. 13. Change under storage days of 0 dose.

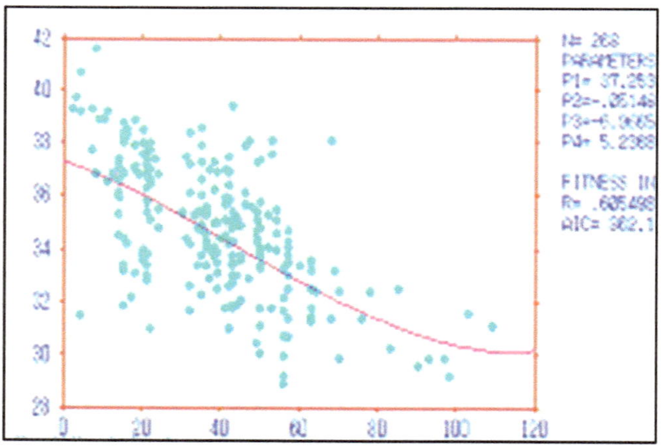

Fig. 14. Change under storage days of 5000 dose

Fig.15 is shown change of SD value under storage days by EDA. In Fig 15, SD value has fallen so that the concentration of reagent as which it is regarded the change united with the biorhythm of 28 diurnal periodicity exists. SD value with reaction capability is falling as to slide with concentration dose. Data of Fig.15 has mean very important to improved accuracy as fault elements The abscissa shows the storage days of every seven days interval in Fig.14. The change interval was a periodical target on the 28 days interval of biorhythm. In the addition, in the domain of low binding capacity, all the unstable of accuracy are as same.

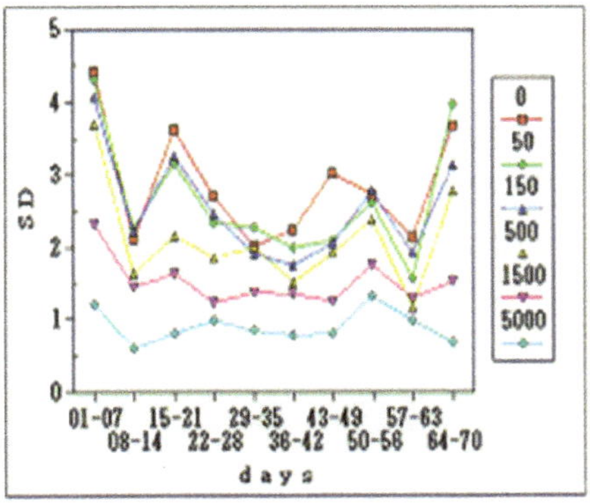

Fig. 15. Storage days variance by type B of EDA

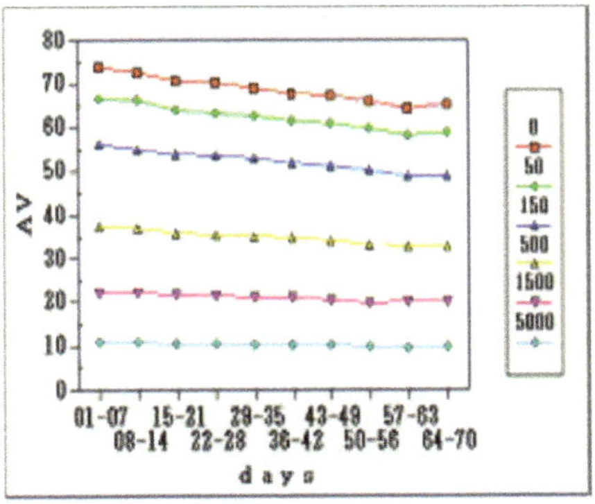

Fig. 16. Storage days variance by type A of ANOVA

Fig.16 is shown change of average value under storage days by the data based on ANOVA, only a flat change of right going down is shown and a periodic change is not seen.

7.6.2 MA data of additional reagents

In this experiment, since it seem that the data of 7.6.1 is important. I verified whether the same result would be obtained with two homeopathic test reagents of the same usage RIA. Furthermore, in order to investigate an information criterion sample number and likelihood test and it used multivariable analysis in the time series. The results are shown Fig.17 and Fig.18.

Fig.17 and Fig. 18 shows the graph which laid change of SD and of the number of samples on the top of storage days. Fig. 17 shows Thyroxin data. Fig.18. shows Testosterone data.

In both figure, although a periodic change is looked at by change of SD for under every storage days of six sorts dose as Standard Deviation (SD) in order to explore the root cause of fault elements, it will become unstable data few samples. The both figures showed the same characteristic results. Thereby, the check of the reproducibility of an uncertainty factor was completed. About the number of samples, it is the information criterion by maximum likelihood theory need to be inquired. This research is under experiment. In Fig.17 and 18, it is unstable area by number of five or less samples. The ordinate shows SD and number of samples. The abscissa shows the storage days of every seven days interval. The ordinate shows SD and number of samples In both graph, the ordinate is the affinity (%) and frequency. The abscissa is shown every seven storage days.

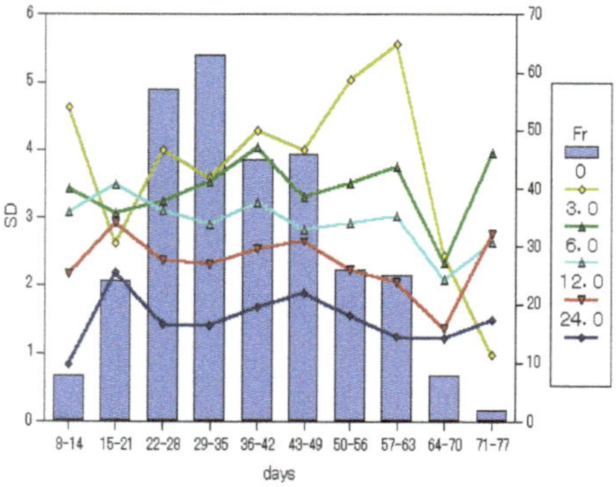

Fig. 17. Shows Thyroxin data.

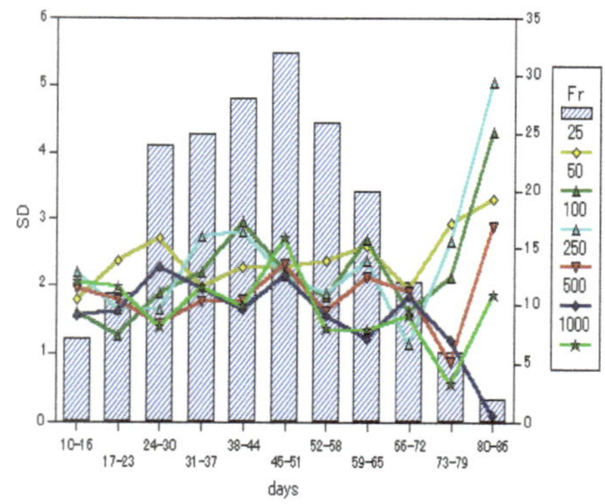

Fig. 18. Shows Testosterone data.

7.6.3 Interaction and allosteric effect in data

The purpose of this chapter is to outline methods for assessing uncertainties related to material inhomogeneous that can be a factor in uncertainty analysis (see chapter 6.2).

One more is found in the uncertainty factor. The interaction effect and the allosteric effect are generating by the source of biorhythm. Binding of antibodies to antigens is reversibly and very specific. The introduction of an immune response depends on the size of antigen. Small molecular weight compounds (<2000 Daltons) such as drugs is unable to induce antibody formation.

The interaction investigated the phenomenon prevented from fundamental reaction principle. The example of calibration curve has generating of an interaction which is shown sample in Fig 19. The interaction and allosteric effect were considered by from the experiment results of in fig.19. Ordinate is affinity (%) of reagent. Abscissa is dose of reagent at table position as from 0 to 5000.

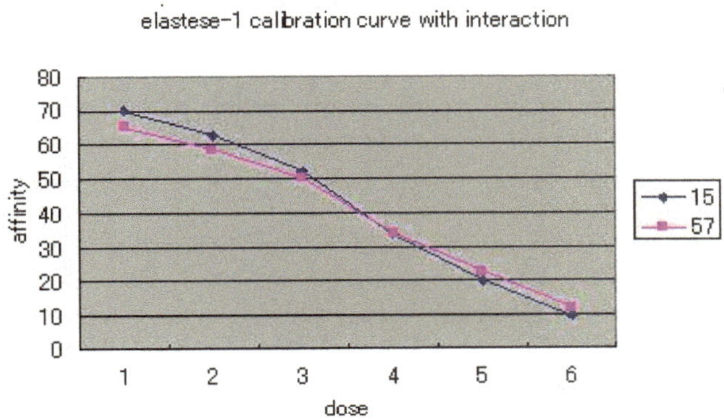

Fig. 19. Interaction is exact between 2 calibration curves as an example

Some one exists in the allosteric effect which inhabits reaction process. Both effects like in the bio- science on theorem of logistic function.

8. Conclusion

QA system of clinical test data by ISO-GUM is made utilization in 2006. It included in IT system plan for medical health care in 2010. This research is continued in order to reliance of QA further. In ISO-GUM it is pursued two or more buried multivariable uncertainty factors. Validity of ISO-GUM is increasing by additional issue many supplements.

This research started for the purpose of preventing the clinical misdiagnosis by ambiguity data on medical care. Therefore, the work is obtaining that of data stable in higher accuracy has been continued. Improve strategy found out by uniting QE and ISO-GUM. The assurance of measurement data which was able to be attained new technology for is main purpose. Clinical data obtained and expected from exact prediction of patient individual's pathological change. After that, a result came to be utilized for the world wider base medical care of EQAS when improvement of ambiguity was obtained. The result of research has been satisfied,

Uncertainty of measurement, traceability and numerical significance are separate but closely related concepts that affect both the format and information conveyed by a quantitative test result. In addition, use of SI units provides a consistent basis for the reporting of clinical laboratory. Katal as a SI unit which evaluates the reaction kinetics of chemical will be used for the near future.

The experimental result sees enable exact diagnosis decision taking in Bayes inference to QE and ISO standard. Medical Laboratory Quality System (MLQS) is essential in laboratory to the correct result for patient and donor by Good Laboratory Practice (GLP)

QA has grown to be equivalent QC which can obtain the same result when and anywhere.

9. References

[1] BIPM. IEC. IFCC. ISO. IUPAC. OIML."Guide to the Expression of Uncertainty in measurement" in 1993, revised and corrected 2nd ISBN 92-67-10 pp188-9, 1995

[2] Joint Committee for Guides in Metrology (JCGM). "Guide of the expression of uncertainty in measurement". Supplement 1 In 2006

[3] Barry N Taylor and Chris E Kuyatt. "Guideline for Evaluating and Expressing the Uncertainty of NIST Measurement result". {NIST Technical report Note 1297 in 1994 edition}

[4] Mario. Plebani "Errors in clinical laboratories or errors in laboratory medicine?" Clin. Chem. Lab. Med. 2006. 44(6)

[5] Barry Taylor. Chries E.Kuyatt "Guidelines for Evaluating and Expressing the Uncertainty of NIST Measurement Results" NIST Guide for Evaluation and Express. Measurement. Uncertainty. In Sept 1994.

[6] Paula. R.G.Couto. Jailton.C.Damascene. Renata. M.H.Borges. "Uncertainty estimation of mechanical assay by iso-gum 95 and Monte Carlo simulation-case study: tensile strength, torqe and brinell hardness measurements" IMEKO September 17-22 2006 in Brazil

[7] Lu Tang and Scott Willi "Uncertainty calculation in Analytical Chemistry" Chemical Metrology, INNS, URCC. February 2001

[8] G.Fanti. A "Possible improvement to the iso-gum", {IMEKO-2000 World Congress, Vol.9, pp.77-82, in Austria, 2000}

[9] Voluntary EMC Laboratory Accreditation Center Inc. "Guideline for Traceability of Measurement" VLAC- VR103 Second Edition in 2002.12.19

[10] K.Gregory. G. Bibbo. and J.E.Pattison. "A standard approach to measurement uncertainties for scientist and engineers in medicine". Australasian Physical & Engineering Sciences in medicine. Vol.28 No 2 P131-139 in 2005

[11] Jerzy M.Korczyski. "Calculation of expanded uncertainty" Joint IMEKO TC-1 & XXXIV MKM Conference 2002. {Wroclaw 8-12 September, 2002.}

[12] U.M.Koczynsky. "Calculation of expanded unceriainty" IMEKO world congress in 2000

[13] Pawel, Fotowics, "Method for calculating the coverage factor in calibration".{OIML Bulletin XLIII No.4 October, 2002}

[14] Dr.Henrik.S. Nielse "Iso-gum 14253-1 Decision Rules –Good or Bad.?" National Conference of Standards Laboratories. Workshops & Symposium In 1999

[15] Mauro. Panteghini. "Traceability, Reference Systems and Result Comparability". Clin Biochem. Rew. Vol.26. p97-104 August 2007.

[16] RODBAD and D,M.Hutt. "Statisticl analysis of immunoraiometory" (Labelld antibody ASSAY.) {IAEA SM-!77/278, 1977

[17] Colin.Selby "Inference in immunoassay" Ann.clin.Biochem. Vol.36. 1999.pp704-721

[18] Dorozhovets Mykhaylo. Warsza Zygmunt Lech. "Methods a upgrading the uncertainty of type A evaluation (2)"

[19] Enginering statistics handbook issue NISTNIST/SEMATECH, issue NIST in 1992

[20] Maurice Cox, Peter Haris "Up a GUM tree? Try the Full Montel" National Physical Laboratory. Tedding Middlesex. UK. TW11. OLW

[21] Greg O Donnell and Robert Geyer "The Estimation of Uncertainty by the Utilization and Quality Control Data" Thornleigh Ver, 1 18/06/2001

[22] Wikipedia, the free encyclopedia "Radioimmunoassay"

4

The Development and Changes of Quality Control in Japan

Kozo Koura
Kozo Koura & Associates
Japan

1. Introduction

1.1 "Investigation and research era of quality control" (1946-49)

In 1945 after the Second World War, we were nearly on the verge of starvation: over 80% of our industrial facilities had been destroyed, and industrial production dropped to a little over 10% of the prewar standard. It may tell that, Japan was staying without natural resources, nothing but "industrious people and its brains"1).

Japanese Standards Association (henceforth JSA) was established under instruction of Industrial Standardization Law in 1945 and Union of Japanese Scientists and Engineers (henceforth JUSE) was organized in 1946 and have become the mother's body of the quality control promotion of our country since then. While Chairman of JUSE, Ichiro Ishikawa (the first chairman of Keidanren: the Federation of Economic Organizations) planned the overseas technical investigation team composed by the members of learning and experience for revival of Japan industrial business in 1948 with approved the grant of investigation research cost of Economic Stabilization Board, and as a result of investigation it was proposed to introduce "Quality Control (henceforth QC)" into Japan. JUSE carried out a QC seminar (Quality Control BASIC course) in 1949. QCRG (QC Research Group) organized by the members of a lecturer of this seminar had exerted them for development of quality control in Japan later.

In "the statement of the first publication" of the first-issue of the "Quality Control" (present "Quality Management") in March 1950, Ichiro Ishikawa proposed as "For reconstruction of a peace country, a cultural country, and a democratic country, the product of our country must enable it to be positive competitiveness in the world market dignifiedly heading toward 'The Figure which should exist' for the future of our country economy and industry (sentence abbreviated)"2), as the vision of QC of Japan.

2. "SQC Ers" (1950-54) and "Era of systematic management reinforcement of QC" (1955-59)

Dr. W.E. Deming was invited and the eight days course of QC was held in 1950. The Deming Prize founded a prize to commemorated Dr. W.E. Deming's contribution and friendship in a lasting way and to promote the continued development of Quality Control in

Japan in 1951, and it became the Driving Force of QC Development of subsequent Japan. Spread and application of statistical quality control (SQC) were prosperous in this time, and it was called SQC Era. Also, the management and improvement by control chart, process control table, or process analysis were advanced by the "Deming Cycle", and it was also called as "Process-Control Priority-Focus Era".

Dr. J.M. Juran was invited in 1954 and his seminar was opened, under Dr. Juran's concept of "QC is a part of business management", and it went into subsequent "the Era of Systematic Management by reinforcement of QC", and progressed "Establishment of the Concept and the Technologies of Management". A Deming Cycle was generalized as a PDCA (Plan-Do-Check-Act) Cycle, and systematization and institutionalization of QC, and the QC Activity System were advocated.

3. "Total quality control (henceforth TQC) era" (1960-69)

By the 1960, "Quality Control magazine" proposed the special issue for "QC Implementation with All Member" by one year's project and the TQC of Dr. A.V. Feigenbaum was also taken into and spreaded, and which were later developed with a next Japanese QC (Japanese TQC). That is, triggered from Dr. Juran's "General management" lecture, the concept of "Management Item" was introduced and further developed into the "Management Item Table by position and rank", "Flag System", and "Policy Management (Management by Policy)".

Further, expanding into substantial strengthening on "Initial Production-Flow Line Quality Control" in new product development and on sale stage and "Vender-Vendee Relation's QC 10 principles", and eventually "Quality Assurance System" were established, and these views were further developed into the improvement of management efficiency, i.e., "Cross-Functional Management" by establishment of management system of quality (Q: Quality), the cost/ price and profits (C: Cost/Profit), and the time for delivery and the quantity of production (D: Delivery/Volume). Then "Policy Management" and "Cross-Functional Management" were established as the fundamentals of business management system. As horizontal management against ordinary vertical management in organization.

On the other hand, QC Circle was born in 1962, the base of "management of humanity respect" was built, and "seven QC tools" and the "QC story for problem solving" were developed as the tools of the activities.

The 1st Quality Control Symposium (henceforth QCS) was held in 1965, and its subject had taken as "Introduction, Promotion and Rooting of QC". The QCS had the role of discussion and discovering right approaches on QC under the collaboration of industrial and academic worlds. "Six Items of the Specific Features of TQC in Japan" was deliberated upon them and announced at the International Conference on Quality Control in 1969 (ICQC'69 TOKYO).

4. "Era of establishment of TQC" (1970-79)

The Japanese Society for Quality Control (JSQC) was established and the Japanese Quality Control Medal was founded in 1970.

In 70/71 years, "QC Circle Koryo (General Principles of QC Circle)", and "How To Operate QC Circle Activities" were published. QC Circle Activities were further had organized nine branches offices in the whole country. Moreover, it began to spread through out the world, and China and Korean version in 76 year, and English version of QC Circle Koryo (word of henceforth each country) were published in the 80-year. The International Convention on QC Circle was held in 1978 (ICQCC1978-Tokyo).

The Deming Prize awarded company came out of the construction industry for the first time in 1979.

The Quality Assurance was compiled into one book as a "Quality Assurance Guide Book", and the concept of quality was also expanded to deal with reliability, PLP (product-liability prevention), environmental management, QC in office work and sales, and conservation of resources and energy problem. Moreover, advanced technologies development, such as "New Seven Management Tools", "QFD (Quality Function Deployment)", and multi-variable analysis using the computer was handled in QC.

So called the "Quality Revolution" to which Dr. Juran was mentioned 3) was attained, and made-in-Japan product was begun to export to a world market, and Japanese TQC had been also come to be accepted abroad.

5. "Era of leap and development of TQC" (1980-89)

Reverse export of TQC of Japan started. Diversification and an advancement of a customer demand had been progressed, "Attractive Quality" was advocated in 84, sensitivity quality is also dealt with. Policy Management and Daily Management had been being substantial and clarifying role of management, the management-strategy problem had also been taken, and Group-Wide QC also progressed. Moreover, TQC was also implemented from manufacturing industry to service industry, the QC of Software develops, and the "Social Quality" lecture by QC magazine had been taken in 86.

Then, the internationalization of the Deming Application Prize was determined in 1984, and Florida Electric-Power and Light Incorporated Company, U.S.A became the first recipient company as an overseas company in 1989. ICQC'87 TOKYO was held in 1987 and "Ten Items of the Specific Features of TQC of Japan" was announced.

The technology transfer of quality prize started in the same year, and the Malcolm Baldrige National Quality Award in U.S.A. was founded by referring to Deming Prize in 1987, and also continued to the European Quality Award founded in 1991.

6. "Era of internationalization of TQC and TQM reconstruction" (1990-99)

While the introduction of the ISO 9000 Quality System / 14000 Environmental Management System to company in our country and the integration with TQM could be considered, it was announced "Declaration TQM" by QCS 1997, and "the Comprehensive Quality Management in the 21st Century" was advocated, and "Stakeholder Relationship Management" with stockholder, customer, employee, society, and environment, etc. came to be considered. Moreover, the systematization of TQM is also proposed and the total quality of a wide sense was defined. On the other side, while "Strategic Policy Management" was advocated, "Global Quality Management (GLQM)" came to be taken into account in connection with the internationalization of TQM, and international competitiveness and TQM (Japan-U.S.A. comparison) were also studied.

As for Japanese economy, stagnation advanced with collapse of bubble. Eventually, the consensus for Quality Management had been slipped of minds, and caused serious quality problems and ethical business problems, which were further apprehensions of deterioration of business management quality, product/service quality, and social quality which were structured for social industrial base in Japan, and quite afraid of losing Japanese industries competitiveness in the international market. In December 1999, "Hakone Declaration" was adopted as "the Design of Establishment of the Japan Organization for Quality Innovation" at QCS, Hakone.

"Deming Prize Application Guide" was also revised in the same year.

7. "Era of TQM innovation" (2000-)

"The TQM Encouragement Prize Guide" as preceding step of the continuous promotion to Deming Prize was founded in 2000. Moreover, based on the "Hakone declaration" in 1999, "the Japan Organization for Quality Innovation (JOQI)" was founded on May 23 2001 by the cooperation with JSQC and participation of five quality related organizations, and started for the activity to the reconstruction of a Japanese Management model.

And while learning to the way of revival of the U.S.A., it is Japanese Evolution of TQM started.

The basic concept and the examination standard (mark system) of Deming Prize Application Guide were also revised in 2002.

A "Quality Control" magazine also renewed its name under "Quality Management" magazine, and began to study for the integration of new management with TQM, such as Top's Viewpoint and Corporate Governance, Value Management, Balanced Score Card and Customer Value Management as a new current of quality business management, began to be studied.

In November 2002 "the Asia Quality Network" was formed as QC Research / Promotion Organization of ten nations of Asia, to promote "activity which raises the quality as a factory in the world" was started.

Moreover, ICQCC2003-Tokyo was held in October 17-20 2003 and 1400 attendants participated from 22 countries contained Asia 13 countries and ICQCC(International Conference for QC Circle) 2011 Yokohama on September 11-14, 2011 had been attended 14 participating nations, 1108 participants and 177 presentations.

JOQI reported its activities on May 28 2004 and finished the Objective and Role of Establishment of New Quality Management (System) Model in 21st Century, the contents of report are as follows;

1. New Commodity Development
2. Business Process Innovation
3. Value Creation

4. Self-assessment of management system
5. Development Program of Management Director of Technology
6. Development Program of Quality Professional Specialist
7. Human Resources Development at Workshop, Evolutional-QC Circle (e-QCC)
8. Structuring of Healthcare Management System based of ISO 9000

ICQ'05-Tokyo (International Conference on Quality'05-Tokyo-) had been held on September 13-16 2005, and attended 51 participating nation, 1066 participants, and 165 presentations.

8. "Toward new age"

In order to recovering from dishonor of the Made-in-Japan product of "poor and cheap", we have walked along the Journey of the QC of our country toward first Keidanren chairman Ichiro Ishikawa's vision realization. And it has come to be exportable the product of "good and reasonable price" to the world, and to be called an economic big country like today.

In "Upcoming Century of Quality, 21 Century" which Dr. Juran is said 3), there should announce new vision, and, probably, it should carry out and cut after this. Fortunately it is stated that the 8th chairman of Keidanren, Shyoichiro Toyoda's lecture dissemination as "To be country, Japan which is a Country be trusted and respected by the World", for a future of Japan 4). We wish to move forward with this as the 21st century vision of "Quality Establishment-of-a-Country Japan" in 2003, 5).

9. References

Kenichi Koyanagi (1963): Quality Emphasis in Japan's Postwar Trade, pp.4, JUSE Publishing Co.

Ichiro Ishikawa (1950): Statement of First Publication --Quality Control, Vol.1, No.1

J.M. Juran: Upcoming Century of Quality (1), ENGINEERS, September, 1994, pp.1-9;

The Keynote Address of ASQ 48th AQC, May 24 1994.

Shyoichiro Toyoda (1998): Quality Month-Special Event, Lecture Gist; Refer to Syoichiro Toyoda (1996): Creation of "Attractive Japan", Toyo Keizai Sinbunnsya, p.17, "Fundamental Idea".

Quality Establishment-of-a-Country Japan, Vol.54. No3, pp.36-38, Quality Management, JUSE Publisher, 2003

5

Quality Assurance in Education

Geoffrey Doherty
Emeritus Professor
UK

1. Introduction

The purpose of this chapter is to give an overview and critique of Quality Assurance (QA), its role, function and effectiveness as practised and researched in education organisations. To place contemporary QA in its historical context, some well-trodden ground will be revisited. There is nothing new about government and other authorities' inspectorial interest (some of it demeaning) in the effectiveness of teaching. Despite this, terms such as 'quality', 'quality assurance' and 'management' are still hotly contested, particularly in Higher Education (HE) since the increased focus on 'accountability' over the last three decades. Much antagonism both overt and covert is shown towards those responsible for quality – amply demonstrated by letters and occasional articles published in the educational press. This is a fact of life in universities, university colleges, further education (FE) colleges and schools, though to a less extent, perhaps, in schools and FE Colleges. What follows is an attempt to come to grips with some of the reasons, historical, conceptual, methodological and cultural.

2. An historical perspective

Generally speaking, when enterprises embark on quality improvement they start with **inspection** – a form of **quality control (QC)** – a reactive approach which identifies a weakness, non-compliance or whatever and endeavours to correct it ensuring it will not happen again. In outcome terms, of course, the damage has been done. The next phase is to move on to **quality assurance (QA)** – a proactive approach which attempts to identify problems and deal with them immediately, or even better prevent them from happening at all. The inspectors in white coats are replaced by problem solving groups, usually with some form of quality leader or leaders. Both QC and QA can be, and often are, led from the top down – i.e. they are managerial initiatives, clearly aimed at bringing down costs, improving processes, profitability and so on.

Obviously, QC is immediately applicable to manufacturing, where the outcomes are products of one kind or another. However, the fact is that quality assessment was applied to education before the industrial revolution was at its peak. In 1833 a Government Grant was given for elementary education provided to poor children by some church and non-denominational bodies. In 1837 the Government appointed the first school inspectors to monitor the effectiveness of the grant. So began Her Majesty's Inspectorate (HMI). Two key characteristics of inspection are clear from this elementary beginning: firstly, the importance of size and, secondly, the concern for value for money. Enough was being invested in schools to make the appointment of two inspectors worthwhile. The passing of the Foster Education Act in 1870 brought state elementary education for all and with it the setting up of School Boards, an enlarged inspectorate and the introduction of the infamous 'Revised Code' - payment by results. Teachers' pay depended on the successful achievement of examination results including tests of reading and mental arithmetic. In the United Kingdom (UK), these small beginnings have developed into The Office for Standards in Education (Ofsted) which is now responsible for Children's Services, Early Years, Primary, and Secondary Schools, Sixth Form Colleges and Colleges of Further Education. It employs over 400 HMIs, though inspections are mostly carried out by some 2000 Additional Inspectors (AIs)commissioned by HMI, but employed by privately owned companies - Regional Inspection Service Providers (RISPs).

The huge expansion of business and manufacturing in Europe and the USA during the nineteenth and twentieth centuries encouraged interest in methods of improving production, the most interesting and influential of which, was the work of F. W. Taylor, an American engineer who invented the concept of 'Scientific Management' (Taylor 1911), which analyses what workers do, how they do it and how long it takes - the basis of contemporary, highly sophisticated 'time and motion studies', aimed at measuring each element of production tasks and reducing them to a minimum of repeatable actions. Taylor became the first independent 'Management Consultant' and his ideas had a tremendous influence on the organisation of manufacturing during the early part of the twentieth century. However, its present relevance is that this can be seen as an early form of quality control, the links between this approach and Statistical Process Control (see below in Quality Tools) being obvious. Incidentally, the law of unintended consequences brought opprobrium to both Taylorism and Payment-by-Results. On the one hand, the resulting de-socialisation and de-skilling of the work force and on the other the negative effects on education of 'teaching-to-the-test' did nothing for the successful achievement of the outcomes.

Consumers have always been interested in the quality of what they consume and the involvement of external bodies in taking responsibility for quality control can be traced back a long way - certainly to the Medieval Guilds. The British Standards Institute (BSI), now a huge international business and owner of the internationally recognised *kitemark* dates back to 1903. These days, the BSI works in close collaboration with the more recently established International Organisation for Standardisation (ISO). Achieving a standard, aimed at ensuring potential customers that quality is assured, is valued by business and manufacturing enterprises which are driven by the imperatives of market survival, competition, and the responsibility of making profits for their shareholders. Improving market share by driving up quality and driving down costs are very attractive goals, particularly for large firms. Per se they have nothing to do with education. These elements - quantitative measurement, control and compliance with their implied threat to autonomy - are what cause academics and teachers to be extremely suspicious of quality systems in general.

The Second World War accelerated the need for effective quality control - it is a good idea to ensure that the bombs explode when they are dropped, that the wings don't fall off your aircraft in a dogfight, that your rifle is a reliable weapon etc. Significant contributions to the effectiveness of the USA's war machine were made by two American scientists and engineers - Walter A. Shewhart and W. Edwards Deming. Their influence on the development of Total Quality Control (TQM) and Total Quality Improvement (TQI) has extended far beyond engineering processes. Shewhart was the originator of Statistical Process Control (SPC) (Shewhart 1939; Neave 1990; 1993) and Deming extended statistical methods to non-manufacturing as well as manufacturing enterprises (Lambert, 1993) and became one of the original quality 'gurus'. Even more importantly, perhaps, he succeeded in conveying the statistical control methods in such a way that the worker on the shop floor could understand and use them. Unfortunately, after the war, American industrialists showed little interest in Shewhart and Deming. However, Japanese industrialists did. They invited them and another American, J.M.Juran a quality management consultant, to visit Japan and teach them how to make high quality products.

The Japanese did not spurn the message. During the following decades, their industrialists honed the tools, becoming leaders of the industrial world and competing annually for their W. Edward Deming quality award. Their best-known leaders are: Ohno, (1992); Ishikawa, (1986); and Taguchi (!986) They outline the quality tools and methods (see below), of which the most frequently used are: control charts, flow charts, paired comparisons, cause and effect analysis, force-field analysis, histograms, pareto analysis, quality function deployment (QDF), kaizen, quality circles and small step progression. The western nations were not swift to emulate the Japanese. However, by the seventies alarm bells were ringing. Firms such as Toyota, Nissan, Honda, Mitsubishi, Panasonic, Technics and Toshiba were making inroads into the automobile and brown goods markets. By the eighties other consultants were following in the footsteps of Deming and Juran - notably: Crosby; Moss Kanter; Peters & Waterman. Large management consultancies (e.g. A. T. Kearney) were becoming involved, and at least one firm, Rank Xerox, was mounting comprehensive and sophisticated TQM internal training programmes. Development over the last two decades has been exponential. Heavily based on TQM and TQI concepts and methodologies are Lean Management Systems and Six Sigma (originally developed at Motorola in 1986). The British Quality Foundation (BQF) was founded in 1943, more or less at the same time as the European Quality Foundation (EQF). Both institutions actively promote the European Quality Foundation Model (EQFM), a business excellence model to which all enterprises, large or small can aspire - education included. There is a much prized annual award, not unlike the Baldridge Award, founded in the USA in 1981 and subsequently (since 1991) administered by the American Society for Quality (ASQ). Thus the development of quality assurance systems is now thoroughly embedded in private sector enterprises throughout the world.

There have been to some extent similar developments in state education systems but nothing quite so dramatic. Education organisations are notoriously anxious to preserve the status quo. Change is slow, regarded with deep suspicion and usually resisted - the time is never ripe. Nevertheless, change does happen, particularly when governments are footing the bill. During the 20thCentury in the UK, the school leaving age has been raised from 12 to 14; 16; and now 18. There are thousands of primary and secondary schools in the UK, the annual cost of which has risen to billions of pounds. This is replicated throughout the civilised world: hence government and, indeed, taxpayers' interest in having some assurance that their monies are being well spent. Ofsted-style school inspections are still generally the norm. Though there are considerable variations in responsibility and style between nations - as, for instance, between the UK, Government controlled (via Ofsted) and RSIP conducted and the USA, where school inspection is contracted out to wholly private enterprises such as the American Society of Home Inspectors or even Tribal, a UK based provider.

In contrast, Higher Education (HE) remained relatively untouched by Government interference until some years after World War 11. In comparison with today, universities were tiny. Apart from Oxbridge, London and the Red-bricks (the large civic universities), undergraduate numbers were counted in hundreds, not thousands. Less than three percent of the post-eighteen population attended universities, making them truly elitist institutions. They were fee-paying, though some scholarships were available, including highly sought after State Scholarships. Government interest was represented by the University Grants Committee (UGC), but it was not until 1946, after the Education Act of 1944 encouraged significant increases in student numbers, that it assumed some responsibility for planning and development. As education costs rose, so did Government concern. In 1969 the then Minister of Education, Shirley Williams, put down 13 discussion points with the intention of encouraging economies in the university system. They were rejected out of hand by the Committee of Vice-Chancellors and Principals. This was the beginning of erosion of trust between Government and the HE system. The argument that the UGC had acted as an effective buffer between the Government and universities, mitigating financial cuts and more extreme Government policies, was well put by Elton (1992). However, 'erosion of trust' could be applied to the education system as a whole, not just the universities. In 1976 James Callaghan, the Prime Minister, made his famous Ruskin College speech initiating the 'Great Debate' about education. The root concerns were the maintenance of educational standards, education including HE and employability and the provision of 'value for money'. These concerns motivated and, indeed still motivate, both left- and right-wing governments to extend more and more direct control via the use of performance indicators, quality assurance and audit to HE. The UGC was replaced by the Higher Education Funding Council (HEFC) in 1987. This became the Higher Education Funding Council for England (HEFCE) in 1992.

In 1997 The Quality Assurance Agency (QAA) was set up as an independent, not-for-profit company. It is responsible for academic standards and quality in England and N. Ireland and is also separately contracted to Scotland and Wales and will accept advisory roles and take on overseas contracts. The QAA methodology will be discussed in a later section; however, its role and function is not loved by academics. For a detailed and comprehensive critique of the perceived flaws in the methodology, see Laughton (2003). Government concerns have not changed very much over the first decade of the 21st Century. Suspicions that the QAA lacks the necessary teeth to require improvements have been rife since the 1970s.

There are variations between national approaches to QA. The USA favours accreditation by private agencies as do many Nordic countries, including Germany. Others have a higher degree of Government involvement, sometimes, as with the UK and Ireland, through Government funded, but independent agencies such as the UK's Quality Assurance Agency (QAA) (Kis, 2005: Williams, 1993). Occasionally other stakeholders, e.g. students, graduates, employers, are involved as committee members or observers. Governments in the West, particularly in the UK and USA remain critical about school and HE performance, being particularly concerned about declining academic standards, whilst emerging economies like China, Korea, Brazil and Chile commit more and more resources to education, especially HE, which they perceive as essential to the continued growth of their economies. For instance, despite their political and economic difficulties, the Arab States, identifying knowledge as a key element in overcoming poverty, improving peoples' capabilities and developing a competitive economy, have embarked on a comprehensive QA programme for universities. They are using QAA methodology (UNDP, 2006).

Negative reactions from academics and teachers in general to Government inspired external quality control have been consistent over the years. They are not confined to the UK. Academics repeatedly refer back to the loss of trust between academia and Government (Brennan and Silver 1992; Johnston 1992; Loder 1992; Hodges 1993) and also to the intractable conflict between academic values and the managerial ideology which underpins the QA approach (Gorbutt *et al* 1991; Becher 1992) The conflict is ongoing and regularly aired in *The Times Higher Education (THE)* for instance McNay (2006) on the counter productive results of QA processes or Ashworth (2009) on the irrelevance of the QAA. Thus, whilst QA methods are embedded in industry and commerce, they are heavily criticised and at best only reluctantly tolerated in education, especially in HE in the UK. Educationalists have become increasingly antagonistic towards what they perceive as the crude materialist and managerial values of policy makers who, in turn, have become increasingly exasperated by what they perceive as the educationalists' endless capacity for talk without action, rationalisation without results. Fred Inglis, Emeritus Professor of Cultural Studies at the University of Sheffield encapsulated the mood:

> "There are no books on my desk, only quality papers. These are the dry thoughts of their dry season. Let virtues be forced upon us by their impudent crimes." (1993).

3. Concepts and definitions

Before discussing quality systems, it would seem important to know what we mean by quality. Unfortunately this is not as straightforward as it may seem. One of the reasons for the slow development of and resistance to QA in education is that academics are, of course, trained to ask questions, to be sceptical, if not challenging. In academia, quality has been a contested concept since ancient times, for instance:

> "Goodness is not the same as being, but even beyond being, surpassing it in dignity and power." (Plato c. 380 BC)

V.

> "Any kind of excellence renders that of what it is the excellence *good* and makes it perform its function *well*." (Aristotle c. 380 BC)

Plato's definition is utopian - something which has 'quality' is closer to its 'ideal' form than a similar thing that is of poorer quality. Aristotle is defining quality as fitness to purpose. The first is metaphysical: it cannot be measured; the second is realistic: it can be measured, so Aristotle might be regarded as the father of modern quality systems. This is not to say that 'fitness to or for purpose' is not still contested. A favourite writer of academics on quality is Pirsig (1976). His most quoted claim is:

> "Quality is, how do you know what it is, or how do you know it even exists? If no one knows what it is, then for all practical purposes it doesn't exist at all. But for all practical purposes it does exist."

This is followed by a couple of hundred pages of fascinating metaphysical discussion attempting to gain some grasp of this elusive concept. Metaphysics, unfortunately, is of little immediate, practical use in assuring the 'quality' of a product whether it is an automobile, a school curriculum or a research degree; which is not to say that quality experts totally reject the elusive, subjective element in the experience of satisfaction. For instance, a past Chief Inspector of Schools commented:

> "Quality is quite important and I, like others can recognise it when I see it - Maradonna's second goal against England in the 19th World Cup..." (Melia, 1990)

or, perhaps even more surprising, given his considerable influence on the successful development of the Japanese model of excellence in the West, Tom Peters (1990): Quality is all about:

> "...getting the customer to say WOW!"

This simplistic, apodictic, definition, 'I-instinctively-know-it-when-I-see-it' might be attractive but, like the utopian definition, it is subjective and therefore unmeasurable.

Fitness to purpose, however, is a much more useful concept, especially if 'purpose' can be defined as 'satisfying the customer'. This is an easily acceptable concept for manufacturers. Generally speaking, if consulted, as market researchers aim to do, customers universally want well-designed, well made products at the cheapest possible price. The BSI/ISO Quality Management Standard (ISO9001 Series) quite clearly defines quality as "...satisfying customer wants and needs,* A company, of course, must further clarify this in the context of its mission, aims and objectives. The Japanese were clear from their original approach to Deming and Juran that they wished to transform the quality of their manufacturing industries so as to gain a world-wide reputation with customers for excellence. This they successfully achieved and the rest of the world has followed their lead. It is interesting to note in passing, that the Japanese seem to have avoided philosophical arguments about *quality* by concentrating on *outcomes*. 'Improve the process and the quality will improve itself' is the Japanese approach - TQI, perhaps, rather than TQM. This is not to say that their 'gurus' do not have interesting contributions to the 'quality debate'. Taguchi (1986), for instance, exhorts designers to concentrate on the differences between *species* and *product* quality. There is no point in arguing about *species* when we should be focussed on *products* - in simple terms to attempt argue that a Bentley is a better quality motor car than a Mini shows a total misunderstanding of quality. The designer should be aiming to make the Bentley the 'best' high-quality motor-car of its kind, and the Mini the 'best' popular motor car. This places the Japanese approach firmly in the 'fitness to purpose' camp. The implications for different kinds of qualifications should be obvious.

However, educationists at every level have, understandably, rejected out of hand the idea that children or students can be regarded as 'products' and continue to resist the customer paradigm. They ask: Who is the customer? Claiming that it is impossible to identify any one, single customer as everything depends on the context. Can children or students be described as customers? Educationists are agreed that teaching and learning is a transaction to which the learner makes a significant contribution, indeed, where deep learning takes place, the most important contribution, so that in the most successful transactions, the teacher is also a learner. Is the teacher the student's customer? What about the post-graduate researcher who provides his/her professor with data etc. for the next paper or book? Similarly, a parent, a Professional Body, a Local Authority, a Government, a Research Council or some other contractor could all be regarded as customers. Clearly, they have different and conflicting requirements. There is no shortage of literature on the subjects of either quality in education or the nature of the customer or consumer, see : the whole of the *British Journal of Educational Studies(1992);* Bookman (1992); Barnett (1992); Pollitt (1992); Harvey *et al* (1992); Harvey and Green (1993); Richards, (1994); Kis *op cit.*(2006); Hackett (2011); Watson (2011). The basic reservations about the intangible, elusive non-measurable characteristics of quality in education and the range of different 'customers' have changed very little since the 1990s. Harvey and Green, whose work has had considerable influence both nationally and internationally opted for an 'excellence' model of quality and preferred the descriptor 'stakeholder' to customer. Kis *op cit.* (2006) comments that (a) the 'excellence' model sometimes used in conjunction with 'fitness-to-purpose', 'zero defects' or 'value for money' is now widely used, globally and that, whilst 'consumerisation' is equally generally resisted, in regimes where institutions have a high degree of autonomy (e.g. USA, UK, Germany, Scandinavia) 'accountability' is becoming an increasingly important issue and (b) that the descriptor 'stakeholder', presumably because it is just about acceptable to academics, is also more less or less universally used.

The answer to the questions to whom and for what are academics, education institutions (public or private), or manufacturing or commercial enterprises accountable is not straightforward. It might seem obvious that the latter are accountable to their customers who want value for money and will go elsewhere if they do not get it. However, there is nowadays a growing pressure on private enterprises to be ecologically and socially accountable. There are BSI and ISO standards for such matters. As we have already noted, education organisations have a multiplicity of stakeholders frequently with conflicting interests, not only to society for its future wellbeing and prosperity through the education of its children, teenagers and and students but also to subject disciplines, other academics and professional bodies. For 'society' read government, especially for state institutions where governments or their agencies are the paymaster. Though there may be other paymasters or sponsors – for research projects, for instance. In developed and more and more in emerging countries, governments are investing vast sums of money in pre-, primary, tertiary and higher education. In the UK the Government is pouring over ten billion £ Stirling into state education. Tax-payers' and hence governments' anxieties to be ensured that 'value-for-money' can be demonstrated are hardly surprising.

In the private sector, standards are not problematic since aiming for ever higher standards of quality is seen as encouraging the customer to trust in the quality of the product. Achieving a recognised standard improves market credibility. In contrast, academics express deep concerns about academic standards, in respect of both schools and universities and schools are also held accountable for the development of the appropriate moral and social values of young people. In addition, standards require *compliance* and compliance entails loss of *autonomy*, an intolerable proposition to academics and to a less extent schoolteachers, hence the reservations about standards:

> "The difficulty in talking about standards is that the concept is like 'truth', or 'goodness', or 'beauty', both logically indispensible and impossible to define without, considerable philosophical elaboration." (Pring, 1992).

Furthermore, at a more practical level, slavish adherence to a 'standard' can be a serious barrier to innovation and progress. This is particularly true of normative standards or descriptive statements which are prescriptive, stating how things 'ought' to be done, like a subject curriculum or a medical procedure. Ironically, apocrypha has it that 'ought' is the most overworked word in the schoolteachers' vocabulary.

Criterion-referencing, in contrast, sets out what must be achieved to reach a standard. The Council for National Academic Awards (CNAA - 1965-1992), set up by the UK Government in 1965 as the degree awarding body for the new Polytechnics in the UK was also the custodian of the degree standard. To gain accreditation the institution had to demonstrate that it could meet and sustain a threshold academic level. This did not preclude later changes and improvements. One of the reasons the 'new' - post 1992 - universities in the UK were less neurotic about the role and function of the QAA (see below) was because for over 25 years they had been part of the CNAA system, which included peer validation, institutional review and annual monitoring. The uneasy relationship between universities and governments is neatly summed up in a THES editorial in 2001:

> "Universities, have sought over more than a decade, to do as little as possible consistent with keeping governments off their backs. Governments have, in consequence, become increasingly frustrated and meddlesome." (THES editorial, 2001).

4. Doing quality

As outlined in the first section, the Japanese are credited with initially developing the philosophy and methodology of continuous improvement with the aim of achieving product excellence. This approach thy called *kaizen* which translates as 'making things better' (Ohno; Ishikawa. *op cit.*). It is founded on the principle of the total involvement of the whole workforce - that is to say everyone, from the top down. To succeed, the enthusiastic support of the senior management is essential. One of the key elements is the *quality circle*, a small group of 6 - 9 workers, which engages in *problem solving* of issues related to their work. The idea - 'if you want to improve something, ask the people who do it' - seems pretty simple and straight forward. Nevertheless, it is an anathema to top-down management systems of the sort which developed post-Taylor. Another key principle is to take small steps. It is proposed that consolidated small steps achieve more than attempting to change in leaps and bounds. Of course, senior managers have to listen to what the quality circles suggest: another characteristic not common to Western managers, at the time.

4.1 Quality tools for TQM, QA, QC and QI

The *quality tools* originally developed to support kaizen are aimed at identifying and solving problems and improving processes. The most commonly used are:

- **The PDSA cycle**: This is the basic methodology of continuous improvement:

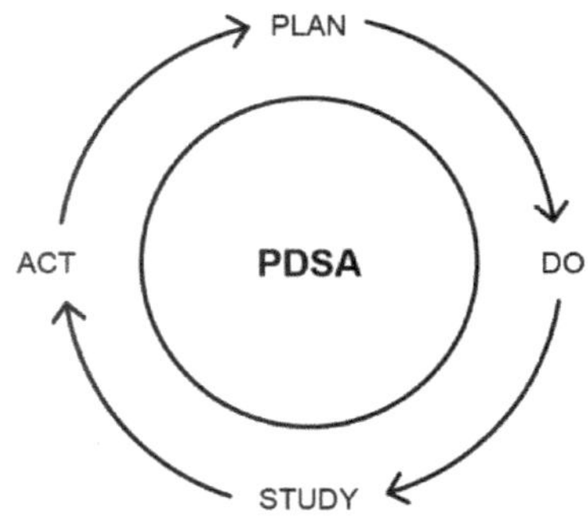

Plan = a change or a test aimed at improvement; Do = carry out the change or the test (preferably on a small scale); Study = what did we learn? What went wrong? Act = adopt the change, or abandon it, or run the cycle again (Deming, 1993).

- **The Problem solving cycle**: is a version of PDSA specifically focussed on problem solving:

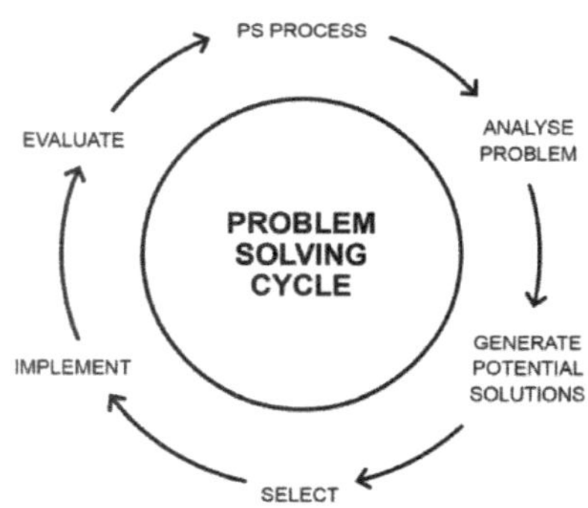

- **Brainstorming**: An effective and enjoyable method of getting a huge number of ideas related to a given issue from a small number of people in a short space of time. It is a 'feel free' meeting, which has some important rules: e.g. no one should be criticised; anything goes, no matter how outlandish. Brainstorming sessions are not without structure: there must always be a leader; all ideas should be written down on a flip chart; time must be apportioned to brainstorming the issue and to reflect on and evaluate the ideas generated.
- **Flow Chart**: A relatively simple way of describing or mapping a process, mapping from start to finish. **Analytical Flow Charts**, which are rather more complicated, are so commonly used throughout the world that the international symbols used can be found in the 'insert' section of the Windows Vista control bar (Bank, 1992; Implementation Group of the BDA,1995)
- **Histogram**: is a graphic illustration of data distribution (also in the Vista 'insert' tool bar).

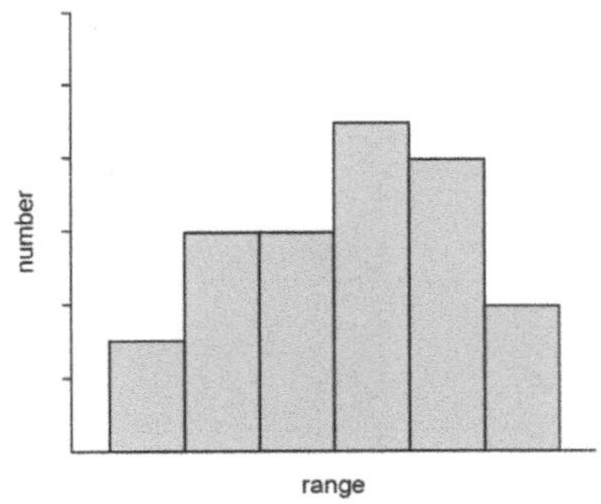

- **Run Chart**: Another effective tool for presenting data, a line graph illustrating how a process changes over time.

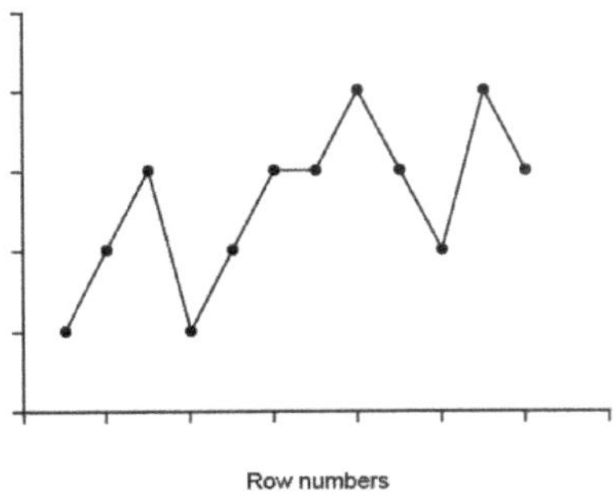

- **Statistical Process Control (SPC)**: A much more sophisticated chart which demonstrates the variation of a process between upper and lower control limits. It enables distinction to be made between *'special'* and *'common'* causes of variation (Neave, 1990) This is one of the basic tools for straightforward product improvement but it is very difficult to apply to outcomes which are not numerically measurable, e.g. in education, health and social services.

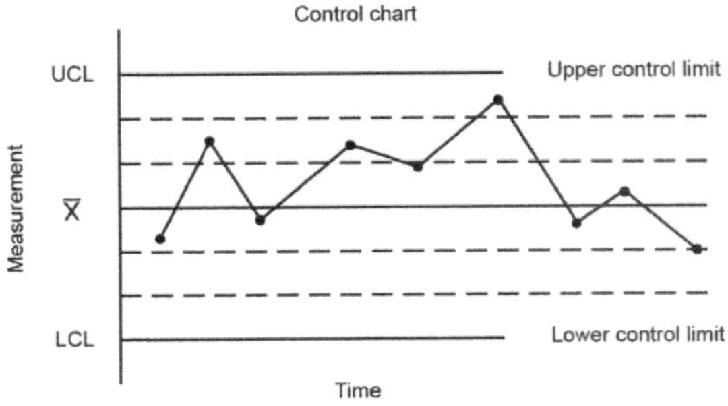

- **Cause and effect or fishbone diagrams**: conceived by Ishikawa, this diagram aids in analysing a problem in the context of a complete process:

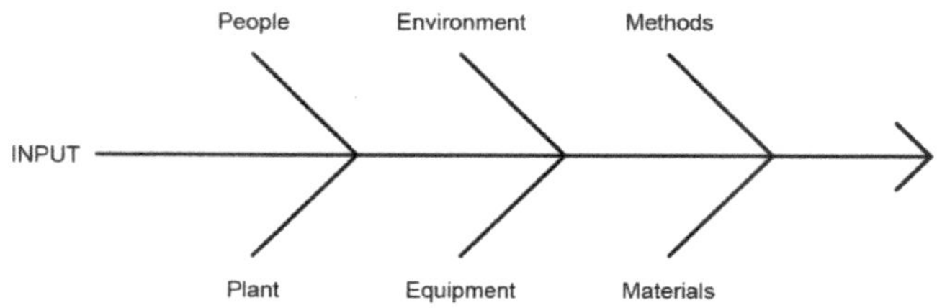

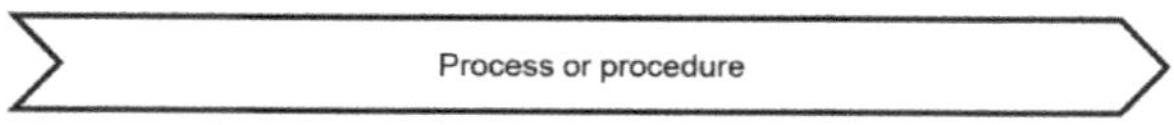

- **Pareto analysis**: Pareto was an Italian economist who noticed that in many unequal distributions about 80% of a problem could be attributed about 20% of its causes. This has become known as the Pareto Principle. If the 20% can be identified, problem solving can be focused more effectively on the 80% – student or staff absenteeism, for instance.

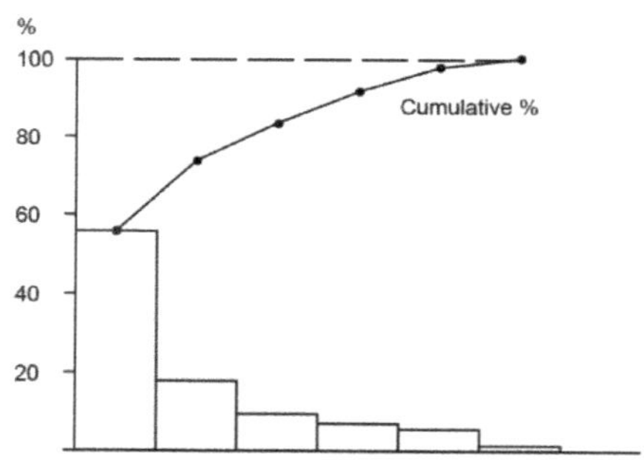

- **Paired comparisons**: a method of helping a group to quantify preferences. Each option (e.g. potential solution) goes head-to-head with every other option. In each 'face-off' members vote for the option they prefer. Votes are totalled when all the comparisons have been made.
- **Customer chains:** a frequent cause of 'hold-ups', poor communication and loss of quality is the breakdown of customer chains. Most outcomes involve several linked processes. When one process is completed, the next process is the 'customer', so to speak, of the former process:

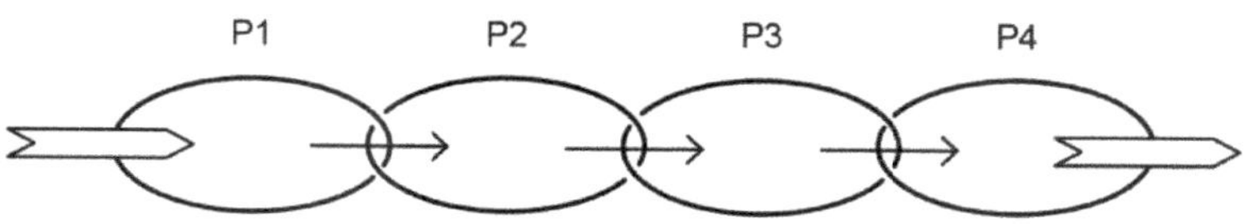

This represents an internal customer chain, where the P1 team is responsible for the quality management of what they hand over to P2 and so on. A breakdown anywhere in the chain can have serious consequences; hence: 'we are only as strong as our weakest link'.

- **Questionnaires**: common sense suggests that the obvious way to find out what the 'customer' wants is to send out a questionnaire, so this is a favourite resort of market researchers and even, in large organisations, internal QA. Of course, in matters of scientific, empirical research common sense is a poor guide. Questionnaires are an essential tool for sociologists. There are serious statistical problems in respect of sample size, response rate, validity, reliability and construct validity. Devising and testing an acceptable questionnaire for a piece of social science research can be quite expensive and time-consuming. However, many QA and market 'researchers' have a much more pragmatic, broad brush interest in gaining a quick turn-around. In the UK and many other countries *Student Satisfaction Surveys*, constantly contested by academics have been introduced. School Inspectors send standard questionnaires to parents - one size fits all wherever or whatever the type of school. There are also more carefully constructed questionnaires available such as *SERVQUAL*, devised in the 1980s to measure satisfaction with services, which has been applied to student satisfaction and variations such as *PERFSERV* and *HEdPERF*.
- **Benchmarking**: requires researching 'best practice' in the field of activities an organisation is involved in. There is a useful overview of its use in HE in Jackson (2001). The aim is to identify specific activities where improvement is perceived as necessary. Usually, it is external, which is likely to lead to the best results, especially where innovation is sought, though it can be internal, where some part of the organisation is outperforming the others. It can be an effective, though costly and time-consuming, contribution to quality improvement.
- **Force-Field Analysis**: a means of identifying forces helping and hindering the institution or section to get from where it is to where it wants to be. This can be a very helpful tool for achieving aims or circumnavigated intractable opposition e.g.

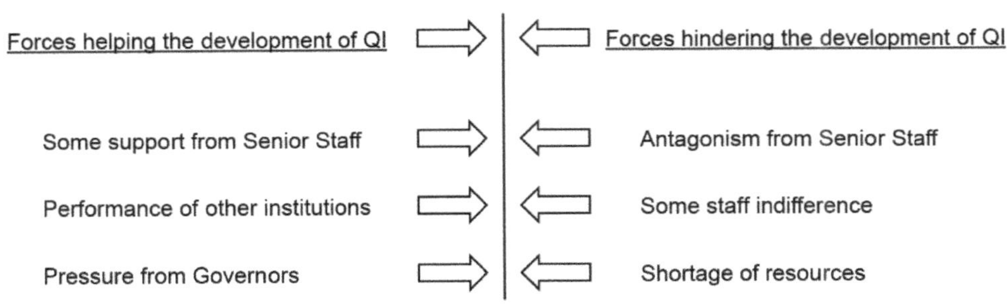

- **Quality Function Deployment (QFD)**: is a quite complex method of identifying customer needs. 'Listen to the voice of the customer' is a cornerstone of TQM. First, the customer's equirements for x are agreed, x is delivered and then both the evaluation and the provider's evaluation of how it was delivered are compared. There is a very interesting account of the application of QFD in Thakkar; Deshmukh; and Shastree (2006).

There are a huge number of books and journal articles elaborating these tools. A very clear account of the most frequently used can be found in Banks, J.(*op cit.*1992). For a really down-to-earth account of how to use most of these (and other) quality tools, see Rank-Xerox (1993). A really positive effect of training people to use these tools is that everyone can become involved - cleaners or senior managers. They are relevant and useful in indentifying straightforward or complex problems, whether the process is simple or complicated: extensively used by quality circles (see below) or improvement teams they encourage the sense of involvement or 'empowerment' across whole organisation.

4.2 Performance indicators

Performance indicators, of which there are many, have been around for a long time in education. We have already noted that the first school inspectors were implementing 'payment-by-results' in UK state schools through performance outcomes, including tests in the 1830s. Some, e.g. financial indicators are not necessarily directly related to quality (though procurement might be). The following list considers only those that are obviously relevant. Also, as far state schools, FE and HE Colleges and Universities are concerned, many resources are either centrally required by statutory regulations or in the UK and USA Local Authority (LA) and District requirements:

- **Provision**: area per pupil/student; staffing budget; building maintenance; capital expenditure; staff development budget; procurement; unit costs; marketing expenditure.
- **Process**: student/staff ratio; average student hours (HE and FE), average teaching staff hours(full and part-time); average class size (schools); % pupil/student attendance; staff absenteeism; quality of teaching and learning; space utilisation; consumables; pupil/student/staff support.
- **Progression**: examination results; progression ratio (schools =secondary to tertiary; FE and HE = annual retention rate); progression to post-graduate study (FE and, mostly, HE); successful employment; added value (AV) and contextual added value (CAV); post-graduate degrees, including doctorates, awarded; research grades awarded (UK).
- **Targets:** very popular where the pressure to make improvements in outputs is strong in either state or private sector enterprises. They are now extensively used by government in education (and in other public services) as a means of exerting pressure on schools, colleges and universities to meet priorities and improve and/or monitor such matters as academic progress, admissions, progression through the system, and employment. They are also used internally for similar purposes and are thus perceived by many staff as instruments of external or internal managerial coercion. Hitting or falling short of targets contributes to league tables, much hated by the profession, but loved by governments and the media as a beating stick. The attraction of targets and league tables to the paymasters can be contrasted to their rejection, as much on philosophical as statistical grounds, by TQM gurus such as Tribus, M. (1994) and Deming (*op cit.* 1993). Deming regards numerical goals (both in industry and education) as futile and counterproductive, concluding: "ducation "should be a system in which pupils on up from toddlers through the university take joy in learning, free from fear of grades and gold stars, and in which teachers take joy in their work, free from fear in ranking."

- **Contextual value added**: in education value added is usually calculated by measuring performance (using Standard Attainment Tests ((SATS) or some equivalent) at the beginning of a stage (primary, secondary etc.), working out by means of complex (and contested) statistical methods what the expected attainments, weighted for social and economic background should be and giving it a numerical score. Thus 100 = expected attainment achieved and anything above 100 = *positive value added* and anything less = *negative value added*. Apart from the dubious methodology, the fact that this is an 'industrial' *input – output* process does not endear it to educationists. It is the 'holy grail' of their paymasters.

Where performance indicators can be measured, even when the outcomes are somewhat intangible, they are regarded as 'hard' indicators or 'metrics': in the UK, for instance, Ofsted inspectors grade teachers and QAA Subject Review teams used to grade teaching quality, though individual scores were never divulged. It must be emphasised that QA, QC, QI and TQM are not synonymous. Although many of the tools were designed specifically to facilitate the development of TQM, which *is* a system and will be discussed below, for the most part in themselves they are neutral. They can be utilised to improve quality from a variety of conflicting ideological stances. Command economies have as great a need to produce high quality goods and services as capitalist economies. Different organisations will have differing mindsets and values and therefore differing priorities. Coercive governments and managements can, and do, make very effective use of targets and other quantitative measures (all of which encourage the development of league tables). Essentially, performance indicators are impersonal can easily be used to provide a stick with which to intimidate the uncooperative or ineffective employee. In contrast other organisations, inspired by human relations organisation theory, will use quality tools as they were designed: to encourage and facilitate worker involvement and empowerment.

It is not the purpose of this chapter to venture into the domain of sociology. However, a frequent criticism of QA, QC, QI and 'quality' in general is that they require a 'tick-box mentality' and encourage the development dull and brutish bureaucracies. On the contrary, if an organisation's management style is creative, positive and entrepreneurial so will be its approach to QA. State education is now mass education at all levels, primary, secondary and tertiary. Schools with 1300+ pupils and colleges and universities with 20,000+ students are so huge that some level of bureaucracy cannot be avoided, nor can management and leadership even if the management style is 'collegiate' – more about this later.

4.3 Systems

'Systems' is used here in the sense of a coherent way of doing something - Quality Assurance - not in the sense of everything that goes on in the black box. Organisation theory and organisational development is a different and much larger subject including, for instance: Normative Strategic Management; Management by Objectives; Human Relations Theory, etc. QA of some kind will be part of the larger enterprise, even if it is a small or medium-sized firm, a small Primary School or a University with thousands of students. TQM is one quality system which can be described as a complete organisational system in itself, so we will deal with this first.

- TQM

TQM was the name given to W. Edwards Deming's concept of transformational quality management. It inspired the Japanese industrialists and was eventually embraced in the West. It is nowadays peddled, in various forms, by a multitude of management consultants. TQM reaches far beyond QA, its underpinning philosophy being grounded on a passionate belief in a positive attitude to human nature. Deming and his followers call for a radical change not only from closed to open management and leadership styles but also to Government attitudes and policies, especially to education. It is a holistic management system requiring a system-wide quality culture in which everyone in the organisational hierarchy from the bottom up takes responsibility for her/his contribution to the whole: hence 'total'. Taking a risk is encouraged; mistakes are tolerated and regarded as learning experiences' rather than blameworthy; suggestions for improvement are encouraged, evaluated and adopted where appropriate. This is what is meant by 'empowering your people'. Unfortunately, the meaning of the word 'empowerment' has been debased through casual usage. In *The New Economics* (1993, *op cit.*), Deming describes his version of TQM in detail in the context of current practices in Western Industry. Government and Education: all three he puts to the sword:

> "What they do is to squeeze out from an individual, over his lifetime, his innate intrinsic motivation, self-esteem, dignity. They build into him fear, self-defence, and extrinsic motivation. We have been destroying our people."

The cornerstone of his system is the famous key 14 management points. These call for: commitment to focus on quality - both of incoming resources and the continuous improvement of the quality of outcomes: active participation of all members of the organisation, improved communications (bottom up and the top down) and cooperation and coordination: the importance of meeting customer needs: the driving out of fear: the substitution of inspection by procedures which detect errors and encourage continuous quality improvement. To transform the product, whether it be manufactured or a service, the culture of the entire organisation needs to be transformed - an aim which is neither easy nor swift to achieve.

Given its positive, optimistic view of human nature, transparent and easy communications up and down the organisational hierarchy, commitment to the mission and shared values, it might be thought that TQM would be attractive to academics. To some extent this has proved correct. Sallis and Hingely (1992) give a clear account of TQM and its take up in education and an abundance of books and journal articles have been published since. There are several journals in the UK and USA devoted to quality e.g. *Quality Assurance in Education,* and the *Journal for Quality and Participation*. There is more in the American journals about TQM in schools than in those in the UK, where the concentration is heavily weighted towards HE and FE. Professor Geoffrey Alderman put the case for TQM in HE rather persuasively, if not rationally:

> "...(the university approach) is grounded unashamedly in a total quality management (TQM) philosophy. Because TQM represents a holistic approach to the achievement of quality because it stresses the part everyone in an organisation has to play for quality overall and because it places a premium on the ideas of constant improvement, cultural evolution, quality circles and team work. Its approach strikes a chord with academics, and its philosophy has found a number of academic champions - even if some of them practise TQM without being aware of it." (Alderman 1996).

However, the Sallis and Hingely report shows that the impact of developments in TQM was greater and more widespread in FE than in HE, probably because of the vocational orientation of FE. A comprehensive guide to the implementation of Strategic Quality Management in FE Colleges (Miller & Inniss 1992) was commissioned by the UK Department of Employment. By 2000, the BEM (Business Excellence Model) approach was well established in some UK FE Colleges (McAdam & Walsh (2000)).This tendency is even more marked in 2011 than it was in the 1990s. Currently, the European Union provides strong support for the development of harmonised QA processes for vocational education across its member states through EQAVET (European Quality Assurance in Vocational and Educational Training). EQAVET provides a wide range of support and training and there is now a European Centre for the Development of Vocational Training one aim of which is to establish a European framework for quality assurance based on EQFM.

- Lean Six Sigm

Six (or 6) Sigma is a potent quality control metric, originally developed from Shewhart's process control statistics at Motorola in the 1980s. Sigma (σ) is the symbol for standard deviation in statistics and Six Sigma is a means of driving down deviations (errors) in a process to an incredibly small per cent. Over the next twenty years it progressed to a much wider application in an organisation. The method is known as DMAIC - Define an opportunity; Measure performance; Analyse opportunity; Improve performance; Control performance. Recently it has been linked to *Lean Manufacturing*, generally referred to as **Lean.** This is a process method derived from the Toyota Production System (TPS), dating back to the 1950s. This aims at eliminating waste thus maintaining value whilst reducing work i.e. it links back to 'zero defects', 'Taylorism', 'Time-and-Motion' and 'Fordism'. These links with the attitudes and values of mechanism and mass production would not seem very attractive to academics; nevertheless, there are examples of its application in HE (e.g. Dorman 2011).

European Quality Foundation Model (EQFM)

The European Quality Foundation Model (EQFM) and the British Quality Foundation Model (BQFM) are effectively the same. They are sometimes jointly referred to as the *Business Excellence Model* (BEM). These systems can be applied equally well to Service Industries or Business and Manufacturing. The model is now well established across Europe with over 30,000 firms participating, including the education sector. The model comprises nine criteria:

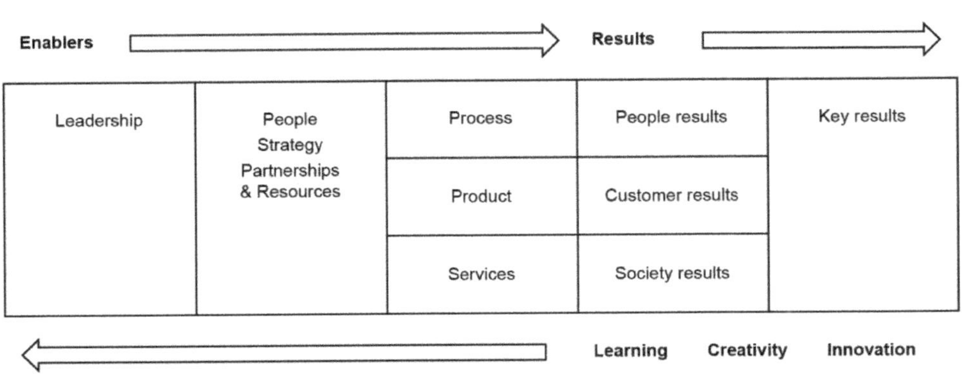

The various elements are differently weighted. Firms assess themselves against the nine criteria. Based on their self-assessment, they can then apply to be *recognised* as a firm having achieved sustainable excellence. The highest accolade is to win the EQFM *Quality Award*. The Foundation trains assessors from member firms, so the assessment method is essentially based on self-assessment followed by peer-group assessment. Firms are encouraged to use the RADAR improvement process:

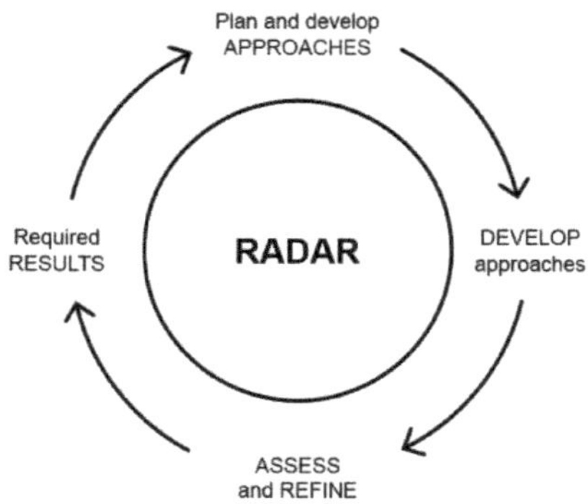

Essentially this is the same process as PDSA and any of the QA and QI tools can be appropriately applied. The model is NOT a standard. It does not lay down how to do anything, only the criteria by which the achievement of excellence will be judged. The links between the EQFM and TQM are obvious. Like TQM, the model allows academics to retain their autonomy, though it does require external, third party assessment. The American parallel is the annual Baldridge Awards. Unlike the EQFM awards, Baldridge is nationally administered by the National Institute of Standards and Technology. The Awards are presented by the President. They were officially set up by public act in 1987 and, since then the scope has been widened from a focus on manufactured products to include services and education; Baldridge is now referred to as the Performance Excellence Model. Criteria for the awards are remarkably similar to EQFM - including such elements as: leadership: improving performance and results; strategic planning; customer focus; measurement and analysis; developing a learning organisation. Schools have figured regularly since 2001 in the Baldridge Awards but less frequently in BQFM and EQFM awards.

ISO 9000 (series)

ISO 9001 was derived directly from the British Standard BS 5750 (Series) - 'series' means that there is more than one version of the standard, e.g. an enterprise opts for the basic management standard or for the inclusion of an extra clause covering product design. Opting for product design made achieving the standard significantly more difficult. BS5750 was immensely influential and regarded as a general guarantee of 'quality'. By the 1990s firms outside the UK were seeking registration. In the 1980s the American National Standards Institute (ANSI) adopted the standard unchanged and in1987 the International Standards Organisation (ISO) adopted BS5750 as the basis of ISO 9000 (now 9001 Series). The ISO keeps its standards under constant review through a system of technical committees and the standard has been refined over the years. The most important developments were made in the 2000 version which placed increased importance on process management and continuous improvement. Nowadays, well over one million firms are ISO 9001 registered.

Like BS5750, which it has replaced, ISO 9001 is a documented system, so that evidence of its effective implementation can be produced at any time. Certification, which can only be given by an Accredited Firm, is not based on self-assessment, but on the production of a *Quality Manual* containing an account of how each of the clauses is interpreted and how it will be implemented. Every member of the company, whatever her/his position in the hierarchy is expected to know the contents of the *Quality Manual*. The manual must also give a clear account of the enterprise's interpretation of quality (BS 5760 and ISO 9001 both define quality in terms of customer wants and needs), its management policies, how the system will be continuously improved and how it will be internally audited. Certification is not permanent. Regular auditing by an accredited auditor is also a requirement. Auditing, internal or external, is not inspection of the quality of products. Error free delivery of the quality system as set out in the *Quality Manual* is the criterion. Auditors are trained professionals. No preparation for an audit is needed. Auditors ask to be shown the appropriate documented evidence: either it is there or it is not. Minor errors, if there are not too many of them, are acceptable - provided the enterprise later produces evidence that they have been corrected. Systems failure results in loss of Certification - obviously a dire consequence for a firm's reputation for quality. *Compliance* is thus a key element in certification.

ISO 9001 is not an immediately obvious choice for education, especially schools and universities. Compliance and autonomy are seen as opposites, though the compliance required by ISO 9001 is compliance with what you say you have already decided to do rather than compliance with what someone else has told you to do, also educationists claim not to know whom their customers are. It is also perceived as a 'tick box', mindlessly bureaurocratic and costly system - resources are always in short supply in state schools. Only one university in the UK (Wolverhampton) has successfully pursued the BS5750/ISO 9001 route. The process and the pros and cons are well documented (Doherty, 1993, 1995; Storey, 1993). Much more interest is shown at departmental and subject level, particularly where quality systems are taught as part of the curriculum in Engineering Departments and Management and Business Schools (Thorhauser, T. & Passmore, D.L. 2006). This contrast in attitude between departments and institutions is remarked on and criticised by Matthews, (1993) as an example of double standards in academia, resulting in the "real danger" of universities and colleges teaching one set of values while having a different set for themselves. FE Colleges in the UK and Technical Colleges in the USA and elsewhere (CEDEFOP, 2010) still show more interest in implementing the BEM than schools and universities.

4.4 Inspection

As we have seen, inspection by 'men in white coats' has long been regarded as counter-productive in industry and commerce though quota sampling might be used as a quantitative quality measure. Inspection is used here as a 'third party' and overtly impartial examination of a product, company, institution or part of an institution to assure the quality of what is being delivered. The most common are School and FE inspection and HE Review for which the Office for Standards in Education, Children's Services and Skills (Ofsted) the Quality Assurance Agency (QAA) respectively are responsible in the UK. Ofsted is a quite small government department that contracts out its inspections to various private education consultancies. Developed countries have similar arrangements. The QAA is a not-for-profit limited company. It is a statutory body in England and a registered charity in Scotland and Wales. Agencies, many modelled on the QAA have sprung up, particularly in the Pacific Ring and Australia. In the USA and South America all QA is contracted to private agencies, although there are national guidelines provided by such organisations as the Council for Higher Education Accreditation (CHEA) and the Central American Quality Assurance System (CAQAS). In addition to nationally orientated agencies, there are also Professional Bodies, some which have stringent codes of practice and QA requirements, for instance to name but three of dozens: Law; Psychology; Engineering. Again, professional bodies are established in all developed countries.

QA must have something to assure, there must be a product. As we have noted, there is some difficulty in defining the product of schools, colleges and universities.

Currently, the nearest to a consensus is to define the product as 'teaching and learning' or 'learning experiences', where 'teaching' is subsumed into 'learning' as one of the experiences, as it were. Practically all are agreed for philosophical and ideological reasons that the pupil/student cannot be the product. Furthermore, she/he is an active participant in a process where transactions between the teacher and learner are happening all the time. In addition, the product must have a 'purpose'. Purpose generates as much disagreement as product, among educationists, that is. Sometimes purpose is defined by Government 'policy'. In the UK, for example, the 1944 Education Act was quite clear that the purpose of education was to ensure equal opportunities for children to receive free education up to age 15 and later 16, suited to their abilities, allowing them to achieve their full potential. Subsequent to the Robbins Report of 1963, this principle was extended to HE. The Report, incidentally, was also quite clear about the purposes of HE which included "...the promotion of the general power of the mind so as to produce not mere specialists but, rather cultivated men and women." There have been many Acts since, but this principle remained more or less firm until quite recently. In the 1990s, UK Governments have toyed with the idea of 'marketising' education with a view to raising performance through 'market' competition. University fees were re-introduced in 1998 and have been a political bone of contention, sometimes quite violent, ever since. The political, ideological and financial arguments surrounding these developments are not relevant here, but the practical implications are. Charles Clarke, when he was Secretary of State for Education in 1992 claimed that the purpose of HE was to contribute to the development of the UK's economy and society in the context of global change. This purpose is out of kilter with the Robbins' principle and is a clear example of political pressure on both education providers and those responsible for assuring the quality of their products.

Ofsted: Ofsted's methodology has evolved over the years, becoming leaner in the process. Prior to 2005, two months' notice was given of an impending visit. This was followed by a large team of inspectors who spent a week in the school. Every teacher was observed and the management of everything from finance and resources to the curriculum was scrutinised. Post 2005, two days' notice via phone call is given and a much smaller team visits the school for a few days. The key document a school must complete is the *self-*

assessment form, in which it is required to provide a detailed self-assessment of such elements as Senior Management, Curriculum Content and Management, Teaching and Learning, Student Achievement, Resources, Student Behaviour and School Governance. The touch is lighter, considerably lighter for high performing schools. Elements, including teaching and learning are graded: 1 (outstanding); 2 (good); 3 (satisfactory); 4 (inadequate). Schools are graded overall as: 1 (outstanding); 2 (satisfactory); 3 (failing). *Failing Schools* are put in *Special Measures*, usually resulting in the replacement of the Head Teacher and Senior Management team and other draconian measures designed to turn the school round. The criteria for all these grades are clearly set out at length in widely available Ofsted documents.

This current methodology is destined to change in 2012. The self-assessment is to be dropped and the inspected elements narrowed to five. The method is intended to be 'proportionate', so that for instance, 'outstanding' schools will not be inspected at all, unless certain alarm bells begin to ring. The new procedures are currently under discussion, but will be agreed and advertised in time for a September 2012 start. Some variation of this methodology is common to most other school and college inspection systems. Generally, following Ofsted visits, schools either set themselves, or are set targets for improvement by other bodies with responsibilities for school performance - such as Local Authorities.

Quality Assurance Agency: like Ofsted, the QAA has changed and evolved since its predecessor was set up 1992 by HEFCE. This was the Quality Assessment Division, which first embarked on Subject Reviews. It was fairly quickly replaced (1997) by the Quality Assurance Agency (QAA), an independent agency partly funded by subscriptions from the universities and partly by government and other contracts. Like Ofsted, its methodology has evolved over the years. Originally, there were two arms to QAA methodology - *subject review* and *institutional review (or audit)*. Subject departments were reviewed in turn and then the cycle started again. Departments were required to produce a self-assessment, which was then used as the basis for the visit by a small team of reviewers (usually 3 - 4). Review teams were led by a Review Chair chosen from a fairly small cadre of self-employed professional quality consultants together with peer-group reviewers all of whom had completed a QAA training course. A cross-section of teaching was observed and a grade (1,2,3,4) from good to unsatisfactory awarded. No lecturers were named. A distinction was drawn between *academic standards* - those aspects concerned with curriculum content, student numbers and attainment - and *quality of student experience* - resources, student progression and student support. In contrast, Universities and Colleges were reviewed by a similarly composed, but larger, Audit teams.

With various adjustments over the years, this two-pronged approach lasted until 2001, when regular *subject review* was discontinued in favour of periodic *institutional reviews*, with closer scrutiny of subjects only where there appeared to be cause for alarm. After lengthy consultation with the institutions, a new, even leaner and more flexible version of review has been agreed for implementation in 2011. To support the review process, the QAA publishes a series of booklets which have evolved over the years through consultation and collaboration with universities and colleges, setting out the *Academic Infrastructure*: *Frameworks for HE Qualifications; Subject Benchmark Statements; Programme Specifications; Code of Practice.* The Agency is a full member of ENQA (the European Network for Quality Assurance) and its methodology has had considerable, world-wide influence (Mok, K. 2005).

Both the *inspection* and the *peer review* models attract often cogent criticism, especially from dissenting academics. Even peer observation, except when it is voluntary, is opposed on the grounds that the observer warps the learning experience and that it is impossible to grasp and properly measure the elusive 'quality' factor. Other common criticisms are that they are costly and disruptive, diverting resources away from supporting educational objectives. Preparation for a major inspection or QAA visit would begin months in advance. Writing and editing the self-assessment or completing the self-assessment form was a major task and everyone involved in the event needed to be thoroughly conversant with it: all of which diverted concentration away from the core activity of teaching. To these negative effects can be added: lowering staff morale; the encouragement of 'teaching to the target'; departmental and institutional gamesmanship; increased bureauracy and managerialism. A further limitation, is that both approaches are heavily weighted towards teaching and learning. They are not coherent holistic systems looking in equivalent detail at the contribution of administration, procurement or the learning environment although as they have evolved over the last two decades, what counts as quality of experience has widened considerably. The progress made towards leaner methods and lighter touch in inspection methods in the UK can, perhaps, be partly attributed to the influence of the Business Excellence Model. In 1998 the Cabinet Office set up a task force for Modernising Government Quality Schemes which produced a report, *The Next Steps Report*, recommending the encouragement of three initiatives: The Charter Mark; Investors in People; and the BEM. The influence of TQM methods can be seen in the development of Ofsted criteria, the increased range of qualitative elements and the collection of data related to 'customer satisfaction' - parents and students.

Over the last two decades there has been a steady stream of research papers focused on specific quality tools, for instance: Thakkar *et al.op ct.* (2006); Leadership (Osseo-Assare, *et al* 2005); Service Quality (2005); SERQUAL and other service tools (Deshmukh, *et al.* 2006). The most notable development since the 1990s is the wide spread, global interest in QA systems. For example, a superficial content analysis of the contributions to the journal QA in E reveals that in 2001 c. 80% were from UK sources and most of the remaining few were from either the USA or Australia, whereas in 2010 90% were from non-UK sources, in which contributions from far eastern institutions and Spain, Portugal, Australia and the USA were the most frequent.

We have already noted that a wider perspective (Kis, *op cit.)* has shown that, while there has been some increase in TQM, EQFM and Baldridge activities in educational institutions including schools (in 2009 there were six school winners of BQF Quality Awards, three of EQF Awards and there have been regular school winners of Baldridge Awards for many years), it is neither extensive nor consistent. In their more specific research into ISO 9001 in US and UK universities (Thorhausen & Passmore *op cit.*) report on the generally favourable responses to the implementation of ISO 9001 but, interestingly, that many of those who had achieved registration were voluntarily relinquishing it, the most frequent reasons cited being cost and QA fatigue. Given the ever mounting pressure from governments, competition between institutions and the interest of other stakeholders - students, parents, employers - (not to mention the well known barriers to any kind of radical change in organisations) this is hardly surprising:

> "There are multiple layers of overlapping audit, assessment, accreditation and external examining. However, that is only part of the burden; there is also the internal quality

> procedures that have to be adhered to, including for many academic staff, annual module or programme reports, reporting on research activity and publication, periodic programme revalidation, occasional departmental or faculty reviews, possibly internal teaching assessments, internal audits, facilitating and responding to student feedback and individual performance review or staff development procedures..." (Harvey, 2005).

There seems some reluctance for social scientists to show interest in or commit resources to the effects of the BEM on education. Leonard, & McAdam, (2001) argue that there is a need for much more in-depth, qualitative, ethnological research in the field and suggest a methodology. Ethnological and anthropological research call for longitudinal studies: costly and attractive neither to quality assurers nor paymasters, both of whom like quick results. Moreover, governments, providing agents and their 'inspectors' tend to trust numerical data like league tables (however dubious their reliability) rather than qualitative data. Other factors militating against large scale quality research programmes are, again, the lack of a generally agreed theory of quality and the sheer size of TQM and BEM initiatives, which involve the whole organisation and its people, not just bits of it. Ironically this makes it easier to restrict research activities to one department rather than the whole institution. Finally, there is the general aversion to, and lack of trust in any approach derived from a manufacturing source.

5. Culture shock and conflict

There is pretty well universal agreement among writers on quality about one thing at least, namely that quality initiatives of any kind are doomed to failure without the development of a quality culture in the organisation, something very difficult to achieve. Superficially, at least, 'culture' is not such a contested subject as 'quality'. Applied to people, its dictionary definitions which include: common beliefs; characteristic ways of doing things; behaviours; prejudices; customs and values, appear to be generally acceptable. Thus developing a TQM 'quality culture' involves developing a commitment from every member of an organisation to place a high value on error free, continuous improvement.

Organisation theory is an important element in Business and Management Programmes. The literature on organisations is extensive and beyond the scope of this chapter. However, both TQM and BEM call for an open, cooperative, consultative management style. There is also general agreement that the culture of an organisation will stem from top management and that the development of a quality culture is impossible without the total commitment of senior managers. This is difficult enough in manufacturing and business, where the product and production processes are clear. As we have already seen, clarity is not a characteristic of education's purposes, processes or outcomes and the officially favoured 'quality' methods are *Inspection* and *Peer Review*, neither of which requires open, cooperative, consultative management styles. Atkinson (1991), a quality consultant not an academic writing for business and industry not academia, sets out an admirably clear account of the relationship between leadership, TQM and cultural change. Effective leadership is inspirational, charismatic even, generating in followers enthusiasm for and commitment to the task, whatever it might be. Leaders 'do right things', whereas managers 'do things right'. This is standard doctrine in the management and business studies curriculum. There is a strong tendency, therefore, for managers to cherish the status quo, especially in education, where the pace of change is notoriously slow. Two other problems he emphasises are the difficulty of developing a change to commitment to quality in the workforce and the necessity for comprehensive training programmes before trying to introduce TQM methods.

The 'workforce' in the context of education is not a straightforward entity. There are two if not three quite different 'workforces' namely, academics and the support staff, which include administrators, and the rest. Administrators, even in small primary schools, have a culture of their own. They have been more favourably disposed towards quality systems, especially those which focus on 'error free' delivery, than academics. The Conference of University Administrators published its own TQM handbook (Doige & Whitchurch 1993). As already noted, there is not a great deal of research into institutional overviews of QA ad TQM. Ehlers (2009) considers several currently popular theories of organisational behaviour and the importance of values in organisational development, concluding that TQM is the most appropriate approach to quality in the context of the culture of academic organisations. Harvey (*op cit.*) refers to the wide range of different QA agencies, reporting that, despite this variety nearly all of them use the same method - performance indicators and self-assessment followed by peer review. Other researchers, however, are more negative. Poole (2010) postulates that the cultural divide between QA professionals and academics is as wide as the rift between Snow's (1963) literary v. scientific cultures. The introduction of QA, in the shape of the QAA and perceived by academics as a time wasting irrelevance, is linked to Thatcherism and the rise of the New-Right ideology. He suggests that this cultural divide could be narrowed by consultation, participation and more emphasis on quality enhancement (all characteristics of TQM). Davies (2007) in a case study research reports that academic staff see EQFM as a 'managerial' attempt to become involved in 'collegiate' issues and have problems in trying to translate the 'business' language of EQFM into their terms. To involve academics, Davies suggests there should be more emphasis on teamwork, self-improvement and creating a supportive environment. In another attempt to assess academic attitudes based on interviews, Cartwright (2007) identifies a big discrepancy between the rhetoric of quality managers and QAA practice. Staff display cynicism, regarding the QAA methodology as bureaucratic, time wasting, and part of management intentions to exercise more control over academic autonomy, leading to a culture of fear and suspicion. These studies are based on small sample sizes, but they all report similar negative reactions.

The same themes have become a liet motif over the last two decades clearly demonstrating that academics perceive quality management as currently delivered by the QAA as a consistent threat to their collegiate culture, a term which is loosely defined as an organisation made up of equals: professionals, sharing the same academic values and, where decisions have to be made, they are consensus decisions made after appropriate, open discussion. In reality, collegiums are idyllic dreams. Anyone who doubts this should consult the *Micro-cosmographia Academica: being a guide for the young academic politician* (Cornford 1908). It contains some priceless pearls of wisdom - The Principle of the Wedge; the principle of the Dangerous Precedent; The Principle of Unripe Time. The key management aim is to do nothing whatsoever and power is exerted through influence. The *Micro-cosmographia* is a young academic's handbook of how to acquire it. 'Academic culture' is a carpetbag term: an oversimplification. Academics tend to be individualists and egalitarian. They resent any form of control or rules (except their own, which are quite stringent) because the rules of others involve the hated concept of 'compliance', which is regarded as incompatible with academic freedom and innovation (autonomy). Also, there are 'cultures' within the culture. Disciplines, departments and institutions vary. Any ex-auditor or ex-subject reviewer will testify to the differences in ethos or culture between old, redbrick, the 1960 generation, 'new' universities and ex-colleges of education. The 'new' ex-polytechnic universities were already accustomed to a higher degree of 'managerialism' than the others and the redbricks and ex-colleges of education regarded themselves as 'collegiate' in the style of the old universities. The same goes for school inspectors who will claim to gain a clear impression of a school's culture as soon as they walk through the door.

The strength of commitment to this collegiate culture explains to a large extent the depth of cynicism, resistance and lack of trust with which a wide range of academics, educationists and other public services regard QA systems as they experience and perceive them: "Naming and shaming is a central part of quality assurance procedures throughout the public services in Britain...The act of inspection itself is riven with the potential for shame and humiliation." (Morley 2005) One wonders what Deming, Juran, Ishikawa *et al.* would make of that. Overall, the state of Quality Assurance in Education is not very different in 2011 from what it was in 1991: subject to conflict and dispute. On the one hand governments mistrust the education system's willingness to commit to raising standards (schools) and the post-secondary (particularly the universities') willingness to commit to providing qualifications aimed at improving the economy and on the other, educationists particularly academics regard government motives as politically motivated and QA methodology as time-wasting, bullying and coercive. This varies, of course, across the huge spectrum of third world, emergent and advanced economies, but the criticism appears to be stronger in the developed countries than the rest. In the UK, the QAA and Ofsted have developed a 'lighter' touch and the evidence suggests this is the result of the influence of TQM and the Excellence Model. In both schools and tertiary education, there is a growing interest in TQM and BEM and researchers have become much more active in the field, though much more in small scale, in-depth rather than large scale projects.

6. Conclusion

The claim that education is 'different' from business and manufacturing is well substantiated and not mere 'whinging' as governments and other observers would like to believe; though this is not to say that educationists, particularly academics, do not frequently present themselves as obtuse to the rest of the world. In the UK alone, the Government annually pours billions of £ Stirling into the education system. Not unreasonably, whatever its political ideology, a government expects to receive value for money. Constant reports that standards have deteriorated rather than improved: dumbed-down A levels, ridiculously high percentages of 1st class and 2.1. degrees, a fall from 7th to 29th in the world's school performance league table are produced as 'evidence'. The education profession clings to the 1944 Act and Robbins' Report principle that free education up to first degree standard is everyone's right, despite the fact e this is no longer a 'given' in the public at large. Hence the official reliance on *inspection* and the taste for hard, metric data, whatever its reliability, and easily absorbed league tables, whatever their reliability.

TQM and BEM systems offer a much more positive approach to both QA and management style, than *inspection*. TQM goes beyond QA and has social and economic implications. However, Deming's vision of a society focussed on people committed to cooperation, intrinsic motivation and without the unethical greed of capitalism seems as idealistic as the academic collegiate dream. Restricted to QA in education institutions, both TQM and the BEM offer autonomy in the sense that definition of quality and application of the QI system is the institution's own and not imposed from without. The price is that the institution has to produce evidence that it is doing what it said it was going to do. This must now be seen in the context of globalisation, the introduction of HE fees in England and the rapid rise of much more serious competition for HE students from universities and colleges in the emerging economies. Both the USA and the UK who are undisputed market leaders at the moment, will have to struggle to maintain this position. More and more, despite the reluctance of academics to accept the fact, students are becoming customers and will be looking for 'value for money'. In this context of international competition, TQM and the BEM may well prove to be much more 'market orientated' than *inspection*. This may require a change of tactics for the paymasters. Ironically, as far back as 1983, Moss Kanter (*op. cit.)* describing employees' enthusiastic attitudes to the open, entrepreneurial style of an 'exciting' place to work, reported their statements that: working in the company was like being in "...a family, a competing guild, a society on a secluded Pacific island ...a university, a theocracy..." Let us hope that in the current application of the QAA's and Ofsted's outdated methodologies the paymasters do not lose the baby with the bathwater.

7. References

Alderman, G. (1996) *Self-regulation v. Inspection: the quality debate in British higher education* Quality World Technical Report IQA: Bedford.

Aristotle J. A. K.*et al* (1976) *The Ethics of Aristotle. The Nicomachean Ethics* Hammondsworth: London.

Ashworth, A. (2009) *Is this Watchdog Needed?* THE 16 April

ATKearney (undated) *Creating the Environment for TQM* and *Total Quality: Time to Take Off the Rose-tinted Spectacles* in conjunction with *the TQM Magazine.*

Atkinson, P. (1991) *Leadership; Total Quality and Cultural Change* Management Services, June.

Banks, J (1992) *The Essence of Total Quality Management* Prentice Hall: London.

Barnett, R. (1992) *Improving Higher Education Total Quality Care* OUP: Buckingham

Becher, T. (1992) *Making audit acceptable: a collegial approach to quality assurance* Higher Education Quarterly 46 (1).

Brennan, J. & Silver, H. (1992) *Who will lead the Charge or Sound the Retreat?* THES (25). *British Journal of Educational* 1992 40 (1).

Brookman, J (1993) *Bordered by Red Tape* The Higher, 5 March.

CEDEFOP (European Centre for the Development of Vocational Training (2010) *Accreditation and Quality Assurance in Vocational Education and Training: Selected European Approaches*: Luxenbourg.

Davies, J. (2007) *The effect of academic culture on the implementation of the EQFM Excellence Model in UK universities* QA in E 15 (4).

Deming, W. E. (1993) *The New Economics,* MIT Centre for Advanced Engineering Study, Cambridge: MA.
Doherty, G. D. (1993) *Towards total quality management in higher education: a case study of the University of Wolverhampton* Higher Education 25 (3).
Doherty, G. D. (1995) *BS5750 parts 1 and 2/ISO9000 (series) 1987: do they fit and is it worth it?* QA in E 3 (3).
Dorman, M.S. (2011) *A new lean paradigm in higher education: a case study* QA in E 19 (3).
Ehlers, U. D. (2009) *Understanding quality culture* QA in E 17 (4).
Editorial (2001) Times Higher Educational Supplement 30 March 2001.
Elton, W. (1992) *Quality Assurance*
Elton,W. & Partington, P. (1993) *Teaching Standards and Excellence in Higher Education* 2nd. Ed. CVCP, USDU: Sheffield.
Gorbutt, D, *et al* (1991) *Quality Support: Measurement and Myth* in Assessment and Evaluation in Higher Education 16 (1).
Hackett, L. (2011) *The Big Picture Show* Times Higher Education 30 July.
Harvey, L. *Et al.* (1992) *Criteria of Quality* Quality in Higher Education Project, Birmingham Central University of Central England: Birmingham.
Harvey, L. (2002) *Evaluation for what?* Teaching in Higher Education 7 (3).
Harvey, L. (2005) *A history and critique of quality evaluation in the UK* in QA in E 13 (4).
Harvey, L. And Green, D. (1993) *Defining Quality* in Assessment and Evaluation in Higher Education 18 (1).
Ishikawa I. (1985) *What is Total Quality Control?* , Harper and Row: NY.
Implementation Group of the BDA (1995) *Flowcharting: How and Why,* British Deming Association: Salisbury, UK.
Jackson, N. (2001) *Benchmarking in UK HE: and Overview* QA in E 9 (4).
Kis, V. (2005) *QA in Tertiary Education: Current Practices in OECD Countries and a Literature Review on Potential Effects* OECD Tertiary Review, www.oecdorg/data oecd/53/30/38006910pdf
Lambart I.T.G. (1993) *Out of the Crisis with George, Mises and Deming British Deming Association, Salisbury,* UK.
Laughton, D. (2003) *Why was the QAA approach to Teaching Quality Assessment Rejected by Academics?* Assessment & Evaluation 28 (3).
Matthews, W. E. (1993 *The missing element in higher education* Journal for Quality and Participation Jan/Feb.
McNay, I. (2006) *We sacrifice our souls on the altar of employability* THE 27 October.
Melia, T.(1990) *The Role of HMI in Quality Assurance,* Address to the CVCP Academic Audit Unit 7 December.
Miller, J. & Inniss, S. (1992) *Strategic Quality Management (3 booklets: Guide for Senior Managers, Guide for Middle Managers; Guide for Teachers and Teams),* Consultants at Work: Ware.
Mok, K. (2005) *The Quest for World Class University: Quality Assurance and Benchmarking in Kong,* QA in E 13 (4).
Morley, L. (2005) *Opportunity or exploitation? Women and quality assurance in education* Higher Rducation: gender and education 17 (4).

Moss Kanter R. (1984) *The Change Masters: Corporate Entrepreneurs at Work,* Allen & Unwin: London.

Neave H.R. (1990) *The Deming Dimension,* SPC Press Ltd: Knoxville TN.

Neave H.R. (1993) *How SPC?* British Deming Association, Salisbury: U.K.

Neave, H. R. (1993) *The Experiment on Red Beads* British Deming Association, Salisbury: UK. *Nest Steps Report 1998 (1999)* Cmnd. 4273 Cabinet Office, Stationary Office: London.

Osseo-Assare, A. E. *et al.* (2005) *Leadership best practices for sustaining quality in UK higher education from the perspective of the EQFM Excellence Model* in QA in E 13 (2).

Peters, T. (1990) *Liberation Management: Necessary Disorganisation in the Nanosecond Nineties* Macmillan: London

Peters T. J. (1988) *Thriving on Chaos* Macmillan: London.

Peters T. J. & Waterman R. H. J. (1982) *In Search of Excellence* Harper&Row: NY.

Pirsig, R. M. (1976) *Zen and the Art of Motor Cycle Maintenance* Corgi: London.

Poole, B. (2010) *Quality, semantics and the two cultures* QA in E 18 (1).

Pring, R. (1992) *Standards and Quality in Education* British Journal of Education 40 (1).

Richards, S. (1994) *The Consumer Paradigm in Public Management* in Doherty, G. D. Ed. Developing Quality Systems in Education Routledge: London and NY.

Plato, trans. Cornford, F.M. (1994) *The Republic* OUP: London.

Pollitt, C. Ed. (1992) *Considering Quality an Analysis:Guide to the Literature in the Public Services*

Rank-Xerox (1988) *Leadership through Quality (Reference and User Manuals),* Xerox, Rochester: NY.

Rank-Xerox (revised 1996) *Leadership through Quality (Problem-solving Process: Users' Manual),* Rochester: NY.

Shewhart, W.A.(1934) *Statistical methods from the viewpoint of quality control* The Grad. School, University of Michigan.

Storey, S.(1992) *Laying the Foundation for Total Quality Management at Wolverhampton Polytechnic* in Ellis, R. (ed.) Quality Assurance for University Teaching, OUP: Buckingham.

Taguchi, G. (1986) *Introduction to Quality Engineering* Asian Productivity Association: Tokyo.

Taylor F. W. (1911) *Principles of Scientific Management,* Harper Bros. New York and London.

Thakkar. J;Deshmukh, S. G; and Shastree, A. (2006) *Total quality management (TQM) in self-financed technical colleges. A quality function deployment (QFD) and force-field analysis approach.* Quality Assurance in Education 14 (1).

Thorhauser,T. and Passmore, D.(2006) *ISO 9000 in Education: a Comparison between the United States and England,* Research in Comparative and International Education 1 (2).

Tribus. M. (1994) *Total Quality Management in education: the theory and how to put it to work* in Developing Quality Systems in Education, Routledge 1994: Londondon and NY/

United Nations Development Programme (UNPD) & Regional Bureau for Arab States (RBAS) (2006) *Quality Assessment of Programmes in the Field of Education in Arab Universities: A Regional Overview Report* UNDP, NY. Kulwer: Netherlands.

Watson, D. (2011) Who would want to be in the Driving Seat on the Road to Nowhere? *Times Higher Education 7 July.*

Williams, G (1993) Total Quality management in education: panacea or placebo? *Higher Education* 25 (3).

6

Implementing Quality Management Systems in Higher Education Institutions

Maria J. Rosa[1,2], Cláudia S. Sarrico[1,3] and Alberto Amaral[1,4]
[1]*CIPES – Centre for Research in Higher Education Policies,*
[2]*DEGEI, Universidade de Aveiro,*
[3]*ISEG, Universidade Técnica de Lisboa,*
[4]*A3ES – Agência para a Avaliação e Acreditação do Ensino Superior,*
Portugal

1. Introduction

In the last decades, several factors have contributed to raising public concern over higher education institutions' quality, leading to the emergence of quality measurement and improvement devices such as performance indicators, accreditation, programme and institutional assessment and quality audits, and there have been attempts to import models from the private sector into higher education systems and institutions (Sarrico, Rosa, Teixeira and Cardoso, 2010).

This has led to the emergence of a debate on the applicability of quality management principles, methodologies and tools to the higher education sector. As reported in the literature on higher education, several voices have been heard about the non-applicability at all of those management theories, especially because they derived from industry and had nothing to do with the higher education ethos (Harvey, 1995; Maassy, 2003; Birnbaum, 2000; Kells, 1995; Pratasavitskaya and Stensaker, 2010). Other authors gave a more nuanced view on the subject, claiming that although higher education institutions were not companies some of the basic principles and tools could be applied as long as they were instruments at the service of institutions and their governance and management boards, subject to the institutions academic mission, goals and strategies (Williams, 1993; Harvey, 1995; Dill, 1995).

Although the debate is old, no firm conclusions have been reached so far. It seems, nevertheless, that in Europe, due to the developments on quality assurance that followed the Bologna Declaration, higher education institutions are now being "forced" to implement internal quality assurance systems based on the European Standards and Guidelines (ESG). Apparently the way these systems should be organised and function is not that specified, apart from the need to comply with the seven standards established in the ESG Part I, being up to each institution to define and implement its own quality assurance system in accordance with its mission, goals and institutional culture (Santos, 2011). So maybe this is the time to look again at quality management principles, tools and frameworks and see if they can be of some help to the development of these quality assurance - or management - systems.

The aim of this chapter is to revisit the debate on the applicability of quality management to the higher education sector, discussing its possibilities and impossibilities, in the light of the new developments in quality assurance in European higher education, namely the guidelines put forward in ESG Part I, since these provide indications for the institutions to set up their own systems. Some of the most well known quality management frameworks, such as the ISO 9001 standard, the EFQM Excellence Model and the Balanced Scorecard will be discussed on the basis of their usefulness for implementing quality management systems in higher education institutions. More specifically one will try to highlight how these quality management frameworks may be used to implement the seven standards for quality assurance established in the ESG Part I within HEIs, through an analysis of the criteria/indicators/requirements of the frameworks that accommodate each one of the standards. According to Sarrico *et al.* (2010) all these models propose to assess higher education institutions as a whole, including not only its teaching and research missions, but also other activities and, notably, institutional management, which is in the authors' opinion the area of quality management that needs to be developed and improved further in higher education. This discussion is particularly relevant since according to Pratasavitskaya and Stensaker (2010), the analysis of models and approaches of quality assurance at the institutional level has been rarely addressed in the literature, which is considered by the authors as an unfortunate situation since

> "quality management, at least theoretically, can have potential benefits; for example, with respect to identifying available options higher education institutions may choose from in order to respond to increasing external pressures for demonstrating academic output" (Pratasavitskaya and Stensaker, 2010: 38).

2. Total quality management (TQM) and higher education: A long debate

Rosa and Amaral (2007) argue that it is difficult, maybe even impossible, to find a unique and unequivocal definition of TQM and that the better one can do is to put forward a set of principles that underlie most TQM approaches. To Campatelli *et al.* (2011: 696), TQM is "an approach to management characterised by the definition of some general and inspiring guiding principles and core concepts that represent the way the organisation is expected to operate in order to obtain high performance." Both the ISO 9001 standard (ISO, 2005) and the EFQM Excellence model (EFQM, 2010) have established these principles and core concepts, which are usually used as the rationale to develop quality management systems within organisations (see Figure 1).

Goldberg and Cole (2002), referred in Calvo-Mora, Leal and Roldán (2005) consider TQM as being a "strategic option and an integrated management philosophy for organisations, which allow them to reach their objectives effectively and efficiently and to achieve a sustainable competitive advantage" (Calvo-Mora *et al.*, 2005: 742). According to Campatelli *et al.* (2011), TQM has been initially applied with good results in private sector organisations, being latter on used also in the public sector, namely in health care and education organisations.

Williams (1993: 229) refers to the rise of TQM in higher education as a "product of the market ideologies of the 80's and of the managerialism that accompanied it". From the very beginning its emergence in higher education led to heated debates on its applicability, originating what may perhaps be designated as two "schools of thought". On one side the quality management literature presented the idea that education, including higher education, was another field where quality management principles, methodologies and tools could perfectly be applied leading to better schools, where more motivated teachers and students contributed to an improved teaching/learning process, being capable of doing more with the same or even with less (Sahney, Banwet and Karunes, 2004; Campatelli *et al.*, 2011; Cruickshank, 2003): "It is inevitable for quality management processes, which have helped to transform business and overcome their quality problems, to be transferred to the field of education" (Peak, 1995, referred in Calvo-Mora *et al.*, 2011: 741). According to this "school of thought" the reasons behind the adoption of quality management, namely TQM, in higher education lied on the need higher education had to react to the increasing pressures and demands from its stakeholders, finding itself in a market-oriented environment where internal and external customers needed to be satisfied and even delighted. Furthermore in a time of declining resources, combined with higher education rising costs and an increasing pressure to provide high quality education (Cruickshank, 2003), there was the need of 'doing more with less'. The identification and application of "the relevant concepts of TQM to each and every aspect of academic life; that is, to the teaching, learning and administrative activities" (Sahney *et al.*, 2004: 145) appeared then as a viable solution.

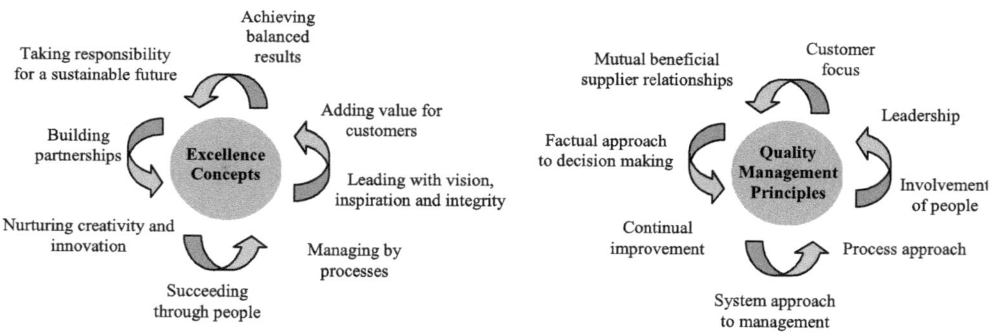

Fig. 1. The TQM principles that underlie the ISO 9001: 2008 standard (quality management principles) and the EFQM Excellence Model (concepts of excellence) (adapted from ISO (2005) and EFQM (2010))

On the other side, the higher education literature generically tended to put forward the idea that schools (and higher education institutions) were a very different type of organisation, with a different ethos and particular characteristics that made difficult, or even impossible, to implement a management philosophy derived from industry (Birnbaum and Deshotels, 1999, and Seymour, 1991, cited in Pratasavitskaya and Stensaker, 2010; Massy, 2003; Birnbaum, 2000; Kells, 1995). In fact, although not much publicised, it is known that in several institutions the application of quality management did not contribute to internal quality improvement (Harvey, 1995). Quality management is not a management approach easily applied to higher education institutions, especially because the academic culture of these organisations is quite strong and resistant to its concepts, principles and practices. And this resistance begins with terminology. Terms such as product, client, empowerment or even strategy, not to mention TQM or reengineering do not easily resonate in higher education institutions. For Massy (2003:165) "The greatest resistance to quality process improvement comes from professors who think it's just another business-oriented fad. The language of some TQM advocates contributes to this view (...) Customer, scientific method and removal of all forms of waste are sure to raise the hackles of academics." According to Houston (2008), even the definition of quality that prevails in industrial/business environments, based on the idea of satisfying customers' needs and expectations, is problematic in higher education. The fact is that higher education has a multitude of interested parties, with different perspectives about its role. Houston (2008) identifies interlinked environments and expectations in which universities operate that correspond to looking at these institutions from three different perspectives: economic, societal and educational. Labelling any of the groups included under these perspectives, or in their connections, as the customers that define quality makes higher education demands over-simplified: "Customer-focused definitions of quality fit the context of higher education poorly." (Houston, 2008: 63).

Seymour (1991, referred in Pratasavitskaya and Stensaker, 2010) mentioned the following factors as reasons for the unsuccessful application of TQM to higher education: resistance to change; insufficient administration commitment; high time investment due to personal training; difficulty in applying TQM tools to the HEIs environment; little experience of team leaders and staff in team-work; and the concerns HEIs have with their own results not being sufficient enough. Rosa and Amaral (2007) add the absence of effective communication channels; the difficulty in measuring an HEI's results; the co-existence of several purposes and objectives for the HEI; an emphasis in the individualism and a significant degree of internal competition; the bureaucracy affecting decision-making circuits; and the absence of a strong leadership, highly committed to the ideas and principles it wants to implement and capable of involving all the institution's members.

To Birnbaum (2000) the most relevant barrier to quality management implementation in higher education has to do with the need for a compromise between it and the traditions, values and purposes of higher education institutions. According to the author, total quality management has probably been the first management tool capable of provoking a serious discussion not just about its technical merits and demerits, but also about its educational and social implications. For Kells (1995: 458): "An extremely important question is the extent to which managerial innovations can be successfully adapted to the environments they seek to serve, rather than, as is feared by many in the higher education world, there being the expectation that the institution must comply with the method."

In the higher education literature there are nevertheless more nuanced views on the subject, claiming that although higher education institutions are not companies some of the basic principles and tools can be applied as long as they are instruments at the service of institutions and their governance and management boards, subject to the institutions academic mission, goals and strategies (Rosa, 2003). Williams (1993) considers that continuous quality improvement; quality consistency; participation of academics, students and non-academic staff; satisfaction of the clients' needs; and the existence of management procedures that reinforce quality are a number of quality management principles that nobody would consider irrelevant within the higher education context. In the author's opinion all these principles can significantly contribute to the development of massified higher education systems and their institutions, either explicitly oriented towards the market or not. Harvey (1995) considers that some TQM aspects can be adapted to higher education institutions and defends a new collegialism that emphasises professional accountability and cooperation, reflecting two quality management key elements: delegation of responsibility for quality and teamwork. The new collegialism emphasises the continuous improvement within the existent academic framework. Dill (1995) assumes a not very different perspective but replaces the notion of social capital for Harvey's new collegialism. For Dill (1995) there are also some important lessons that higher education institutions can learn from quality management, the most relevant being the central place that social capital should occupy inside organisations. Dill (1995) argues that efforts directed at enhancing quality should then be put into identifying networks and integration mechanisms that promote social capital development, leading to increased academic cohesion, communication and integration. Following this line of reasoning, examples of empirical studies aimed at probing the possibility of applying quality management tools and models for evaluating and improving the quality of management, services and processes of higher education institutions have been put forward. Examples include the application of the EFQM Excellence Model to higher education institutions, as illustrated by the works of Rosa (2003), Rosa and Amaral (2007), Campatelli, Citti and Meneghin (2011), Osseo-Asare Jr. and Longbottom (2002) and Calvo-Mora, Leal and Roldán (2005); the use of the Balance Scorecard (see, for example, Sahney, Banwet and Karunes, 2004; Asan and Tanyas, 2007 and O'Neill *et al.*, 1999); the application of the 14 Deming Principles (Redmond, Curtis, Noone and Keenan, 2008); the development of a TQM educational excellence model (Sakthivel and Raju, 2006); or the conduction of benchmarking exercises to improve institutions practices and performance (Jackson and Lund, 2000).

3. New developments in quality assurance in European higher education – The need for quality management systems

The quality concept in higher education can be said to be as old as the medieval ages (Rosa and Amaral, 2007), being possible to distinguish already in the 13th century two models of quality assessment: the French model of vesting control in an external authority - the archetype of quality assessment in terms of accountability - and the English model of a self-governing community of fellows - an example of quality assessment by means of peer review. These two models address two important dimensions of quality, extrinsic and intrinsic, the last one being dominant over the centuries in the academe. In fact the extrinsic dimension only has become relevant in the 1980's when the concerns with higher education quality, its assessment, management and improvement started to be a central policy issue for governments and society (Liaison Committee of Rectors' Conferences, 1993). This change in the approach to quality in higher education can be linked to a number of factors, such as (Rosa and Amaral, 2007: 183):

> "the massification of higher education, changes in the relationship between higher education institutions and governments (from a model of state control to a model of state supervision), the increasing role of market regulation, increasing institutional autonomy and the problems of the principal/agent, and the loss of trust in universities associated with new public management."

The 1980's and the decades that followed represented then a movement in the balance between two distinctive objectives higher education quality assessment has: accountability and quality improvement - the first objective being mainly pursued by governments and the second one by higher education institutions and academics. In fact there has been a rising prevalence of the accountability objective over the improvement one, leading to an increased attention by European governments and other institutions over higher education qualities, this is the qualities found in the services provided to society by higher education institutions (Rosa and Amaral, 2007).

From the 1980's onwards the development of quality assurance in Europe was fast. According to Schwarz and Westerheijden (2004) in the early 1990s fewer than 50% of the European countries had initiated quality assessment activities at supra-institutional level, while in 2003 all countries except Greece had entered into some form of supra-institutional assessment. Furthermore if state approval and accreditation schemes in the years 1998 and 2003 are compared, an overwhelming movement towards accreditation is observed. In fact, all recently implemented quality systems are based on accreditation (e.g. Austria, Germany and Norway), while old systems based on quality assessment were replaced by accreditation systems under the aegis of independent accreditation agencies (Flanders, The Netherlands and Portugal) (Amaral and Rosa, 2010). A key contributor to this movement has been the Bologna Declaration, signed by 29 European countries in 1999, and the process that followed. One of the main objectives of the Declaration was to encourage European cooperation in the quality assurance of higher education with a view to developing comparable criteria and methodologies. According to Schwartz and Westerheijden (2004: 36), the Bologna process has been an important "driver for change with regard to quality in steering mechanisms."

Following the Bologna process, the European Ministers of Education adopted in 2005 the Standards and Guidelines for Quality Assurance in the European Higher Education Area (ESG) and in 2007 established the European Quality Assurance Register for Higher Education (EQAR). The ESG are a set of standards, procedures and guidelines that higher education institutions and accredited agencies (the ones responsible for assessing and accrediting higher education programmes and institutions) should follow in order to implement, assess and accredit quality assurance systems in the European Higher Education Area (EHEA). According to ENQA (2007) they constitute a first step to the establishment of a widely shared set of underpinning values, expectations and good practices in relation to quality and its assurance, by institutions and agencies across the EHEA, aiming at providing a source of assistance and guidance to both HEIs and agencies, while contributing to a common frame of reference. The ESG are divided in three parts: Part 1, referring to standards and guidelines for internal quality assurance within higher education institutions; Part 2, referring to standards for the external quality assurance of higher education; and Part 3, referring to standards for external quality assurance agencies.

Although ENQA (2007: 13) claims "it is not the intention that these standards and guidelines should dictate practice or be interpreted as prescriptive and unchangeable", the truth is that the mere existence of the ESG Part 1 led governments and higher education institutions to start paying more attention to the implementation of quality assurance systems within higher education institutions. In fact Santos (2011) refers that in all countries that have signed the Bologna Declaration institutions are obliged to implement internal systems for quality assurance, in accordance with the fundamental idea that quality and quality assurance are primarily their responsibility. The author also mentions that in most countries the way these internal quality assurance systems are organised and function is not that specified, being up to each institution to define and implement its own system.

Furthermore another relatively recent development may also promote the implementation of quality management systems within European higher education institutions. It is the so-called quality enhancement movement, which can be seen as an attempt by universities to regain trust from society by restating that quality is their major responsibility and that the role of outside agencies should be limited to quality audits. The idea is not that new, since at the beginning of the nineties a similar approach was proposed in the US by academics, the basic idea being that accreditation should be transformed into audits of the institution's internal quality control procedures and self-examination schemes (Trow, 1996).

Even if in the UK the Quality Assurance Agency (QAA) has defined quality enhancement as "the process of taking deliberate steps at institutional level to improve the quality of learning opportunities", this still remains a not well-defined concept, with institutions still looking for their own definition of it (HEA, 2008). Nevertheless there are a number of common patterns to every institutional approach. A Higher Education Academy report (HEA, 2008) and a study conducted by Fillipakou and Tapper (2008) reveal some of the common quality enhancement characteristics taken from replies from diverse higher education institutions. These include institutions having the responsibility for the quality of the learning process; the existence of a flexible and negotiated evaluation model, non-mandatory and shaped by the participants in the teaching/learning process; and external vigilance of HEIs relying on institutional audits.

These late developments in quality assurance in European higher education, and the consequent emphasis that is being put on the need for HEIs to build their own mechanisms to assure and improve teaching and learning quality, lead us to think that maybe this is the time to look again at quality management principles, tools and frameworks and see if they can be of some help to the development of these quality assurance - or management - systems.

4. Quality management frameworks: Revisiting the debate on their applicability

Harvey (1995) once mentioned that there was time to go beyond TQM and that rather than "debate suitability it is time to look to practice and determine the worthwhile aspects of TQM and relocate them in the higher education context" (Harvey, 1995: 123). Also Joseph Juran, one of the leading quality management gurus, has observed that "the prime need is to discover the realities under the labels, i.e. the deeds, activities, or things which the other person is talking about..." (Juran and Gryna, 1988: 2, 13, cited in Houston, 2008).

Besides gurus' ideas on what quality management is and how it should be implemented, managed and improved, there are several methodologies to measure and guide quality assessment and improvement in organisations. Within these the ISO 9000 standards, the

Balanced Scorecard and the EFQM Excellence Model deserve to be noticed due to their international recognition and previous validation. Discussing the possibilities of using them and/or their logics within higher education institutions to frame quality assurance/management systems is made in this section of the paper, taking into consideration what the ESG establishes as standards for internal quality assurance. Furthermore one will try to go a step further by showing that these quality management models, while encompassing the ESG, are in fact broader than the standards, allowing, if adequately applied, for a truly quality enhancement approach within higher education institutions and not just a quality assurance one as, in our opinion, is the case with the application of just the ESG.

As previously referred there is a long debate about quality in higher education. This debate has repeatedly shown the difficulty in reaching a consensus not only about what quality is, but also about its implications for higher education. Sarrico *et al.* (2010) consider that quality can have multiple meanings in higher education and that this variety has had important consequences in the development of methods and instruments of assessing quality. According to the authors,

> "one of the most important issues refers to the multidimensionality of higher education institutions, which perform multiple and very different missions. Thus, higher education institutions face multiple stakeholders that have different expectations and different priorities regarding higher education. These differences will necessarily reflect in their specific approach to the evaluation of higher education, which should reflect the multiplicity of interests and aspirations that converge into each higher education institutions. The development of several models of institutional evaluation has often been an attempt to reflect those differences and nuances into an integrated evaluative framework (Sarrico *et al.*, 2010:51).

Furthermore Sarrico *et al.* (2010: 52) claim that

> "there are already some examples in the literature that account for the application of institutional quality assessment models that provide this integrated view on higher education quality, providing frameworks for better institutional management (...) leading to continuous quality improvement. These are the cases of the EFQM Excellence Model, the Balanced Scorecard, Benchmarking exercises or the EUA Institutional Evaluation Programme. All these models propose to assess higher education institutions as a whole, including not only its teaching and research missions, but also other activities and, notably, institutional management."

The ESG (ENQA, 2007: 16-19) establishes seven standards for quality assurance within HEIs, complemented with guidelines for their implementation:

1. Policy and procedures for quality assurance

   Institutions should have a policy and associated procedures for the assurance of the quality and standards of their programmes and awards. They should also commit themselves explicitly to the development of a culture which recognises the importance of quality, and quality assurance, in their work. To achieve this, institutions should develop and implement a strategy for the continuous enhancement of quality. The strategy, policy and procedures should have a formal status and be publicly available. They should also include a role for students and other stakeholders.

2. Approval, monitoring and periodic review of programmes and awards

   Institutions should have formal mechanisms for the approval, periodic review and monitoring of their programmes and awards.

3. Assessment of students

   Students should be assessed using published criteria, regulations and procedures which are applied consistently.

4. Quality assurance of teaching staff

   Institutions should have ways of satisfying themselves that staff involved with the teaching of students are qualified and competent to do so. They should be available to those undertaking external reviews, and commented upon in reports.

5. Learning resources and student support

   Institutions should ensure that the resources available for the support of student learning are adequate and appropriate for each programme offered.

6. Information systems

   Institutions should ensure that they collect, analyse and use relevant information for the effective management of their programmes of study and other activities.

7. Public information

   Institutions should regularly publish up to date, impartial and objective information, both quantitative and qualitative, about the programmes and awards they are offering.

Can these standards be implemented resorting to the quality management frameworks proposed in the quality management literature? Would it be wise to do so?

The Balanced Scorecard

The Balanced Scorecard (BSC) was developed by Kaplan & Norton (1992) and aims at attaining a balance between several dimensions of performance (see Figure 2). These include the following ones: financial, internal business processes, customer and the needs of learning and growth. The critical point is that in order for the organisation to be successful it needs to understand the interactions between actions (operational and developmental) and results (external indicators and financial ones). This apparently simple idea has had a tremendous impact in recent years in encouraging managers, at various levels, to develop balanced indicators of performance. Furthermore, this tool has produced other important changes by exposing managers to both indicators of determinants and results. The development of the BSC attempts to understand the interaction between both, helping organisations to develop a performance measurement mechanism that links strategy and operations.

Those organisations that are capable of reflecting their strategy in their systems of performance measurement seem to be better equipped to implement the former. By using knowledge on the interactions between indicators on operations, funding, external (customers), and development (learning and growth), organisations can become smarter. Those more able at exploring those interactions are likely to perform better. This analysis is not trivial since the interactions between operational decisions and organisational performance can be extremely complex.

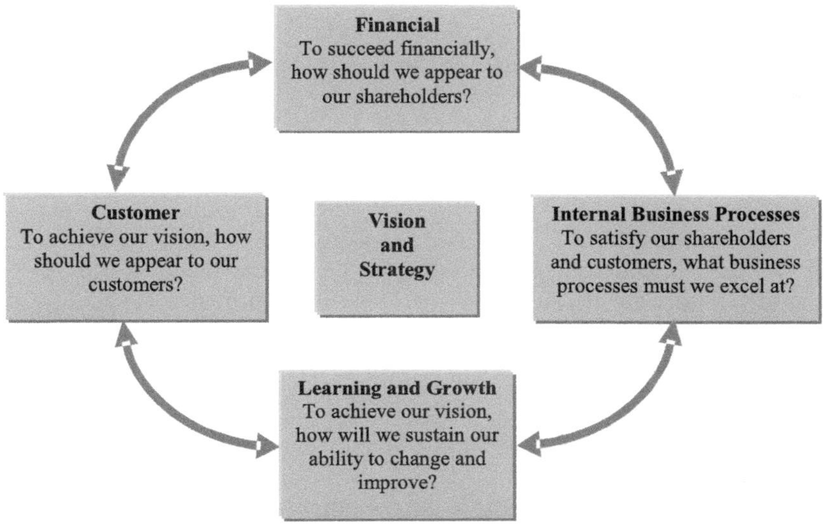

Fig. 2. Balanced Scorecard (adapted from Kaplan and Norton, 1992)

Several studies reported in the literature demonstrate the application of the BSC to the educational context (see, for example, Asan and Tanyas, 2007; O'Neill *et al.*, 1999; Cullen *et al.*, 2003; Karathanos and Karathanos, 2005), on the basis that "educational institutions also need to be managed through strategic concepts, in order to meet demands and keep up with change" (Asan and Tanyas, 2007: 1006). Furthermore, if one looks at the ESG Part1 one can see that standards could be met through a proper definition of the different performance indicators included in the BSC. Starting with *Policy and procedures for quality assurance* which should be part of the Vision and Strategy at the core of the model. *Approval, monitoring and periodic review of programmes and awards*; *Assessment of students*; *Quality assurance of teaching* staff and *Learning resources and teaching staff* could be met when defining the Internal Business Processes indicators; *Public information* would go well under the Customer indicators and *Information systems* resorting to Learning and Growth. Moreover the philosophy behind the BSC, of linking actions and results, helps to make sense of the links between the seven standards put forward in the ESG and that otherwise may be looked at as independent ones, adding a developmental character to a quality assurance system that otherwise would look as mainly operational. Interestingly nothing is said on financial sustainability in the ESG, something which is already addressed in some national systems of quality assurance.

The EFQM Excellence Model

Another model that attempts to improve the capacity of organisations to incorporate the interactions between actions and results in their decision-making process is the EFQM Excellence Model, a creation of the *European Foundation for Quality Management* (EFQM).

According to Campatelli *et al.* (2011: 693) the EFQM "promotes the use of a standard management model (...) capable of bringing the organisation to excellence level and a standard evaluation process that could be applied to all types of organisations, regardless of sector, size, structure or maturity."

The EFQM model attempts to analyse how satisfaction can lead to excellence in organisational results. It was devised as a non-prescriptive framework, which recognises that excellence may be sustainably achieved through the adoption of different approaches (Calvo-Mora *et al.*, 2005). The model states that satisfaction of customers and employees, and societal impact are achieved through leadership, which facilitates and stimulates institutional strategies and the management of personnel, resources and processes (see Figure 3). The institution's capacity to deal successfully with those aspects will depend ultimately on their organisational excellence. The model states that the learning processes within the institution will help it to improve its performance and further development. The proponents of this model consider that first-rate organisations have a better institutional performance due to better management practices, i.e., better management of processes, skills, people (either employees or customers) and structure. On the other hand, this behaviour impinges on a clear vision, a service-oriented concept, well articulated strategies and an adequate culture stimulated by effective leadership. First-rate organisations go beyond intuition and not only understand the links between operational decisions and organisational development, but also use that understanding effectively.

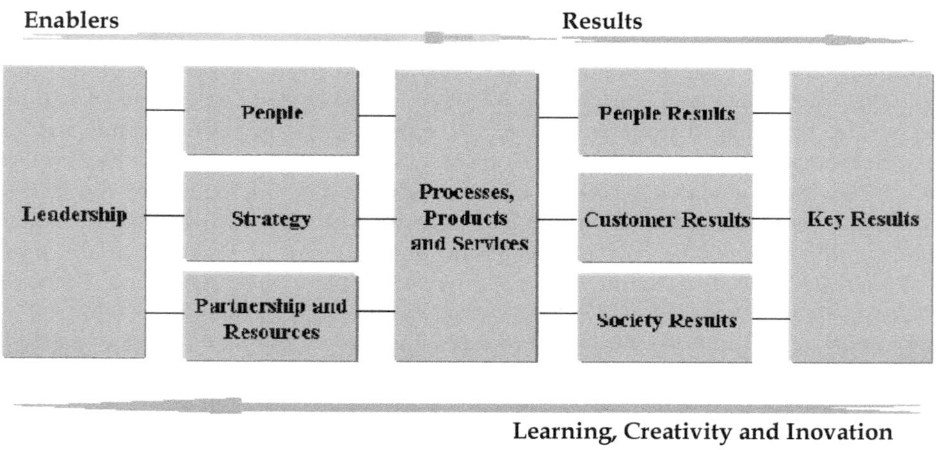

Fig. 3. The EFQM Excellence Model (adapted from EFQM, 2010)

According to Rosa and Amaral (2007), since the EFQM Excellence Model is associated with business, it has been resisted by academics and did not receive broad-based support inside HEIs. But the authors also alert to the fact that "the growing concern with quality, the need to be accountable towards society and the increasing presence of the market in higher education systems have made quality assessment, management, assurance and improvement an unquestionable reality, covering teaching, research, services and institutional-level approaches." (Rosa and Amaral, 2007: 194). So it is not that surprising that some higher education systems and institutions have started to consider its application, as has been reported in the literature (Saraiva, Rosa and Orey, 2003; Farrar 2000; Osseo-Asare Jr. and Longbottom, 2002; Hides and Davies, 2002; McAdam and Welsh, 2000; Cortadellas, 2000; Schmidt and García-Legaz, 2003; Calvo-Mora, Leal and Roldán, 2005, Rosa and Amaral, 2007; Campatelli, Citti and Meneghin,2011), since it "provides a unique framework for the global evaluation of an organisation and strongly promotes the implementation of a continuous improvement system." (Campatelli *et al.*, 2011: 693).

Looking at the model criteria it is relatively easy to see that they cover the standards for quality assurance predicted under the ESG (2005): *Policy and procedures for quality assurance* is addressed under the Leadership and Strategy criteria; *Approval, monitoring and periodic review of programmes and awards* and *Assessment of students* under the Processes, Products and Services criterion; *Quality assurance of teaching staff* under the People criterion; *Learning resources and student support* is covered by the Resources and Partnerships criterion; and finally both the *Information systems* and the *Public information* standards may be addressed through the different enablers and results criteria, depending on institutions local circumstances as referred in the ESG (ENQA, 2007). Furthermore, the arrow in the model, representing its feedback philosophy, also contributes to the implementation of the *Information systems* standard, since this one has to do with institutional self-knowledge, being the starting point for effective quality assurance (ENQA, 2007). Additionally, as was the case with the BSC, the EFQM model allows to uncover the links between the seven standards, since it is also based on the idea that quality actions (enablers) lead to quality results.

The ISO 9001: 2008 Standard

Another possible quality management framework that HEIs may be willing to consider is the ISO 9001:2008 standard, which is probably the most well known of the referred frameworks. This standard sets the requirements for implementing a quality management system in an organisation, independently of its dimension or type of activity, including education institutions.

According to the ISO (2005), a quality management system is the management system that directs and controls an organisation with respect to quality. The ISO 9001:2008 standard establishes the minimum requirements to set up such a system, which are organised in five main blocks: *Quality Management System; Management Responsibility; Resource Management; Product Realization;* and *Measurement, Analysis and Improvement* (see Figure 4).

According to the ISO 9000:2000 developing and implementing a quality management system comprehends several phases, among which:

- determining the needs and expectations of customers and other interested parties; establishing a policy for quality and the organisation quality goals;
- defining the processes and responsibilities needed to attain the quality goals defined;
- determining and making available the resources needed to attain the quality goals defined;
- establishing the methods to measure each process' efficiency and efficacy;
- applying these measures to determine each process efficiency and efficacy;
- identifying the means to prevent non-conformities and eliminate its causes;

- establishing and applying a process to the continuous improvement of the organisation's quality management system.

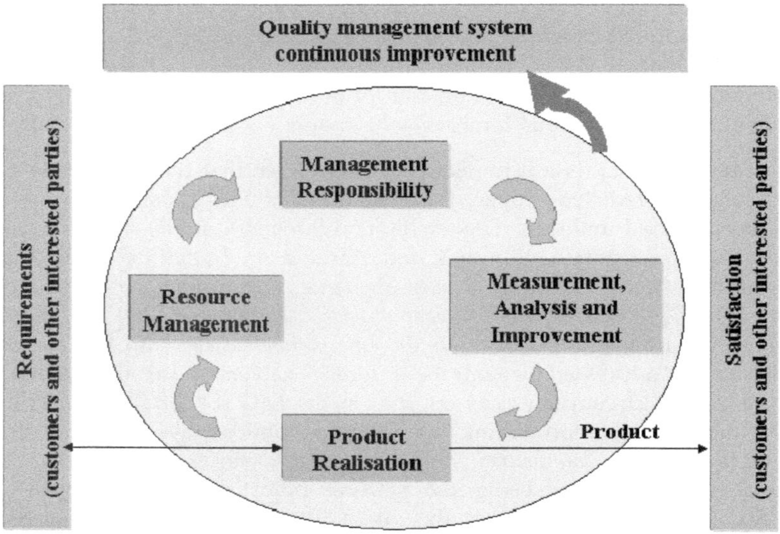

Fig. 4. The ISO 9001:2008 standard model (adapted from ISO, 2008).

A lot of criticism has emerged around the application of this standard to higher education institutions, namely because the "ISO approach entails too general a view of the 'production process' of higher education" (Csizmadia, 2006, cited in Pratasavitskaya and Stensaker, 2010:39), implying a high degree of processes' standardisation that is incompatible with HEIs' nature (Rosa, 2003). It is true that there are not many references in the literature to the ISO 9001 implementation in higher education institutions; nevertheless Rosa (2003) argues that this standard can in fact be applied if the HEI so wishes and believes that implementing a quality management system according to it will allow embarking in a continuous improvement process (that does not end with the standard implementation, although the implementation can constitute a good departure point). And the truth is that some HEIs have indeed decided to implement such quality management systems (Sohail *et al.*, 2006; Welsh and Dey, 2002), obtaining benefits such as a cost effective method for accountability, the development of an improvement-driven focus through re-focusing core processes to improve both productivity and service levels, to take into account a broader number of stakeholder views, to enhance the use of data for quality assurance purposes and improvement in inter-departmental working conditions and student enrolment (Brookes and Becket, 2007). Nevertheless the truth is that the application of the ISO 9001 in higher education has been generically limited to the institutions' services and not to their core functions, namely teaching and learning.

Looking at the ISO 9001 requirements shows their implementation will answer to the ESG. Addressing Management Responsibility will lead to the implementation of a *Policy and procedures for quality assurance*, while the *Approval, monitoring and periodic review of programmes and awards* and the *Assessment of students* are covered under the Realisation of the

Product. When implementing the Resource Management requirements, the *Quality assurance of teaching staff* and the *Learning resources and student support* standards are addressed. Finally the requirements put under Measurement, Analysis and Improvement allow for the implementation of the *Information systems* and *Public information* standards. Again the idea underlying the ISO 9001 standard that there is the need to continuously improve an organisation quality management system, based on the application of the PDCA (Plan-Do-Check-Act) cycle, contributes to establish a link between the seven standards established in the ESG, allowing for the implementation in the HEIs of a quality assurance system with a truly developmental character.

From the analysis just presented it seems than that the ESG may be applied in European HEIs if these decide to opt for the implementation of a quality management system based on one of the most well known quality management frameworks. Additionally, when doing so institutions are establishing interrelationships among the standards, allowing them to really take into consideration the need for *feebdback* processes that will lead to higher education institutions and their functions, namely teaching and research, continuous improvement. In a sense if HEIs will go for the implementation of the ESG *tout court* they risk being implementing a rather operational quality assurance approach, while if they opt for their implementation via the adoption of a quality management model there is the possibility that quality assurance will become more developmental, hopefully leading to true quality enhancement.

5. Conclusions

In 2007 Brookes and Becket (2007) made a review of the literature on quality management in higher education and came to the conclusion that a number of environmental forces were driving change within and across countries, leading to a firm emergence of the quality management issue on the agenda of many higher education institutions. The review revealed that the most popular response HEIs gave was the testing or implementation of quality management models developed by industry. From the empirical studies reported in the literature and analysed by the authors, benefits related to the implementation of these models as well as limitations were identified. The benefits included the adoption of a strategic approach to quality measurement and management and the identification of quality enhancement priorities; limitations related largely to the dilemma of applying business models to higher education. Furthermore the authors refer that "the models are reported to have far greater applicability in measuring administrative or service functions within HEIs rather than the quality of research or teaching and learning" (Brookes and Becket, 2007: 105-106).

This paper argues that at least in Europe the need for higher education institutions to develop their own internal quality management systems has become a reality, for which European standards and guidelines have even been defined. So maybe it is again the time to look at existent quality management models, going beyond the debates about whether quality management is or is not suited for higher education, focusing less on the label and paying more attention to the content and substance of such models (Harvey, 1995; Pratasavitskaya and Stensaker, 2010). Because it may just be as Birnbaum (2000: 104) points out: "TQM was sound; it was the implementation that was at fault". Furthermore the discussion presented in this paper leads us to conclude that quality management models not only have full potential to cover the standards and guidelines established by ENQA in 2005 to higher education, but additionally they may allow HEIs to go a step further, opening the possibility for them to really move towards quality enhancement. It is just a matter of adequately implementing them, so that they cover what lies at the heart of higher education: teaching and learning.

Interestingly, and certainly worth further research, is the fact that the adoption of the ESG by the European HEIs may finally allow for *benchmarking* exercises, which so far have been difficult to implement due to the rather different nature of quality assurance practices existent in HEIs all over Europe. Benchmarking is a process that can be used to improve higher education institutions' practices and performance (see Jackson & Lund, 2000). It is a self-evaluation method that allows a better knowledge and measurement of the organization's performance in order to promote continuous improvement. *Benchmarking* has been used in higher education especially in supporting areas such as administrative, technical and management activities. The effective management of those activities becomes critical for higher education institutions' performance, especially since the latter face an increasingly complex context of tight financial resources and rapidly changing environment. *Benchmarking* not only allows for the comparison of indicators, which makes possible the review of proposed goals, but also allows for the sharing of best practices within functional communities, identifying more intelligent ways of performing the same tasks and new solutions for common problems.

6. References

Amaral, A., & Rosa, M. J. (2010). Recent Trends in Quality Assurance. *Quality in Higher Education, 16*(1), 59-61.

Asan, S. S., & Tanyas, M. (2007). Integrating Hoshin Kanri and the Balanced Scorecard for Strategic Management: The case of higher education. *Total Quality Management and Business Excellence, 18*(9), 999-1014.

Birnbaum, R. (2000). *Management fads in higher education. Where they come from, what they do, why they fail.* San Francisco: Jossey-Bass Inc.

Birnbaum, R., & Deshotels, J. (1999). Has the academy adopted TQM? *Planning for Higher Education, 28*, 29-37.

Brookes, M., & Becket, N. (2007). Quality management in higher education: a review of international issues and practice. *International Journal of Quality Standards, 1*(1), 85-121.

Calvo-Mora, A., Leal, A., & Roldán, J. L. (2007). Relationships between the EFQM model criteria: a study in Spanish universities. *Total Quality Management and Business Excellence, 16*(6), 741-770.

Campatelli, G., Citti, P., & Meneghin, A. (2011). Development of a simplified approach based on the EFQM model and Six Sigma for the implementation of TQM principles in a university administration. *Total Quality Management and Business Excellence, 22*(7),691-704.

Cortadellas, J. (2000). *Aplicação do modelo EFQM no ensino – Universidade Politécnica da Catalunha*. Paper presented at the Sistemas de Avaliação da Qualidade no Ensino Superior.

Cruickshank, M. (2003). Total quality management in the higher education sector: a literature review from an international and Australian perspective. *Total Quality Management and Business Excellence, 14*(10), 1159-1167.

Cullen, J., Hassal, T., & Broadbent, M. (2003). Quality in higher education: from monitoring to management. *Quality Assurance in Education, 11*(1), 5-14.

Dill, D. (1995). Through Deming's eyes: a cross-national analysis of quality assurance policies in higher education. *Quality in Higher Education, 1*, 95-110.

EFQM. (2010). *EFQM Excellence Model*. Brussels: EFQM.

ENQA. (2007). *Standards and Guidelines for Quality Assurance in the European Higher Education Area*. Helsinki: ENQA - European Association for Quality Assurance in Higher Education.

Farrar, M. (2000). Structuring Success: A Case Study in the Use of the EFQM Excellence Model in School Improvement. *Total quality management, 11*(4/6), 691-696.

Filippakou, O., & Tapper, T. (2011). Quality Assurance and Quality Enhancement in Higher Education: Contested Territories? . *Higher Education Quarterly, 62*(1/2), 84-100.

Harvey, L. (1995). Beyond TQM. *Quality in Higher Education, 1*, 123-146.

HEA. (2008). *Quality enhancement and assurance: – a changing picture?* York: Higher Education Academy.

Hides, M., & Davies, J. (2002). *Implementation of the EFQM Excellence Model in the UK Higher Education Sector. A Comparison With Other Sectors*. Paper presented at the 7th World Congress for Total Quality Management - Business Excellence, Make it Happen!

Houston, D. (2008). Rethinking quality and improvement in higher education. *Quality Assurance in Education, 16*(1), 61-79.

ISO. (2005). Quality management systems - Fundamentals and vocabulary: Geneve: International Organisation for Standardisation.

ISO. (2008). Quality Management Systems - Requirements: International Organisation for Standardisation.

Jackson, N. & Lund, H. (2000). *Benchmarking for higher education*. Buckingham: Open University Press.

Kaplan, R. S., & Norton, D. P. (1992). The balanced scorecard - measures that drive performance. *Harvard Business Review, Jan-Feb*, 71-79.

Karathanos, D., & Karathanos, P. (2005). Applying the Balanced Scorecard to Education. *Journal of Education for Business, 80*(4), 222-230.

Kells, H. (1995). Creating a culture of evaluation and self-regulation in higher education organisations. *Total quality management, 6*(5/6), 457-467.

Liaison Committee of Rectors' Conferences (1993). *Quality Assessment in European Higher Education. A Report on Methods and Mechanisms, and Policy Recommendations to the European Community*. Brussels: Liaison Committee of Rectors' Conferences.

Massy, W. F. (2003). *Honoring the trust. Quality and cost containment in higher education.* Bolton: Anker Publishing.

McAdam, R., & Welsh, W. (2000). A Critical Review of the Business Excellence Quality Model Applied to Further Education Colleges. *Quality Assurance in Education, 8*(3), 120-130.

O'Neil, H. F. J., Bensimon, E., Diamond, M., & Moore, M. (1999). Designing and implementing an academic scorecard. *Change, 31*(6), 32-40.

Osseo-Asare Jr, A. E., & Longbottom, D. (2002). The need for education and training in the use of the EFQM model for quality management in UK higher education institutions. *Quality Assurance in Education, 10*(1), 26-36.

Pratasavitskaya, H., & Stensaker, B. (2010). Quality management in higher education: towards a better understanding of an emerging field. *Quality in Higher Education, 16*(1), 37-50.

Redmond, R., Curtis, E., Noone, T., & Keenan, P. (2008). Quality in higher education. The contribution of Edward Deming's principles. *International Journal of Educational Management, 22*(5), 432-441.

Rosa, M. J. (2003). *Definição de Bases Estratégicas e de Excelência para o Desenvolvimento do Ensino Superior em Portugal.* Universidade de Aveiro, Aveiro.

Rosa, M. J., & Amaral, A. (2007). A self-assessment of higher education institutions from the perspective of the EFQM excellence model. In D. F. Westerheijden, B. Stensaker & M. J. Rosa (Eds.), *Quality assurance in higher education: trends in regulation, translation and transformation* (Vol. 20, pp. 181-207). Dordrecht: Springer.

Sahney, S., Banwet, D. K., & Karunes, S. (2004). Conceptualizing total quality management in higher education. *The TQM Magazine, 16*(2), 145-159.

Sakthivel, P. B., & Raju, R. (2006). Conceptualizing total quality management in engineering education and developing a TQM educational excellence model. *Total Quality Management and Business Excellence, 17*(7), 913-934.

Santos, S. M. d. (2011). *Análise comparativa dos processos europeus para avaliação e acreditação de sistemas internos de garantia de qualidade.* Lisboa: A3ES - Readings.

Saraiva, P., Rosa, M. J., & Orey, J. (2003). A Large-scale Application in Portuguese Schools of the EFQM Excellence Model. *Quality Progress, 36*(11), 46-51.

Sarrico, C. S., Rosa, M. J., Teixeira, P. N., & Cardoso, M. F. (2010). Assessing quality and evaluating performance in higher education: worlds apart or complementary views? *Minerva, 48*, 35-54.

Schmidt, A., & García-Legaz, F. (2003). *Análisis del poder predictivo del modelo EFQM para la gestión y mejora de la calidad en la universidad pública española.* Paper presented at the XIII Jornadas Hispano-Lusas de Gestión Científica - La empresa familiar en un mundo globalizado.

Schwarz, S., & Westerheijden, D. (2004). *Accreditation and Evaluation in the European Higher Education Area.* Dordrecht: Kluwer Academic Press.

Sohail, M., Daud, S., & Rajadurai, J. (2006). Restructuring a higher education institution: a case from a developing country. *International Journal of Educational Management, 20*(4),279-290.

Trow, M. (1996). Trust, markets and accountability in higher education: a comparative perspective. *Higher Education Policy* 9(4), 309-324.

Welsh, J., & Dey, S. (2002). Quality measurement and quality assurance in higher education. *Quality Assurance in Education, 10*(1), 17-25.

Williams, G. (1993). Total quality management in higher education: panacea or placebo? *Higher Education, 25*(3), 229-237.

7

Challenges for Quality Management in Higher Education – Investigating Institutional Leadership, Culture and Performance

P. Trivellas[1], P. Ipsilantis[2], I. Papadopoulos[2] and D. Kantas[2]
[1]*Technological Education Institute (TEI) of Chalkis,*
[2]*TEI of Larissa,*
Greece

1. Introduction

In an era of global economic recession, Higher Education Institutions (HEIs) are experiencing severe pressure from budget reduction. As a result, they have been forced to develop more competitive ways in order to tap on resources and capabilities stemming from contemporary rapid technological and organisational changes. Thus, quality management issues have drawn the interest of academics and practitioners in order to build a sustainable competitive advantage battling against economic recession (Altbach et al., 2009).

Nowadays, Greek economic arduous position triggered off downsizing and cost reduction at unprecedented levels in the pubic sector, forewarning a similar orientation of retrenchment of higher education. Hence, resource scarcity and decline will guide inevitably to the corrosion of institutional effectiveness accompanied with lack of innovation, rigidity, dissatisfaction, conflict, reduced quality and turnover, unless HEIs adapt to the vulnerable environmental conditions and fiscal recession (Cameron & Smart, 1998).

Almost six years ago, a reform act was initiated for the adoption of necessary metrics and processes in order to assure the quality of services provided by Higher Education Institutions (HEIs) in Greece. This national quality assurance (QA) system aims at improving transparency, comparability and accountability of the Greek higher education system, fostering quality culture throughout the HEIs. Teaching and administrative staff as well as students are expected to be the main participants and contributors in this process. Evidence from other countries have shown that the introduction of an assessment system, QA procedures, and long range planning leading to cultural change has met the opposition and resistance of the majority of HEIs' stakeholders (Morley, 2003; Van Damme, 2002). The success of the quality management systems' change and the necessary transition in quality culture of HEIs depends on the ability of academic leaders to handle crisis and to build a strategy supportive culture with the contribution of all participants.

This chapter aims to give an overview of the QA system deployed at the Technological Education Institute of Larissa (TEIL) and to describe the transition process through the investigation of leadership, culture, student satisfaction, graduates assessment and teaching performance evaluations.

First, the chapter provides the general framework for QA initiated in higher education in Greece, as the foundation of the quality system TEIL adopted. Then the chapter describes the profiles and the emergence importance of leadership and institutional culture in the implementation stage of QA system, to engage faculty and administrators in the evaluation of service quality provided. Thereafter, this study presents the findings of a large scale survey conducted by the QA unit of TEIL, investigating the quality level of teaching processes and supporting services offered to students. Finally, this chapter concludes with the perceptions of graduates about various aspects of quality in order to provide insight into the strengths and weaknesses of the established system and offer valuable suggestions for areas of improvement.

2. Quality in higher education

Though a plethora of meanings and connotations of 'quality' term has been proposed, its subjective nature has contributed to confusion and vagueness in the existing literature leading to the lack of a universally accepted definition. In addition, the discipline of quality in higher education suffers from ambiguity due to the interchangeably adoption of conceptual different terms such as *quality*, *accountability*[1] and *assessment*[2].

Yet, it might be more fruitful to provide an insight into many views that formulate a fuzzy construct referred to as quality through theoretical consensus rather than defining it.

Initially, institutions of higher education have adopted the internally focused resource view determining quality by the assessment of their internal resources, such as the number of books in its library, the number of faculty with terminal degrees, size of the endowment and reputation neglecting the influence of the changing external environment and the emergence of sophisticated higher education customers (Seymour, 1992). Increased competition, cost-efficiency, accountability and service orientations forced higher education to gradually swift its focus on a value added or performance approach of excellence, where quality is determined by its outcomes, such as efficient allocation and use of resources and producing highly satisfied and employable graduates (Koslowski, 2006).

Originated in ancient Greece, Aristotle[3] rejects that quality is an act, and favours that it is a habit. In another way, Crosby (1979) described quality as conformance to requirements based on customer expectations. From a related but different perspective, Juran (1945) established the fitness for use view as determined by the customer. However, Drucker (1985) determined quality as customer's willingness to pay in relation to what he/she values.

---

[1]Accountability as an overarching principle has frequently been related to external forces such as accrediting bodies, governmental agencies and the public (Koslowski, 2006).

[2]Assessment as one of the many components of quality reflects the further refined tangible process resulting to the evaluation of individual instruction, specific curriculum, and student learning as a way to deal with and enhance the quality of higher education (Palomba & Banta, 1999).

[3]Aristotle of Stagira (384 BC -322 BC) was an ancient Greek philosopher who first devised a comprehensive system of Western philosophy, encompassing morality and aesthetics, logic and science, politics and metaphysics.

In an attempt to synthesize diverse quality perspectives, Seymour (1992) interpreted in the higher education context the five different types of quality proposed by Garvin (1988), as follows:

- *Transcendent* quality is a result of educator's reputation and expertise, however, it is internally focused ignoring the role of external agencies as well as the public.
- *Manufacturing-based* quality is based on service conformance to specifications and implies that the provider is capable of appropriate self-regulation. Seymour (1992) advocates that this definition is useful in an educational context, since it counterbalances the resource view of quality. Still, educational institution is responsible to make the critical decisions about service design (degree offerings, curriculum etc.) abolishing external environment and neglecting consumers' wants and needs.
- *Product-based* quality reflects student learning, produced by the curriculum and faculty. It is closely linked with assessment due to its foundation on measurable and objective indices and once agreed upon by administration and academia can advance teaching and learning. Notwithstanding, it suffers from imprecision and it might be deceptive, because several metrics such as retention rates and graduate entrance examination scores fail to shed light on what students have learned and the competences gained during their studies.
- *Value-based* quality refers to acceptable or above expectations performance at an acceptable price, meaning that higher education customers may consider the expected salary after graduation in comparison with tuition fees.
- *User-based* quality is determined by the customer's needs, wants, desires, and preferences, but in the educational context, they are highly idiosyncratic and partially subjective. This approach is antithetical to transcendent, and resource views.

Regardless of the debate on quality perceptions which are inevitable, legitimate and evolving, there is some agreement that quality has to be determined by stakeholders (Koslowski, 2006). Thus, higher education has a number of stakeholders such as students, graduates, their parents and family, academic and administration staff, employers, governmental agencies, local community, and society, all of whom experience different aspects of quality.

For example, **academics** characterized by professional preparation and autonomy in the delivery of the educational process are more inclined to define quality in resource rather than performance terms, such as individual reputation, number of research publications produced, and number of courses taught. However, **administrators** favour concrete and definable measures of success or failure of administration and coordination processes, in relation to abundant (and sometimes potentially contradicting) institutional goals and obligations (Koslowski, 2006).

Bonvillian and Dennis (1995) recognize that quality in higher education is determined by three related ingredients, namely (a) market forces as a result of fierce competition and socio-economical environment of higher education, (b) the political context referring to accreditation and public funding in order to efficiently accomplish more with less[4], and (c) student's experiences and expectations.

---

[4]Recognizing the significant impact of global financial crisis on OECD countries, the Institutional Management in Higher Education entitled its last general conference: *Higher Education in a World Changed Utterly: Doing More with Less*, (IMHE, 13-15/9/2010, Paris, France) in order to highlight and reinforce initiatives toward sustainability and effectiveness.

In this regard, the application of the marketing paradigm in higher education has advanced the view that tertiary institutions serve customers endorsing service quality rather than students. Thus, their satisfaction of their learning experience is crucial to develop institution's reputation and gain competitive advantage over other institutions.

Moreover, on the grounds that students may not be in a position to recognize exactly what is better for them at that point in life, Sitkin et al., (1994) distinguish that tertiary institutions should educate customers in what they need rather than just aiming to satisfy what they want.

Further, the term '*students*' is commonly tightly linked with a one-way approach of learning, limiting the potential for greater collaboration and participation, while the concept of '*customers*' indicates a sense of sharedness, interaction and feedback leading to knowledge co-construction where involvement promotes knowledge acquisition and experiential learning (Yeo, 2009).

In a similar vein, Rinehart (1993) discerns two different approaches of students regarded as customers: (a) internal customers actively involved in the input and output of the learning process, and (b) external customers playing the role of future employees and employers. The latter type promotes the idea that education should prepare students for the future by developing curricula suited to employment needs, even though students hardly envision of what they need to learn.

3. Employees' perceptions of service quality in higher education

Despite the various quality views have been put forth, there is some consensus that quality has to be determined by stakeholders, especially in the services sector. Service quality enhancement requires a sustained improvement in the clarity, accuracy and reliability of services delivered under a holistic perspective. The imbalanced focus of quality attempts on only external customers' perceptions, ignoring internal customers would stimulate the resistance among the latter. On the other hand, it may not be feasible to simultaneously meet all stakeholders' criteria, due to limited resources or conflicting demands.

In this regard, several scholars verified the link between customer-perceived service quality and employees' satisfaction, enforcing the definition of service quality as a perceived judgment (Snipes et al., 2005; Brown & Lam, 2008; Schlesinger & Heskett, 1991). Nevertheless, most researchers in higher education have exclusively investigated students' view of service quality, neglecting employees' perceptions of teaching processes as well as administration services. In our study of *faculty and administration staff*, we bridge this gap by adopting employees' view of the service quality they provide, following Slatten's et al. (2011) recommendations.

Though most of the instruments[5] developed to evaluate customer satisfaction in educational settings have gained acceptance in USA, Europe and Australia, their immoderate focus on the teaching aspect of quality rather than students' holistic experience has been proved to be

[5]For example: Classroom Environment Scale (CES), My Classroom Inventory (MCI), Individualized Classroom Environment Questionnaire (ICEQ), College Student Experiences Questionnaire (CSEQ), Course Perception Questionnaire (CPQ), Student Evaluation of Education Quality Questionnaire (SEEQ).

their crucial drawback (Cuthbert, 1996; DiDomenico & Bonnici, 1996). In this regard, the various *"educational instruments"* aiming at evaluating specific aspects of higher education may fail to determine the level of *'satisfaction'* as a holistic experience of students which requires an integrative and interpretive paradigm (Yeo, 2009).

Among the several measuring instruments have been developed aiming to capture and explain a holistic view of service quality, SERVQUAL (e.g. Parasuraman et al., 1994), has proved to be the most popular, as acknowledged even by its critics (e.g. Asubonteng et al., 1996). The 22 items of this instrument are categorised into the reliability, tangibles, responsiveness, assurance, and empathy service quality dimensions.

In our study of faculty and administration staff, two frameworks of service quality measurement based on SERVQUAL were synthesized referring to quality of teaching and administration quality: First, Owlia and Aspinwall's (1996) theoretical framework of service quality with an emphasis on teaching aspects of education (academic resources, competence, attitude, content), and second, Waugh's (2001) model of administrative and supportive services quality (tangibles, reliability and responsiveness, assurance and empathy).

In the context of the sustained growth and diversification of higher education systems, OECD[6] launched a program on Institutional Management in Higher Education (IMHE) putting special emphasis on innovation, quality of teaching and learning improvement, the measurement of performance and learning outcomes, access and regional competitiveness. In a student-centered approach of education promoted by IMHE, teaching (bringing learning about), assessment (finding out what has been learned), and university management (organizing learning) should be aligned to be achieved the intended learning outcomes.

4. A conceptual framework of service quality in higher education

The conceptual framework of this chapter is based on Yeo's work (2009) and links research findings with the three interrelated aspects of service quality in higher education, namely customer orientation, course design and delivery, and support services.

4.1 Customer orientation

In this study, customers are classified into three distinct yet interrelated groups: employers of prospective HEI graduates, graduates, current students and parents. Performance of this strategic objective is evaluated by the 'graduate survey' and 'faculty and administrator survey' (figure 1). The latter field research investigates institutional culture, leadership, service quality, organisational commitment, job satisfaction and job performance of academic and administration staff.

4.2 Course design and delivery

Performance in this area is mainly assessed by the 'student survey', 'graduate survey' and 'faculty and administrator survey'.

---

[6]OECD, Assessment of Higher Education Learning Outcomes (AHELO) programme, Access on 25/8/2011, Retrieved from
<http://www.oecd.org/document/22/0,3746,en_2649_35961291_40624662_1_1_1_1,00>.

4.3 Support services

Satisfaction in this area is evaluated by the 'support service survey'. Quality expectations include the availability of facilities such as computer and technical laboratories, library services, internet facilities, as well as administrative and technical support.

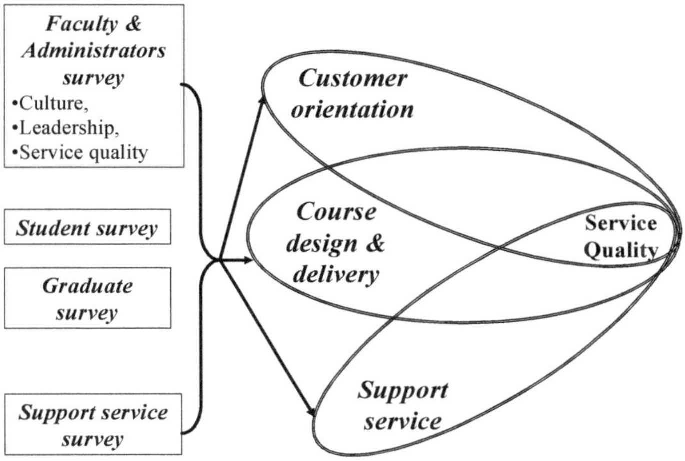

Fig. 1. Conceptual framework of the chapter.

5. Quality assurance framework for higher education in Greece

Higher Education in Greece is provided by 38 HEIs through a binary system, as in other European countries, of Universities (22) and Technological Education Institutes (16) all of which are state entities operating under a regulation framework put in place in the early eighties and is under reformation nowadays. Institutional autonomy of the Greek HEIs is limited mainly to educational and research activities as shown in Table 1 based on the 6 criteria describing institutional autonomy as defined by OECD (Santiago et al., 2008).

A tremendous expansion of Higher Education in terms of institutions and number of students emerged between 1999 and 2004 leading to 40% increase in student intake and two fold increase in the number of academic departments/programmes offered (1998: 238 departments /56,000 annual student intake, 2010: 488/80,000), making Greece the top OECD country in terms of increases in higher education expenditures as well as number of students, and among the top three countries (Spain, Turkey and Greece) in terms of increases in public spending per student (OECD, 2007). However, this expansion was not guided by quality, as most decisions regarding expansion of specific institutions were taken under the pressure of local and regional politics (viewed as transfer of income to regional economies for boosting growth), without any formal accreditation procedures in place, and mainly without a corresponding increase in academic staff.

INSTITUTIONAL AUTONOMY OF GREEK HEIs *Bold letters indicate areas of autonomy in HEIs,* *Italics denote areas controlled by the state or non-existent in the Greek HE*					
INSTITUTIONAL GOVERNANCE	STAFF	STUDENTS	FINANCE	EDUCATION	RESEARCH
Legal Status *Own infrastructure* *Commercialization of activities* **Parameters for internal decision - making** *including freedom to set up internal governance structure*	**Selection appointment** *and dismissal of academic staff* *Academic career structure* **Career advancement** *Working conditions (e.g. salaries)*	*Selection of students* *Number of students enrolling.*[a]	*Set and differentiate tuition fees* *Borrow funds on capital markets* *Allocate funds as the institution sees fit.*[b] *Income generating activities* *Right to build up a portfolio of assets and to accumulate financial capital*	**Supply of Programmes,** *including their accreditation* **Design curriculum** **Contents of courses** *Modes of instruction and delivery.*[c]	**Design research** **Decide the priorities of research**
[a]Institutions Proposition. Final decision made by the ministry of Education (MoED) [b] Must be approved by the MoED [c] No part-time studies or non-degree studies. Open & Distance Learning offered only by the Hellenic Open University.					

Table 1. Autonomy of Greek HEIs

Furthermore, current issues such as internationalization, flexibility in curricula and institutional diversity were not addressed leading to a situation where most of HEIs lack any strategic planning and cannot exploit market opportunities while a significant number of students show a low interest in their studies as it is evidenced by the fact that more than 50% of the total student population in universities do not complete their studies on time (required years +1), according to the Greek national statistics service (EL. STAT., 2008).

Several reforms were attempted in the regulatory framework for HEIs, addressing mainly peripheral and minor issues. Significant improvements that were introduced lately (since 2006) were neither welcomed by the academic community (i.e. introduction of quality assurance - currently implemented at 50% of academic units), nor could be applied because of the economic crisis (i.e. contracts between Government and HEIs on four year operational plans) nor have taken effect yet (i.e. limiting the maximum time required to complete degree requirements - to take effect in 2012).

The establishment, in 2006, of the Hellenic Quality Assurance Agency for Higher Education (HQAA, 2009)[7], an independent agency governed by a board of academics, nominated by the rectors and presidents of HEIs, marks the formal introduction of a QA system in Greek HEIs. The role of the HQAA is to oversee and co-ordinate the QA process and external assessments at HEIs, inform and advise the government on issues related to quality in higher education and promote public awareness. In this way, accountability at department,

---

[7]Greek Republic (2005) Quality assurance in higher education. Credit transfer and accumulation system - diploma supplement. Greek State Law 3374/2005.

institution and government level is promoted, transparency is increased, and quality of education and learning is improved.

6. Quality assurance at the Technological Education Institute of Larissa

The Technological Education Institute of Larissa (TEIL) is the largest HEI in Central Greece. Established in 1983, comprises four faculties (Management and Economics, Agriculture and Food Technology, Engineering, Health Studies, and Forestry & Wood/Furniture Technologies) offering twenty (20) undergraduate and nine (9) postgraduate programs of studies in applied sciences to a student population exceeding 17,000.

The institution had identified the need to introduce QA procedures long time before QA became mandatory for HEIs. In 2002, TEIL invested in the introduction of quality processes through the implementation of a pilot project, exploiting funding of the 2nd Community Support Framework[8]. Self-evaluation reviews, followed by external audits and reviews for the first time were generated at the departmental and institutional level.

As of 2007, the QA procedures pertaining to higher education institutions became mandatory. TEIL was one of the first HEIs in the country to introduce the QA procedures, where all of its departments (except four newly established departments that first had to complete 5 years of operation) managed to deliver their first self-evaluation report by the end of 2009, as compared to a national average of about 15% of all academic departments.

7. Quality Assurance System at TEIL

7.1 Establishment of quality units & committees within the institution

TEIL developed the necessary organizational structure to establish QA processes in accordance with ENQA guidelines (ENQA 2005):

'Each university should formulate a policy and procedures for the QA in their programs and commit themselves to the development of a quality culture. To this end, a strategy should be devised with a role for students and other stakeholders.'

In particular, the following bodies were formed to plan, introduce and implement QA policies and procedures:

- Institutional *Quality Assurance Unit (QAU):* Headed by the Vice-President of Academic Affairs, four academic staff members nominated by the General Assembly of the institution, one member of the administrative/technical staff, one representative of the undergraduate students, and one of the postgraduate students.
- Departmental *Internal Evaluation Work Teams (IEWT):* A three to five member committee of academic staff, nominated by the Assembly of Academic Staff of the corresponding department, the main task of which is the issuance of the annual departmental reviews, the release of the Internal Assessment Report (self-evaluation review) every four years.

[8]Community Support Framework (CSF) is a set of programmes funded jointly by the European and National governments of member states in different sectors (i.e. education, transportation, environment) to achieve economic and social cohesion in Europe.

7.2 Defining areas of evaluation and establishment of metrics

HQAA provided to all HEIs a framework of evaluation criteria which mainly address the programme of studies, the teaching and learning process, interaction and cooperation with industry society, internationalization, and research. The list of criteria was adapted to the specific institutional environment of TEIL, and the sources and processes of data collection were identified including the development of focused questionnaires addressed to students, academic and administrative staff. Data are then processed to meaningful information through the derivation of a series of performance/assessment metrics defined by the QAU with the consensus of the academic departments. In addition, a set of metrics was established for the evaluation of the performance of institutional-wide student related services such as ICT, library, registry and administration facilities, athletic facilities, dormitories, and student restaurant/cafeteria. Metrics were defined in such a way to qualify the following characteristics:

- To ensure systematic and uninterrupted collection of data so that time series are produced
- To provide the ability to aggregate from course level to institutional level where appropriate
- To produce comparisons within each department but also across faculties and departments possible, and finally
- To maintain compatibility with the national metrics defined by HQAA.

7.3 Quality assurance process

Following the general guidelines of HQAA, QA process at TEIL works on a four-year cycle, as it is illustrated in fig. 2. The internal processes lead to the compilation of self-evaluation reports, while the HQAA controls the external review process (i.e. maintaining registry of reviewers, setting up the review teams, compiling the external review reports).

The QAU prepared proper templates for the academic departments and organised a series of workshops to train the members of the departmental IEWT.

The process of compiling the first self-assessment report was a rather long one (on the average one and a half year) requiring laborious involvement from academic staff, and especially those who staffed the self-evaluation team due to lack of administrative personnel. The fact that historical data were not readily available, and had to be collected from different sources and checked for validity, along with the requirement to build a five year retrospective view of the main metrics, so that trends can be identified resulted in many person hours to be allocated in this process. The provisions of MIS support for the next rounds will facilitate the process to a large extend.

7.4 Early experiences of a quality odyssey

Introducing a QA system in an organization is a difficult task on its own since it affects the "business as usual" status and introduces additional burden to personnel, especially in an academic environment where most of the necessary quality processes are seen as non-academic, taking away valuable staff time from research and other academic activities. Establishing a quality assurance system in a period where the vast majority of the HE community is against formal QA processes (by the middle of 2009 - one and a half year after the release of the HQAA guidelines only 10% of all academic departments had submitted a self-evaluation report) makes the implementation of such a project even more difficult.

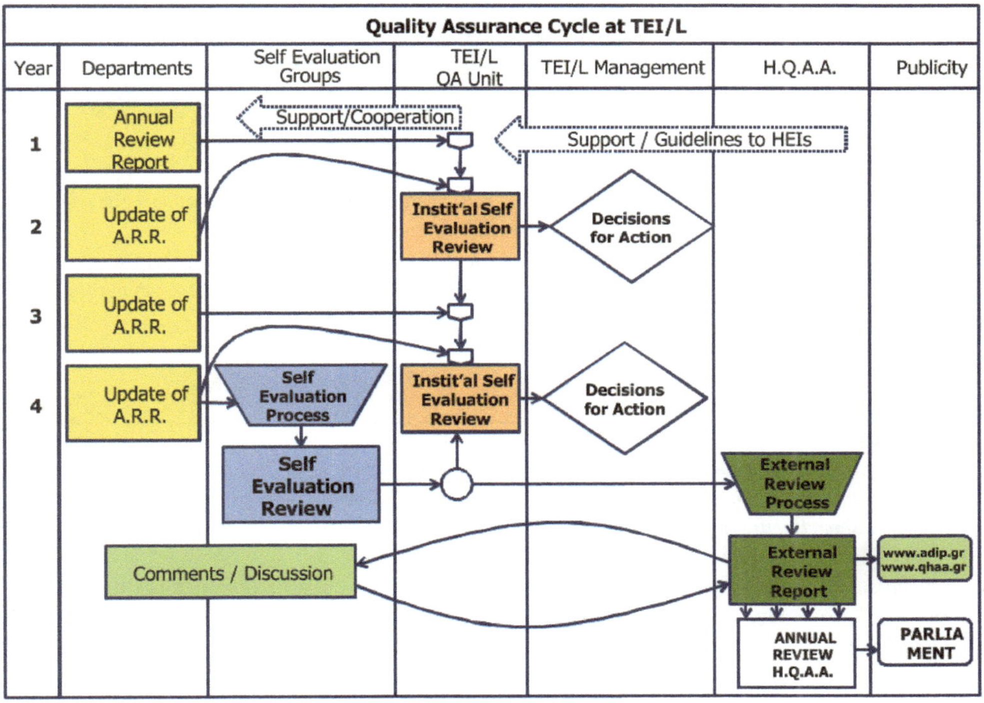

Fig. 2. Quality assurance cycle of TEIL (proposed by the HQAA)

On the contrary, nowadays the majority of the faculty have recognised the need to standardize and improve procedures related to the basic operations involved in the teaching and learning processes. As a result, several academic departments and the QAU are currently committed to contribute to the development of the core elements of the Quality Management System, compatible with the ISO 9001:2008 requirements.

As many stakeholders have had different attitudes, interests, concerns regarding the implementation of QA at TEIL, the task of recognizing and managing stakeholder's interest becomes vital for the success of the project. At TEIL, stakeholders are categorized into four groups: institution's top management, institution's middle management, academics and students, whose interests and actions prescribed by QAU are listed in table 2:

Institution's top management: *Interests*	**(institutional level)** *Actions*
Promote excellence at institution – wide level *Increase funding opportunities since non-conformance with QA requirements may lead to reduction in funding* *Become more competitive*	*Make Quality a strategic issue for the institution. Mention it in the mission statement* *Introduce institution-wide policies (e.g. QMS-ISO9001) across all departments. Link results to allocation of funds* *Commit resources to support QA process* *Extend QA to administrative processes* *Introduce QA in research*
Institution's middle management: *Interests / Attitudes*	**(departmental level)** *Actions*
Acceptance. A "chance" to reform curricula and teaching *Opportunity. Identify weaknesses and areas of improvement with much less internal conflicts, since "we are obliged to it"* *Rejection. Problems lie with the institution not with the department* *Necessary evil. Will do it because otherwise could lose funds. Exploit the system.* *Reluctance. Bureaucratic burden on academic staff must be minimized*	*Support, diffuse and promote best practices* *Diffusion of knowledge from departments that do it successfully in regular meetings (one per semester)* *Build information systems to support the processes and reduce burden on staff* *Even those who do it just due to their obligations will realize some benefits. Encouragement of efforts.*
Academic Staff: *Interests / Attitudes*	*Actions*
Opportunity: A "chance" to address / provide input / become involved / discuss issues at departmental level *Fears: Concerns for low ratings by students, and/or low research output. Increased insecurity.* *Doubts: QA added value questionable. reliability of student assessments*	*Increase awareness for low performance* *Provide relative assessment reports (compared to department median and quartiles) to all staff, while protecting privacy* *Discussions of annual report at departmental meetings with the presence of students* *Private meetings with Dept. Head for exceptional cases. Plan corrective action* *Take student assessment into account in contract renewal of part-time staff*
Students: *Interests / Attitudes*	*Actions*
Enthusiasm: It is our turn to evaluate you *Expectations: Improvements to the teaching / learning process, Better infrastructures* *Eagerness: No real changes happen. Too many surveys no results.*	*Keep students informed. Discuss departmental self-assessment review results in the presence of all students.* *Make results of external reviews known to students.* *Assign responsibilities. Make clear to them that some may participate in the external review* *Publicize results* *Utilize input from student evaluation forms in module reviews.*

Table 2. Main stakeholders of TEIL services

8. Institutional culture and leadership in higher education

In an era of economic crisis, HEIs are experiencing severe pressure from budget reduction, and they are obliged to establish new rigid systems of quality assurance, new rules and regulations and tight monitoring. As a result, academic leaders have been forced to develop more competitive ways to explore and embrace new roles in order to tap on institutional resources and core competences. Furthermore, the study of leadership in higher education faces many difficulties due to the dual control systems, since leaders have to excel in different contexts including administrative and academic departments and to deal with mixed expectations (Lewis & Smith, 1994).

Even though HEIs may be considered as service institutions, as adaptive or entrepreneurial entities, or as learning organisations (Askling & Kristensen, 2000) pursuing different strategic orientations, a convergence on a sense of strong institutional leadership prevails (Askling, 2001). Strong leaders are supposed to instigate change processes, set overarching objectives and formulate the necessary strategies to accomplish them. Nevertheless, empirical evidence support that some strong leaders with clear and predefined objectives about the outcome of the evaluation process such as well-structured self evaluation reports may have a negative impact, because they actually suffocate motivation and involvement from the academic staff (Stensaker, 1999).

Thus, the lack of a participative culture, stimulating discussions and analysis for current and future actions may conclude to disappointment and alienation among the staff or even resistance to change. Institutional culture may be defined as the collective personality of a HEI and reflected at the shared values, beliefs and behaviours of its members (Quinn, 1988). Resistance to change from its internal members such as faculty and administrators is reasonable, because change stimulates uncertainty, fear and suspicion in the stakeholders engaged (Thomas, 1998). Elements of culture such as autonomy from external control, adaptation, morale, conflict resolution, goal achievement, and formalization modify the degree to which faculty and administrators accept policy or changes associated with QA initiatives. Thus, the major challenge academic leadership has to deal with is to empower members to discard old ways and obsolete values; essentially to 'unlearn' and espouse new values (Elwood & Leyden, 2000).

Many countries have established some sort of external evaluation system like ISO 9000 as well as accreditation systems such as the Accreditation Board for Engineering and Technology (ABET) and the European Quality Improvement System (EQUIS). Though, literature review supports a positive effect of the aforementioned systems on organisations as a whole, empirical evidence reveal that resistance to accreditation by employees may be attributed to increased workload and bureaucracy, negative emotions of stress, insecurity and distrust, low level of commitment, autonomy restraint, lack of knowledge and experiences, and limited acceptance of the system (Van Kemenade & Hardjono, 2009). Hither, institutional leadership and culture may act as catalysts to reap the benefits of accreditation.

8.1 Competing Values Model

Competing Values Model (CVM) is adopted in the study of faculty and administration staff, for the operationalization of the institutional culture and leadership constructs (Deshpande et al., 1993; Quinn 1988; Trivellas & Dargenidou, 2009a; 2009b). CVM has gained wider acceptance among researchers as it has been validated not only as a model of culture (Kwan & Walker, 2004; Smart 2003; Zammuto & Krakower, 1991), but also as an instrument for other organizational phenomena such as organisational effectiveness, leadership (Quinn 1988; Hart & Quinn, 1993; Smart 2003; Trivellas & Geraki, 2008; Trivellas & Dargenidou, 2009a) and MIS effectiveness (Trivellas et al., 2006; Trivellas & Santouridis, 2011). CVM may also be used as a tool for mapping organizations' culture and leadership profiles and conducting comparative analysis.

CVM emphasizes the competing tensions and conflicts across two axes, which form a four-cell model (Fig. 3). The vertical axis extends from change and flexibility to control and order. The horizontal one reflects the conflict between internal focus and external orientation to the competitive environment. The intersection of these two dimensions determines four quadrants, which establish four archetypes of organizational culture which correspond to four leadership styles, namely adhocracy and adaptive leadership, clan and people leadership, hierarchy and stability leadership, market and task leadership (fig. 3):

Adhocracy culture and *adaptive* leadership style values entrepreneurship, creativity, proactiveness, resource acquisition, and innovativeness in discovering new markets and directions for growth. They are characterized by flexibility, adaptability and external orientation. The broker, who acquires resources for the institution and the innovator who actively supports adaptation to changes are the two roles assigned to this orientation.

Clan culture and *people* leadership addresses employee commitment, empowerment, morale, participation, teamwork, personal involvement and cohesiveness. They revolve around trust, while they promote conflict resolution and confine resistance to change. Mentoring subordinates and facilitating teamwork are the core activities attached to this style.

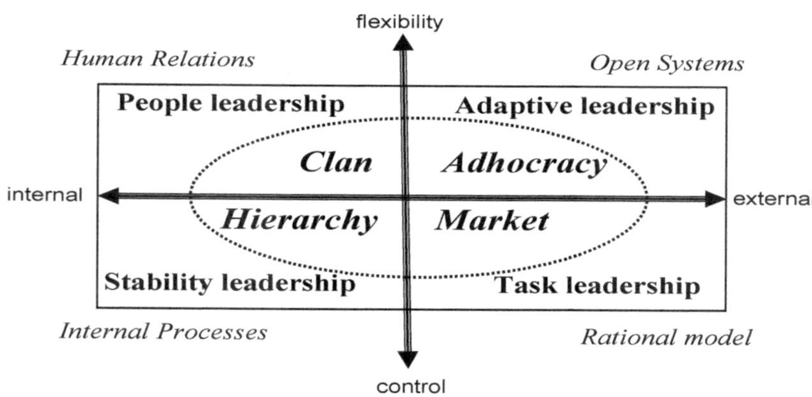

Fig. 3. CVM of culture and leadership styles

Hierarchy culture and *stability* leadership focuses on rules and regulations, order, stability, control, documentation, information management, centralization of decision making, standardization, dependability and reliability reflecting inward orientation and formalized structures. This leadership style consists of the monitor role striving to ensure compliance, track progress, and analyse results; and the coordinator one who maintains order and flow of the system.

Market culture and *task* leadership emphasize goal achievement, productivity, task accomplishment, planning and setting objectives, and efficiency (Cameron & Freeman, 1991; Deshpande et al., 1993), reflecting their extrovert and control orientation. This leadership orientation covers the producer role that motivates individuals to take actions and the director one, who clarifies expectations and establishes objectives.

There is growing evidence to support the view that institutional effectiveness as well as leaders' effectiveness depends on their cognitive and behavioural complexity, in order to respond successfully to a wide range of situations that may in fact necessitate seemingly conflicting and opposing behaviours (Smart, 2003; Denison et al., 1995; Hooijberg, 1996; Hart & Quinn, 1993; Quinn et al., 1992). In similar vein, managers who balance competing leadership roles are found to be more successful than those who adopt a restricted number of roles (Hooijberg, 1996; Hart & Quinn, 1993). Accordingly, institutional success lies on HEIs capability to develop an overall organizational culture that comprises a healthy balance of the four archetypes proposed by CVM (Smart, 2003).

8.2 Leadership and culture profiles of faculty and administration staff

The field survey of faculty and administration members of TEIL, comprised 134 valid questionnaires (response rate about 85%), and was based on a structured questionnaire measuring leadership, culture, organisational commitment, job satisfaction and service quality with 5-point Likert–type scales. Tapping on the advantages of CVM, the institutional culture and leadership instruments adopted were validated by several researchers (Trivellas & Dargenidou, 2009a, 2009b). In the same regard, higher education service quality was operationalised by adopting both the quality dimensions emphasised on teaching aspects proposed by Owlia and Aspinwall (1996), and Waugh's (2001) measures of administration quality. Affective and continuance components of Allen and Meyer's (1990) organizational commitment scale were also utilised. Job satisfaction construct was built upon Warr et al. (1979) recommendations.

Regarding leadership styles, faculty considered people leadership as the most frequently adopted one, while administration staff ranked task leadership as the dominant one. The CVM based instrument applied as a diagnostic tool, reveals that TEIL is deficient in roles emphasizing innovativeness, creativity, risk taking, monitoring, and networking with external constituencies reflected on adaptive leadership.

	Faculty	*Administration*	*Sig.(t-test)*
Leadership styles			
Adaptive leadership	3,79	3,87	n.s.
Task leadership	4,19	4,65	$p<0.05$
Stability leadership	4,12	4,26	n.s.
People leadership	4,23	4,17	n.s.
Valid N	66	68	

Table 3. Results of paired t-test analysis among leadership styles

Besides, t-test analysis was used to asses the statistical significance of the differences between faculty and administration members of TEIL. Results summarized in table 3, indicate that administration staff assigns higher priority to task leadership, in comparison with academics. In fig. 4, the leadership profiles of faculty and administration staff are illustrated.

Examining organisational culture, faculty preferred hierarchy and clan archetypes, while administrators considered hierarchy as the dominant one (Trivellas & Dargenidou, 2009b). The CVM suggests that TEIL is deficient in adaptability and growth potential, as well as in market orientation. Moreover, findings reveal that administrators characterised less by a clan culture reflecting loyalty, involvement and cohesiveness and more by a hierarchy one related to formalized structures, rules and regulations, control and decision making centralization, in comparison with academics (Trivellas & Dargenidou, 2009b).

These findings are in alignment with other researchers' conclusions that public institutions are characterised by bureaucratic cultures emphasizing on order and control and their leaders adopt conservative and stability oriented roles (Hooijberg & Choi, 2001).

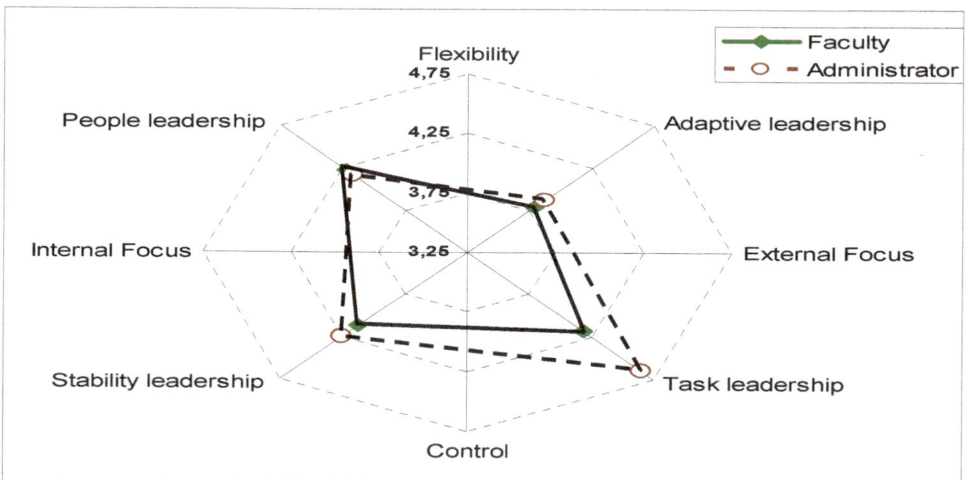

Fig. 4. Leadership profiles of faculty and administration members of TEIL

8.3 Culture, leadership, commitment and job satisfaction as antecedents of higher education service quality

Education may be considered as a transformative process promoting the enrichment of students' knowledge, the empowerment of their ability to think critically and to challenge their worldviews and assumptions (Yeo, 2009).

Extending this argument, a transformation from centralised hierarchical structures to decentralisation, employee involvement and effective leadership is a prerequisite for TEIL's adaptation towards the establishment of QA processes. Given that these transformations frequently cause employees' reluctance or opposition to change, the successful introduction of a QA system depends on crucial factors such as **culture** (Trivellas et al., 2006), trust and long term **commitment** to the organization (Zammuto & O'Conner 1992; Trivellas, 2009); participation in decision making (Franz & Robey, 1986), and **leadership** (Smart, 2003). In this way, leadership come to play a decisive role in the transformation of attitudes, and management has to adopt the suitable human resource practices in order to facilitate changes towards enhanced service quality. Furthermore, an adhocracy culture facilitates critical thinking and experimentation to transcend taken-for-granted frames of reference, leading this transformation process, and elevates effectiveness resulting to improved customer satisfaction. Thus, the role of institutional culture, leadership, organisational commitment and job satisfaction as antecedents of higher education service quality was explored at TEIL.

Regarding institutional **culture**, higher education literature has consistently supported a three-tier order, with clan or adhocracy dominated institutions being the most effective, followed by those characterised by market culture, and lastly, hierarchical culture archetypes are associated with low performers (Cameron & Ettington, 1988; Cameron & Freeman, 1991; Smart et al., 1997). This research (Trivellas & Dargenidou, 2009b) provides supporting evidence in this argument, as adhocracy culture significantly contributes in the improvement of all aspects of higher education service quality. In other words, intrapreneurship, experimentation, creativity, proactiveness, adaptation and innovativeness are values conducive to enhanced quality of teaching and administration.

Quality of teaching building on academic resources, teaching expertise, theoretical and practical knowledge, attitude of academic staff, and curriculum content, requires creative spirit, experimentation, receptiveness to radical new ideas, tolerance to ambiguity and aptitude to change. In a similar vein, flexibility, adaptation and proactiveness are the foundations for the improvement of administration quality, referring to administration contact, reliability, confidence, understanding and caring.

Examining **leadership**, the innovator (adaptive leadership) and the monitor (stability leadership) roles were found to be the most powerful predictors of higher education teaching quality (Trivellas & Dargenidou, 2009a). For example, the improvement of quality of teaching competence (teaching expertise, theoretical and practical knowledge) requires not only experimentation, risk taking, creativity, adaptation, and tolerance to ambiguity reflected on the innovator but also compliance with rules and regulations, and documentation as the monitor role prescribes. The latter role also fosters quality of teaching attitude which emphasizes on the availability of academic staff for guidance and advice. Nevertheless, the producer (task leadership) and the mentor roles (people leadership) were negatively related to teaching attitude. Given that teaching attitude also refers to educator's empathy and understanding of students' academics needs, sometimes in expense of achieving the predefined goals, the producer's objective to maintain productivity may cause alienation and emotional detachment. On the contrary, too much intimacy and emotional engagement with students under the mentor role, may lead to complications that disturb the necessary political balance in human relations to ensure objectivity and fairness in decision making.

Regarding administrators, the broker (adaptive leadership) and the facilitator (people leadership) roles which are considered to be flexible in the resolution of problems and supportive in building consensus towards its practical application, were strongly associated with both dimensions of administration quality (reliability and responsiveness, assurance and empathy). Reliability and responsiveness refer to administrative contact, provision of administrative material, confident and dependable administrative advice, and advanced notice of administrative changes. Accordingly, assurance and empathy concern courteousness and confidence, individual contact and understanding, caring and secure contact.

Investigating the relationships among **job satisfaction, institutional commitment** and service quality (Trivellas & Dargenidou, 2011), academics' teaching attitude, and satisfaction stemming from rewards and recognition were proved to be strongly associated with affective commitment, while availability of resources and academicians' competence were found to be related to continuance commitment. In a similar pattern, satisfaction with rewards and recognition as well as reliability and responsiveness of administration quality were linked with affective commitment of administrators.

9. Quality assessments by students

9.1 Questionnaire design and sample

The instrument for the students' survey included 43 items referring to students' self-evaluation and the evaluation of course elements, course teachers and laboratory workshop teachers. The structured questionnaires were answered by all students who were present in the class session in the day the survey for each course took place.

The questionnaires were specifically designed for the aim of this research, according to the basic principles of social sciences research (Gordon & Langmaid, 1988; Tull and Hawkins, 1990; Doyle, 1998; Aaker et al. 2004). Finally, 22657 valid questionnaires were analyzed, filled by students of 15 departments of the TEIL, during the academic year 2009-2010.

9.2 Results

Considering course content, the research revealed the strengths and the weak areas which need improvement of all courses evaluated. In particular, the 3 strongest points include:

1. The material covered, met the objectives of the course
2. The subjects of the work-papers were given on time
3. The objectives of the courses are made clear to the students

On the contrary, the 3 weakest points of all courses evaluated are the following:

1. The usefulness of the exercise workshops is limited
2. The educational material (book, notes, extra bibliography), were not delivered on time
3. The quality of the exercise workshops is low.

Minor differences were detected among TEIL departments mainly due to peculiarities associated with their relevant disciplines.

Moreover, the correlation analysis (Pearson correlation) brought to light some interesting relationships:

- The early delivery of the course material ($r = 0.716$, $p<0.01$), as well as the use of the course materials ($r= 0.651$, $p< 0.001$) are associated with better understanding of the course objectives and content.
- Teachers' crucial component of successful guidance through the course is the provision of edifying and analytical feedback to students ($r = 0.795$, $p< 0.001$).
- Work papers examination contributes more than other assessment methods to the better understanding of the subject by the students ($r= 0.653$, $p < 0.001$).
- The level of difficulty of the course, is positively related to the workload and the credits (ECTS) assigned to it ($r = 0.373$, $p< 0.001$).
- The usefulness of the exercise workshops is strongly related to the assessment of their overall quality ($r = 0.749$, $p< 0.001$).

Regarding faculty, the students of TEIL are quite satisfied from the responsiveness of the academic staff towards their duties such as: their attendance to the courses, the prompt correction of their project work and the time spent for collaboration with the students.

Exploring differences among students, senior students (at the 7th and 6th semester) are evaluating higher their educators than the junior ones perhaps because, students become more cognizant during the last years of studies, given that the modules they attend are more specialized and practice-oriented.

The correlation analysis verified that the more the academic staff organize their teaching materials, the better they succeed in stimulating the interest of the students for the course ($r= 0.788$, $p< 0.001$) and the better they can analyze and present the concepts of the course in a simple way, with interesting examples ($r = 0.769$, $p < 0.001$).

10. Support services

10.1 Sampling

The support services' field survey adopted the stratified random sampling method. In other words, the sample is not drawn at random from the whole population, but separately from a number of disjoint strata of the population in order to ensure a more representative sample. Stratification, that is the process of dividing members of the population into homogeneous subgroups before sampling, was based on two criteria: (a) the department and (b) the year of study of the student. The strata are mutually exclusive and collectively exhaustive. Then random sampling is applied within each stratum. This often improves the representativeness of the sample by reducing sampling error. Thus, almost 200 students from 14 departments was the initial population (2800 in total), resulting to 2114 valid structured questionnaires (75.5% response rate).

10.2 Results

The students of TEIL are found to be moderately satisfied from the suitability of the classrooms and from the teaching equipment with significant differentiation among the departments. The degree of overall satisfaction regarding the student dormitories is rather low, while much higher is the degree of satisfaction regarding the library of the Institute. However, the frequency of use of the Institute's library services is considered low (average 1.53 times per month per student), while the existence of the Career Office is not as known (only to 54.7% of the respondents) as it was expected to be. There is moderate to low satisfaction on the services provided from the student restaurant, while the satisfaction regarding the campus cafeteria is higher. The sports facilities are considered as satisfactory, while the students use them 2.08 times per month on average. The degree of the awareness of the internet and network services of the Institute is high (75.4%), while the number of students who have an email account at the Institute is considered rather small (67.2%).

High enough is the awareness of the students regarding the existence of the e-class platform (69.4%), while finally, the web sites of the Institute is considered very useful for the information of the students.

Among the most important students' suggestions on the upgrade and improvement of support services are the following:

- The renovation and upgrade of infrastructure (classrooms, laboratory equipment).
- The incitement of the students to use the books of the library more as well as the enrichment of the library collection.
- The more effective promotion of the role of the Career Office.
- Incitement of the students to exploit ICT facilities (e.g. web-page) as well as the upload of more courses on the e-class platform(departments should demand this by all teaching staff whether permanent or not).
- The renovation of the Halls of Residence Complex for student accommodation and the introduction of more efficient administration.
- The improvement of the service quality provided at the student restaurant.

11. Quality assessments by graduates

11.1 Questionnaire design and sampling

One additional initiative of the Institutional Quality Assurance Unit (QAU) of TEIL in 2010, concerned the carrying out of a survey on the graduates of all the Departments for professional and educational issues. So, a structured questionnaire was developed to be filled from the graduates of all departments during the graduation day in December of 2010. It consisted of 28 questions referring to the choice of the students to study in the institute and in the specific Department they graduated from, as well as to the evaluation of the graduates' educational experiences, employment issues and their immediate professional goals. The 12 out of the 14 departments at TEIL participated and 499 valid questionnaires were collected.

11.2 Results

The fact that the Department of TEIL was one of the first choices of the graduates 4 to 6 years ago, in a percentage of 69.3%, is very encouraging. Of course, this choice depends on the specific discipline of each department (Pearson $chi^2 = 40.865$, $p<0.001$). In particular, the Departments that were on the list of graduates' first choices, are: Infrastructure Engineering (86.8%), Business Administration (82.9%), Electrical Engineering (83.3%), Business in Tourism (82.1%), Mechanical Engineering (80.0%). On the contrary, the least preferred correspond to the Departments of: Animal Production (37.5%) and Forestry and Natural Environment Administration (47.1%). Certainly, this does not imply that the above preferences are also in effect today. The most important reasons for students' selection of the specific Department are the attractiveness of the subject of their studies (45.1%), and their employment prospective (33.9%).

The fact that 88.9% of the graduates would recommend the Department that they have studied in, to their friends, is very heartening. It is worth noticing that departments, which were not the first preferences of the graduates at the time of their entrance to higher education (e.g. Project Management, Electrical Engineering), will be strongly recommended by them at the time of graduation to other candidates.

Finally, if the graduates had the chance today to choose their department, the vast majority of them would choose the same one that they graduated from (e.g. Accountancy (93.9%), Infrastructure Engineering (94.7%)).

Furthermore, the graduates from the Departments of TEIL, state that they are satisfied on the most educational issues, ranking work placement and dissertation thesis on the top of the list. Regarding correlation analysis, the following relationships were confirmed:

- The increase of the quality of the workshops, leads to an increase on the sufficiency and the quality of the acquired knowledge of the graduates ($r = 0.614$, $p<0.001$). The sufficiency and the quality of the knowledge acquired in turn ($r=0.714$, $p<0.001$) as well as educators' relation with students ($r = 0.551$, $p<0.001$) determine to a large degree the perceived quality of the course.
- The suitability and quality of the curriculum is strongly related to the sufficiency and the quality of the studies ($r = 0.609$, $p<0.001$), the effectiveness of the faculty ($r=0.609$, $p<0.001$), but also the better link with the requirements of the labour-market ($r=0.646$, $p<0.001$).
- The strong association between the services provided by the library and faculty effectiveness ($r=0.516$, $p<0.001$), verified the crucial role of the library in facilitating the educational process by supporting not only educators and but also students as highlighted in the previous session (10.2).
- A proper work placement for the compulsory industrial training required in all programs of study at TEIL, contributes positively to the higher quality of dissertation thesis ($r=0.588$, $p<0.001$).
- Moreover, as it was expected, the total score of the graduates found to be significantly associated with the quality of the theory classes (ANOVA, $F=3.190$, $p<0.001$), the quality of the textbooks (ANOVA, $F=4.106$, $p<0.001$), educators' effectiveness (ANOVA, $F=2.759$, $p<0.005$), the efficient link between curriculum and market labour demands (ANOVA, $F=2.707$, $p<0.005$) and the quality of the curriculum (ANOVA, $F=2.307$, $p<0.05$).

Finally, the three most important benefits that graduates of the TEIL have acquired from their practical exercise are (a) the in depth understanding on the relevant subject (73.1%), the practical application of knowledge (65.3%) and the experience gained (63.7%).

12. Conclusions

This chapter aims to synthesize the experience gained and the conclusions drawn from the investigation of leadership, institutional culture, students' & graduates' satisfaction and teaching performance based on multiple stakeholders' surveys (faculty, administration staff, students and graduates) at the TEIL, in which a QA system is deployed in the last five years.

Regarding *institutional culture,* though adhocracy culture reflecting intrapreneurship, creativity, proactiveness, adaptation and innovativeness significantly contributes in the improvement of all aspects of higher education service quality, TEIL's academics are dominated by hierarchy and clan archetypes and administrators favour hierarchical values. Thus, TEIL is suffering from a shortage in adaptability and growth potential, as well as in market orientation, which may facilitate initiatives towards enhanced quality of teaching and administration. In a similar vein, administrators should depart from values tied with formalized structures, rules and regulations, control and decision making centralization in order to be actively involved in QA procedures.

According to OB literature, institutional success lies on TEIL's capability to develop an overall organizational culture that comprises a healthy balance of the four archetypes proposed by CVM, since *adhocracy* stimulates creativity and adaptation, *clan* triggers conflict resolution, morale and cohesion, *hierarchy* boosts reliability, standardisation and documentation, and *market* culture facilitates goal achievement and efficiency, all integral components for a holistic view of QA.

A similar pattern emerged investigating *leadership*. Although, the innovator (adaptive leadership) and the monitor (stability leadership) roles were found to be the most powerful predictors of higher education teaching quality, faculty considered people leadership and administration staff ranked task leadership as the dominant ones, both of them neglecting the crucial role of adaptive leadership. In alignment with management literature, managers who balance competing leadership roles are found to be more successful than those who adopt a restricted number of roles. Consequently, effective faculty and administration leaders should emphasize on all four leadership styles in order to improve quality of teaching and administration building on a QA system.

Investigating the relationships among *job satisfaction, institutional commitment and service quality,* academics' teaching attitude, and satisfaction stemming from rewards and recognition were proved to be strongly associated with affective commitment. Similarly, satisfaction with rewards and recognition as well as reliability and responsiveness of administration quality were linked with affective commitment of administrators.

Thus, effective HR practices and appraisal systems at TEIL should put special emphasis on rewards and recognition of faculty and administrators in order to improve their affective commitment and service quality provided.

In a different aspect, students of TEIL are quite satisfied from the responsiveness and commitment of the academic staff towards their duties, while senior students are evaluating higher their educators than the junior ones. As it was expected, the well organized teaching material by faculty stimulates the interest of the students for the course and facilitates the presentation of the relevant concepts in a simple way, with interesting examples. Furthermore, students' better understanding of the course objectives and content is associated with course material quality, teachers' successful guidance and feedback as well as the adoption of essays examination as assessment method.

Regarding support services, students stress the requirement of renovation and upgrade of infrastructure (classrooms, laboratory equipment, and student restaurant), as well as the necessity to be motivated in order to take advantage of library, Career Office and ICT services.

Under a different point of view, graduates are so contented with their studies that the vast majority of them would recommend the Department that they have studied in, to other candidates, even though this might not be their first preference at the time of their entrance to higher education. Moreover, sufficiency and quality of the acquired knowledge as well as teachers' relation with students are associated with higher course quality as perceived by graduates. Furthermore, quality of the curriculum, faculty effectiveness, and studies connectivity with market requirements promotes the overall quality of the studies. Also, the crucial role of the library in improving the educational process performance by supporting not only educators and but also students was verified. Likewise, graduates' total score found to be significantly associated with the quality of the curriculum, the theory classes and the relevant textbooks, as well as educators' effectiveness and the associated links of studies with market labour demands.

In the light of the above findings, QAU prioritise the following actions:

- To raise quality of education and learning as a top policy item in the strategic agenda of the institution, and make the continuous improvement of educational provisions and quality of student learning a matter of the institutional mission statement.
- To promote all stakeholders' consciousness of QA processes as integral and effective part of internal procedures overcoming impediments due to bureaucracy and useless documentation.
- To assure the reliability and transparency of QA processes.
- To nurture a more balanced culture as proposed by CVM, with a special emphasis on adhocracy and clan archetypes (e.g. instigating trust about private data protection, publicity).
- To suggest motivational and appraisal systems aiming to facilitate continuous improvement at the institutional, departmental and individual levels.
- To support institution as a whole, as well as departments to formulate operational plans, set clear objectives related to learning objectives, design suitable curriculum under multidisciplinary perspective, make decisions and take actions towards their accomplishment.
- To assist the alignment or conformance with guidelines imposed by Bologna declaration, EHEA, OECD, European and Greek Qualifications Framework etc.

Setting the agenda for the future in alignment with OECD considerations, TEIL has to deal with the following challenges (Altbach et al., 2009):

- The phenomenon of *massification*, responding to mass demand as a result of the rise of service industries and the knowledge economy.
- *Inequalities in access*. Although, policy initiatives have focused on widening the undergraduate participation in recent years, has not benefited all sectors of society equally.

- *Increasing student mobility and internationalisation.* Greece is a champion in the number of students studying abroad regarding its population, with 4784 international students per million of Greek residents (2009).
- *Teaching, learning and curricula.* An increasingly diverse student body necessitates the establishment of new systems for academic support and innovative approaches to pedagogy taking into consideration indigenous philosophies, cultures, languages and histories. In addition, several key demographic trends will modify the education landscape such as female majority, the mix of the student population (e.g. international students, older students, part-time and working students), different social groups, disadvantaged groups, educators with varied employment contracts and part-time staff.
- *Quality assurance, accountability and qualifications frameworks.* Quality assurance in higher education, certification of institutions and the qualifications they award are tightly linked with postsecondary education mission to provide graduates with new skills, a broad knowledge base and a range of competencies to enter a more complex and interdependent world.
- *Financing higher education and the public good-private good debate.* Greek economic arduous position poses a challenge to the traditional view of postsecondary education as a public good and a 'social contract', contributing to society through educating citizens, improving human capital, encouraging civil involvement and boosting economic development. In response to these financial pressures, tertiary education has sought solutions by initiating tuition fees on postgraduate studies and lifelong professional training programs.
- *The private revolution.* Private higher education is the fastest-growing sector worldwide.
- *The academic profession.* Despite that possibly up to almost 50% of university teachers have only earned a bachelor's degree globally, in Greece all permanent academic staff hold a Ph.D. However, bureaucracy and administration often struggle their autonomy.
- *The research environment.* Teaching, research and public service as primary higher education objectives create constant tension and trade off with each other at different levels leading to shift their interest and resource allocation away from research.
- *Information and communications technology.* Distance education and other technology-induced innovations render traditional HEIs obsolete though they can promote quality offered.

13. Acknowledgment

This chapter is based on results and experiences of the project "Quality Assurance Unit of TEI Larissa" which was co-financed by the EU (European Social Fund) and National funds in the framework of the Operational Programme "Education and Long Life Learning".

14. References

Allen, N.J. & Meyer, J.P. (1990). The measurement and antecedents of affective, continuance and normative commitment to the organization. *Journal of Occupational Psychology*, vol.63, pp. 1-8

Altbach Ph., Reisberg L. & Rumbley L. (2009), *Trends in Global Higher Education: Tracking an Academic Revolution. A Report Prepared for the UNESCO 2009 World Conference on Higher Education*, United Nations Educational, Scientific and Cultural Organisation, UNESCO, Paris, France.

Askling, B (2001). In search of new models of institutional governance: Some swedish experiences. *Tertiary Education and Management*, Vol.7, No.2, pp. 197–210.

Askling, B. & Kristensen, B. (2000). Towards the 'Learning Organisation': Implications for institutional governance and leadership. *Higher Education Management*, Vol.12, No.1, pp. 17- 42.

Asubonteng, P.; McCleary, K.J. & Swan, J.E. (1996). SERVQUAL Revisited: A critical review of service quality. *The Journal of Services Marketing*, Vol.10, No.2, pp. 62-81.

Bonvillian, G. & Dennis, T.L. (1995). Total quality management in higher education: opportunities and obstacles, in Sims, S.J. & Sims, R.R. (Eds), *Total Quality Management in Higher Education: Is it Working? Why or Why Not?*, Praeger, Westport, CT, pp. 37-50.

Brown, St.P. & Lam, S.K. (2008). A Meta-Analysis of Relationships Linking Employee Satisfaction to Customer Responses. *Journal of Retailing*, Vol.84, No.3, pp.243–255.

Cameron, K. & Smart, J.C. (1998). Maintaining effectiveness amid downsizing and decline in institutions of higher education. *Research in Higher Education*, Vol.39, pp. 65-86.

Cameron, K.S. & Ettington, D.R. (1988). The conceptual foundations of organizational culture. In: Smart, J. C. (ed.), *Higher Education: Handbook of Theory and Research* (Vol. 4), Kluwer Academic Press, Dordrecht, The Netherlands, pp. 356–396.

Cameron, K.S. & Freeman, S.J. (1991). Cultural congruence, strength, and type: Relationships to effectiveness. *Research in Organizational Change and Development*, Vol.5, pp.23–58.

Crosby, P. (1979). *Quality is free: The art of making quality certain*. McGraw-Hill, New York, USA.

Cuthbert, P.F. (1996). Managing service quality in HE: Is SERVQUAL the answer? Part 1. *Managing Service Quality*, Vol.6, No.2, pp. 11–16.

Denison, D. R.; Hooijberg, R. & Quinn, R.E. (1995). Paradox and performance: Toward a theory of behavioral complexity in managerial leadership, *Organization Science*, Vol.6, No. 5, pp. 524–540.

Deshpande, R.; Farley J.U. & Webster F.E.Jr (1993). Corporate Culture, Customer Orientation, and Innovativeness in Japanese Firms: A Quadrad Analysis, *Journal of Marketing*, Vol. 57 (January), pp. 23-27.

DiDomenico, E. & Bonnici, J. (1996). Assessing service quality within the educational environment. *Education*, Vol.116, No.3, pp. 353-360.

Drucker, P. (1985). *Innovation and entrepreneurship*. Harper & Row, New York, USA.

EL.STAT. (Hellenic Statistical Authority), (2008), *Universities (students, Institution, academic staff) (beginning & end of the academic year) – 2008* [online]. Available from http://www.statistics.gr/portal/page/portal/ESYE/

Elwood, L.P & Leyden, V.M. (2000). Strategic planning and cultural considerations in tertiary education systems: the Irish case, *Scandinavian Journal of Educational Research*, Vol.44, No.3, pp. 308-323.

ENQA, (2005). *Standards and guidelines for quality assurance in the European higher education area*. Helsinki, European Association for Quality Assurance in Higher Education.

Franz, C.R. & Robey, D. (1986). Organizational Context, User Involvement and the Usefulness of Information Systems, *Decision Sciences*, Vol.7, pp. 329-355.

Garvin, D. A. (1988). *Managing quality*. The Free Press, New York, USA.

Hart, S.L. & Quinn, R.E. (1993). Roles executives play: CEOs, behavioral complexity and firm performance, *Human Relations*, Vol. 46, No. 5, pp. 543–576.

Hooijberg, R. & Choi, J. (2000), Which leadership roles matter to whom? An examination of rater effects on perceptions of effectiveness, *Leadership Quarterly*, Vol.11, No.3, pp. 341–364

Hooijberg, R. (1996). A multidirectional approach toward leadership: An extension of the concept of behavioral complexity, *Human Relations*, Vol. 49, No. 7, pp.917–946.

HQAA, (2009). Report on the quality of Greek higher education in 2008 [online]. Available from: http://www.adip.gr [Accessed 13 June 2011].

Juran, J. (1945). *Management of inspection and quality control*. Harper & Bros, New York, USA.

Koslowski III, F. A. (2006). Quality and assessment in context: A brief review. *Quality Assurance in Education*, Vol.14, No.3) pp. 277-288.

Kwan, P. & Walker A. (2004). Validating the Competing Values Model As a Representation Of Organizational Culture Through Inter-Institutional Comparisons, *Organizational Analysis*, Vol.12, No.1, pp. 21-37.

Lewis, R.G & Smith, D.H. (1994). *Total Quality in Higher Education*, Delray Beach, FL: St. Lucie Press.

Morley, L. (2003). *Quality and Power in Higher Education*, SRHE & Open University Press, ed. Heather Eggins, McGraw Hill, England.

OECD, (2007), Education at a Glance: OECD Indicators 2007, OECD, Paris, France.

Owlia, M.S. & Aspinall, E.M. (1996). A framework for the dimensions of quality in higher education, *Quality Assurance in Education*, Vol. 4, No. 2, pp. 12-20

Palomba, C.A. & Banta, T.W. (1999). *Assessment Essentials: Planning, Implementing, and Improving Assessment in Higher Education*, Jossey-Bass, San Francisco, CA.

Parasuraman, A.; Zeithaml, V. & Berry, L. (1994). Alternative Scales for Measuring Service Quality: A Comparative Assessment Based on Psychometric and Diagnostic Criteria, *Journal of Retailing*, Vol.70, No.3, pp. 201-30.

Quinn, R. E. (1988), *Beyond Rational Management*, Jossey-Bass Publishers, San Francisco.

Quinn, R. E.; Spreitzer, G.M. & Hart, S.L. (1992), Integrating the extremes: Crucial skills for managerial effectiveness. In: Srivastva, S., and Fry, R. E. (eds.), *Executive and Organizational Continuity: Managing the Paradoxes of Stability and Change*, Jossey-Bass, San Francisco, pp. 222–252.

Rinehart, G. (1993). *Quality education: Applying the philosophy of Dr. W. Edwards Deming to transform the educational system*. Milwaukee: ASQC Quality Press

Santiago, P.; Tremblay, K.; Basri, E. & Arnal E. (2008). *Tertiary Education for the Knowledge Society*, vol.1, OECD 2008, ISBN 978-92-64-04652-8, pp.80-82.

Schlesinger, L.A. & Heskett, J.L. (1991). Breaking the cycle of failure in services, *Sloan Management Review*, Vol. 32, No. 3, pp. 17-28.

Seymour, D.T. (1992), *On Q: Causing Quality in Higher Education*, Macmillan, New York, NY.

Sitkin, S.B.; Sutcliffe, K.M., & Schroeder, R.G. (1994). Distinguishing control from learning in Total Quality Management: A contingency perspective. *Academy of Management Review*, Vol.19, No.3, pp. 537–564.

Slatten, T.; Svensson, G. & Sværi, S. (2011). Service quality and turnover intentions as perceived by employees, Antecedents and consequences, *Personnel Review*, Vol.40, No.2, pp. 205-221,

Smart, J.C. (2003). Organizational Effectiveness of 2-year Colleges: The Centrality of Cultural and Leadership Complexity, *Research in Higher Education*, Vol. 44, No. 6, pp. 673-703

Smart, J.C.; Kuh, G.D. & Tierney, W G. (1997). The role of institutional cultures and decision approaches in promoting organizational effectiveness in two-year colleges. *Journal of Higher Education*, Vol.68, pp. 256–281.

Snipes, R.L.; Oswald, Sh.L.; LaTour, M. & Armenakis, A.A. (2005). The effects of specific job satisfaction facets on customer perceptions of service quality: an employee-level analysis, *Journal of Business Research*, Vol.58, pp. 1330– 1339

Stensaker, B. (1999). In Charge? (Absent) academic leadership during external assessments, *International Journal of Educational Management*, Vol.13, No. 6, pp. 266–271.

Thomas, H. (1998). Cultural leadership. In Packwood, T. and Turner, C. (Eds), *Hierarchy, Anarchy and Accountability*. New York, NY: McGraw-Hill.

Trivellas, P. & Dargenidou, D. (2009a). Leadership and Service Quality in Higher Education: The case of the Technological Educational Institute of Larissa, *International Journal of Quality and Service Sciences*, Vol. 1, Issue 3, pp. 294-310, ISSN 1756-669X

Trivellas, P. & Dargenidou, D. (2009b). Organisational Culture, Job Satisfaction and Higher Education Service Quality, The case of Technological Educational Institute of Larissa, *the TQM Journal*, Vol. 21, No. 4, pp. 382-399, ISSN 1754-2731

Trivellas, P. & Dargenidou, D. (2011). Service quality, Job Satisfaction and Organizational Commitment in Higher Education. An Empirical Study of Faculty and Administration Staff. *Proceedings of ICQSS 2011, 14th QMOD conference on Quality and Service Sciences*, pp. 1755-1774, ISBN: 84-8081-211-7, San Sebastian, Spain, August 29-31.

Trivellas, P. & Geraki, A. (2008). Investigating Principals' Leadership Profile in Secondary Education, *Proceedings of ICMMS 2008 International Conference on Management & Marketing Sciences,* pp. 599-602, ISBN 13 978-1-84816-509-0, Imperial College Press Conference Proceedings, Athens, Greece, May 23-25.

Trivellas, P. & Santouridis, I. (2011), The impact of Management Information Systems' effectiveness on task productivity, task innovation, customer focus and management control-The case of the Greek Banking Sector, *proceedings of 3rd IEEE International Conference on Information Management and Engineering (IEEE ICIME 2011)*, pp. 257-261, ISBN: 978-1-61284-240-0, Zhengzhou, China, May 21-22, 2011

Trivellas, P. (2009). Impact of Strategic Human Resource Management on organisational commitment and employee satisfaction in ISO 9001 certified firms: An empirical study in the Central of Greece, *Proceedings of ICESAL 2009, 6th International Conference on Enterprise Systems, Accounting and Logistics*, pp 1-16, ISBN 978-960-287-092-1, Thessaloniki, Greece, May18-19, 2009

Trivellas, P.; Reklitis, P. & Santouridis, I. (2006). Culture and MIS effectiveness Patterns in a Quality Context: A case study in Greece, *International Journal of Knowledge, Culture and Change Management*, Vol. 6, No. 3, pp 129-144, ISSN 14479524, 14479575.

Van Damme, D. (2002). *Trends and models in International Quality Assurance and Accreditation in Higher education in relation to trade in education services*, OECD/US Forum on Trade in Educational Services, Washington, D.C. USA, 23-24 May.

Van Kemenade, E. & Hardjono, T.W. (2009). Professionals freaking out: the case of accreditation in Dutch higher education, *the TQM Journal*, Vol. 21, No. 5, pp. 473-485.

Warr, P.; Cook, J. & Wall, T. (1979). Scales for the measurement of of some work attitudes and aspects of psychological well-being. *Journal of Occupational Psychology*, Vol.52, No.2, p. 129-148.

Waugh, R.F. (2001), Academic staff perception of administrative quality at universities, *Journal of Educational Administration*, Vol. 2, No. 2, pp. 172-188.

Yeo, R.K., (2009), Service quality ideals in a competitive tertiary environment, *International Journal of Educational Research*, Vol.48, pp. 62–76

Zammuto, R.F. & Krakower, J.Y. (1991). Quantitative and Qualitative Studies of Organisational Culture, *Research in Organisational Change and Development*, Vol.5, pp. 83-114.

Zammuto, R.F. & O' Connor, E.J. (1992). Gaining advanced manufacturing technologies' benefits: the roles of organisation design and culture, *Academy of Management Review*, Vol. 17, No. 4, pp. 701-28.

8

Using a Class Questionnaire for Quality Improvement of Engineering Ethics Instruction During Higher Education

Yuji Okita
Kanazawa Institute of Technology
Japan

1. Introduction

This report explains quality improvement of higher education engineering ethics classes. As the sophistication of modern technology increases, its positive and negative influences on society also expand, and the importance of engineering ethics comes to be considered intensively. Therefore, the importance of engineering ethics as a class topic during higher education increases, because the achievements of the numerous graduated students who find positions in the professional engineering field will unquestionably affect the quality of life we enjoy in our modern society.

Because only a few decades have passed since engineering ethics education has come to be regarded as important, its pedagogy has not been well established. The accumulation of good teaching materials and good methods remains insufficient. Regarding today's situation, it can be said that the lecturer in charge is devoting great effort to the improvement of instruction, but mostly through application of trial-and-error methods individually and independently. Consequently, the educational quality of engineering ethics classes is unstable, with great variation. Evaluation and improved quality are desirable.

The author, believing that a questionnaire survey administered to students can be useful to evaluate and improve the engineering ethics class quality, found several lecturers with similar views. Therefore, this joint execution of the questionnaire survey was realized. This paper reports details of the idea, execution, and results of the survey.

2. Current state

What are the current conditions related to evaluation, accreditation of higher education, and engineering ethics education? What are the current conditions of the use of questionnaire surveys in engineering ethics education? These topics are overviewed in this chapter.

2.1 Current conditions of educational evaluation and accreditation

Evaluation of educational outcomes has persisted as a challenging issue for years. In the primary and secondary education field, pedagogic research on educational effects has been performed for a long time, probably because those levels of instruction have constituted nationally compulsory education.

The Organization of Economic Cooperation and Development (OECD) Programme for International Student Assessment (PISA) has supplied much data and information. The PISA Web site[1] reports that "through its surveys of 15-year-olds in the principal industrialised countries. Every three years, it assesses how far students near the end of compulsory education have acquired some of the knowledge and skills essential for full participation in society."

Then how about higher education? Track records of credit unit authorization of class subjects exist, as do those of academic degree authorization. Nevertheless, little information has been supplied from the higher education side to society at large. Consequently, societal confidence in higher education has been shaken.

Under such circumstances, ABET[2], the American accreditation organization for college and university programs, has recently devoted great effort in reforming their accreditation criteria and their reviewing performance, where the outcomes-based assessment is regarded as highly important. The ethical requirement is included in their criteria: their Engineering Accreditation Commission's Criterion 3 (f) says "Engineering programs must demonstrate that their students attain the outcomes of an understanding of professional and ethical responsibility." In Japan, the Japanese Accreditation Board of Engineering Education (JABEE), influenced by ABET and other overseas boards, is striving similarly toward its goals.

The various activities of OECD and UNESCO are also noteworthy. Their key programs and principles would be, for example, quality assurance (quality provision), cross-border higher education, Assessment of Higher Education on Learning Outcomes (AHELO), and the Programme for the International Assessment of Adult Competencies (PIAAC).

It might be readily apparent that quality assurance and improvement of higher education are regarded world-wide as more important than ever, and that academic staff should positively seek adaptive solutions for these issues.

2.2 Current conditions of engineering ethics education

Next, the present conditions of engineering ethics education in higher education are considered.

A fundamental question is whether the degree of achievement and evaluation adapt themselves to morality or ethics. In other words, "can ethics be taught?" Still, higher education institutions which provide class subjects to students hold some responsibility to give them high-quality educational opportunities and reliable credit units as well as degrees. Improving their educational quality is a serious issue.

At the Kanazawa Institute of Technology, a research project titled "The Formation of Ethics Crossroads and the Construction of Science and Engineering Ethics" was promoted during

---

[1] http://www.pisa.oecd.org/

[2] http://www.abet.org/

2004–2007 under funding support of the Japan Science and Technology Agency (Fudano, 2007). How to grasp engineering ethics education effect was a main issue to be tackled. Further consideration of the trial effort to measure the educational effect at engineering ethics was also reported (Honda et al., 2009). A result of the survey that was performed by the Engineering Ethics Investigating Committee of the Japanese Society for Engineering Education was also reported (Kobayashi et al., 2010). The report, described the problem that common understandings related to contents, educational techniques, and assessment methods remain unclear, despite the spread of engineering ethics education during the last decade in Japan.

2.3 Current conditions of questionnaire survey of ethics class

Does a questionnaire survey help to clarify and improve engineering ethics class?

The author, together with other researchers, holds experience in applying questionnaire surveys in research related to corporate ethics. The surveys were administered as a part of "The Formation of Ethics Crossroads and the Construction of Science and Engineering Ethics" described above. The results and continuing research reveal that the questionnaire survey is useful to ascertain the business ethics condition of an enterprise. In other words, it is effective to know what is strong and weak about an enterprise's behaviour judged from an ethical perspective. It offers several leads in improving the effort of business ethics (Okita et al., 2010).

Kageyama and others, including the author, reported a questionnaire survey application result to an engineering ethics class of The University of Tokushima (Kageyama et al., 2009). The questionnaire survey is useful for engineering ethics classes offered at institutes of higher education.

The author has conducted several engineering ethics classes at universities, teaching freshmen to master course students. Questionnaire surveys have been used; they have invariably helped the author to improve class lessons. It follows naturally that if the survey questionnaires were administered by several lecturers to various classes and the results were shared, then the quality of engineering ethics education could be improved more effectively.

3. Assessment of classes by the joint implementation of questionnaire survey

The author and other members performed earlier research of engineering ethics class by the joint implementation of a commonly prepared questionnaire survey (Okita et al., 2010; Shimizu et al., 2010). The Education Working Group of the Ethics Committee of the Institute of Electrical Engineers of Japan (IEEJ) worked well as a basis for activity. The basic views and concrete examples are described in this chapter.

3.1 Basic views

Basic views of the applied questionnaire survey are the following. The surveys were administered to the class students twice: at the beginning and ending of the course. The contents of the two questionnaires are fundamentally identical, so that results can be compared. Data obtained at the beginning period show the initial conditions of the class. Therefore, they help a lecturer to prepare later lessons. The difference between ending and beginning data are expected to show the change of students attributable to learning in the class. Comparison of the data obtained from two or more classes would provide more fruitful information for the improvement of engineering ethics education.

Presume four sets of questionnaire survey data as presented in Table 1.

	Questionnaire Survey Period	
	Beginning	Ending
Class X	Data Xb	Data Xe
Class Y	Data Yb	Data Ye

Table 1. Questionnaire Survey Data.

The variance of Xe and Xb, as with that of Ye and Yb, shows the change of students. It results from the lecturer. Therefore, the student can find hints of which part of the student's own lessons causes, or does not cause, the change if the contents of questionnaire are properly set.

In cases where classes X and Y are covered by the same lecturer, say classes of last year and this year, then the lecturer can know the effects of efforts at improvement under the conditions where groups of students of the two succeeding years are similar. In case classes X and Y are covered by different lecturers, they would be able to improve their teaching methods not only by sharing data, but also by sharing their syllabi, teaching methods, and related information.

3.2 Examples of concrete topics

Questionnaire items of two types are presented as examples in this clause. First is the issue of whether students have knowledge related to topics of engineering ethics. Second is the ability to do something related to those ethical issues.

3.2.1 Knowledge acquisition

We consider two ethics-related question items of questionnaires related to the knowledge of students. These are:

"Do you agree or do you disagree with the following sentences?"

i. I know the meaning of compliance.
ii. I know the meaning of CSR.

Students are asked to answer whether they agree to the given sentences, where so called "Likert 7-point scale" is adopted.

Both compliance and Corporate Social Responsibility (CSR) are presumed to be important in society, especially in enterprises. Compliance is old and passive, but it is still important. However, CSR is new and positive: a corporation should positively perform not only their business matters to acquire profit but also seek to improve human rights, the natural environment, and other matters related to sustainability[3].

Those who live in a modern society, especially who work in a social organization as engineers in professional field, should know these principles well. Nevertheless, university students, at least in Japan, generally have insufficient knowledge about them. By presenting a Likert scale of 1–7 representing seven choices, strongly disagree, disagree, weakly disagree, neither agree nor disagree, weakly agree, agree, strongly agree, the obtained data of the questionnaire can be converted in quantitative figures, so that they can be handled mathematically to obtain the mean of the class, standard deviation (SD), and so on. These might be useful to inspect the following hypothetical thinking.

a. Probably the mean point mark at the time when class starts will be very low, and standard deviation will be small, because students do not, in general, know the word "compliance." (Same as "CSR")
b. The mean score at the end of a class will become high because students will learn in cases where compliance is taken up by a lesson. (Same as "CSR")
c. In a class where students study seriously, the final grade will become higher (the difference from beginning to end of the class will become larger).
d. To teach better will raise the end point mark higher and make the change larger.
e. Some class students are only lightly involved. They might take naps during lessons, or run from or otherwise avoid the lesson. The effect is to widen the standard deviation of the final course grades.
f. A widened standard deviation might also result from poor teaching of the instructor, not by students.
g. The difference of the scores of compliance and CSR suggest a difference of efforts on these two issues currently made by the companies in society.

These hypotheses might sometimes be verified by a single instructor alone, but the cooperative work of plural instructors makes the verification analysis much easier and more effective.

3.2.2 Acquisition of communications skills

Upon graduation, many students want to hold positions in business organizations. Then, regarded from the company side, on what points do they put importance at the employment of freshmen? Keidanren (Japan Business Federation)[4] has conducted annual questionnaire surveys related to employment activities of member companies since 2001. According to it, communications skills are invariably the most important item. The member companies choose five important items from the 25 list of items in the questionnaire survey. In 2010, 81.6% of the respondent companies raised communication skills, the highest of 25 items.

---

[3] The International Standard Organization (ISO) published a new standard ISO 26000 (Guidance on Social Responsibility) in November 2010, which is based on the idea that not only corporations but also social organizations of every type hold social responsibility. CSR is widened to SR.

[4] http://www.keidanren.or.jp/english/

Then, what communication skills does the industrial world expect of university graduates and engineering education? Is it the entertainer's ability to elicit laughter of the audience, or the ability to take orders of guests without mistake at a restaurant? A company ought to ask engineering graduates to have engineering communication skills, which include smooth communication at a team meeting, business-related talk, academic conference presentations, and so on, which are based on common sense and a general understanding of the society in which their work will create values of some kind for the company and society.

Following items are, for example, considered to be useful in the questionnaire to know the interest and attitude of a student.

"Do you agree or do you disagree with the following sentences?"

i. Social, cultural, political, and economic forces will continue to shape and affect the success of technological innovation[5].
i. My area of study has a "nerdy" image.
ii. My area of study is difficult for the general public to understand because it is very complex and sophisticated.

It is doubtful that our society can have the benefits of the sound and solid technologies, unless those who do studies and engineering create new products and services with little consideration of the society they live in. The engineering ethics class must revive and encourage engineering students to adopt a wide interest in various fields such as culture, economy, humanities, law, politics, other engineering disciplines, and the relation between technology and society.

4. Practiced questionnaire survey

In this chapter, actually practiced questionnaire surveys are discussed from the viewpoints of design, execution, and results.

4.1 Design of questionnaire

Based upon the considerations described in the previous chapter, a questionnaire for joint work was designed.

4.1.1 Structure

The questionnaire sheet structure is presented in Fig. 1. A notice to respondents is placed at the beginning, followed by sections soliciting the respondents' attributes and question responses. Question items are classified into three parts: the first is common and mandatory for users who apply this questionnaire; the second is common and optional; the third is arbitrary and optional, where users of this questionnaire can place arbitrary question items as they like.

---

[5] This sentence is taken from the description of "The Engineer of 2020" , p.53, published by the National Academy of Engineering, 2004

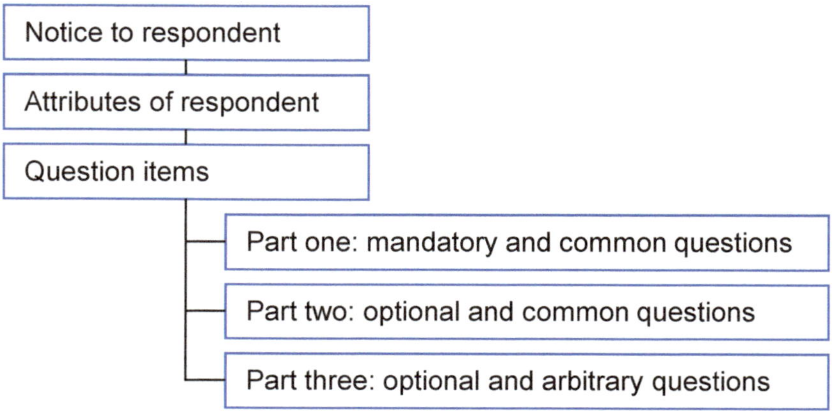

Fig. 1. Structure of questionnaire sheet.

4.1.2 Question items

A notice to respondents includes the following items.

- Purpose of this questionnaire survey
- This is an unsigned survey
- Request to answer instinctively
- Privacy policy

Attributes of the respondent are requested: the name of the institution, name of department, class name, grade year, age, past experience in study engineering ethics in higher education, etc.

Part one of the questions includes 32 items, some of which were explained in Section 3.2. Respondents were asked to choose one of seven alternative answers for each sentence, arranged as "strongly disagree" to "strongly agree". The complete list of the question items is shown below. Arrangement of the items is randomized with small exceptions.

- Q1) My major field of engineering is inconspicuous.
- Q2) I hold a strong sense of ethics (compared to an average person).
- Q3) I know the meaning of CSR.
- Q4) Engineering ethics is important.
- Q5) Engineering ethics is gloomy.
- Q6) I have experience of feeling familiar with engineering ethics.
- Q7) Social, cultural, political, and economic forces will continue to shape and affect the success of technological innovation[6].
- Q8) I know the meaning of compliance.

[6] This sentence was taken from "The Engineer of 2020."

- Q9) In all cases, the freedom of scientific research should be guaranteed.
- Q10) To save someone's life, telling a lie is allowable.
- Q11) If an action troubles nobody, then I can continue doing it.
- Q12) My major field of study has a "nerdy" image.
- Q13) The image of the word "ethics" is formal.
- Q14) My study speciality is difficult for the general public to understand because it is very complex and sophisticated.
- Q15) The pace of the technological innovation will continue to be rapid[7].
- Q16) The possibility of destruction of future generations can occur if the generation of today uses up natural resources such as oil and coal.
- Q17) I am reluctant to study anything by myself other than during class hours, if possible.
- Q18) A bright touch exists in ethics.
- Q19) The presence of technology in our everyday lives will be seamless, transparent, and more significant than ever[8].
- Q20) I should interpret engineering ethics positively.
- Q21) I wish to devote study effort that is barely sufficient to receive a credit unit of a class.
- Q22) To kill people is evil under any set of circumstances.
- Q23) Euthanasia should not be permitted in any case.
- Q24) Environmental destruction is unavoidable if it is necessary for the survival of human beings because no natural life other than humans has a right to exist.
- Q25) The ethical consciousness of a professional and that of a general person should be similar.
- Q26) The image of the word "ethics" is gloomy.
- Q27) I often read newspapers.
- Q28) I am interested in daily social affairs.
- Q29) I am interested in daily economical affairs..
- Q30) I am interested in daily political affairs.
- Q31) I am interested in daily international affairs.
- Q32) I am interested in daily sports affairs.

Part two of the question items are to ask whether they have received course credit unit of humanity and social science classes.

Part three is optional and arbitrary. The questioned items might consist of, for example, working experience, experience in joining academic activities other than that related to formal coursework.

4.2 Joint practice

The attributes of the university which jointly conducted the questionnaire survey are presented in Table 2.

[7] Same as the footnote of Q7.

[8] ibid

University	Major	Grade Year	Number of Students	
			Beginning	Ending
A	Electrical	4	100	97
B	Electrical	3	23	18
C	Electrical	4	68	N.A.
D	Electrical	3	105	104
E	Electrical	3	152	98
F	Chemical/Electrical/Mechanical/System	M1	16	17
G	Chemical	4	62	64
H	Chemical	2	76	77
I	Nuclear	M1	17	10
J	Electrical	3	121	95

Table 2. Attributes of classes contributing the joint practice of the questionnaire survey.

4.3 Results

Data obtained from questionnaires are analyzed from three points of view: interest, knowledge, and attitude. Probably, to be successful in academic life, it is important for a student to be interested in it. Similarly, to study engineering ethics, it is important to have some interest in it. The same view is applicable to knowledge and attitude. Obtaining knowledge is important for students. Studying with a positive attitude is important as well. Some representative results are presented in the following sections.

4.3.1 Interest

Question item Q4) of the Part One of the questionnaires is the following.

- Q4) Engineering ethics is important.

This item asks the respondent about the degree of interest in engineering ethics.

Figures 2 and 3 portray the results. Note that level 4.0 corresponds to the choice "neither agree nor disagree," the area higher corresponds to agree, and lower to disagree.

From these figures, we can infer that students generally show increased interest in engineering ethics as a result of class learning. The only exception is Class I, where the ending period data were not sufficiently obtained because of the time constraint of the academic calendar: some students had to move to the next class before answering the questionnaire. Furthermore, the questionnaire of the ending period was not conducted for Class C: corresponding data were unavailable.

Although students showed increased interest in engineering ethics, the extent differs considerably with classes, perhaps because of the contents of educational materials, handled topics, and lecturing methods. Further analysis is expected.

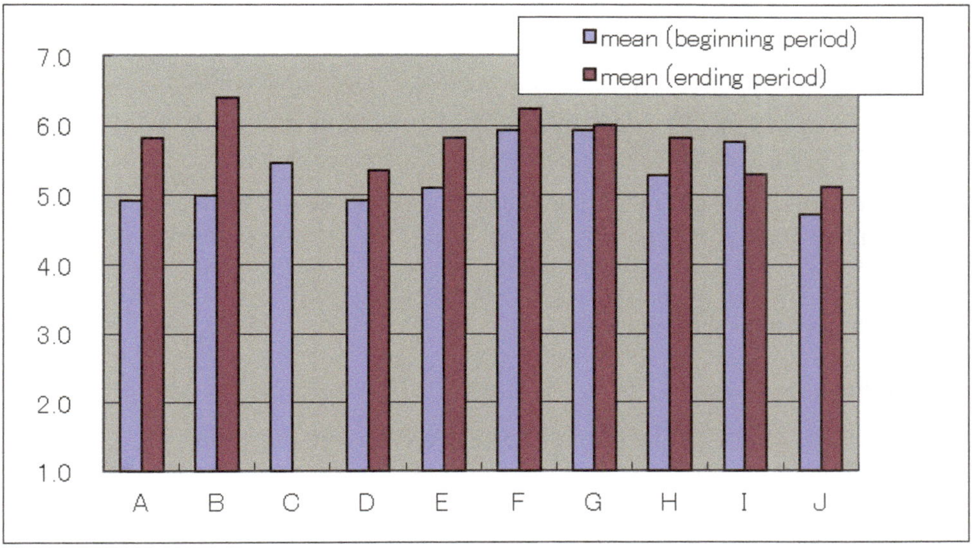

Fig. 2. Mean of replies to Q4: Engineering ethics is important.

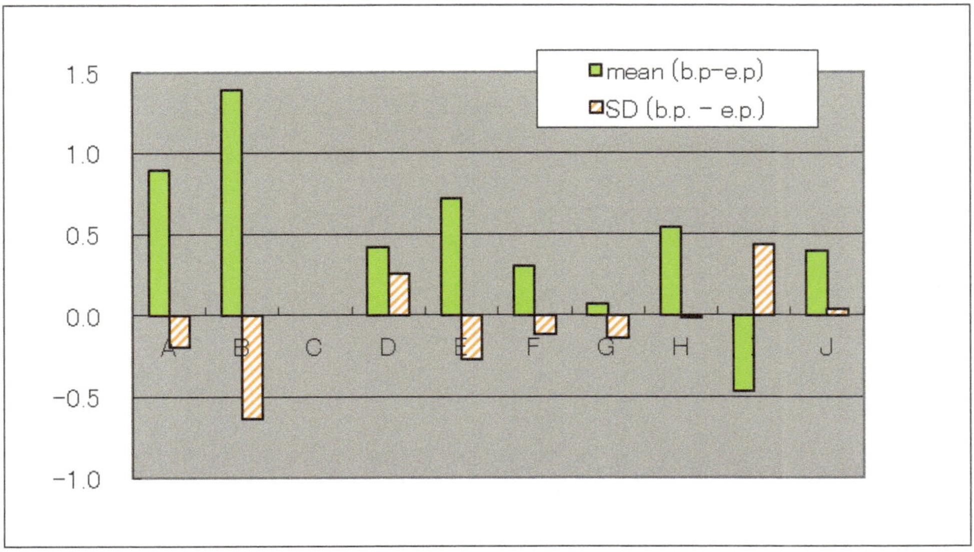

Fig. 3. Difference between beginning and ending scores: mean and standard deviation (SD) of replies to Q4.

4.3.2 Knowledge

Question items Q3) and Q8) of Part One of the questionnaire are as described below.

- Q3) I know the meaning of CSR.
- Q8) I know the meaning of compliance.

Implications of compliance and Corporate Social Responsibility (CSR) were reported in Section 3.2.1. Data of the joint questionnaire survey are shown below.

Compared to the mean data of Q4 (Importance of engineering ethics), those of Q3 (CSR) and Q8 (compliance) vary widely. This variation results from the difference of topics that lecturers handled in their respective classes. Nevertheless, the mean difference data (Fig. 5) illustrate that students of most classes have knowledge related to CSR and compliance through the study of engineering ethics.

4.3.3 Attitude

Question items Q17) and Q21) of Part One of the questionnaire are as follows.

- Q17) I am reluctant to study anything by myself other than during class hours, if possible.
- Q21) I wish to devote study effort that is barely sufficient to receive a credit unit of a class.

The obtained data from the joint survey are presented in Figures 6 and 7.

The data of Q17 and Q21 show good similarity. Both reveal that the engineering ethics class lesson has the effect of improving the study attitude, although that effect might differ among classes and might not be strong.

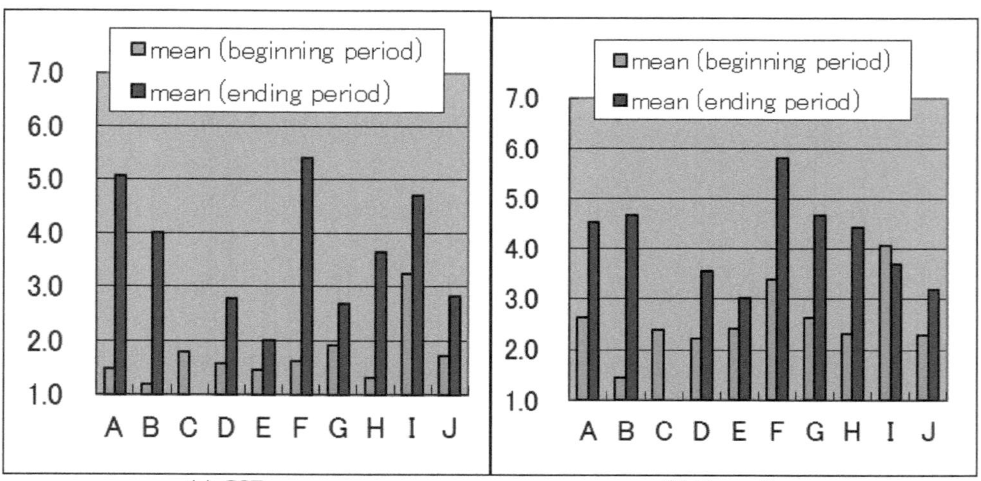

(a) CSR (b) Compliance

Fig. 4. Mean of replies to Q3 (CSR) and Q8 (compliance).

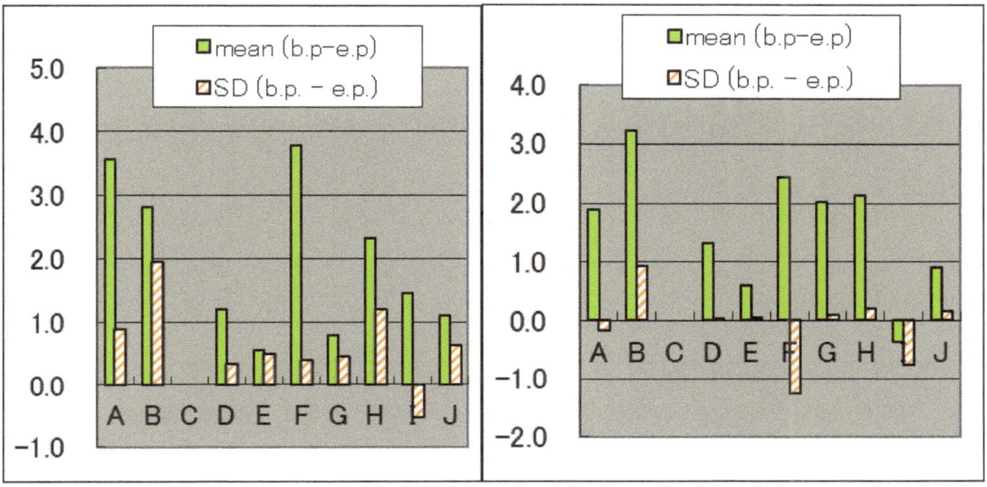

(a) CSR (b) Compliance

Fig. 5. Difference between beginning and ending scores: mean and standard deviation of replies to Q3 and Q8.

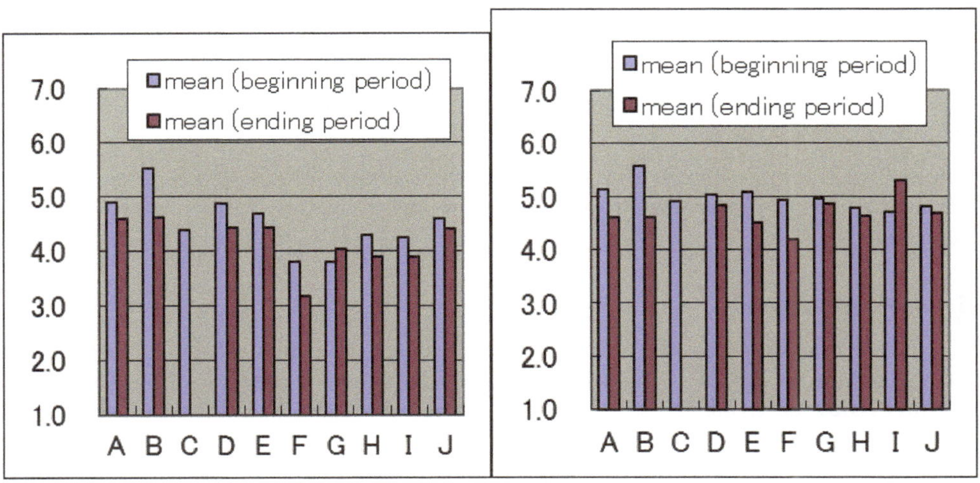

(a) Study attitude (b) Effort to receive credit unit

Fig. 6. Mean of replies to Q17 (study attitude) and Q21 (effort to receive credit unit).

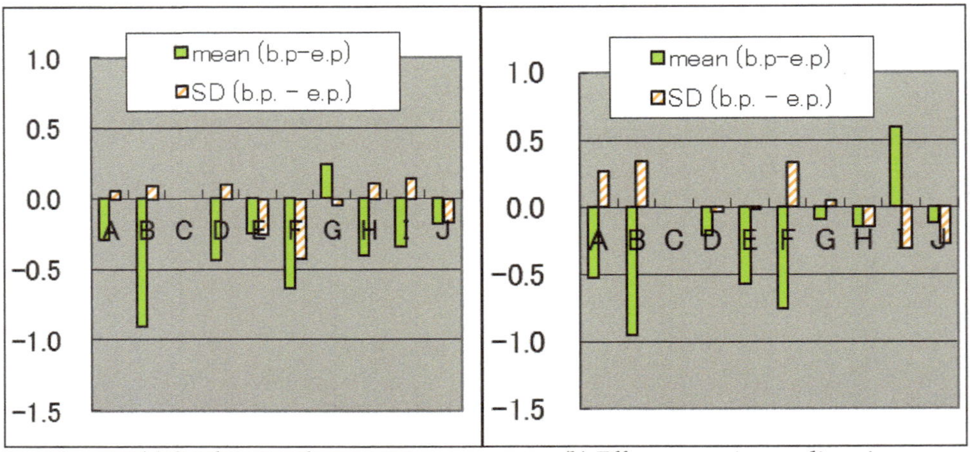

(a) Study attitude (b) Effort to receive credit unit

Fig. 7. Difference of beginning period and ending period: mean and standard deviation of replies to Q17 and Q21.

5. Conclusion

Useful data to improve the quality of engineering ethics education are obtainable through questionnaire surveys administered to students, in cases where the survey is conducted at both the beginning and ending of the class with fundamentally identical contents. Moreover, joint investigation by lecturers increases the benefits of the survey considerably.

The results of the questionnaire surveys showed the student competency development on the interest, knowledge and attitude through the engineering ethics learning.

Further analyses of the obtained data and improvement of the questionnaire surveys are expected. To widen the participation in joint surveys is expected to be useful to improve engineering classes further.

Although the author believes that the statistical analysis on the scored data of the Likert items is useful if it remains in a same set of questionnaires, further study might be necessary.

6. Acknowledgment

The author extends his gratitude to the cooperation of the joint study members; Manabu Aoyagi (Muroran Institute of Technology), Yoshio Kataoka (Kansai University), Junya Matsuki (University of Fukui), Kyoko Oba (Kanazawa Institute of Technology), Susumu Shimamoto (Seikei University), and Kazuo Shimizu (Shizuoka University).

This work was supported by a *kakenhi* grant (23501028) promoted by the Japan Society for the Promotion of Science (JSPS)[9]

[9] http://www.jsps.go.jp/english/e-grants/index.html

7. References

Fudano, J (2007). *The Formation of Ethics Crossroads and the Construction of Science and Engineering Ethics,* Research report submitted to the Japan Science and Technology Agency (in Japanese, partially in English)

Honda, K et al. (2009). Assessment and Evaluation of Educational Effect in Engineering Ethics Education. *The Paper of Technical Meeting on Frontiers in Education,* FIE-09-46, December 2009, pp. 13-23, Japanese Institute of Electrical Engineers (JIEE) (in Japanese, partially in English)

Kageyama et al. (2009). Discussion of Engineering Ethics Education - Questionnaires and Opinion from Students at the University of Tokushima. *The Papers of Technical Meeting on Frontiers in Education,* FIE-09-60, December 2009, pp. 59-64, Japanese Institute of Electrical Engineers(JIEE) (in Japanese, partially in English)

Kobayashi, S et al. (2010). A Study on the Objectives, Methods, and Assessment of Engineering Ethics Education (1) : Toward a Model Sylabus for the Engineering Ethics Course in Japan, *The Papers of Annual Meeting of the Japanese Society for Engineering Education,* No. 4-222, August 2010, pp. 308-309, Japanese Society for Engineering Education (JSEE) (in Japanese)

Okita, Y et al. (2010). A New Concept for a Business Ethics Program and the Development of a Monitoring Methods for the Engineering Ethics Environment of a Corporation. IEEJ Trans. FM, Vol. 130, No. 6, June 2010, p. 11 (in English), pp. 591-600 (in Japanese) , Japanese Institute of Electrical Engineers (JIEE)

Okita, Y et al. (2010). Evaluation of Engineering Ethics Education by Student Questionnaire (Part One: Planning). *The Papers of Technical Meeting on Frontiers in Education,* FIE-10-46, December 2010, pp. 43-48, Japanese Institute of Electrical Engineers(JIEE) (in Japanese, partially in English)

Shimizu, K et al. (2010). Evaluation of Engineering Ethics Education by Student Questionnaire (Part Two: Analysis of Questionnaire). *The Paper of Technical Meeting on Frontiers in Education,* FIE-10-47, December 2010, pp. 49-55, Japanese Institute of Electrical Engineers (JIEE) (in Japanese, partially in English)

9

Quality Assurance in Chile's Municipal Schools: Facing the Challenge of Assuring and Improving Quality in Low Performing Schools

Luis Ahumada*, Carmen Montecinos and Alvaro González
Pontificia Universidad Católica de Valparaíso,
School of Psychology
Chile

1. Introduction

This chapter will focus on the issue of quality assurance and improvement in the Chilean school system, particularly on the challenges faced by public municipal schools in the context of recent educational reform. Building on previous studies on the implementation of a quality assurance system in our country, we outline two interdependent ways to approach this challenge: recognising the schools' current characteristics and improvement needs, and providing appropriate support to build internal capacities for improvement in these schools.

Following the research evidence described in this chapter, suggestions are advanced to influence education policy regarding quality assurance and improvement issues in Chile. We propose the need for closer collaboration and joint work among different levels of the school system (national, municipal and schools), as well as among school stakeholders (principals, students, teachers, parents) to develop internal and system capacity for assuring and improving educational quality. Consequently, to assist schools in building the necessary internal capacity and promote learning and innovation, we highlight the relevance of distributed leadership amongst school stakeholders and the development of a learning-oriented culture to facilitate processes of self-evaluation in schools, which in turn would promote a culture of continuous improvement.

2. Education reform and policy context in Chile

Since the 1980s a market-driven model has operated for the provision of educational services in Chile. The role of the government has been to subsidise demand as parents "choose" among three different types of schools. Municipal schools are administered by the country's 341 municipal governments and are totally financed through a per-pupil voucher system based on student attendance. Private subsidised schools are financed through the same voucher system and, in most cases, charging parents an additional fee. The third type

*Corresponding Author

of schools is private fee-paying schools, which are fully funded by parents. This funding formula has led to a highly stratified educational system (Donoso & Donoso, 2009). The Organisation for Economic Cooperation and Development (OECD) wrote that the educational system in Chile is "consciously class structured" (OECD, 2004). The report further states:

> "The rules of the game are different - and unjustly so - for municipal and private schools. Private schools can both select and expel. Municipal schools - with the exception of the few prestigious ones that are in high demand - are obliged to accept all students asking for access. Under these circumstances, results can be expected to differ in favour of private subsidised schools." (p. 255)

In the case of Chile, research shows that students' performance in national examinations is directly related to their socio-economical status (SES) and the type of school they attend (McEwan, 2001; Raczynski and Muñoz, 2007). This situation illustrates the inequities of the Chilean educational system (Mizala and Romaguera, 2000). A problem that has been hard to resolve through initiatives such as programmes of instructional support for underperforming schools, strengthening of teacher training and assessment, and school management support programmes (Raczynski & Muñoz, 2007). Therefore, demands for improving quality and equity in the school system have been in the front of the policy agenda for the last two decades, but with few positive results so far (Donoso & Donoso, 2009; Belleï, Contreras, & Valenzuela, 2008; Weinstein & Muñoz, 2009).

In an effort to address this situation, the Chilean school system is currently undergoing a deep transformation, with the passing of several legal initiatives aiming to change its structure and functioning (Preferential School Subsidy Act, General Act of Education; National System for Quality Assurance in Education). Weinstein and Muñoz (2009) have noted that the introduction of these new legal frameworks represent a 'breaking point' because it purports to modify the structure of the school system in order to push changes that can redress the differences in achievement among municipal public schools, private subsidised schools and private fee-paying schools (Mizala & Romaguera, 2000). These legal frameworks constitute reform efforts aimed to ensure a high quality and equitable school system for all students.

These reforms have influenced primary and secondary education on issues such as finance, accountability and governance of schools and local administrators (Brunner & Peña, 2007), setting particular challenges to those low-performing municipal schools in vulnerable contexts (Sotomayor, 2006). However, these reforms have also safeguarded the government's role of control and assessment of curricular issues (Muñoz & Vanni, 2008). This is similar to what happens in the rest of the Latin American educational reforms (Ball, Fischman, & Gvirtz, 2003). According to Ball et al. (2003), education policies in Latin American countries have been the subject of many processes of centralised decentralisation, similar to the Chilean case. All of them have followed a similar trend of reform characterised by a 'local empowerment' style of governance (Glatter, 2002), which emphasises the locality of decision-making. These policies and strategies have been widely supported by international multilateral agencies promoting reforms alongside with neoliberal principles (Ball et al., 2003). Additionally, these policies have promoted quality assurance programmes as part of reform efforts in Latin American countries, like Chile (Carnoy, 1999).

One aspect to consider in any system of quality assurance is the relationship between quality and equity. Equity refers to fairness in access to educational opportunities so that resources are distributed in accordance to differential students´ needs and outcomes do not depend on students' social-cultural characteristics. That is, independent of their family backgrounds, in an equitable school system all students complete their education with comparable levels of skills and opportunities to succeed in life (Belfield & Levin, 2002). Quality without equity becomes a mere compliance with standards without consideration of the contextual and local aspects mentioned above.

3. Research evidence from the implementation of a quality assurance system in Chile's municipal schools

Education policy demands the various stakeholders at the national level to take for not only monitoring commitment with established goals and the use of resources, but also to promote the coordination of necessary networks to enable each school to decide and systematically develop efforts to improve and deliver a quality service (MINEDUC, 2005). As new quality assurance systems will be required for Chilean schools, it may be useful to examine lessons learned from a prior quality assurance policy.

The System for Quality Assurance of School Management (SACGE) was implemented between 2003 and 2007 and consisted of an integrated model of four key areas of processes and one area of results. The system promoted a cycle of continuous improvement based on four phases: institutional self-evaluation, external review, school improvement plan and public account (MINEDUC, 2005).

Our research group conducted a study between 2006 and 2008 in order to analyse the influence of SACGE in schools' and their leadership teams' performance. Through a mix-methods approach data collected in six schools, showed that, although all schools evidenced important changes in their practices, not all schools managed to assure quality and develop a culture of continuous improvement (Ahumada, Galdames, González, & Herrera, 2009). The results also revealed several difficulties in the implementation of SACGE (i.e. inadequate training; contradictory information; inconsistent support from local administrators), which in turn hindered the possibility for schools and their school leaders to engage and commit to a process of improvement.

The quality model underpinning SACGE as an assurance system is not prescriptive. It recognizes, values and promotes diversity of schools and it assumes that the implementation of each phase ought to consider the context, particularly the school's culture and history (MINEDUC, 2005). Planning for improvement is one of the most important aspects that SACGE promotes, giving schools the autonomy to identify and define the appropriate management practices for their context and needs. School leaders (principal and management team) are responsible for giving viability to the planned actions. Moreover, SACGE aims to promote continuous improvement in schools, specifically, it seeks to improve the quality of management processes, students' learning (scores on national achievement tests) and the satisfaction of the community. Thus, SACGE invites schools and school stakeholders to take responsibility for the results that the proposed improvement plan aims for and is capable of achieving.

In short, SACGE aims to promote responsibility in principals and their management teams to conduct a systematic review and ongoing development of management processes and outcomes, developing a learning-oriented culture of continuous improvement which involves various school stakeholders. The community the school serves should be part of the decision-making processes as schools and districts define how educational quality will be attained and measured.

4. SACGE's four phases for quality assurance

As described above, SACGE considers four phases for quality assurance in schools: institutional self-evaluation, external review, school improvement plan and public accountability. The institutional self-evaluation phase aims to produce base-line information about the school's management process and results to determine the quality of its daily educational practices and identify those areas that could be improved (MINEDUC, 2005).

Research on institutional self-evaluation (Navarro & Jiménez, 2005; Soto, 2006), indicates that the process of self-evaluation facilitates the awareness of the prevailing leadership style in the organization, the characteristics of the organizational culture and the degree of satisfaction of teachers and school leaders. Navarro and Jimenez (2005) note that to maximize the positive effects of the self-evaluation, it requires prior development of teamwork skills, with strong emphasis on participative management. Soto (2006), however, points out that many principals have an individual conception of the self-evaluation process, which contrasts with the theoretical foundations that foster collaborative work. In this sense, research evidence highlights the need for the principal and his team to assume accountability for the evaluation of educational processes and outcomes (Westhuizen van der, Mosoge, Swanepoel, & Coetsee, 2005), which would facilitate the implementation of a quality-oriented school culture (Cano, 2003; Casanova, 2004).

In a study about the functioning of schools' management team during the implementation of SACGE in ten primary and secondary public municipal schools conducted by our research team, we evidenced four factors that accounted for a good teamwork: trust, sense-making, efficiency and effectiveness, and ineffectiveness (Ahumada, Montecinos & Sisto, 2008). Trust appears to be the key factor in determining a well-functioning team. In schools with a higher level of trust, people felt very comfortable working together, supported each other strongly and had great appreciation for the work they did. Furthermore, they were willing to share and to listen to other members' ideas freely and there was a strong identification with the team and the school.

Management teams, whose functioning was of higher quality, were able to initiate the self-evaluation phase with a clear idea of the workload involved and how it should be managed. In the case of teams whose functioning was of lower or poor quality, they exhibited problems organising and delegating tasks, such as collecting data, interpreting evidence and managing small groups of teachers. It is worth noting that in these cases teachers' participation was instrumental and limited only to data collection (Ahumada, Galdames, Gonzalez & Herrera, 2009).

Building a shared meaning and direction was another important factor that affected the functioning of these schools' management teams. In those teams with a higher quality functioning there were systematic conversations about the school's vision and how to achieve it. According to our research, most management teams faced the self-evaluation phase without delving into the challenges of developing a process such as this one. However, during the implementation of the self-evaluation phase, some teams were able to develop a shared vision about the challenging and meaningful work they had to do in this phase (Ahumada, Galdames, Gonzalez & Herrera, 2009).

During the self-evaluation phase, some aspects related to the construction of meaning and the process for organizing this task emerged, as well as specific aspects associated with organizational learning and the journey of continuous school improvement (Collison & Cook, 2007; Hallinger & Heck, 2011; Weick, Sutcliffe & Obstfeld, 200). However, these aspects were seldom capitalized as time became a scarce resource for the management team and for the teachers engaged in this self-evaluation. These actors lacked the time to reflect on the aim and importance of self-evaluation. Moreover, in many schools it was perceived as an externally imposed task (Ahumada, Galdames, Gonzalez & Herrera, 2009).

The management teams that spent more time and resources to identify problems and found ways to solve them were the same ones that stood out for their internal relationships of trust and mutual knowledge about its members' capacity. These teams, whose functioning was of higher quality, saw self-evaluation as worthwhile and constantly monitored their work, developing strategies to deal with data collection and analysis. In several of these teams, these changes meant redistributing the work schedule, and in some cases, deferring or eliminating other activities; even increasing the number of hours they worked. One of the consequences of these strategies was that some management teams improved their relationship with the rest of the school members, especially with teachers. By contrast, in teams with poor or lower functioning quality the self-evaluation process created a distance between the management team and teachers, negatively affecting the climate within the school (Ahumada, Galdames, Gonzalez & Herrera, 2009).

The second phase, external review, intended to validate the institutional self-evaluation. For the Ministry of Education, the aim of the external evaluation should be providing objective information to the school about their self-evaluation, as a way of contributing to their improvement process (MINEDUC, 2005). This external evaluation was led by a team of professionals - External Panel- who did not belong to the school, but were related to it in their capacity as ministry supervisors or municipal officers. Research evidence indicates that the usefulness of the external evaluation will depend on how the school will assume this instance to generate a learning process, and to ensure commitment and participation of all of the members of the school community (Muijs, Harris, Chapman, Stoll & Russ, 2004).

The External Panel's visit to the school marked a turning point in the process. In general, management teams, due to coordination and relationship problems with the Panel members, developed a negative opinion of this phase. These negative perceptions translated into the idea that SACGE was one of the many ill-implemented and abruptly replaced programs that the Ministry of Education has imposed to schools (Ahumada, Galdames, González & Herrera, 2009). This finding was reported for schools that had high functioning as well as low functioning management teams.

The following phase, developing and implementing an Improvement Plan, consists of designing, implementing and evaluating actions that will generate changes in the school's management or will improve management practices already in place. Thus, the school quality improvement model puts the focus on the school´s and its community´s capacity manage change (Fullan 1999, Canton, 2004). This would involve the establishment of a culture oriented to foster organizational learning and continuous improvement (Stoll, 2009; Voulalas & Sharpe, 2005; Martinez, 2007).

Finally, the Public Accountability phase asks the principal to share, with the community, the school´s outcomes and progress on matters related to the quality of teaching, learning and management processes. The literature has emphasized the importance of schools' responsibility for reporting back to the community their results (Earl, 2004; Glatter, 2002). In this phase, the goal is to promote accountability, seek transparency and facilitate access to the results and management processes to the school community (Marks & Nance, 2007; Ryan, 2005; Schweigert, 2006). However, accountability not only implies to give an account but also to take responsibility for appropriately representing the rights and needs of the community. From this point of view, accountability means building with the community a series of performance indicators that are relevant to them and the school, being part of their success. Accountability as a concept is more about performance than a mere democratic control of processes and outcomes that must be present at school (Anderson, 2009).

5. Discussion and conclusions

The study suggested that systems like SACGE should consider a differentiated approach to quality 'assurance' (compare the schools' performance against a standard) and to quality 'improvement' (review of schools' processes to influence their performance). From this perspective, it is important to consider schools' entry characteristics or base line regarding their school management processes and results, as well as characteristics of the leadership team in charge of this process (Ahumada, Montecinos, & Sisto, 2008). In this regard, findings from our studies indicate the importance of distributed leadership. This leadership is not just that exercised by the principal, as transformational or instructional, but one in which various members of the school community feel responsible for significant improvements in processes and outcomes. As Anderson (2009) points out, it involves leadership exercised by all actors in the school community, ensuring the rights of children who are served by the school.

The concept of distributed leadership is understood here as the one that emerges from everyday actions, involving people responsible for tasks related to the design and implementation of the improvement plan. In our view, this is particularly interesting for understanding the ways of organising work as well as the social relationships involved in performing the tasks entailed in a quality assurance system such as SACGE. On the other hand, distributed leadership is concerned with the contextual variables associated with performing this task, and it seems particularly important to strengthen schools' internal capacity for organizational learning, understood at the individual, team and organizational levels.

We have learned from our research that, for a quality assurance system to be effective, it requires: a culture that facilitates learning, teamwork and strategic clarity where members of the organization share the organizational mission and goals. Strategic clarity must be reflected in the schools' strategic orientations and should guide the development and implementation of the Improvement Plan. In summary, we believe that both the distributed leadership and organizational learning at different levels are central to the current educational reform that seeks to generate a change in the management of educational institutions.

Based on the differences observed between low and high functioning leadership teams, we conclude that capacity building for school improvement is one of the necessary aspects to consider when developing a policy for quality assurance. There are certain skills that should be installed in the educational communities in order to achieve autonomy for improvement: data-driven decision making, distributed leadership, communication and transparency between stakeholders, ability to innovate, to generate new practices and a participatory culture, are just a few.

International research evidence is consistent with our findings regarding the importance of identifying internal characteristics of schools for assuring and improving school quality (Anderson, Leithwood, & Strauss, 2010; Leithwood & Louis, 1998; Scribner, Cockrell, Cockrell, & Valentine, 1999; Stoll, 2009). Collectively, these studies have noted that building internal capacities for data based decision- making, distributed leadership and organizational learning are critical for initiating and sustaining processes for quality assurance and improvement in schools.

To develop these skills it is necessary to articulate the different levels at which a policy for quality assurance and quality improvement is designed and implemented. Greater coordination is necessary among the central (Ministry of Education), intermediate (regional ministry and municipal administrators) and local level (primary and secondary schools). A joint effort among the various stakeholders that make-up the school organization (administrators, students, teachers, and parents) is crucial to achieving a quality education. Quality and equity in education is not a problem that can be resolved at the school level; political and social organizations linked to education must also join in this effort.

Regarding the definition of 'good quality', it is important to consider the interaction between internal and external contextual environments. According to Tikly (2011) there are three overlapping contextual environments that define what good quality education means: policy, school and home/community. The author states that 'creating a good quality education involves paying attention to the interface between each environment' (Tikly, 2011, p.11). In other words, considering the external policy demands, internal capacities of the school and the social and intellectual capital of the community.

The interplay among these factors as well as the school's base line, creates particular configurations cautioning against the 'one size fits all' approach to quality assurance and improvement initiatives (Barrett, Chawla-Duggan, Lowe, Nikel, & Ukpo, 2008; Sayed & Ahmed, 2011). The same goes for the external support that might be provided to low-performing schools, as the literature advises against transferring 'what works' in some school contexts to a different one (Harris & Bennett, 2001).

According to findings from our studies, the school principal and his/her management team, in conjunction with all school´s stakeholders, must work to develop a learning-oriented organizational culture that engages in a continuous improvement of its processes and outcomes. This is certainly hard work that can be facilitated to the extent that there is a shared understanding of where the school stands on the indicators for educational quality and the capacities that need to be developed in order to improve.

6. Acknowledgments

This manuscript was funded by a grant from the Chilean National Fund for Scientific and Technological Development (Fondecyt # 1090570) and by the Centre for Advanced Research in Education (PIA-CONICYT Project CIE-05).

7. References

Ahumada, L., Galdames, S., González, A., & Herrera, P. (2009). El funcionamiento del equipo directivo durante un proceso de autoevaluación institucional en el marco de políticas de aseguramiento de la calidad de la gestión escolar en Chile. *Universitas Psychologica, 8* (2), 353-370.

Ahumada, L., Montecinos, C., & Sisto, V. (2008). Desarrollo y Validación de una Escala para Evaluar el Funcionamiento del Equipo Directivo en los Centros Educativos. *Revista Interamericana de Psicología / Interamerican Journal of Psychology, 42* (2), 228-235.

Anderson, G. (2009). *Advocacy Leadership: toward a post-reform agenda in education*. New York: Routledge.

Anderson, S., Leithwood, K., & Strauss, T. (2010). Leading Data Use in Schools: Organizational Conditions and Practices at the School and District Levels. *Leadership and Policy in Schools, 9* (3), 292 - 327.

Ball, S. J., Fischman, G., & Gvirtz, S. (2003). *Crisis and hope: the educational hopscotch of Latin America*. London, New York: Routledge Falmer.

Barrett, A. M., Chawla-Duggan, R., Lowe, J., Nikel, J., & Ukpo, E. (2008). The concept of quality in education: A review of the 'international' literature on the concept of quality in education. *EdQual Working Paper, 3.*

Belfield, C. R., & Levin, H. M. (2002). Educational Privatization: Causes, Consequences, and Planning Implications. *Fundamentals of Educational Planning, 74*. Paris: UNESCO.

Belleï, C., Contreras, D. and Valenzuela, J. P. (2008). *La agenda pendiente en Educación. Profesores, administradores y recursos: propuestas para la nueva arquitectura de la educación chilena*. Santiago: Universidad de Chile / UNICEF.

Brunner, J. J., & Peña, C. (Eds.). (2007). Reforma Educativa: Aportes para el debate. Santiago: Universidad Diego Portales.

Cano, E. (2003). *Organización, calidad y diversidad*. Madrid: La Muralla.

Casanova, M. A. (2004). *Evaluación y calidad de centros educativos*. Madrid: La Muralla.

Carnoy, M. (1999). Globalization and Educational Reform: what planners need to know. In IIPE (Ed.) *Fundamentals of Educational Planning*. Paris, France: UNESCO.

Collison, V., & Cook, T. (2007). *Organizational learning*. USA: Sage Publications.

Donoso, S., & Donoso, G. (2009). Políticas de gestión de la educación pública escolar en Chile (1990 -2010): una evaluación inicial. *Ensaio: Avaliação e Políticas Públicas em Educação, 17*, 421-448.

Earl, L. M. (2004). *Measuring effectiveness and improvement in schools: why accountability is not about painting by numbers*. Victoria: Jolimont.

Fullan, M. (1999). *Change Forces*. Philadelphia: Falmer Press.

Glatter, R. (2002). Governance, Autonomy and Accountability in Education. In T. Bush & L. Bell (Eds.) *The Principles and Practice of Educational Management*. London: Sage.

Hallinger, P., & Heck, R. H. (2011). Exploring the journey of school improvement: classifying and analyzing patterns of change in school improvement processes and learning outcomes. *School Effectiveness and School Improvement: An International Journal of Research, Policy and Practice, 22* (1), 1 - 27.

Harris, A., & Bennett, N. (2001). *School effectiveness and school improvement: alternative perspectives*. London: Continuum.

Leithwood, K. A., & Louis, K. S. (1998). *Organizational learning in schools*. Lisse, Netherlands ; Abingdon: Swets & Zeitlinger.

Marks, E. M. & Nance, J. P. (2007). Contexts of accountability under systemic reform: Implications for principal influence on instruction and supervision. *Educational Administration Quarterly, 43*, 3-37.

Martínez, E. (2007). La significación en la cultura: concepto base para el aprendizaje organizacional. *Universitas Psychologica, 6* (1), 155-162.

McEwan, P. J. (2001). The Effectiveness of Public, Catholic, and Non-Religious Private Schools in Chile's Voucher System. *Education Economics, 9* (2), 103 - 128.

MINEDUC. (2005). *Sistema de Aseguramiento de la Calidad de la Gestión Escolar: Calidad en todas las Escuelas y Liceos*. Santiago: Ministerio de Educación de Chile.

Mizala, A., & Romaguera, P. (2000). School Performance and Choice: The Chilean Experience. *The Journal of Human Resources, 35* (2), 392-417.

Muijs, D., Harris, A., Chapman, C., Stoll, L., & Russ, J. (2004). Improving Schools in Socioeconomically Disadvantaged Areas -A Review of Research Evidence. *School Effectiveness and School Improvement, 15* (2), 149-175.

Muñoz, G., & Vanni, X. (2008). Rol del Estado y de los Agentes Externos en el Mejoramiento de las Escuelas: Análisis entorno a la Experiencia Chilena. *Revista Iberoamericana sobre Calidad, Eficacia y Cambio en Educación, 6* (4), 47-68.

Navarro, G. & Jiménez, J. (2005). La Autoevaluación Escolar y su impacto en el comportamiento docente, individual y grupal, en la organización educativa. *Estudios Pedagógicos, 31* (2), 57-69.

OECD (2004). *Revisión de políticas nacionales de educación: Chile*. [Reviews of national policies for education: Chile]. Paris: Organisation for Economic Cooperation and Development.

Raczynski, D., & Muñoz, G. (2007). The Chilean Educational Reform: The Intrincate Balance between a Macro and Micro Policy International *Handbook of Urban Education 19*, 641-664. Netherlands: Springer.

Ryan, K. E. (2005). Making educational accountability more democratic. *American Journal of Evaluation, 26* (4), 532-543.

Sayed, Y., & Ahmed, R. (2011). Education quality in post-apartheid South African policy: balancing equity, diversity, rights and participation. *Comparative Education, 47* (1), 103 - 118.

Scribner, J. P., Cockrell, K. S., Cockrell, D. H., & Valentine, J. W. (1999). Creating Professional Communities in Schools through Organizational Learning: An Evaluation of a School Improvement Process. *Educational Administration Quarterly, 35* (1), 130-160.

Soto, M. (2006). Percepción directiva y autoevaluación institucional: el caso de la educación municipalizada en la provincia de Concepción. *Ciencias Sociales Online, 3* (3), 56-70.

Sotomayor, C. (2006). Programas públicos de mejoramiento de la calidad de escuelas básicas en contextos urbanos vulnerables: evolución y aprendizajes de sus estrategias de intervención (1990 -2005). *Pensamiento Educativo, 39* (2), 255-271.

Stoll, L. (2009). Capacity building for school improvement or building capacity for learning? A changing landscape. *Journal of Educational Change, 10,* 115-127.

Schweigert, F. J. (2006). The meaning of effectiveness in assessing community initiatives. *American Journal of Evaluation, 27* (4), 416-436.

Tikly, L. (2011). Towards a framework for researching the quality of education in low-income countries. *Comparative Education, 47* (1), 1 - 23.

Voulalas, Z. & Sharpe, F. (2005). Creating schools as learning communities: Obstacles and processes. *Journal of Educational Administration, 43* (2/3), 187-208.

Weick, K., Sutcliffe, K. & Obstfeld, D. (2005). Organizing and the process of sensemaking. *Organization Science, 16* (4), 409-421.

Weinstein, J., & Muñoz, G. (2009). La Reforma Educacional en el punto de Quiebre. In C. Bascuñán, G. Correa, J. Maldonado & V. Sánchez (Eds.), *Más acá de los sueños, más allá de lo posible. La Concertación en Chile.* Santiago, Chile: LOM.

Westhuizen van der, P., Mosoge, M., Swanepoel, L. & Coetsee, L. (2005). Organizational Culture and Academic Achievement in Secondary Schools. *Education and Urban Society, 38* (1), 89-109.

10

Towards Learning-Focused Quality Assurance in Chinese Higher Education

Yuan Li[1] and Houyi Zhu[2]
[1]*Liaoning University of Traditional Chinese Medicine*
[2]*The University of Manchester*
[1]*China*
[2]*United Kingdom*

1. Introduction

The concept of 'quality' has been contemplated throughout history and continues to be a topic of intense interest today (Reeve & Bednar, 1994). A concern about quality and standards is not new in higher education. Since the mid-1980s, the concept of quality has increasingly influenced discussion around the globe about the role and future of higher education institutions and the academics that constitute those institutions (Watty, 2003). Quality has become a universalising metanarrative (Morley, 2003). Quality and quality assurance issues in higher education have risen to prominence both nationally and internationally (Dunkerley & Wong, 2001). Quality parades as a universal truth continually extends its domain (Morley, 2003). The concern for quality is articulated by university managers themselves, by external agencies deliberately established to assess and reward quality and, increasingly, by the 'clients' of higher education - the students, the employers and, importantly, the state (Dunkerley & Wong, 2001). Despite the prevailing use of the concepts and models of 'quality', many researchers believe that the language and tools of industry-born quality models are an imperfect fit to higher education (Houston, 2008) and the concept of customer-defined quality is problematic (Eagle & Brennan, 2007; Houston 2007; Meirovich & Romar,2006).

There is a great number of studies and publications on the issue of quality assurance in higher education, but most of the studies have difficulties in providing substantial evidence that the core processes of higher education - teaching and learning - are improved as a consequence (Stensaker, 2008). The intuitive answer is that most studies have not reached the needed level of sophistication, but as Stensaker (2008) points out, this is not necessarily a problem solely related to methodology, but to the underlying assumptions of quality assurance and the standard top-down implementation approach. This study is conducted against the background of the prevailing quality culture, and focuses on student learning rather than the widely adopted top-down scrutiny of teaching as the main component of quality assurance schemes. Both systematic literature review and document analysis are adopted to explore how student learning experiences can be integrated into the quality assurance systems in Chinese universities for continuous quality improvement. Based on the discussion of the literature on quality and quality assurance in higher education and a detailed analysis of the current quality assurance practice in Chinese higher education, a learning-focused quality assurance is proposed to offer insights into integrating student learning generically into the quality assurance process for the purpose of the continuous improvement of higher education quality.

2. Quality in higher education

Quality had by tradition been seen as an implicit and natural element of university-level learning and research and an integrated part of academics' professional responsibilities. This changed in the 1990s, with a requirement that higher education institutions should demonstrate, through their institutional leaders and expressed in comparable measures, the quality of their activities (Harvey & Askling, 2003). There is substantial agreement that the quality imperative in higher education was based on pressure from the market and from governments to adapt to an external political agenda (Dill, 2000; Harvey, 1998; Salter & Tapper, 2000). Brooks and Becket (2007) point out that the introduction of the quality imperative in higher education is mainly an externally driven process related to increased demand for accountability and efficiency in the sector.

In fact, the concept of quality is not always made explicitly, though it is used so often by so many people inside and outside higher education. 'Quality' is a highly contested concept, which has multiple meanings for people from different tracks of higher education. Barnett (1992) argues that there is a logical connection between concepts of higher education and different approaches to quality. In his opinion, what we mean by, and intend by, 'quality' in the context of higher education is bound up with our values and fundamental aims in higher education. We cannot adopt a definite approach toward quality in this sphere of human interaction without taking up a normative position, connected with what we take higher education ultimately to be. In turn, what we take higher education to be will have implications for how we conceive of quality, how we attain it, how we evaluate our success in achieving it, and how we improve it. So if we want to offer a particular view on quality we should be prepared to declare where we stand on the key purpose of higher education (Barnett, 1992). He categorises concepts of higher education into two groups:

Group 1:

Four dominant concepts of higher education underlie contemporary approaches to, and definitions of, quality: 1) higher education as the production of qualified manpower; 2) higher education as training for a research career; 3) higher education as the efficient management of teaching provision; 4) higher education as a matter of extending life chances. This group of concepts reflects the thinking about higher education of the national policy makers, funders and institutional managers, and other national interest groups. These concepts are external to the process of higher education, but are driving national debate and development work in quality assessment and are informed by a systematic approach to education (Barnett, 1992). If higher education is perceived as a process of filling particular slots in the labour market with individuals who are going to be 'productive', then one way of assessing quality might be to examine the destinations of the students. Under this conception, students take on value as, and are described in the vocabulary of, 'products' of the system (Barnett, 1992).

Group 2:

This group of concepts is concerned with the students' development, or the educational process to which students are exposed. Such concepts include higher education seen as: 1) the development of the individual student's autonomy, with students acquiring intellectual integrity and the capacity to be their own person; 2) higher education as the formation of general intellectual abilities and perspectives; 3) the enhancement of the individual student's personal character; 4) the developing of competence to participate in a critical commentary on the host society. The concern of this group of concepts is with the educational process that students undergo, not with inputs and outputs and their relationship. This group is not obviously reflected in contemporary debate over quality assurance in higher education. It contains ideas about higher education that do not lend themselves to institutional practice easily captured by system-wide and systematic evaluation procedures such as numerical performance indicators. But their not fitting the standard model of performance assessment does not affect the validity of such conceptions of higher education (Barnett, 1992). As pointed out by Barnett, if we believe that the quality of higher education is more demonstrated in the nature of the intellectual development that takes place in students' minds, in the depth and breadth of understanding that students achieve, in their ability to be self-critical, and in their capacity to apply that understanding and self-critical capacity to all they experience and do, then 'quality' of higher education takes on a quite different character. Under this conception of higher education, the appraisal of quality will not rest content with economic indicators of output, but will turn to exploring the educational process within our institutions. Since there is a logical connection between the development of a worthwhile state of mind and the experiences and educational processes to which students are exposed in their course, a conception of higher education of this kind will prompt an examination of the types of intellectual challenge presented to students, and that in turn will begin to produce an illumination of the internal life of our institutions.

3. Quality assurance in higher education

There are many definitions of 'quality assurance' in the literature (e.g. Ball, 1985; Birnbanum, 1994; Frazer, 1992; van Vugh and Westerheijden, 1993; Woodhouse, 1999). The term 'quality assurance' refers to 'systematic, structured and continuous attention to quality in terms of quality maintenance and improvement' (Vroeijenstijn, 1995). Girdwood (1997) defines the term 'quality assurance' as the policies, systems, and processes designed to ensure the maintenance and enhancement of quality within a programme or institution. Quality assurance is about ensuring accountability, which gives assurance that it is good quality. Harman (1998) suggests that in essence, quality assurance refers to the systematic management and assessment procedures adopted to ensure achievement of specified quality or of improved quality, and to enable key stakeholders to have confidence in the management of quality and the outcomes achieved. Quality assurance may in other words be seen in the context of the regulation of higher education.

Quality assurance is not new. It was originally an integral part of craftsmanship and professionalism (Morley, 2003). Before the quality imperative prevailed globally, the concern for quality and development of quality management arose within higher education institutions. Teachers and administrators within universities and colleges identified what was right for them to teach and made sure it was taught in the accepted way. In the past two decades, the quality imperative in higher education has come from the market and from government (Houston, 2008). More recently, it has been disaggregated from the professions, and formalised and transformed into an object of inquiry (Hart, 1997).

The changes, as analysed by Harvey and Askling (2003), occurred for both pragmatic and ideological reasons. Quality had by tradition been seen as an implicit and natural element of university-level learning and research and an integrated part of academics' professional responsibilities. In the 1990s, universities were required to demonstrate the quality of their activities. Universities were used to seeing excellence or transformation as the self-evident key indicator of higher education quality, but now a self-evident property of higher education became transformed into a mechanism of control.

Systematic procedures for quality assurance and improvement through formal evaluation have been in place in Western Europe since the mid 1980s (Bornmann, et al., 2006). Quality assurance is slowly but steadily becoming an integrated part of higher education (Stensaker, 2008). Quality assurance is by no means a new idea in higher education. For many years, most major higher education systems have had in place various mechanisms of review and assessment. What is new, however, apart from the language, is a more systematic and far-reaching approach to ensuring that institutions and systems have in place mechanisms for review and assessment, and for renewal and improvement (Harman, 1998). Compared to past approaches, the new mechanisms also put much more emphasis on external scrutiny, seeking the views of employers and graduates and, in various ways, making the results of assessment more widely available (Harman, 1998). Stensaker (2008) summarises this process as:

- In the beginning emphasis was given to design issues and the relationship between quality assurance systems and the governance of higher education (Neave, 1988);
- There was a period with greater interest given to methodological issues;
- Much attention was drawn to the human factors (Vroeijenstijn, 1995; Neave, 1996);
- Interest in quality stimulated by leadership and the ways to stimulate staff and student involvement and ownership (Brennan & Shah, 2000);
- More and more governments and quality assurance agencies, and higher education institutions, are held accountable for the impact and outcomes of all this (Stensaker, 2003, 2008; Westerheijden, et al., 2006);
- Currently, higher education is entering an era in which a more nuanced understanding of what quality assurance and quality processes can or cannot prevails (Stensaker, 2008).

3.1 The rationale of quality assurance

Harvey and Askling (2003) argue that from the start quality has been used as a vehicle for delivering policy requirements within available resources. Quality assurance operates as a mechanism to encourage policy driven change. It makes higher education more relevant to social and economic needs, widening access, expanding numbers and doing it with a decreasing unit cost. The rationale for quality assurance is often opaque (Harvey & Newton, 2007). Quality assurance has two underlying broad rationales: accountability and improvement. The perpetual debate about accountability and improvement is as old as quality assurance in higher education (Harvey & Newton, 2007).

Accountability

The term 'accountability' has been widely used in higher education ever since the 1990s. Accountability relates to processes which assess whether minimum standards are in place in a higher education institution or programme. Lewis et al. (2001) defines accountability as demonstrating the worth and use of public resources. Campbell and Rozsnyai (2002) define accountability as the assurance of a unit to its stakeholders that it provides education of good quality. Harvey & Newton (2007) identify accountability as a dominant rationale in quality assurance, but they argue that what exactly accountability is, or requires of the sector, and how that is related to the quality of higher education is less clear. In their view, the often stated reason for the rise of accountability include the cost and potential problems of expansion, the concomitant need to account for increasing amounts of public money, the need to ensure value for both private and public monies, lack of clear lines of accountability within higher education systems, globalisation and the need to keep control of an increasingly unrestricted market (Harvey, 2002). Accountability is seen as a major purpose of external quality processes. Harvey (2002) suggests that accountability has five main functions:

- To ensure that the institution or programme is accountable for the money it receives;
- To ensure that the core principles and practices of HE are not being eroded;
- To ensure that the programme is organised and run properly and that an appropriate educational experience is both promised and delivered;
- To provide proper public information for funders to aid funding allocation decisions, and for prospective students and graduate recruiters to inform choice;
- To ensure compliance with policy.

Improvement

Quality improvement focuses on developmental processes. Improvement potentially depends on the development of definitions and interventions that reflect the interests and concerns of those in the sector (Houston, 2008). Continuous improvement aims at continual increase of performance by emphasising learning and adaptation as keys to success of an organisation, which is also one of the core values of quality management (Deming, 1994; Evans & Lindsay, 2001). Harvey and Askling (2003) point out that improvement has been a secondary feature of most quality assurance systems despite the claims of most external reviews to encourage improvement.

Temponi (2005) suggests the adoption of a continuous improvement approach requires not only upper administration commitment, but also uncovering the current underlying culture and examining the appropriateness of objectives to adopt continuous improvement. Creating a quality culture and long-term commitment to continuous improvement within an academic institution means engaging the administrative and academic systems, and all stakeholders of higher education institutions.

The tension between accountability and improvement

Harvey (2002) points out there has been increasing uniformity of practice for quality monitoring in higher education. This is a pragmatic response to government requirements to demonstrate value for money and fitness for purpose. Nevertheless, what purpose and what constitutes fitness is unclear. The links between accountability mechanisms and quality improvement are rarely clear. Vroeijenstijn and Acherman (1990) point out the tension between accountability and continuous quality improvement. Arguably, accountability is about value for money and fitness for purpose, while continuous improvement in teaching and learning is about enhancement of the student experience, and empowering students as life-long learners. They also argue that the improvement essence of quality is sidelined in the assurance process by a focus on demonstrating compliance. Thune (1996) argues that accountability and improvement are based on different methods based on the ownership of the evaluation system. He identifies that the improvement process has a different form and is independent of control. He argues from his Danish case that accountability and improvement may be combined in a balanced strategy. Middlehurst and Woodhouse (1995) explore whether it is feasible to combine the function of accountability and quality improvement in national arrangements for quality assurance in higher education. They identify that accountability and improvement must be conceptually and practically distinct with separate resourcing. The failure to address different purposes will damage the quality and the integrity of higher education by imbalances of power.

The accountability-led view sees improvement as a secondary function of the monitoring process. Following this approach, external monitoring of quality will lead to improvement as a side effect (Harvey and Newton, 2007). In other words, requiring accountability will lead to a review of practices, which in turn will result in improvement. Harvey (1994) questions this accountability-led view. First, facing a monitoring system demanding accountability, academics will tend to comply with requirements and to minimise its interruption in their existing practice. Second, improvement comes from a changed culture and local ownership, which is in conflict with the principle of compliance in accountability. Third, the extra burden of responding to external scrutiny leads to the feeling of lacking trust, which will demotivate staff who are already involved in innovation and quality initiatives. Harvey and Newton (2007) suggest a view counter to the accountability-led one, which will result in quality improvement: improvement is its own accountability. In other words, if an organisation continually improves it is accountable. This returns the ownership of the quality assurance system to academics. In their view improvement is not something regulated but something attained through critical engagement.

Harvey and Askling (2003) point out the most effective improvement occurs when external processes mesh with internal improvement activities. It is more difficult for external quality assurance to encourage the learning-teaching interface. They argue that the improvement function of quality assurance procedures is to encourage institutions to reflect upon their practices and to develop what they do. Therefore, quality assurance needs to be designed to encourage a process of continuous improvement of the learning process and the range of outcomes.

3.2 The approaches to quality assurance

In one of the earliest classifications of the different approaches to quality assurance, Dill (1992) distinguishes between three forms: the reputational approach, the student outcome approach, and the total quality (management) approach. The reputational approach uses peer review to assess the quality of higher education institutions or programmes. The student outcome approach measures student achievements both when attending higher education and after graduation. The total quality management approach is based on participation, customer orientation, organisational learning and coordination. Over time, approaches to quality assurance are widely discussed and analysed at theoretical level and vary widely among countries.

Billing (2004) explores international comparisons of the purpose of quality assurance in higher education and the extent to which the main national quality assurance frameworks meet this. He concludes with a general model, which summarises the purposes of quality assurance into:

- Improvement of quality
- Publicly available information on quality and standards
- Accreditation (i.e. legitimisation of certification of students)
- Public accountability for standards achieved and for use of money
- To contribute to the higher education sector planning process

Van Vught and Westerheijden (1993) summarise the common elements of quality frameworks in European countries:

- A national agency to co-ordinate and support quality assurance within institutions, which is independent of government;
- Self-evaluation as the vital focus of the external quality assurance process;
- External peer review to explore the self-evaluation with the higher education institution (normally by a site visit);
- Public reports of these evaluation activities;
- No direct relationship of the results of external quality assurance to the funding of higher education institutions.

Thune (2002) summarises the important procedural elements shared among European quality assurance systems: internal self-evaluation; visits by external expert review panel; external evaluation; and public reporting. Harvey (1998) summarises the approaches of quality assurance into: accreditation and evaluation of institutions, audit of procedures within an institutions, accreditation of programmes of study, assessment of teaching quality in subject areas or of programmes, research assessment, and standards monitoring. Harvey and Askling (2003) point out that external quality monitoring takes many forms, ranging from accreditation and institutional audit, through subject assessment and standards monitoring to customer surveys. They have varied objects, foci and purposes and relate to different notions of quality and standards. Harvey and Askling (2003) summarise the object, focus, rationale, approach and mechanisms for quality evaluation under four headings: accountability, control, compliance and improvement.

In Harvey and Askling's (2003) model, the main objective of the quality monitoring process may be the provider in institutional review, medium of delivery, output in programme review, or learners in some cases. The focus may be governance and regulation, curriculum design and administration, learning experience or qualification. The specific purposes of quality monitoring fall under four broad headings: accountability, control, compliance and improvement. External quality monitoring takes several forms, ranging from accreditation and institutional audit, subject review and standards monitoring to customer survey.

Objective	Focus	Rationale	Approach	Mechanism
Provider	Governance & regulation	Accountability	Accreditation	Self-assessment
Medium of Delivery	Curriculum Design, Administration	Control	Audit	Performance Indicators
Output	Learning Experience	Compliance	Assessment	Visit
Learner	Qualification	Improvement	Standards Monitoring	Customer Surveys

Table 1. Object, focus, rationale, approach and mechanisms for external evaluation (Source: adapted from Harvey & Askling, 2003)

Accreditation

Accreditation is a public statement that a certain threshold of quality is passed (Campbell et al., 2000). The formal public recognition embodied in accreditation is seen as being based on agreed, pre-defined standards or criteria (El-Khawas, 1998). Accreditation has two nuances: first, the abstract notion of a formal authorising power, enacted via official decisions about recognition and; second, the quality label that institutions or programmes may acquire through certain accreditation procedures (Haakstad, 2001, as cited in Harvey & Askling, 2003). Accreditation may be of an institution or a programme of study. Accreditation tends to focus on inputs, for example, resources, curriculum and staffing. It may address the teaching process but not focus on outcomes of education, for instance, graduate attributes and employability (Harvey & Mason, 1995). In principle accreditation is based on recognition that the institution has in place appropriate control and monitoring processes to ensure satisfactory quality and standards. Accreditation is usually based on an evaluation of whether the institution meets specified minimum (input) standards such as staff qualifications, research activities, student intake, and learning resources (Harvey & Askling, 2003).

Audit

Quality audit is the process of checking to ensure externally or internally specified practices and procedures are in place (Harvey & Askling, 2003). Audits ensure that the institution has clearly defined internal quality monitoring procedures linked to effective action. An audit is often considered as having the potential of meeting many of the expectations of external control at the same time as it might support improvement (Dill, 2000).

Assessment

Quality assessment sets out to measure the level of quality of inputs, processes and outputs (Harvey & Askling, 2003). Quality assessment may be a judgement of the overall quality of an institution or programme, or of specified component elements. Many assessments are supposedly of fitness for purpose. Institutions or programmes are assessed against mission based criteria. Assessment might include a complex grading system or might be based on a simple satisfactory/non-satisfactory dichotomy (Harvey & Askling, 2003). Assessments may benchmark against other institutions, national norms, or against oneself over time. Assessment may also focus on inputs (for example, teaching staff, learning resources), or process (for example, teaching, learning, support services), or outcomes (for instance, students' academic standards of achievement or professional competence, employment rates, students' perception of their learning). Assessment evidence includes statistical indicators, observation, direct evaluation of research outputs, student and graduate views, employer views, student performance, self-assessment and other documentation, discussion and interviews with teachers, students and managers, and perceptions of other agencies, such as professional bodies (Harvey & Askling, 2003).

Standards monitoring

Standards monitoring makes use of external examiners and has a longer history than external quality evaluation. Harvey and Askling (2003) summarise two main foci of standards monitoring: first, academic standards of a programme of study, identified by the academic work produced by students; second, standards of professional competence, identified through the ability or potential to undertake professional practice. They also argue that standards monitoring may specify standards that are appropriate, or it may endeavour to ensure that standards are at appropriate levels, possibly by checking or grading student work or performance. The purpose of standards monitoring may also be ensuring comparability of standards across the sector or across specific subjects within subject disciplines. Sometimes external examiners grade directly but usually standards are inferred by scrutiny of a sample of work or by monitoring award statistics.

Customer surveys

Quality evaluation often includes participant or client satisfaction with service provision which is at institutional, programme or module level in the higher education context. The feedback from students, graduates or employers is collected to enhance the normal process of self-assessment, statistical indicators and peer review (Harvey & Asklilng, 2003).

Different quality assurance procedures affect universities in many ways. These procedures exert both direct and indirect impact on universities. In the process of implanting external quality assurance, new management and self-regulation, as alternatives to the former models of quality management are institutionalised. Westerheijden (2001) argues that external quality monitoring leads to uniformity rather than diversity. External quality monitoring actually inhibits innovation because of the application of conservative or rigid evaluation criteria. Dano and Stensaker (2007) argue that the role and function of external quality assurance is of great importance for the development of an internal quality culture in higher education. Harvey and Askling (2003, p. 81) argue that:

> *Individual teachers within fields of teaching and learning and didactics have inspired each other and also challenged university teachers to make powerful contributions to improve university teaching. These researchers and teachers contribute in turning the quality issue, which was originally imposed by governments, into something empowering teachers and students.*

3.3 The role of students in quality assurance

In recent years, the role of students in the quality assurance of higher education has been recognised across the world. Across the world, students increasingly play their role in the quality assurance process through providing feedback on the courses they have taken and on the general satisfaction with their educational experiences. Reviewing the literature relating to this topic, we may find that students are more and more involved in measuring quality and in improving their own learning experiences. Student voice is increasingly heard through providing feedback, contributing to the development of learning and teaching, participating in the university decision making process, and presenting student views in a number of ways (Alaniska et al., 2006).

Giving feedback is the most common way through which students participate in quality assurance. The increasingly competitive environment in higher education leads universities to monitor levels of their student satisfaction (King et al., 1999). There is a wide diversity in how, when and what kind of feedback students give. It is typical that feedback is given after each course or at least once in a term. It is believed that student feedback can be used as an effective tool for quality improvement. Harvey (1995) suggests that student satisfaction goes hand in hand with the development of a culture of continuous quality improvement. Rowley (2003) identifies four main reasons for collecting student feedback:

- To provide students opportunities to pass comments on their courses and to collect information for improvement;
- To encourage student reflection on their learning;
- To allow institutions to benchmark and to provide indicators that will contribute to the reputation of the university in the marketplace; and
- To provide students with an opportunity to express their level of satisfaction with their academic experiences.

Both student feedback on courses and these national student surveys are increasingly used by higher education institutions across the world as an important component of quality assurance processes. Students are playing a more and more important role in quality assurance through these surveys. Though student learning experiences are internal issues inside higher education institutions, the publication of survey results and league tables produced accordingly make the internal things external. These surveys provide a means for students, their parents, employers and other stakeholders to assess the quality of these institutions. Therefore, student surveys have become a very useful tool for higher education institutions to benchmark themselves in the higher education market and to monitor the quality of higher education provision.

4. Quality assurance in chinese higher education

Over the last two decades the landscape of Chinese higher education has changed greatly through a process of profound restructuring, decentralisation, introduction of market incentives, university mergers, internationalisation, and enlarging student enrolment. The higher education sector in China has expanded and become more differentiated, especially vertically through reputational differences and funding differentiation. Most universities have expanded their campuses and have shifted to a more market-led culture. The student body has become more diverse with the greatly and rapidly increased numbers, which means students enter higher education with different entry levels and for a greater variety of purposes. Universities compete strongly for their teaching and research funding, and are expected to be more accountable for the funds they receive and tuition fees they charge, and to be more relevant to the economic and social needs of the nation. Along with the expansion of Chinese higher education, the issue of quality has become a concern and has attracted a lot of attention in the Chinese higher education sector. Developing quality assurance schemes has been given priority in the agenda of most Chinese higher education institutions. The nationwide implementation of quality evaluation since 2002 is the main means used by the Chinese government to address the potential quality decline and to realise a macro level control over Chinese higher education institutions.

4.1 The structure of quality assurance in chinese higher education

In the past decade, the introduction of quality assurance regimes into Chinese higher education has covered a broad spectrum of initiatives, from national policy, methodologies of quality evaluation, institutional adoption of quality assurance schemes, to a matrix of quality evaluation systems (with teaching quality evaluation and discipline-based evaluation as the main focus supplemented by a range of other evaluations). In China, at the national level, the responsibility for quality assurance lies with the National Education Evaluation Centre, the specialised agency set up by the Ministry of Education in 2004. It is mandated to coordinate evaluation processes, develop appropriate methods for future quality assessment, guide institutions in their quality assurance development, and compile and publish information on higher education quality. Overall, the structure of the quality assurance system in Chinese higher education can be summarised into external quality assurance processes, and systems within higher education institutions (Figure 1).

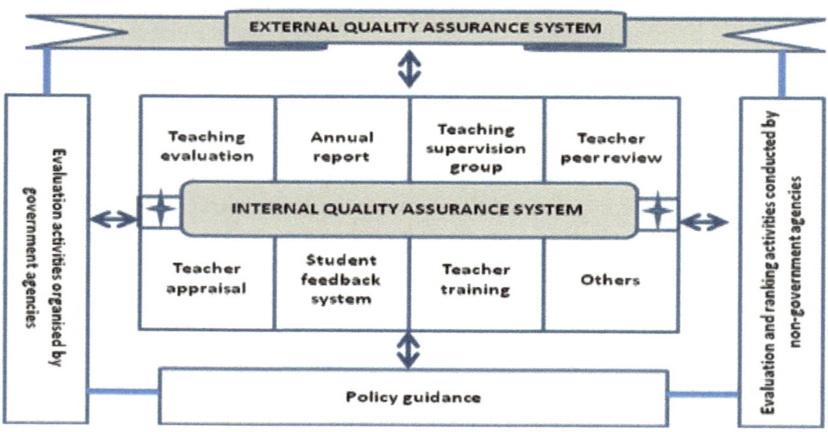

Fig. 1. The structure of quality assurance in Chinese higher education (Li, 2010, p. 66)

The external system can be characterised by three main components: the government's supervision through policy guidance; the government's monitoring through various evaluations carried out by government agencies, among which the most influential ones are the national teaching quality evaluation and the discipline based reviews; the newly emerging non-governmental evaluation agencies and university rankings produced by various non-governmental institutions.

Various measures have been adopted to enhance the quality of education and research activities in Chinese higher education, especially the launch of *Project 211* and *Project 985.* These have had a great impact on quality enhancement in Chinese higher education (Huang, 2005). In addition to these two big projects, other efforts have also been made to assure and improve the quality of education in all Chinese higher education institutions since 2002, including:

- Requiring all professors to teach undergraduate courses and encouraging senior professors to teach core courses and undergraduate courses;
- Setting grants for learning resource renovation;
- Setting grants for developing courses of excellence;
- Setting grants for compiling textbooks of excellence;
- Selecting and awarding the "national outstanding professors in teaching";
- Establishing the Higher Education Evaluation Centre to coordinate various kinds of quality evaluation activities.

Up to now, the education evaluation network in China has been based on evaluation agencies at both national and regional levels (Ding, 2008). At the national level, the Evaluation Office of the Higher Education Department, Ministry of Evaluation, is the government administrative unit in charge of education quality evaluation. The Higher Education Evaluation Centre and the China Academic Degrees & Graduate Education Development Centre are agencies affiliated to Ministry of Education, specialising in conducting evaluations in the Chinese higher education sector. At the regional level, most provincial governments have established their own education evaluation agencies, responsible for education quality evaluation in their provinces.

Currently higher education quality evaluation in China is compulsory and operated by evaluation panels appointed by Higher Education Evaluation Centre. The process of higher education evaluation in China, as in many other countries, includes five basic elements:

- Standards and guidelines issued by the quality evaluation agency and an evaluator panel appointed by the government agencies;
- An institutional self-review report is provided;
- The evaluation panel conducts on-site visits;
- The panel reports back to the institution and Ministry of Education;
- Higher education institutions write their self-improvement report and carry out their self-improvement activities.

In the past decade institutions have been faced with an increase in levels of legislation and involvement from national and local governments, especially in attempts to assure the quality of higher education through formal evaluation techniques and accountability processes. The development of internal quality assurance systems in Chinese universities is the current main emphasis of the quality movement in the Chinese higher education sector. Following the first five-year cycle of national teaching quality evaluations (2002-2007, with an extension to 2008), higher education institutions in China are now encouraged and required by the Ministry of Education to develop their own institutional based quality assurance systems. The common features of internal quality assurance systems in Chinese higher education institutions are (Ding, 2008; Li et al., 2008; Shi et al., 2008):

- The establishment of institutional teaching evaluation centres. These centres are affiliated to or in cooperation with the teaching management office of the institution. Those universities without independent teaching evaluation centres have their own sub-section playing similar roles under the supervision of the teaching affairs/management office. The main responsibilities of these centres are developing and operating the internal quality assurance system.

- The formation of teaching supervision/steering groups. This is also a common practice among Chinese higher education institutions. The group members are the senior teaching staff or retired senior staff with expertise in teaching and teaching management. They are under the supervision of the Vice President for teaching. They are expected to carry out their work directly with teachers and students by observing classes, talking with teachers and students after class. Their responsibilities are to find problems in teaching, and to provide advice on the solutions.
- Peer review. Class teaching observation is also a common practice in most Chinese higher education institutions. Teachers are required to observe each other's classroom teaching, which is considered as a useful way for teachers to learn from each other and to monitor each other's teaching. Besides peer teaching observation, leaders at different levels inside the higher education institutions are required to observe teachers' teaching.
- Student feedback, which is considered as one of the most important quality assurance components, is conducted through surveys, individual and group interviews, student representative reporting, etc. Student surveys are the most commonly used form of student feedback collection, which covers course evaluation, teaching evaluation, and other fields of interest.
- Annual report. Annual institutional self-review report is also a component of internal quality assurance systems. Though the main purpose of such annual self-review reports is not specifically for the sake of quality assurance, it indirectly contributes to realising quality assurance in those institutions.
- Teacher training. Teacher training includes pre-work training, in-service training, and other types of training for teachers. Teachers' pre-work training is considered as the most important of all training schemes. All new teachers are required to attend such training in all Chinese higher education institutions.

4.2 The discrepancy between the current quality assurance and student learning

There has been a tendency to take a 'top-down' approach to the identification and classification of quality assurance issues in Chinese higher education. As a result of the nationwide teaching quality evaluations from 2002 to 2008 and of international exchange, there is a growing body of experiences and knowledge of quality and quality assurance paradigms and cultures. Building up internal quality assurance systems was initiated by the Ministry of Education after the first round nationwide of external teaching quality evaluations. A new round of experimentation and exploration has followed. Various theories, paradigms, and models are being adopted in the Chinese universities' exploration of the quality jungle.

From analysing quality and quality assurance in both Chinese higher education and across the world, we may find that the objectives of quality assurance are largely external influence driven and targeted at efficient and effective management of teaching. The problem observed is that the ultimate aim of higher education, namely the intellectual development and growth of students, has not been given enough emphasis. Most quality assurance policies are generally initiated by the government. The external quality assurance is controlled by the government agencies, and the internal quality assurance is in the hands of teaching administrators. This top-down implementation remains dominant and controls quality assurance activities. The problem observed in this top-down quality assurance is that it has not involved teachers and students actively. Teachers are used as the objects of quality scrutiny and students as the providers of information for teacher and teaching appraisal.

The areas covered by external quality assurance in Chinese higher education can be seen from the institutional quality assurance regulations and procedures. It is the product of the universities' compliance with the national education policies. The areas covered by internal quality assurance mainly include the systematic regulations on teaching, teaching management, teacher appraisal, and students' feedback on teachers and teaching. We may see that the current quality assurance does not address directly the processes of student learning.

The procedures of quality assurance establishment and development in Chinese higher education institutions are a top-down, external influence driven process. They are initiated by the Ministry of Education, and their implementation is guided and supervised by the Higher Education Evaluation Centre. This requires the compliance of higher education institutions in establishing and developing elaborate and comprehensive internal procedures to support the assertions of quality. Such top-down procedures have very little effect on the actual process of student learning and development. In addition, the information collected through quality assurance procedures focuses on resources, teacher qualifications, teaching management, and performance indicators measuring learning outcomes, for example, graduate employment rates. Such information may indicate institutional resources and management and provide a snapshot of what is happening at a certain point in time. However, it is not sufficient to inform quality improvement and student learning development.

Across the world, the exploration of 'quality' in higher education has been dominated by compliance with external agencies' definition of 'quality' as - assurance, accountability, audit and assessment (Houston, 2008). Compliance with the definition imposed from outside the university largely ignores the views of academics, students and others inside who are positioned as the affected but not involved (Ulrich, 2001). We may see the discrepancy between student learning quality and what the current quality assurance assures from the above summary. If we intend to improve the quality of higher education, we should consider how we can incorporate the student learning experience into quality assurance systems to inform the university, teachers, administrative staff, students what and where to work on for improvement. The key issue here is how to incorporate student learning into quality assurance and how to make sure the direct factors influencing student learning quality are managed in a meaningful way, so that continuous quality improvement becomes realisable.

5. Learning-focused quality assurance

Many problems analysed above are linked to a teaching oriented conception of education. The efforts to improve student learning have been underpinned by the belief that learning quality can be assured by teaching quality. Subtly but profoundly it is time to shift higher education institutions from the conception of being an institution complying with external requirements and providing instruction to the ones producing quality learning. It is necessary and urgent to construct a learning-focused concept of education and a way of ensuring the quality defined under this concept.

This learning-focused concept is put forward against the background that the prevailing teacher centred concept of teaching fails in producing quality learning, and the context that the current quality assurance play a very limited role in assuring and promoting student learning. As previously noted, in current quality assurance systems, quality improvement procedures stated in the institutional follow-up reports are mainly about improving teaching profiles and teacher qualifications, but in reality the responsibility for quality enhancement is left to the sense of responsibility or to the priorities of individual teachers. Teachers tend to view external imposition of evaluation and changes as a burden. The quality enhancement initiative, mostly in the form of teaching projects, will only work for those teachers who have an interest in it. Their research usually ends in their publication. It is unlikely to initiate an overall improvement in teaching and learning. Most teachers still teach in the same way, and the quality of learning still relies on who the students are.

Learning-focused quality assurance, as shown in its name, shifts the focus of quality assurance, away from scrutinising institutional compliance with external requirements and scrutinising teaching, towards focusing on improving student learning quality. Learning-focused quality assurance, shown in Figure 2, adopts a more student-oriented approach to learning and teaching and is more sensitive to student learning development. It appreciates student learning experiences, emphasises the interface of teaching and learning, and focuses on student learning activities and approaches.

To improve education quality, universities need to identify and address the characteristics, needs and expectations of students, to respond to different levels of student preparedness while maintaining academic standards, to re-conceptualise teaching and learning in the new paradigm of higher education, to reposition its knowledge functions, and to managing multiple external forces influencing them.

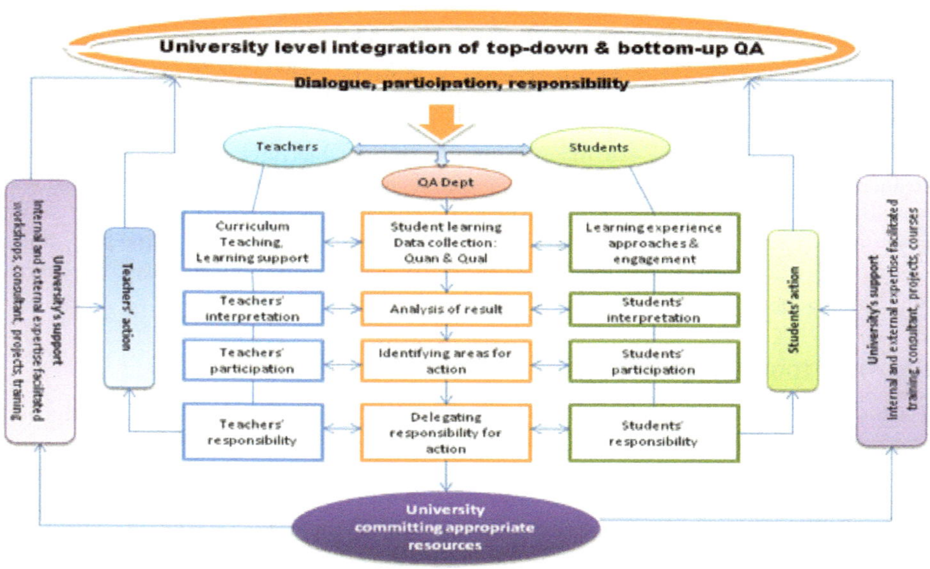

Fig. 2. Learning focused quality assurance

Quality improvement is a complicated process, which needs a holistic system and the engagement of teachers, administration staff, and students. Students' learning experiences need to be integrated into a regular and continuous cycle of data collection, analysis, reporting, transforming into feasible enhancement plans and action, and integrating into learning and teaching practice. Enhancing student learning quality requires a system which may collect appropriate learning and teaching data, identifying the areas for improvement after analysis, delegating responsibility for action to the agents involved, encouraging the ownership of improvement action through the facilitation of the appropriate institutional support of both internal and external resources that can be used, exchanging experiences among students, teachers and administration staff, and stabilising positive experiences into learning and teaching practice.

Learning-focused quality assurance is student learning focused, quality improvement oriented and research informed. It is a holistic system and requires the engagement of teachers, administration staff, and students. In learning-focused quality assurance, students' learning experience are integrated into a regular and continuous cycle of data collection, analysis, reporting, transforming into feasible enhancement plans and action, and integrating into learning and teaching practice. Establishing this is not an easy task, and needs a holistic system of dialogue, participation and responsibility. Learning-focused quality assurance is student learning oriented and supported by the principle of dialogue, participation and responsibility.

Dialogue is the first principle in learning-focused quality assurance. The agents of quality improvement are teachers, students, quality assurance administrative staff, and other staff related in universities. ***Dialogue*** is the communication between them. This dialogue will enhance students' understanding and motivation to learn and teachers' understanding of student learning and their support for this process. The dialogue among teachers will help them be more reflective in their teaching, improve the curriculum, and give better teaching to students. The dialogue between teachers and quality assurance administrators and other staff will help teachers communicate their expected support to their teaching and student learning and keep the university informed of resources needed. The dialogue between university and students through quality assurance systems will help students to understand the vision and values of their universities and help the staff understand student learning needs so that they may configure the necessary support appropriately.

Participation is the second principle in learning-focused quality assurance. Quality improvement is unlikely without the active participation of the three main agents inside higher education institutions: teachers, students, administrators. In almost every step in learning-focused quality assurance, their participation is necessary. Quality learning at any universities requires the active participation of both teachers and students. This process is guaranteed by quality assurance staff's action of involving both teachers and students in designing how to collect student learning data, to best understand and grasp the information about student learning and to disseminate the information to related departments and people. This participation at the stage of data collection can be in multiple forms: by contributing ideas in designing questionnaires and themes of qualitative enquiry, giving feedback to the designed questionnaires, communicating opinions and insights in discussion groups, and providing reliable information in any student learning data collection processes. The participation at the stage of analysing results is mainly in the way that all related agents contribute their interpretation, which will make the next step of identifying necessary action more evidence based and will better address the needs in reality. The participation at the stage of identifying areas for action is to discuss what and how to improve quality and to reach a consensus, so that the appropriate quality enhancement procedures will be established. Participation at this stage of delegating responsibility for action implies that all involved agents are to take their respective responsibilities in taking action.

Responsibility is the third principle in learning-focused quality assurance. It is essential for any quality enhancement initiatives to have effects. Teachers' responsibility is to undertake the quality enhancement initiatives actively, to reflect on their own teaching, to understand student learning, to explore how to better align their teaching to student learning, to organise a better learning environment through their teaching and their dialogue with other stakeholders, to develop their curriculum regularly, to provide appropriate feedback to students, and to encourage them to take their own responsibility for learning. Students' responsibility is to become the owners of their own learning by thinking and developing their learning objectives, participating actively in learning, managing their learning time and activities appropriately, reflecting on their own learning approaches and engagements regularly, communicating their learning needs and questions to teachers and other staff promptly, participating in the quality assurance process, providing feedback actively, and making good use of resources and support to reach the best learning outcomes. University administrators are the facilitators and coordinators in the whole process. Their responsibility is to manage data collection and analysis, to coordinate the participation of teachers and students in the process of data collection, analysis, interpreting results, and identifying areas for action, to ensure the quality of their participation and the feasibility of the enhancement plan, to identify and provide effective support to both teachers and students in the process of their undertaking the quality enhancement actions by organising training, workshops, courses, forums, consulting sessions with the support of internal and external experts and resources.

In learning-focused quality assurance, the responsibility for quality enhancement will not be left to the sense of responsibility or to the priorities of individual teachers, as the current quality assurance does. Universities will take their responsibilities in providing the incentives and support structures for teachers to enhance their teaching, and for students to become the owners of their learning. Both teachers and students are not staying passively at the receiving end of quality assurance; instead, they are the key agents and drivers of continuous quality improvement. It is also important to approach quality holistically and combine cultural elements, structural dimensions and competencies into one holistic framework, in order to enable stakeholders to develop visions, shared values and beliefs, and to delegate the ownership of learning and improvement at all institutional levels.

6. Conclusion

To summarise, improving quality is about a change in culture, which involves a slow process of evolution (Harvey, 1998). Quality assurance should not be the synonym of formalism and conformism; instead, it can be used as catalyst for change from teaching-focused higher education to a new paradigm with more focus on student learning.

Learning-focused quality assurance is built on a more holistic understanding of the relationship between quality assurance and learning enhancement. The role of learning-focused quality assurance involves monitoring and managing the complex learning situation, and improving the quality of student learning. The core of learning-focused quality assurance is learning: student learning quality and how universities and teachers facilitate students to improve their learning.

We are clear that 'quality' in the context of higher education is bound up with our values and fundamental aims in higher education. It is never unnecessary for higher education institutions to ask themselves what the term 'quality' means to them. In this chapter, we have taken the view on quality from student learning and development. Therefore, the implications we draw here are based on the viewpoint that student learning quality should be the focus of a university's quality.

- A sophisticated quality assurance system does not mean students' learning quality can be assured. It is not necessarily a problem related simply to methodology, but to the underlying assumptions of quality assurance and the standard top-down implementation approach. The quality assurance mechanisms should focus on the quality of student learning.
- It is necessary to create a structure for teachers, students and administrators to have dialogue on learning, to actively participate in the quality improvement actions, and to take their responsibility in assuring and improving learning quality. Keeping the ongoing dialogue, sustaining active participation of students, teachers and administrative staff, and encouraging them to take responsibility for learning and improving quality play a significant role in assuring and improving learning quality.
- The student learning experiences need to be integrated into a regular and continuous cycle of information collection, analysis, reporting, transforming into feasible enhancement plans and action, and integrating into learning and teaching practice. Integrating individual learning into a learning-focused culture will strengthen quality learning on campus and nurture the learning of individuals involved.
- Universities should also encourage and support teachers with all possible resources to reflect on their own teaching, to understand student learning, to explore how to better align their teaching to student learning, to organise a better learning environment through their teaching and dialogue with other stakeholders, to develop their curriculum regularly, to provide appropriate feedback to students, and to encourage them to take their own responsibility for learning.

Institutional support in sustaining the quality engagement of teachers, administrators, and students is critical because quality improvement is not something regulated but something attained through critical engagement of all stakeholders involved. The learning-focused quality assurance proposed in this study encourages care for quality at all levels in institutions through care for learning. Such a quality assurance system requires higher education institutions to establish clear learning orientation, to actively involve students, teachers and administrators, to ensure dialogue among teachers, students, and administration staff, to support teachers to research and improve their teaching, and to create and sustain an environment which may enable students to be the owners of their own learning, and further to realise the continuous quality improvement in higher education institutions. At the end of the chapter, we need to say that the learning-focused quality assurance proposed in this study is very conceptual. Further studies investigating how to translate this conceptual model into real world practice to motivate teachers, students and administrators to commit themselves in improving learning quality might be needed.

7. References

Alaniska, H., Codina, E. A., Bohrer, J., Dearlove, R., Eriksson, S., Helle, E., & Wiberg, L. K. (2006). *Student Involvement in the Processes of Quality Assurance Agencies*, European Association for Quality Assurance in Higher Education, Helsinki, Finland

Ball, C. (1985). *Fitness for Purpose: Essays in Higher Education*, SRHE & NFER-Nelson, Guildford

Barnett, R. (1992). *Improving Higher Education*, SRHE & Open University Press, Buckingham

Billing, D. (2004). International Comparisons and Trends in External Quality Assurance of Higher Education: Commonality or Diversity? *Higher Education, 47*, 113-137

Birnbaum, R. (1994). The Quality Cube: How College Presidents Assess Excellence. *Journal of Higher Education Management, 9*(3), 71-82

Bornmann, L., Mittag, S., & Daniel, H. D. (2006). Quality Assurance in Higher Education - Meta-evaluation of Multi-stage Evaluation Procedures in Germany. *Higher Education, 52*(4), 687-709

Brennan, J., & Shah, T. (2000). *Managing Quality in Higher Education: An International Perspective on Institutional Assessment and Change*, OECD, SRHE & Open University Press, Buckingham

Brookes, M., & Becket, N. (2007). Quality Management in Higher Education: A Review of International Issues and Practice. *International Journal of Quality Standards, 1*(1), 1-23

Campbell, C., Kanaan, S., Kehm, B., Mockiene, B., Westerheijden, D. F., & Williams, R. (2000). *The European University: A Handbook on Institutional Approaches to Strategic Management, Quality Management, European Policy and Academic Recognition*, European Training Foundation, Torino

Campbell, C., & Rozsnyai, C. (2002). *Quality Assurance and the Development of Course Programmes*, UNESCO, Bucharest

Dano, T, & Stensaker, B. (2007). Still Balancing Improvement and Accountability? Developments in External Quality Assurance in the Nordic Countries 1996-2006. *Quality in Higher Education, 13*(1), 81-93

Deming, W. E. (1994). The Need for Change. *Journal for Quality and Participation, 17*(7), 30-32

Dill, D. (1992). Quality by Design: Towards a Framework for Academic Quality Management, In: *Higher Education: handbook of Theory and Research*, J. Smart (Ed.), Vol. VIII, pp. 37-83, Agathon Press, New York

Dill, D. (2000). Capacity Building as an Instrument of Institutional Reform: Improving the Quality of Higher Education through Academic Audits in the UK, New Zealand, Sweden, and Hong Kong. *Journal of Comparative Policy Analysis, 2*(2), 211-234

Ding, X. (2008). *Research on Quality Assurance in Higher Education*, Jiangsu Education Press, Nanjing

Dunkerley, D., & Wong, W. S. (2001). *Global Perspectives on Quality in Higher Education*, Ashgate, Aldershot

Eagle, L., & Brennan, R. (2007). Are Students Customers? TQM and Marketing Perspectives. *Quality Assurance in Education, 15*(1), 44-60

El-Khawas, E. (1998). Accreditation's Role in Quality Assurance in the United States. *Higher Education Management, 10*(3), 43-56

Evans, J., & Lindsay, W. M. (2001). *The Management and Control of Quality,* South-Western/Thomson Learning, Cincinnati OH

Frazer, M. (1992). Quality Assurance in Higher Education, In: *Quality Assurance in Higher Educatio,* A. Craft (Ed.), pp. 9-29, Routledge, Hong Kong

Girdwood, A. (1997). Quality Assurance in Higher Education: Trends and Aspects, *Proceedings of Seminar on Quality Assurance in Higher Education,* Tertiary Education commission, University of Mauritius, October 1997

Haakstad, J. (2001). Accreditation: the New Quality Assurance Formula? Some Reflections as Norway is about to Reform its Quality Assurance System. *Quality in Higher Education, 7*(1), 77-82

Harman, G. (1998). The Management of Quality Assurance: A review of International Practice. *Higher Education Quarterly, 52*(4), 345-364

Hart, W. A. (1997). The Quality Mongers. *The Journal of the Philosophy of Education, 31*(2), 295-308

Harvey, L. (1994). Continuous Quality Improvement: A System-wide View of Quality in Higher Education, In: *University-Wide Change, Staff and Curriculum Development.* P. T. Knight (Ed.), pp. 47-70, SEDA, Birmingham

Harvey, L. (1995). Student Satisfaction. *The New Review of Academic Librarianship 1*(1), 161-173

Harvey, L. (1998). An Assessment of Past and Current Approaches to Quality in Higher Education. *Australian Journal of Education, 42*(3), 263-276

Harvey, L. (2002). Quality Assurance in Higher Education: Some International Trends, *Proceedings of the Norwegian Higher Education Conference,* Oslo, 22-23 January 2002

Harvey, L., & Askling, B. (2003). Quality in Higher Education, In: *The Dialogue between Higher Education Research and Practice,* R. Begg (Ed.), pp. 69-83, KLUWER, London

Harvey, L., & Mason, S. (1995). *The Role of Professional Bodies in Higher Education Quality Monitoring,* QHE, Birmingham

Harvey, L., & Newton, J. (2007). Transforming Quality Evaluation: Moving on, In: *Quality Assurance in Higher Education,* D. F. Westerheijden, B. Stensaker & M. J. Rosa (Eds.), pp. 225-245, Springer, Dordrecht, the Netherlands

Houston, D. (2007). TQM and Higher Education: A Critical Systems Perspective on Fitness for Purpose. *Quality in Higher Education, 13*(1), 3-17

Houston, D. (2008). Rethinking Quality and Improvement in Higher Education. *Quality Assurance in Education, 16*(1), 61-79

Huang, F. (2005). Quality Enhancement and Quantitative Growth: Changes and Trends of China's Higher Education. *Higher Education Policy, 2005*(18), 117-130

Jacques, M. L. (1999). The Call of Quality: Doing Right Things Right. *Quality Progress, 32*(9), 48-54

King, M., Morison, I., Reed, G., & Stacow, G. (1999). Student Feedback Systems in the Business School: A Departmental Model. *Quality Assurance in Education, 7*(2), 90-100

Lewis, D. R., Ikeda, T., & Dundar, H. (2001). On the Use of Performance Indicators in Japan's Higher Education Reform Agenda. *Nagoya Journal of Higher Education, 1,* 67-98

Li, Y. (2010). Quality Assurance in Chinese Higher Education. *Research in Comparative and International Education, 5*(1), 58-76

Li, Y., Li, X., & Qu, Q. (2008). *Practice and Study of Chinese Higher Education Teaching Quality Evaluation*, Zhongshan University Press, Guangzhou

Meirovich, G., & Romar, E. J. (2006). The Difficulty in Implementing TQM in Higher Education Instruction. *Quality Assurance in Education, 14*(4), 324-337

Middlehurst, R., & Woodhouse, D. (1995). Coherent Systems for External Quality Assurance. *Quality in Higher Education, 1*(3), 257-268

Morley, L. (2003). *Quality and Power in Higher Education*, SRHE & Open University Press, Maidenhead

Neave, G. (1996). On looking both ways at once: Scrutinies of the private life of higher education, In: *Inside academia. New challenges for the academic profession,* P. A. M. Maassen & F. A. Van Vught (Eds.), De Tijdstroom, Utrecht, the Netherlands

Neave, G. (1998). The Evaluative State Reconsidered. *European Journal of Education*, 265-284

Reeves, C. A., & Bednar, D. A. (1994). Defining Quality: Alternatives and Implications. *Academy of Management Review, 19*(3), 419-445

Rowley, J. (2003). Designing Student Feedback Questionnaires. *Quality Assurance in Education, 11*(3), 142-149

Salter, B., & Tapper, T. (2000). The Politics of Governance in Higher Education: The Case of Quality Assurance. *Political Studies, 48*(1), 66-87

Shi, H., Liu, X., Chen, W., & Yang, R. (2008). *Theory and Practice in Higher Education Teaching Quality Assurance and Monitoring*, Henang University Press, Kaifeng

Stensaker, B. (2003). Trance, Transparency and Transformation: the Impact of External Quality Monitoring in Higher Education. *Quality in Higher Education, 9*(2), 151-159

Stensaker, B. (2008). Outcomes of Quality Assurance: A Discussion of Knowledge, Methodology and Validity. *Quality in Higher Education, 14*(1), 3-13

Temponi, C. (2005). Continuous Improvement Framework: Implications for Academia. *Quality Assurance in Education, 13*(1), 17-36

Thune, C. (1996). The Alliance of Accountability and Improvement: The Danish Experience. *Quality in Higher Education, 2*(1), 21-32

Thune, C. (2002). Future Relationships and Dilemmas of Quality Assurance and Accreditation, *Proceedings of the 24th EAIR Annual Forum,* Prague, 8-11 September 2002

Ulrich, W. (2001). The Quest for Competence in Systemic Research and Practice. *Systems Research and Behavioural Science, 18*(1), 3-28

Van Vught, F. A., & Westerheijden, D. F. (1993). *Quality Management and Quality Assurance in European Higher Education: Methods and Mechanisms,* Office for Official Publications of the Commission of the European Communities, Luxembourg

Vroeijenstijn, A. I. (1995). *Improvement and Accountability: Navigating between Scylla and Charybdis, Guide for External Quality Assessment in Higher Education,* Jessica Kingsley, London

Vroeijenstijn, A. J., & Acherman, J. A. (1990). Control Oriented versus Improvement Oriented Quality Assessment, In: *Peer Review and Performance Indicators: Quality Assessment in British and Dutch Higher Education,* L. C. J. Goedegebuure, P. A. M. Maassen & D. F. Westerheijden (Eds.), VSNU, Utrecht

Watty, K. (2003). When Will Academics Learn about Quality. *Quality in Higher Education, 9*(3), 213-221

Westerheijden, D. F. (2001). Ex oriente lux?: National and Multiple Accreditation in Europe after the Fall of the Wall and after Bologna. *Quality in Higher Education, 7*(1), 65-75

Westerheijen, D. F., Hulpiau, V., & Waeytens, K. (2006). Lines of Change in the Discourse of Quality Assurance: An Overview of Some Studies into What Impacts Improvements, *Proceedings of the 28th Annual EAIR Forum*, Rome, 30 August-1 September 2006

Woodhouse, D. (1999). Quality and Quality Assurance, In: *Quality and internationalisation in higher education,* H. d. Wit & J. A. Knight (Eds.), pp. 29-44, OECD, Paris, France

11

Integrated Higher Education Management: Summary of Management Approaches

Juha Kettunen
Turku University of Applied Sciences, Turku, Finland

1. Introduction

The great challenge of autonomous higher education institutions is to create and describe their quality assurance systems, which may be based on old traditions and various management approaches. The existing and emerging management approaches have different backgrounds. Therefore, they have to be integrated with each other so that they support each other and ensure that higher education institutions can attain their objectives. Strategic management and quality assurance are the main management theories, and they include many minor approaches and concepts. Each institution has a challenge to put these approaches together and integrate them as a management system of the institution (Kettunen & Kantola, 2007).

Strategic plans include strategic themes and objectives, which describe the route from the perceived present situation to the desired future (Kettunen, 2010b). Higher education institutions which are able to take personnel, students and other stakeholders into account in future planning are able to build understanding and commitment to the strategic plan. Strategic planning is no longer a compliance task of the top-management making sure that the regulatory and statutory requirements are met. The new information technology provides means to have dialogue and enables the most important stakeholders to engage in the strategy process. The higher education institution should be able to describe its strategic plan using a strategy map, which describes the strategic objectives in the perspectives of the customer, finance, internal processes and organisational learning (Kettunen, 2004).

Quality assurance has developed independently from strategic management and it emphasizes the different aspects of management. Even though these management approaches have different backgrounds, they can be integrated into the management system supported by the management information system (Kettunen & Kantola, 2005). The quality assurance system supplements strategic management, because the purpose of quality assurance is to ensure that the institution can achieve its strategic and other objectives. The success of the institution in reaching its strategic objectives can be measured using the indicators of the Balanced Scorecard. Autonomous higher education institutions plan and build their own quality assurance systems and each institution should be able to describe the main elements of the quality assurance system in an understandable way.

The purpose of this chapter is to summarise strategic management, quality assurance and other management approaches, and create the conceptual framework of integrated higher education management. The study explains the importance of dialogue in the strategy process to engage academic people in future planning and definition of the pedagogical profile of the institution, which can be defined in the strategy process. Then the study explains the internal and external evaluation of educational units, which are processes to enhance quality and pedagogical development. The study also explains the internal and external quality audits of processes.

This study is organised as follows. Section 2 explains the strategy and quality maps developed to describe the strategic plan and the quality assurance system of a higher education institution. Section 3 describes the strategy process which has turned to a new era, because Web-based dialogue enables the personnel, students and other stakeholders to engage in the strategy process. Section 4 presents the role of pedagogy in quality assurance and presents innovation pedagogy adopted for the profile at the Turku University of Applied Sciences. Thereafter, Section 5 presents the internal benchmarking of degree programmes as the procedure of quality assurance. Section 6 presents the external evaluation of the centres of excellence. Section 7 presents process management as a means to ensure the achievement of objectives and the continuous improvement of high quality processes. Process management also is an important approach for the information systems to ensure high quality. Section 8 presents the conceptual framework of integrated higher education management. Finally, the results of the study are discussed and summarised in the concluding section.

2. Strategy and quality maps create the framework

Higher education institutions are accountable for their performance not only for those bodies that finance education, but also students, employers and other stakeholders. These institutions are responsible to implement the education policy, their own strategic objectives and quality assurance, and they are obliged to participate in external evaluations and audit their quality assurance systems. Accountability means that the institutions have the responsibility to build their own quality assurance systems. They need a conceptual framework for the integration of a strategic plan and quality assurance, which can be tailored to meet the needs of the institution. The strategy and quality maps are means to describe the integration of strategic management and quality assurance in higher education (Kettunen, 2011c).

The purpose of the quality assurance system is to ensure that the strategic and other objectives can be achieved. Therefore, the management of higher education institutions should have a clear understanding how quality assurance is related to strategic management. The managers need concepts to describe quality assurance, strategic management and main processes. The managers also should have a precise understanding about how the institution can reach its goals with the processes. Strategy and quality maps, which are like road maps, are the needed concepts. They describe the main land marks on the route to the desired future situation illustrated by the vision, but omit all the minor details of specific considerations.

The internal processes can be described at the institutional level by the value chain, which is a broad description of the sequential processes. The main elements are research and development representing knowledge acquisition, support services representing knowledge management and education representing knowledge dissemination, as noted by Lee and Yang (2000). The quality cycle of continuous improvement (Deming, 1986; Tague, 2005) can be combined with the main processes of education. Support services represent the "plan" phase and education represents the "do" phase. Students, employers and other stakeholders give feedback in the "check" phase and finally the processes are improved by research and development in the "act" phase.

At the plan phase, the objectives of education are planned and defined in the curricula and other planning documents, and the processes are inspected if necessary. The plans then are implemented following the process descriptions. The third phase includes the evaluation of processes, objectives achieved and quality deviations based on feedback from students and other stakeholders. Corrective actions are taken if non-conformance is detected. In the final phase, the process descriptions and activities are improved. The procedure of the quality audit is essential to maintain and continuously improve the processes.

The strategy map introduced by Kaplan and Norton (1996, 2004) can be used to describe and communicate the strategic plan of the institution. The strategy map describes the strategic objectives and balances them into four perspectives, which typically are the customer, finance, internal processes and organisational learning. The strategy map also describes the linkages between the strategic objectives. The internal processes perspective describes the main processes of an institution, which are research and development, support services and education. Typically, the linkages between the perspectives are described in higher education so that the financing and organisational learning is required for the internal processes, which achieve customer satisfaction.

The quality map is the graphical representation of the quality assurance system. The quality map helps the management of the institution describe the main characteristics of quality assurance. The global environment, national education policy and regional aspects are taken into account in strategic planning, which defines the strategic objectives. The quality assurance system ensures that the strategic objectives can be achieved. The management of the higher education institution takes responsibility to develop internal processes based on the strategic plan and the feedback from students and customers. The support services of the faculty define the learning objectives and develop curricula. The faculty implements the plan in the education process and collects feedback from students and customers.

Strategic management and quality assurance are different approaches of management, because they have been developed independently of each other. They can, however, be integrated with each other. The strategy process has a forward-looking stance and it produces the strategic objectives, but the purpose of the quality assurance system is to ensure that strategic objectives can be reached. Strategy and quality maps describe the essential characteristics of the management approaches. The strategy process is partly in the quality assurance system, because it produces the important strategic objectives. On the other hand, continuous improvement is a common element in both the strategy and quality maps.

3. Strategy process as a dialogue to achieve commitment

Strategic management adapts to a rapidly changing environment and provides the strategic themes and objectives for the desired future. Autonomy of higher education means that academic people have the power to subvert, constrain or ignore changes they do not accept. Therefore, it is extremely important to engage them in the strategy process. This process has turned to a new phase which enables the personnel, students and other stakeholders to engage in it. The Web-based dialogue enables management to collect ideas, to conduct a content analysis and transform a large amount of data into strategic themes and objectives even in large organisations (Kettunen, 2010b).

The strategy process typically starts with environmental scanning. The environment of higher education institutions includes education policy, networked strategies and local demand for labour among the other things. The strengths and weaknesses of the institution are analysed taking into account the opportunities and threats of the environment. The new age of environmental scanning extends the traditional environmental analysis of strategic planning with the Web-based dialogue (Ilmola & Kotsalo-Mustonen, 2003; Kettunen, 2010b). The dialogue in the strategy process is more important than the written document, because participation in the process supports the commitment to the strategic plan. On the other hand, the environmental analysis is related to the fulfilment of customers' requirements, needs or desires following the fitness-for-purpose principle (Welch & Dey, 2002).

A Web-based dialogue can be designed to collect, evaluate and analyse signals form personnel, students and partners to pass the surveillance, mentality and power filters (Ansoff, 1984). The surveillance filter blocks the information out of the sight and scope of an individual. The mentality filter limits the information to seeing only the short-sighted changes in time. The power filter blocks information from other organisational levels and cultures. The Web-based survey is used to identify especially the opportunities and threats of the institution. The respondents of the survey are able to evaluate the ideas presented by other people and present new ones.

The Web-based survey provides background information for the face-to-face dialogue. The managers of each organisational unit should analyse and discuss the results of the Web-based dialogue. The results of the strategy processes are collected in the strategic plan, which describes the strategic themes and objectives. The strategic plan is implemented using the annual action plans, the human resources plans, the budgets and other plans. The Balanced Scorecard introduced by Kaplan and Norton (1996, 2004) is an approach which is used to measure the achievement of strategic objectives.

The environment is the common element for strategic planning and quality assurance. The fitness-for-purpose principle of quality assurance and environmental scanning of strategic planning are similar. Strategic planning describes the route to the desired better future, but quality assurance ensures that the desired outcome described by the vision can be reached. It can be concluded that quality assurance complements strategic management and provides tools for implementation of the strategic management. Even though the Balanced Scorecard approach has been developed in the context of strategic management, it also is suitable for quality assurance.

4. Innovation pedagogy as a profile of a higher education institution

Higher education institutions are not equal, because they have different profiles. That is especially true in countries which do not have common and strong education policies. Higher education institutions define their profiles to differentiate themselves, create competitive advantage and increase their external impact on their environment. If the education policy and the strategic plans differ among institutions, the quality assurance systems are hardly identical. The Finnish Ministry of Education and Culture asked the higher education institutions to define their profiles in 2009.

Many traditional research universities rely on individual-centred learning. Students listen to lectures, read literature and memorise material for examinations. Individual learning is relevant when the purpose is to disseminate facts, concepts and information. Although individual-centred learning has attracted attention previously, such systems often place an unusual cognitive burden on the learner in social situations and networks. Working life does not function so that people memorise things alone, but rather ask help from other people. Individual learning is necessary but not sufficient in situations in which the purpose is to create capabilities for students to have a positive impact in their working life.

The Finnish universities of applied sciences are part of a wider community and their regional development has a prominent role based on the legislation. Sociocultural theory and the constructivist view of learning developed by Vygotsky (1978) and Piaget (1999) are widely accepted pedagogical starting points at these institutions. Collaborative learning assumes that learning takes place as learners interact. Individual-centred learning found at traditional research universities is extended to group-based learning at the universities of applied sciences. It is relevant to the relatively complex and multidisciplinary problem-solving tasks of applied research and development, where students participate and create capabilities for working life. Collaborative learning takes place, for example, in reflection, negotiation, debating or multidisciplinary study groups.

Networked learning takes place when students are connected to their potential employers and the partners of the institution. This is relevant in professional higher education, where internships, applied research and development and theses are carried out with working life. Internships take at least half a year at the Finnish universities of applied sciences. Networked learning using information and communication technology supports learning with regard to the development of searching, evaluating and understanding information sources, which are necessary skills of scientific inquiry. High learning outcomes need metacognitive skills and abilities to regulate the learning.

The Turku University of Applied Sciences defined its profile as innovation pedagogy based on multi-field activities, where entrepreneurship, applied research and development, and internalisation combine with education to promote innovations in its region (Kettunen, 2011b). Innovation pedagogy provides a pedagogical framework to enhance the quality on the institutional level. The elements of innovation pedagogy can be found in the Act of the Universities of Applied Sciences and the learning practices. Therefore, innovation pedagogy is suitable for all the universities of applied sciences.

The concept of innovation pedagogy includes all the most important elements of the universities of applied sciences and is especially suitable for those institutions that want to increase their external impact on the region and promote innovations. According to literature, innovations are incremental or radical (Tidd et al., 2001; Bessant et al., 2005). The incremental innovations are improvements of existing products, services and processes. Actually, the incremental innovation is a similar concept to the continuous improvement of quality assurance. The radical innovations correspond to the major strategic changes among the new products, services and processes. Radical innovations among the processes mean the reengineering of processes described by Hammer and Champy (1993).

The profile is specified by the focal areas of knowledge at the institution. The research and development programmes are based on the focal areas of faculties, including applied information and communication technology, biocompetence and business know-how, expertise in health care and medication, lifelong well-being services, marine environment and construction expertise and working life-based approaches to creative arts. The institution and faculties allocate financial resources to these focal areas. In addition, external funding is sought to supplement internal funding.

5. Internal benchmarking of degree programmes

The quality assurance system includes, among other things, the evaluation of degree programmes or subjects. In many countries, subject benchmarking is a systematic procedure to evaluate education and academic standards (Haargeaves & Christou, 2002; Yorke, 2002). The procedure of benchmarking for seeing and accepting improvement is a useful tool to avoid the prevailing opposition to change (Calabrese, 2003; Ford & Ford, 2002). Benchmarking is especially useful in enhancement-led evaluation, which aims to improve the activities for better quality.

Benchmarking also is a useful procedure for learning good practices and accepting the necessary changes to develop education and other activities, and it actually is the comparison of activities. At the same time, it is hoped that the comparison leads to enhancement-led evaluation, which aims to support the improvement of activities for better quality. Traditionally, the emphasis has been on the careful selection of benchmarks that represent the best practices. However, the selection of a benchmarking team often is a neglected feature of benchmarking.

Benchmarking includes the search for benchmarks and best, or at least good, practices. The search for benchmarks is close to quantitative research and typically the comparison of indicators, which measure the achievement of strategic objectives. The Balance Scorecard approach is a valuable framework, which can be used in the context of benchmarking taking into account that the profiles and strategic objectives of the institutions usually are different. The search for best or good practices is the other line of evaluation. Both of these lines aim for the continuous improvement of education. When two organisational units are compared, the underlying assumption of benchmarks and best practices is that these units have comparable learning objectives based on similar strategic objectives.

The process of benchmarking includes the formation of an evaluation team, the nomination of a chairman and the selection of the evaluated unit. The evaluated unit is informed to write the self-evaluation report, which the evaluation team analyses and prepares the evaluation visit. The evaluation team analyses the statistical data and self-evaluation report and visits the evaluated unit. During the visit, the evaluation team analyses continuous improvement and how the unit has succeeded in its activities. The evaluation visit includes separate interviews of the management, teachers, students and the advisory board, including representatives from working life. The dissemination of evaluation results includes the seminar and the publication.

If the members of the evaluation team are selected from the same degree programme or subject, they are competent to evaluate the learning objectives, contents and methods. If they are selected from other domestic or foreign institutions, they are able to make sound comparisons about the networked collaboration and international exchange from a larger perspective. If the members of the evaluation team are selected from other degree programmes, subjects, units or institutions, the strength of that cross functional evaluation team is in the sound comparison of learning methods, the processes of education and innovative collaboration among the degree programmes, units or institutions. This kind of "cross-evaluation" favours the multidisciplinary activities and innovations and meets the needs of customers supporting the research and development (Kettunen, 2010a).

The weakness of the international evaluation is the lack of knowledge of the education system of the country where the evaluation is carried out. That may lead to misunderstanding and incorrect evaluation. Even though a member of the evaluation team is from the home country, the evaluation team's knowledge is limited. If the evaluation team suggests good practices from other countries, they are not necessarily useful, because the education policy, educational systems and traditions of the different countries are not equal. Other problems are the culture and language, which do not help the evaluators to understand all the details of higher education.

The internal benchmarking of a degree programme is an important step in the long-lasting development of education. The benchmarking produces recommendations included in the evaluation report. The degree programme takes responsibility to plan these recommendations for development steps included in the annual action plan. Educational development usually takes many years until it is implemented in the classroom practices. The internal benchmarking can be followed in the external evaluation, which may take the form of centres of excellence.

6. External evaluation of the centres of excellence

The evaluation of the centres of excellence is the procedure of the quality assurance agency to enhance the quality of education at higher education institutions. The evaluation is targeted in teaching processes and outcomes that the institution has achieved. The evaluation encourages the institution and its organisational units to improve the quality and effectiveness of education. The evaluation encourages educational units to the long-term development of education, networking and collaboration. The evaluation provides information on pedagogical decisions, which is important in the evaluation of educational quality (Kettunen, 2011a).

The Finnish Higher Education Evaluation Council is a national quality assurance agency which has the responsibility of evaluating candidates for the centres of excellence in higher education. The evaluation of the centres of excellence is targeted to the evaluation of the operation and outcomes of education. The evaluation supplements the quality audits, where the quality assurance system is the object of evaluation. The quality assurance agency also has other types of evaluations, which have been targeted for thematic purposes.

The evaluation target "operation" includes the description of the education unit and linkages with the strategic plan and pedagogical outlines of the institution containing the core competence, collaboration with working life, networking within and outside the institution, the integration of education with research and development and the planning, operation, evaluation and development of the education unit. In addition, the description of the operation includes forecasting and responding to the knowledge needs of the operating environment containing the control of resources, evaluation and development, the planning process of curriculum, the learning process and guidance, learning environments and the procedures of quality assurance. The evaluation target "outcomes" of the educational unit include the outcomes of the education unit in relation to the defined objectives containing outcomes related to students, working life, personnel, economic performance, international activities and regional development.

The evaluation of excellence starts with the participatory self-evaluation, which is supplemented by the external evaluation, in which the participants come from other higher education institutions. The participatory self-evaluation describes the history, presence and future of the units. It also describes the pedagogical approach used and compares the activities with national average figures in higher education. The evaluation is looking for good or best practices and national benchmarks to develop education. The evaluation panel should pay attention to the selection of indicators, which may measure the achievement of national or institutional objectives. The Balanced Scorecard indicators may be different among institutions; therefore, they cannot be fully compared with each other.

The cross functional evaluation panel is collected from other higher education institutions. It studies and scores the written self-evaluation reports. Based on the written reports the evaluation panel makes a decision about the evaluation visits. The aim of the visits is to ascertain that the operations and results are as described in the self-evaluation report and for the panel to arrive at its own view about the quality of education. After the visit, the evaluation panel makes its proposal about the centres of excellence. The results of the evaluation are published to disseminate the evaluation results and good practices of high quality education.

The evaluation of the centres of excellence is a powerful procedure to develop higher education. The evaluation acknowledges the development of education, brings out high quality learning practices and disseminates excellent outcomes. It also emphasises the importance of pedagogical development in quality assurance. The quality assurance of education is not only procedures and systems, but excellent educational units should have pedagogical views which guide them to reach the desired objectives following the outlines of the education policy.

7. Process management as an approach of quality assurance

The common European quality assurance of higher education institutions (ENQA, 2009) has a relatively short history, where the Berlin Communiqué (2003) was a prominent takeoff.

Autonomous European higher education institutions have an obligation to plan and implement their own quality assurance systems. When the quality assurance systems mature over time, they move away from the compliance auditing and focus more on process management and the achievement of objectives (Pollit et al., 2002). Business process management has taken hold in private companies, but recently process management has reached higher education institutions.

The strategy map describes the main processes at the overall level of an institution, including research and development, support services and education. Educational management requires a much more detailed level to guide personnel, students and stakeholders in the processes. It is important to engage personnel in the process description and maintenance. The planning and implementation of information systems require the most detailed descriptions of processes. They usually are too complicated for the ordinary users who are looking for help in performing their tasks. The majority of improvements can be achieved without information systems.

A process comprises the network or a series of value-added activities performed by the collaborator to accomplish the assignment. The lanes of the flow charts describe the responsibilities of those who are performing the processes. The lanes describe the collaborating partners and information systems, and the responsibilities include activities and decisions that have essential connections between each other. From the point of view of process maintenance and improvement, it is necessary to keep the process descriptions as simple as possible for the personnel and students to carry out the necessary tasks. The internal customers come from the institution but external customers and partners come from other organisations. Therefore it is important to extend the concept of quality assurance outside the institution.

Continuous improvement is the main element to learn from experience and to develop the processes (Mehra & Argawal, 2003; Escrig-Tena, 2004). The planning, implementation, evaluation and improvement of processes require active maintenance and audits of the processes. Business process reengineering defined by Hammer and Champy (1993) is the fundamental reconsideration and radical redesign to achieve drastic improvement of performances. Instead of incremental continuous improvement, the processes are redesigned or eliminated altogether.

The processes are audited in the external and internal audits. External auditing is carried out by the quality assurance agency, based on the self-evaluation report, auditing material and the site visit. The evaluation panel meets separately with the different groups and stakeholders of the institutions and publishes the audit results. The external audit is part of the auditing of the whole quality assurance system in Finland, described in the quality manual of the Finnish Higher Education Evaluation Council (Korkeakoulujen arviointineuvosto, 2010). The external quality audit group of the Turku University of Applied Sciences recommended the systematic auditing of internal processes (Hintsanen et al., 2009).

The other maintenance operation is the internal audit, whereby the institution collects the evaluation group that evaluates the need to improve or reengineer the processes and inspects the conformance of processes. It is necessary to focus first on the improvement of the processes and achievement of the objectives, and if no major amendments are found, to look for quality deviations. There is no use to search only for quality deviations if the processes are inefficient and do not produce the desired outcomes when better processes are available. The improvement or reengineering of processes is a great challenge which needs clear understanding about the strategic objectives, competent and dedicated people and support of top-management.

The flow charts of processes describe the different organisational units and stakeholders that have the responsibilities of the actions and decisions. The higher education institutions have to identify not only the internal organisational units but also the customers and partners. The customers include potential, occasional and key customers. The partners include not only the ordinary partners, but also strategic partners who have an important collaborating role in achievement of the strategic objectives. Hence quality assurance should be extended to the external partner if the management and internal processes are connected with the external partners of the institution.

The integration of information systems to quality assurance is another main driver to describe processes in higher education. Most Finnish higher education institutions have the quality manuals and manual binders of evidence for the quality audits. These manuals are passive documents that do not include dialogue, feedback and adequate guidance for students and personnel. Intranet technology provides a solution to reduce or avoid paper-based systems and support process management. It is reasonable to plan the structure of the Intranet to follow the process map and include essential process information and guidance for ordinary users. It also is essential to include the detailed flow charts and process cards to the Intranet for those who need more information. A well-planned Intranet also should have a feedback system to evaluate and improve the processes. The Intranet is supported by the customer management system which collects customer and partner information to increase the quality of processes.

8. Integrated higher education management

Strategic management, quality assurance and many other management approaches have different kinds of backgrounds and they have been developed independently of each other, but they meet in practice in many organisations. It is a great challenge to conceptually integrate strategic management, quality assurance, the Balanced Scorecard, the promotion of innovations, process management, partners, investments and economic success, and customers in higher education institutions to the consistent system of integrated higher education management to ensure high quality outcomes. This challenge is solved in this section.

Table 1 describes the integration of strategic management with quality assurance to achieve the objectives in the perspectives of the Balanced Scorecard. A higher education institution must respond to the changes in its environment, including the international and national economic policy and the local demand for labour, research and development. Strategic management must adapt to the environment changes and therefore is a stronger management approach which overshadows quality assurance. The ideas presented in the table may take slightly different kinds of concepts in other organisations.

Successful strategic management produces radical or incremental innovations which are outcomes of organisational learning. The innovations invented in the internal processes and networks also may have strategic importance. In such a case, the innovations should be taken into consideration in the strategy process. Quality assurance typically produces incremental innovations, even though quality assurance may help management analyse the internal strengths and weaknesses with respect to the opportunities and threats of the environment and consider the possibility for radical innovations.

The core of strategic management is in the internal processes, because the desired new outcomes cannot be reached without renewed actions. Internal processes may produce new or improved products or services with reengineered or improved processes. The reengineering of processes has much potential, but it sometimes is interpreted as disturbing in higher education. The internal processes and the knowledge do not change very quickly among autonomous faculties. Reengineering may cause changes among the strategic or other partners involved in the processes. Quality assurance typically produces continuous improvement in existing processes. The improvement also may cause changes among the partners and their operations.

Strategic management may cause large investment from the financial perspective. If radical innovations cause reengineering of processes and new partners, the changes are risky investments which do not happen rapidly in higher education, because the lengths of study typically take many years, and because students follow their curricula and personal study plans. Reengineering is a human investment which requires research and development and in-house training, but it also may require real investments for new premises and equipment. In a successful case, quality assurance produces increased economic success, which may lead to better learning outcomes and lower costs. Continuous improvement can be used to avoid large investments and organisational changes in the later stages.

Strategic management which produces radical or incremental innovations may increase customer satisfaction among potential, occasional and key customers. High customer satisfaction is the desired goal, which can be described by the success to achieve learning objectives, new knowledge and innovations among other objectives. The scope of quality assurance is limited, because continuous improvement mainly affects key and occasional customers. Continuous improvement may keep customers satisfied, avoid drop-outs and keep the processes sustainable. The concept of customers also can include stakeholders if one wants to study their role in management.

Integrated higher education management can be summarised so that the radical innovations are the reengineering of processes and respectively incremental innovations are continuous improvement seen from the internal processes perspective. This is a result which opens up opportunities to integrate innovations with strategic management, quality assurance and process management with strategic or ordinary partners. These management approaches can be integrated in practice in higher education institutions and also in other organisations. The implementation of the strategic plan means typically investments to achieve economic success, but quality assurance usually has smaller economic consequences.

Strategic management which aims to satisfy potential customers is the great challenge of change management, because it is based on innovative products or services, entails the reengineering or improvement of internal processes, calls for renewed collaboration with partners and involves investments. Quality assurance that aims to increase economic success and customer satisfaction usually is based on incremental innovations and continuous improvement with ordinary partners.

	Strategic management	*Quality assurance*
Organisational learning: innovative management	• Radical or incremental innovations	• Incremental innovations
Internal processes: products, services and renewed processes	• Reengineering or continuous improvement with partners	• Continuous improvement with partners
Finance: sustainable economic success	• Investments and increased economic success	• Increased economic success
Customers: satisfied students and employers	• Potential, occasional and key customers	• Occasional and key customers

Table 1. Integrated higher education management drives innovations, renewed processes, economic success and satisfied customers.

The concept of integrated higher education management can be extended with respect to the competitive strategies in higher education. Higher education institutions can be focused on specific customers and thereafter, the focus strategy can be divided into differentiation and cost effectiveness (Kettunen, 2002). Integrated higher education management also can be studied in different autonomous higher education institutions which have created their own quality assurance systems. The integration of various management approaches represents consistent and efficient management, which can be evaluated in the auditing.

Another possibility to extend the concept of integrated higher education management is to explore its meaning in the different subjects or fields of education. For example, the subject of design has a wide and narrow meaning. The wide meaning of the design of the product embraces, among other things, the planning of the structure and shape, the usage in the cultural context, and the relationship with the strategy and quality assurance. The wide meaning of the design is multidisciplinary, which is common in real life. The narrow meaning in industrial design means the art to provide the outside shape of a product. In the Swedish language, the word for design is *formgivning*, which means 'to give a form'. That interpretation represents the narrow connotation of design, which is more common in education than in industry.

9. Conclusion

Autonomous higher education institutions have the responsibility to build their own quality assurance systems. The purpose of a quality assurance system is to ensure that the strategic and operative objects of the institution can be achieved. Quality assurance is not the only management approach in higher education institutions, which typically have adopted many pedagogical approaches and management theories developed in private and public organisations. The results of this study show that the conceptual framework of integrated higher education management encompasses the necessary management approaches of higher education institutions.

The strategy and quality maps provide a conceptual framework and practical tools for higher education institutions to describe their strategic plan and quality assurance system in a graphical form. They help management communicate the future plans and ways to implement them to achieve the desired objectives. They make the management documents understandable, which increases the commitment of personnel, students and stakeholders to these plans. The conceptual tools also can be used in the external evaluations of the institution.

The weakness of a written and detailed strategy document is the weak commitment at the lower levels of an organisation. Therefore, it is important to engage personnel, students and other stakeholders in the strategy process. The Web-based strategy dialogue can be used to collect, evaluate and analyse the ideas of the stakeholders for a better future and increase the commitment to the strategic plan. It is important that the managers of organisational units analyse and discuss the results of the Web-based dialogue and implement the good ideas not only in the strategic plans but also in the annual action plans, curricula and other necessary documents. Even in a large higher education institution every individual can be involved in the strategy process.

Each higher education institution has its own profile and customers. The profile differentiates the institution, creates competitive advantage and increases the external impact. The profile of the Turku University of Applied Sciences was defined as innovation pedagogy in the strategy process. The breadth of expertise of the institution in applied research and development and education encompasses several focal areas which are planned based on the established knowledge of the faculties. The multidisciplinary faculties of the institution are able to support the creation of innovations and respond to the customer needs of the region with the networked applied research and development and the integrated education.

The internal benchmarking of degree programmes is the enhancement-led evaluation carried by the cross functional evaluation groups. The advantage of the cross evaluation is its comparison of learning methods and processes to promote innovative collaboration among the degree programmes and to support the profile of the institution to lead to the creation of innovations. The external evaluation of the centres of excellence is the procedure of the quality assurance agency to support the long-lasting educational development of the institution. In Finland, the external evaluation encompasses the evaluation of operations and outcomes of the organisational unit. The internal and external evaluations are important elements of the quality culture, which are connected to the profile and other pedagogical development of the institution.

Process management is an emerging management approach in higher education which attains more importance when the quality assurance systems mature over time. The processes and other elements of the quality assurance system must be maintained and improved in a systematic manner. The quality audits and the implementation of information systems are the main drivers to implement process management. The maintenance and internal audits of processes should be designed at the moment of accepting the process descriptions, but not later than after the external audit. Reconsideration and radical redesign of processes is the primary task of the process audit if the strategic objectives are not achieved. If no major improvements are found, the quality deviations must be reported.

10. References

Ansoff, I. (1984). *Implanting Strategic Management*, Prentice Hall International, ISBN 013451808X, Englewood Cliffs, USA.

Berlin Communiqué (2003). *Bologna Process Berlin 2003, Realising the European Higher Education Area*, Communiqué of the Conference of Ministers Responsible for Higher Education in Berlin on 19 September 2003.

Bessant, J.; Lamming, R., Noke, H. & Phillips, W. (2005). Managing Innovation beyond the Steady State. *Technovation*, Vol.25, No.12, pp. 1366–1376, ISSN 0166-4972.

Calabrese, R.L. (2003). The Ethical Imperative to Lead Change: Overcoming the Resistance to Change. *International Journal of Education Management*, Vol.17, No.3, pp. 7–13, ISSN 0951-354X.

Deming, W.E. (1986). *Out of the Crisis*, MIT, Center for Advanced Engineering Study, ISBN 0-262-54115-7, Cambridge, USA.

ENQA (2009). *Standards and Guidelines for Quality Assurance in the European Higher Education Area*, European Association for Quality Assurance in Higher Education, ISBN 952-5539-04-0, Helsinki, Finland, Retrieved from http://www.enqa.eu/files/ESG_3edition%20(2).pdf.

Escrig-Tena, A.B. (2004). TQM as a Competitive Factor, A Theoretical and Empirical Analysis. *International Journal of Quality & Reliability Management*, Vol.21, No.6, pp. 612–637, ISSN 0265-671X.

Ford, J.D. & Ford, L.W. (2002). Resistance and the Background Conversations of Change. *Journal of Organizational Change Management*, Vol.15, No.2, pp. 105–121, ISSN 0953-4814.

Hammer, M. & Champy, J. (1993). *Reengineering the Corporation: A Manifesto for Business Revolution*, Harper Collins Publishers, ISBN 0-06-662112-7, New York, USA.

Hargreaves, J. & Christou, A. (2002). An Institutional Perspective on QAA Subject Benchmarking. *Quality Assurance in Education*, Vol.10, No.3, pp. 187–191, ISSN 0968-4883.

Hintsanen, V., Luukka, M.R., Lounasmeri, T., Majander, M., Renvall, J., Holopainen, H. & Hiltunen, K. (2010). *Audit of the quality assurance system of Turku University of Applied Sciences*, Korkeakoulujen arviointineuvoston julkaisuja 2:2010, Tammerprint Oy, ISBN 978-952-206-134-8, Tampere, Finland, Retrieved from http://www.kka.fi/files/982/KKA_0210.pdf.

Ilmola, L. & Kotsalo-Mustonen, A. (2003). Filters in the Strategy Formulation Process, *Journal of Universal Computer Science*, Vol.9, No.6, pp. 481–490, ISSN 0948-695x.

Kaplan, R. & Norton, D. (1996). *The Balanced Scorecard: Translating Strategy into Action*, Harvard Business School Press, ISBN 0-87584-651-3, Boston, USA.

Kaplan, R. & Norton, D. (2004). *Strategy Maps: Converting Intangible Assets into Tangible Outcomes*, Harvard Business School Press, ISBN 1-59139-134-2, Boston, USA.

Kettunen, J. (2002). Competitive Strategies in Higher Education, *Journal of Institutional Research*, Vol.11, No.2, pp. 38–47. ISSN 1443–2110.

Kettunen, J. (2004). Bridge Building for the Future of the Finnish Polytechnics, *Journal of Higher Education Outreach and Engagement*, Vol.9, No.2, pp. 43–57, ISSN 1534-6102.

Kettunen, J. (2010a). Cross-evaluation of Degree Programmes in Higher Education, *Quality Assurance in Education*, Vol.18, No.1, pp. 34–46, ISSN 0968-4883.

Kettunen, J. (2010b). Strategy Process in Higher Education. *Journal of Institutional Research*, Vol. 15, No.1, pp. 16–27, ISSN 1443–2110.

Kettunen, J. (2011a). Evaluation of the Centres of Excellence in Higher Education. *Tertiary Education and Management*, Vol.17, No.2, pp. 151–161, ISSN 1358-3883.

Kettunen, J. (2011b). Innovation Pedagogy for Universities of Applied Sciences. *Creative Education*, Vol.2, No.1, pp. 56–62, ISSN 2151-4755.

Kettunen, J. (2011c). Strategy and Quality Maps in Higher Education. *US-China Education Review*, Vol.8, No.2, pp. 149–156, ISSN 1548-6613.

Kettunen, J. & Kantola, I. (2005). Management Information System Based on the Balanced Scorecard. *Campus-Wide Information Systems*, Vol.22, No.5, pp. 263–274, ISSN 1065-0741.

Kettunen, J. & Kantola, M. (2007). Strategic Planning and Quality Assurance in the Bologna Process. *Perspectives, Policy and Practice in Higher Education*, Vol.11, No.3, pp. 67–73, ISSN 1360-3108.

Korkeakoulujen arviointineuvosto (2010). *Korkeakoulujen laatujärjestelmien auditointikäsikirja vuosiksi 2011–2017*, Korkeakoulujen arviointineuvoston julkaisuja 16:2010, Tammerprint 0y, ISBN 978-952-206-159-1, Tampere, Finland, Retrieved from http://www.kka.fi/files/1201/KKA_1610.pdf.

Lee, C.C. & Yanc, J. (2000). Knowledge Value Chain. *Journal of Management Development*, Vol.19, No.9, pp. 783–794, ISSN 0262-1711.

Mehra, S. & Agrawal, S.P. (2003). Total Quality as a New Global Competitive Strategy. *International Journal of Quality & Reliability Management*, Vol.20, No. 9, pp. 1009–1024, ISSN 0265-671X.

Piaget, J. (1999). *The Psychology of Intelligence*, Routledge, ISBN 0415210062, London, UK.

Pollit, C.; Girre, X.; Lonsdale, J.; Mul, R.; Summa, H. & Waerness, M. (2002). *Performance or Compliance? Performance Audit and Public Management in Five Countries*, Oxford University Press, ISBN 0-19-829600-2, Oxford, UK.

Tague, N.R. (2005). *The Quality Toolbox*, ASQ Quality Press, ISBN 0-87389-639-4, Milwaukee, USA.

Tidd, J.; Bessant, J. & Pavitt, K. (2001). *Managing Innovation: Integrating Technological Market and Organizational Change*, Wiley, ISBN 9780470093269, Chicester, UK.

Vygotsky, L.S. (1978). *Mind in Society: The Development of Higher Psychological Processes*. In: M. Cole, V. John-Steiner, S. Scribner & E. Souberman (Eds.), Harvard University Press, ISBN 0-674-57629-2, Cambridge, USA.

Welch, J.F. & Dey, S. (2002). Quality Measurement and Quality Assurance in Higher Education. *Quality Assurance in Education*, Vol. 10, No.1, pp. 17–25, ISSN 0968-4883.

Yorke, M. (2002). Subject Benchmarking and the Assessment of Student Learning. *Quality Assurance in Education*, Vol. 10, No.2, pp. 155–171, ISSN 0968-4883.

12

Quality Assurance of Medicines in Practice

Beverley Glass[1] and Alison Haywood[2]
[1]James Cook University,
[2]Griffith University,
Australia

1. Introduction

The quality of medicines in relation to their stability has previously been seen to be entirely the responsibility of scientists within the pharmaceutical industry. Nowadays, however, since pharmacists have an important role to play in the delivery of safe, effective and quality medicines (drug products), they are increasingly being required to make decisions related to the stability of drugs in the day-to-day practice of both hospital and community pharmacy. These decisions range from deciding on whether a liquid dosage form, extemporaneously prepared from a commercial tablet or capsule is stable or whether it is appropriate to repackage a drug product into a dose administration aid (DAA). Pharmacists always consider the stability of the drug concerned and whether it might be preferable to crush a tablet or sprinkle the contents of a capsule over food or mix it in a drink, as is common practice in many nursing homes, when patients are unable to swallow. Pharmacists are also required to consider the implications of the transfer to a non-manufacturer's pack on the stability of a drug product. Despite the widespread use of DAAs, and the common practices of extemporaneous dispensing (compounding) and crushing tablets to be mixed with various liquid media, there is little available data regarding the drug stability following repackaging or alteration, involving compounding liquids or crushing of tablets.

This chapter examines the stability implications of extemporaneously prepared liquids, tablet crushing and repackaging of tablets and capsules into DAAs, by reviewing the literature and highlighting the research undertaken by the authors. Results from their published data have revealed stability concerns with only 7.2% of oral liquids prepared extemporaneously, primarily due to interactions between the drug and the excipients in the formulation. Further research has also provided evidence on the stability of medicines commonly repacked in DAAs to support pharmacists in making appropriate clinical and operational decisions regarding this repackaging process.

These findings will confirm the importance of knowledge of drug stability for the practising pharmacist, who is involved with modifying or altering drug products, whether repackaging into DAAs, compounding oral liquids or advising on the suitability of crushing tablets. Guidelines on appropriate practice will be provided, in addition to highlighting challenges facing the pharmacist in continuing to deliver safe, effective and quality medicines.

2. Dose administration aids

Populations are increasing in age worldwide as are the number of medicines prescribed, with a number of these patients receiving their medication in DAAs due to the benefits in terms of health outcomes and cost of health care. Despite the widespread use of these devices, there is little available data on the quality implications, based on the stability of these drug products when repackaged into such devices (Walker, 1992; Ware, Holford et al., 1994; Nunney & Rayner, 2001; Church & Smith, 2006). Repackaging of a medicine, involving removal from its primary packaging, invalidates the stability guarantee of the manufacturer. In fact, drug manufacturers on the whole, tend to discourage repackaging of medicines and as there is little quality data available to support this process. It is thus the role of the health care team to ensure optimal patient care by making an informed judgment as to the effect on the quality and safety of this repackaging process.

2.1 Compliance aids

Compliance aids, also referred to as Multi-Compartment Compliance Aids (MCCAs) or DAAs (Figure 1), are devices which have been developed to assist patients in managing their medicines by arranging individual doses according to their prescribed dose schedule throughout the day.

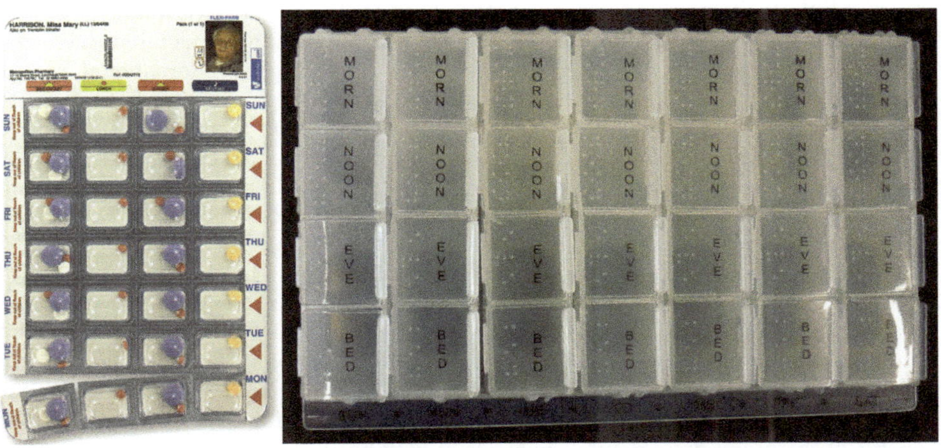

Fig. 1. Examples of Compliance Aids: Webstercare Cold Seal Flexi-Pak® (left), dosette box (right).

These aids have been used to facilitate medication administration to patients for over 30 years and their wider application has been strengthened by various government programs worldwide to facilitate their use (Llewelyn, Mangan et al., 2010). A community pharmacy contractual framework in the UK now places emphasis on assessing and providing practical compliance aids to all patients who fall within the protection of the Disability Discrimination Act 1995 and need help with medicine taking (Chan, Swinden et al., 2007). The Australian Government Department of Health and Ageing has funded a number of professional programs and services under the Better Community Health Initiative of the 4th and 5th Community Pharmacy Agreements, including the DAA Program, where the dollar value was increased from nearly $73 million to $132 million (Australian_Government, 2010). The objective of this program is to identify patients who will derive the most benefit from the supply of a DAA, to develop a sustainable service and payment model capable of meeting the program's aim and to trial the broader use of these aids within the community setting. The provision of a DAA service through community pharmacies is expected to reduce medication related hospitalisation and adverse events through improved medication management. This should result in improved quality of life and health status for patients, and have flow on benefits for the community by reducing the demand on aged care facilities and costs associated with adverse reactions to medication mismanagement.

2.2 Stability of repackaged medicines

The stability of a pharmaceutical product may be defined as the capability of a particular medicine, in a specified container, to remain within its physical, chemical, microbiological, therapeutic and toxicological specifications. The shelf-life of a drug product may be affected by the intrinsic stability of the active pharmaceutical ingredient (API) and the excipients, the potential interactions between them, the manufacturing process, the dosage form or drug product, and the packaging and environmental conditions encountered during their transport, storage and use (Aulton, 2007).

Pharmaceuticals are expected to meet specifications for identity, purity, quality and strength throughout their defined storage period. The stability of manufactured medicines is routinely confirmed by the drug manufacturers according to international regulatory requirements (ICH, 2003), where stability studies on packaged medicines are conducted at real-time long-term and accelerated conditions at specific temperatures and relative humidity (RH). This represents storage conditions experienced in the distribution chain of the climatic zone(s) of the country or region of the world concerned. Manufacturers' packaging is designed to protect drug products from environmental factors encountered during storage, such as light, air (oxygen, carbon dioxide and other gases), and moisture, while limiting interactions between the product and the packaging material. However, this does not guarantee the stability of the API and the drug product on removal and repackaging into a DAA. A recent survey of 392 repackaged products revealed that, although some information regarding the potential stability of solid dosage forms in DAAs can be obtained from manufacturers, there is still a lack of short-term stability data for the transfer of drug products into these devices (Church & Smith, 2006). Thus, although the benefits of the use of these devices, in terms of both health outcomes and cost of healthcare have been reported (Simmons, Upjohn et al., 2000; Lee, Grace et al., 2006), there is little available data regarding the shelf-life of drug products when repackaged. The fact that there is little available stability data is significant considering the value of these aids, the investment in dollars and the extent to which they are being used to aid adherence.

In electing to repackage a drug product into a DAA, healthcare professionals must consider the implications on drug stability of the transfer to a non-manufacturer pack. Pharmacists thus rely largely on individual drug storage recommendations for the medicine in the manufactures packaging, general guidelines for repackaging, e.g. in Australia (PSA, 2007), the UK (RPSGB, 2003) and USA (USP, 2010), and their basic understanding of inherent drug stability to make case-by-case recommendations as to whether repackaging is appropriate. A small number of medicines have been investigated for their stability following repackaging into DAAs, namely aspirin, atenolol, clozapine, frusemide, paracetamol, prochlorperazine, and sodium valproate (Table 1). These studies have contributed significantly to the body of evidence available on repackaging into DAAs and are detailed below. These drug candidates were chosen for study for a number of reasons, including that they were commonly repackaged into DAAs, because of their proven susceptibility to various environmental conditions and/ or because of anecdotal evidence from the practice of problems encountered when these drugs are repackaged either from a patient or health professional perspective.

Active Ingredient	Stability considerations	Literature reference
Aspirin	Moisture sensitive	(Haywood, Llewelyn et al., 2011)
Atenolol	Light sensitive	(Chan, Swinden et al., 2007)
Clozapine	Oxygen sensitive	(Perks, Robertson et al., 2011)
Frusemide	Light sensitive	(Bowen, Mangan et al., 2007)
Paracetamol	Moisture sensitive	(Haywood, Mangan et al., 2006)
Prochlorperazine	Light sensitive	(Glass, Mangan et al., 2009)
Sodium valproate	Hygroscopic	(Llewelyn, Mangan et al., 2010)

Table 1. Drug candidates investigated for repackaging into DAAs.

2.3 Repackaging moisture-sensitive medicines

Three studies (Haywood, Mangan et al., 2006; Llewelyn, Mangan et al., 2010; Haywood, Llewelyn et al., 2011) have reported on the stability considerations of repackaging medicines that are hygroscopic or sensitive to moisture. The medicines were exposed to controlled room temperature conditions and accelerated conditions (elevated temperature and relative humidity). The findings of all studies emphasize the importance of ensuring that these medicines are stored in appropriate well-sealed devices. While well-sealed devices, such as the WebsterPak® are commonly utilised in community and hospital practice, it is not known which devices are used by patients in their homes, since many of these devices that do not have adequate air-tight seals, such as the dosette box (Figure 1), are available in supermarkets and general stores. One of the studies also investigated whether splitting tablets, as is commonly undertaken by patients on low-dose Aspirin, had any adverse effects on the stability and quality of these tablets, when repackaged.

2.3.1 Paracetamol

Haywood and colleagues (Haywood, Mangan et al., 2006) examined the stability implications of repackaging commonly used 500 mg paracetamol tablets (Panamax, Sanofi Synthelabo) in a DAA frequently employed in practice (Multidose WebsterPak®). Paracetamol has the potential to undergo hydrolysis and therefore requires protection from moisture. The samples in this study were stored under controlled long-term (25°C; 60% RH) and accelerated (40°C; 75% RH) conditions as per the ICH (International Conference on Harmonisation of Technical Requirements for Registration of Pharmaceuticals for Human Use) guidelines for a period of three months. Physical characteristics of the tablets, including weight uniformity, physical appearance, hardness, friability, disintegration and dissolution were evaluated at time = 0, directly after heat sealing, 1 month and 3 months. Chemical stability was confirmed by a validated high performance liquid chromatography (HPLC) method. The results were compared to control samples stored in the original packaging.

The results of the study were very favourable for repackaging paracetamol, indicating that all requirements for the physicochemical stability were met with the paracetamol content within the required British Pharmacopoeial (BP) range of 95–105% of the labelled amount, for all storage conditions, even those conditions of high humidity. The results suggest that paracetamol tablets repackaged into a DAA offering sufficient protection from moisture would remain stable for a reasonable in-use period of approximately six weeks (i.e. allowing two weeks for advanced packing and delivery on a four-week supply).

2.3.2 Aspirin

Aspirin (acetylsalicylic acid) in low-doses has been increasingly prescribed for primary prevention of stroke and acute myocardial infarction in the elderly. A recent study has indicated that low-dose aspirin is also a cost-effective option in primary prevention and the majority of healthcare systems are more than willing to pay for any additional Quality-Adjusted Life Years (QALYs) gained (Annemans, Wittrup-Jensen et al., 2010). It is often more cost-effective for patients to purchase standard dose (300 mg) aspirin and to cut the tablet in half to achieve a 'low-dose' equivalent. Tablet splitting or dividing has been an accepted practice for many years as a means of obtaining the prescribed dose of a medication and for cost-saving purposes (Marriott & Nation, 2002). However, the storage of split tablets is not well discussed in the literature and anecdotal evidence suggests that especially elderly patients or their carers split tablets in advance and then store the split tablets in bottles that previously contained the same medication, a different medication or some other substance, or in a compliance aid such as a dosette box (Marriott & Nation, 2002). Due to stability concerns, patients and carers are usually advised that when half a tablet is taken, the unused half should be immediately discarded, particularly with medicines that are known to be unstable when exposed to light and air.

Considering the above practices and the fact that acetylsalicylic acid is rapidly hydrolysed to salicylic acid on exposure to moisture, a recent study assessed the stability of 300 mg aspirin tablets (Solprin Dispersible Tablets, Reckitt Benckiser) when repackaged in a dosette box obtained from a local supermarket and stored under a number of in-use conditions, both as a whole and split tablet (Haywood, Llewelyn et al., 2011). This is an important consideration as it is a decision to be made in practice by pharmacists or at home by patients and carers. The acetylsalicylic acid content remained within BP specifications (95–105% of the labelled amount) for all except the accelerated storage conditions, with 93.4% of the drug remaining in this case, and the salicylic acid content at 0.04% (BP limit 0.0006%). The split tablets did not display any additional degradation of the acetylsalicylic acid or increased amount of the degradant salicylic acid, when compared to the whole tablets under the same conditions. While the acetylsalicylic acid content remained within specifications under standard room conditions, it must be noted that the limit for the degradant was exceeded under all conditions. Additionally there was some colouration and disintegration of the tablets, thus compromising the quality of this medication and suggesting that this practice of repackaging by patients is inappropriate. This study in fact does confirm the instructions given that these dispersible tablets should not be removed from their original packaging and that if tablets are split, that the remaining half should be discarded.

2.3.3 Sodium valproate

Llewelyn and colleagues (Llewelyn, Mangan et al., 2010) undertook a study on the stability of 100 mg sodium valproate tablets (Epilim, Sanofi-Aventis) repackaged into DAAs (WebsterPak®) and stored under various temperature and humidity conditions. Sodium valproate, a commonly used antiepileptic is unstable in the presence of moisture due to its hygroscopic and deliquescent nature. The results revealed that while the sodium valproate content in the tablets remained within an acceptable range (90–110% of the labelled amount) under all storage conditions for eight weeks, the physical stability was not maintained, with unacceptable weight variation in the tablets, changes in their dissolution profiles and significant changes in their appearance, under accelerated conditions, due to the hygroscopicity of the API, even after only three weeks of storage (Figure 2) (Llewelyn, Mangan et al., 2010).

Fig. 2. Sodium valproate tablets after 21 days of storage at accelerated (40°C; 75% RH) (left), refrigerated (2–8°C) (middle) and controlled room temperature (25°C) (right) conditions.

The results of this study highlight the fact that accelerated conditions of temperature and humidity should be taken into account, and that cognisance is taken from the fact that in different countries, and in fact within the same country, the climatic conditions might vary considerably. Medicines therefore may be appropriately repackaged in, for example, a temperate region such as London, Los Angeles or Sydney, but repackaging that same medicine in tropical or desert regions such as Darwin or Dubai may be inappropriate due to increased temperature, humidity and light conditions. These findings are thus significant because these accelerated conditions are not uncommonly encountered in northern Australia and other tropical regions worldwide.

2.4 Repackaging light-sensitive medicines

Three studies (Bowen, Mangan et al., 2007; Chan, Swinden et al., 2007; Glass, Mangan et al., 2009) have reported on the stability considerations of repackaging medicines that are sensitive to light. The medicines were exposed to controlled room temperature conditions (25°C; 60% RH) and accelerated conditions (40°C; 75% RH) and two of the studies also included those light conditions specified by the ICH (ICH, 1996).

2.4.1 Atenolol

Atenolol is reported to be photoreactive when exposed to UVA-UVB radiation with photodegradation increasing with a decrease in the pH value (Aryal & Skalko-Basnet, 2008).

The main photodegradation product at pH 7.4 has been identified as 2-(4-hydroxyphenyl) acetamide (Andrisano, Gotti et al., 1999) (Figure 3).

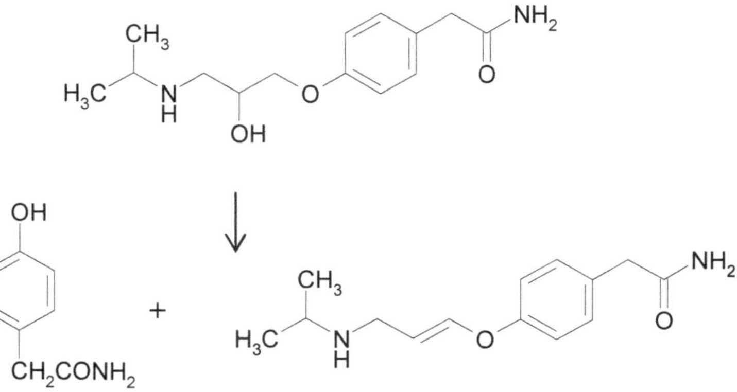

Fig. 3. Photodegradation of atenolol

Chan and colleagues (Chan, Swinden et al., 2007) reported on the stability of atenolol 100 mg tablets in a 28-chamber compliance aid with transparent lids, stored for four weeks at room temperature (25°C; 60% RH) and accelerated (40°C; 75% RH) conditions. Tablets were also stored at room temperature in their original packaging and in Petri dishes. All tablets stored at room temperature conditions passed the necessary physical tests (weight uniformity, physical appearance, hardness, friability, disintegration and dissolution) and demonstrated chemical stability, however the tablets stored at accelerated conditions although chemically stable, showed an increase in disintegration time and decreased dissolution time when compared to the tablets stored in the original blister packaging. A limitation of the study was that only one sampling time point (Week 4) was used with comparison of the results to the tablets stored in the original packaging, as opposed to observing changes from an analytical determination at the outset (time = 0). Further, the brand of the compliance aid or the degree of protection afforded against air, moisture and light is unknown (Glass, Haywood et al., 2009).

2.4.2 Frusemide

Frusemide, a light sensitive drug commonly used in the treatment of cardiac failure and hypertension is often repacked into a DAA. A study by Bowen and colleagues (Bowen, Mangan et al., 2007) reported on the in-use stability implications of repackaging 40 mg frusemide tablets (Uremide, Alphapharm). In addition to storing the medicines under controlled room temperature conditions (25°C; 60% RH), the DAAs (WebsterPak®) were stored for eight weeks under conditions that reflected a 'home-environment' where DAAs were stored blister-side up in a bathroom exposed to a standard 60W tungsten bulb and indoor indirect daylight/ window- filtered sunlight; and a 'pharmacy-environment' where DAAs were stored blister-side up on a bench top exposed to fluorescent lighting and indoor indirect daylight. Photostability studies were also performed according to the ICH guidelines (ICH, 1996). The results confirmed the drug content to be within the BP range of 95–105% for all storage conditions, including the ICH light conditions. Although the physical stability was confirmed by all tests, including weight uniformity, hardness, friability, disintegration, dissolution, under both controlled and in-use conditions, the exposure to light in the pharmacy and under ICH conditions, even after one week, resulted in a yellow colouration of the tablets. The progressive yellow discolouration (Figure 4) (Bowen, Mangan et al., 2007) of the tablets over an eight week storage period was attributed to exposure to fluorescent lighting, which was not encountered under the 'home-conditions'. Although the colour change was noted as having negligible effects on the drug content and other physical parameters, such as dissolution of the tablets, it was considered an unacceptable change, since patients are likely to be concerned about a possible compromise in the quality of the medicine, and this may have an adverse effect on patient acceptance and hence adherence.

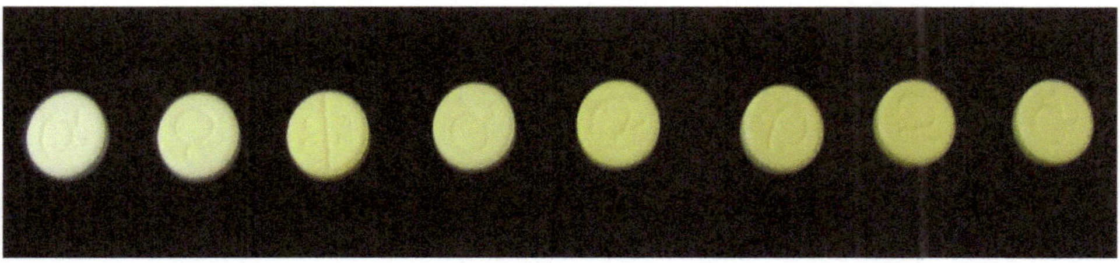

Fig. 4. Progression of colouration of frusemide tablets exposed to fluorescent lighting over an eight week period from left (week 1) to right (week 8)

2.4.3 Prochlorperazine

Prochlorperazine, a phenothiazine drug, widely used as an anti-emetic is susceptible to oxidation to the sulphoxide (a metabolite and a photodegradant) under the influence of light (Figure 5). The main metabolites and degradants of all phenothiazines have been found to be inactive at the dopamine receptors (Nejmeh & Pilpel, 1978; Moore & Tamat, 1980) and prochlorperazine is reported to cause photosensitivity effects in patients. A study by Glass and colleagues (Glass, Mangan et al., 2009) reported on the physicochemical stability of 5 mg prochlorperazine tablets (Stemetil, Sanofi-Aventis) repackaged into a DAA (WebsterPak®). In addition to the controlled room temperature (25°C; 60% RH) and accelerated (40°C; 75% RH) storage conditions, tablets were also exposed to an in-use condition by placing the repackaged DAAs blister-side up on a bench top exposed to fluorescent lighting and indoor indirect daylight. The results showed chemical and physical stability to be within BP limits (drug content within 95–105%) however, there were noticeable organoleptic changes in the tablets stored under the in-use conditions, with a progressive grey discolouration over an eight week period, starting in week 2. The discolouration and the potential for the resulting photodegradants to cause adverse effects in patients suggest that the quality of this medicine had been compromised.

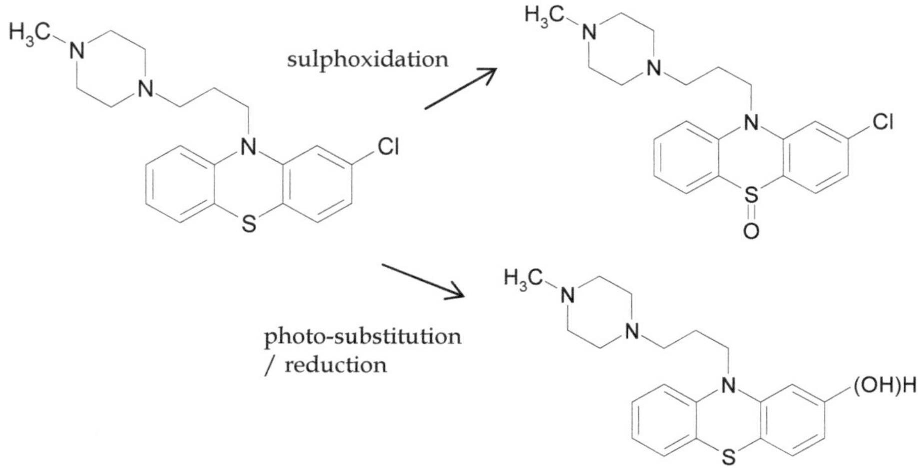

Fig. 5. Photodegradation of prochlorperazine

2.5 Repackaging medicines sensitive to oxidation

Only one study (Perks, Robertson et al., 2011) has reported on the stability of repackaging medicines sensitive to oxidation, with preliminary findings suggesting that clozapine undergoes a photo-oxidative process, since the coloration is not apparent during the exposure period in the absence of light.

2.5.1 Clozapine

Clozapine is an atypical antipsychotic used in the treatment of schizophrenia. A study by Perks and colleagues (Perks, Robertson et al., 2011) was prompted due to anecdotal reports from hospital pharmacy practice about discolouration of returned clozapine tablets that were repackaged into DAAs, and the known susceptibility of clozapine to oxidation (Kohara, Koyama et al., 2002). The study evaluated both the chemical content and physical stability of clozapine tablets (Clopine, Hospira, and Clozaril, Novartis) repackaged into a DAA (WebsterPak®) over a six week period. The DAA's were stored under two environmental conditions, namely accelerated conditions (40°C; 75% RH), and under a simulated controlled room temperature condition (25°C; 60% RH) by placing the DAA with the blister-side facing up on a windowsill in the laboratory, with exposure to both window-filtered sunlight and fluorescent light. The results were compared to control samples stored in the original packaging under the above conditions. The study also investigated the stability of the tablets exposed to ICH light conditions (ICH, 1996). Although the physical stability, namely weight uniformity, hardness, friability, disintegration and dissolution, was confirmed for all tests at room temperature, under accelerated conditions, the disintegration test did not meet the BP requirements. However, the subsequent dissolution test was successful with 85% of clozapine dissolving in 45 minutes. The chemical stability was confirmed for all storage conditions, including under ICH conditions (for light), with clozapine content within the BP range of 90–110%.

Based on the susceptibility to oxidation, and in order to reproduce the colouration noted in the practice of repackaged samples returned by patients, a sample was removed from the original manufacturers packaging and exposed to light and air on the bench beside the simulated controlled room temperature condition samples (Figure 6) (Perks, Robertson et al., 2011). The results for the exposed tablets confirmed the discolouration referred to in the anecdotal reports from practice. These findings show that clozapine, when correctly repackaged, maintained its physical and chemical stability for six weeks. It is therefore assumed that these reports of tablet discoloration were as a result of improper handling of these DAAs by patients. These findings highlight the importance of the role of the pharmacist in providing patient care and advising on the correct handling and storage of their DAAs.

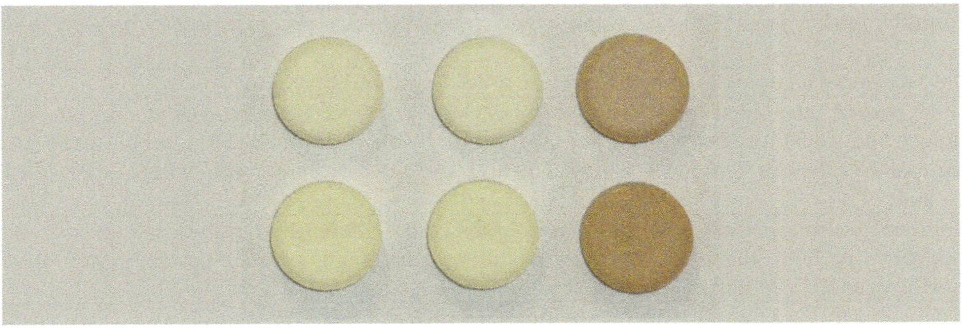

Fig. 6. Photographic comparison of the colour of clozapine tablets from left to right: control sample, sample stored in a DAA under simulated controlled room temperature conditions after 6 weeks, and a sample left out of the DAA packaging on the windowsill after 6 weeks (Clopine® above and Clozaril® below)

2.6 General guidelines for repackaging medicines

Various published guidelines, such as the Pharmaceutical Society of Australia (PSA) Professional Practice Standards (PSA, 2010) and Dose Administration Aids Service Guidelines and Standards for Pharmacists (PSA, 2007) have provided general guidance on the suitability of repackaging of solid dosage (tablets and capsules) forms into DAAs and are summarised in Table 2.

A flow chart for the quality management of DAAs in practice is shown in Figure 7 (Haywood, Llewelyn et al., 2011).

The following practical recommendations for ensuring the quality and stability of medicines repackaged into DAAs arising from the above studies include: (i) selecting an appropriate brand of DAA for repackaging medicines that affords appropriate protection against air and moisture; (ii) protecting the DAA containing drugs susceptible to photodegradation from light in the pharmacy and in patients' homes, achieved by either storing the DAA protected from light and/or placing the DAA into a light-protecting sleeve (e.g. foil, cardboard); (iii) careful removal of tablets to prevent accidental rupture of adjacent blisters, thus exposing tablets to air and moisture; (iv) monitoring the DAA integrity during repackaging, dispensing and throughout the in-use period; (v) consideration of an appropriate location to store the DAA to avoid unnecessary exposure to light, heat, humidity and away from children, if the device is not child-resistant; (vi) counselling patients on correct use and appropriate storage locations for their DAA; and (vii) following appropriate professional practice guidelines (Haywood, Llewelyn et al., 2011).

Medicines generally unsuitable for packing into DAAs include effervescent, dispersible, buccal and sublingual tablets and significantly hygroscopic preparations.
Medicines administered on an 'as required' basis are generally unsuitable for packing into DAAs, since they may be taken unnecessarily on a regular basis or removed for use at an earlier or later stage, thus exposing the remaining contents to the environment.
Cytotoxic preparations or other medicines posing occupational health and safety risks are generally inappropriate, however the risk-benefit of packing must be considered; for example, packing may be appropriate where non-adherence is considered to be a greater risk.
Only devices that are well sealed and tamper evident should be used.
The length of time taken for the end-to-end packing process should be kept to a minimum; tablets and capsules should be removed from the manufacturer's foil or blister pack immediately before the DAA is packed, and the DAA sealed immediately after it is packed.
Any heat sealing methods should be used quickly and efficiently to minimise exposure of medicines to heat, the likely storage conditions (e.g. exposure to heat, humidity, and moisture) and the length of time the DAA will be in transit.
It is useful to maintain a list of medicines / medicine types that should not be removed from their original pack for repacking in a DAA.

Table 2. Summary of PSA guidelines relating to the suitability of repackaging into DAAs

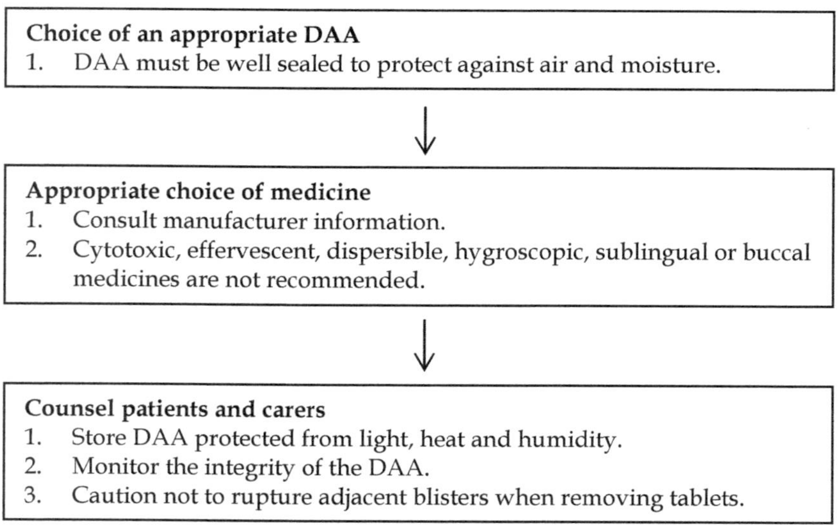

Fig. 7. DAAs – Assuring quality in practice.

3. Crushing tablets

The world's population is ageing and many patients suffer from dysphagia either as a consequence of disease or as part of the ageing process. Elderly patients are more prone to diseases linked to dysphagia, such as Parkinson's and Alzheimer's disease, other dementias, stroke and cancer (Morris, 2005). Further, several age-related changes may contribute to dysphagia, including a decline in salivary gland function resulting in xerostomia, and changes in the sensory function of the aerodigestive tract such as a deterioration of the pharyngo-upper esophageal sphincter (UES) and laryngo-UES contractile reflexes (Kawamura, Easterling et al., 2004). Polypharmacy, involving the use of multiple medicines at the same time, is common among older patients. Dysphagia therefore poses significant challenges for medicine administration, since most adult medicines are available as mainly solid dosage forms (i.e. tablets or capsules).

Tablets and capsules are commonly modified or altered for ease of administration, not only in traditional care settings such as residential aged care facilities and in hospital wards, but also by patients experiencing swallowing difficulties at home. Opening a capsule or crushing a tablet before administration will in most cases render its use to be 'unlicensed'. In addition to the legal implications involved in modifying medicines, there are important safety, stability and bioavailability considerations. Pharmacists thus have a role to play in advising not only on safe medicine use but on any stability issues which may occur on modifying oral solid dosage forms and thus contribute to improving the quality use of medicines in older patients.

3.1 Medicines that should not be crushed

There are certain tablets and capsules that should not be modified. The potential risks associated with crushing tablets and opening capsules and the impact on the API is summarised in Table 3 (Haywood & Glass, 2009). Important considerations include (Haywood & Glass, 2009):

- Controlled or sustained release medicines are designed to release the active ingredient in a controlled manner over a defined dosing period. Crushing or splitting these drug products will interfere with the release characteristics and usually cause an unintended large bolus dose resulting in toxicity and increased incidence of adverse effects. For example, crushing sustained release verapamil (e.g. Veracaps SR®) will result in an increased risk of hypotension and bradycardia.
- Enteric coating is used to protect the active ingredient from the acidic environment of the stomach and film coating affords protection from light. Tablets, capsules and capsule contents such as pellets can contain these coatings. For example, nifedipine (e.g. Adalat®) is extremely susceptible to light and even brief exposure to ambient light will result in degradation of the API. Omeprazole (e.g. Losec®) is degraded by exposure to acid and whilst the tablets may be dispersed in yoghurt or orange juice, crushing such tablets will expose the API to acid in the stomach resulting in a loss of activity before the site of absorption in the upper small intestine.
- Delayed release medicines are designed to release the API at a defined site in the gastrointestinal tract. For example, mesalazine (e.g. Mesasal®) is formulated as a resin coated tablet designed to dissolve and release medication in the lower small intestine, where it exerts a local anti-inflammatory effect.

- Enteric coated tablets, for example Cartia®, are designed to minimise the irritant effect of aspirin. If these tablets are crushed, they may be irritant to the oesophagus and the stomach (upper gastrointestinal tract).
- Sugar or film coats are designed to disguise APIs with an unacceptable taste. For example, quinine (e.g. Quinate®) is coated to disguise the bitter taste. Crushing these tablets, although not altering their effectiveness, may compromise patient adherence.
- There are occupational health and safety (OH&S) concerns with not only cytotoxic medicines, but also teratogens (e.g. isotretinoin) and those medicines with the potential to cause contact dermatitis (e.g. chlorpromazine). It is important to remember that the clinical indications for the use of some cytotoxics extend beyond the treatment of malignancy, for example, cyclophosphamide (e.g. Cycloblastin®) and methotrexate (e.g. Methoblastin®) are used to treat some inflammatory disorders. Health care workers and carers need to take appropriate measures to avoid exposure to the powder from these crushed tablets.

Dosage forms that should not be crushed	Potential risks
Extended/ sustained-release product	Increased toxicity/ adverse effects
Enteric coat protecting an acid-labile API	Decreased stability/ efficacy
Film coat protecting a light-sensitive API	Decreased stability/ efficacy
Delayed-release coat designed to release API at a defined site in the GIT	Decreased efficacy
Enteric coat protecting the upper GIT from the API	Increased local irritant effect
Sugar/ film coat disguising a poor tasting API	Unacceptable taste/ poor compliance
OH&S of cytotoxic/ teratogenic APIs	Potential hazards to health care workers and careres

Table 3. Risks associated with modifying tablets and capsules

Information on the appropriateness of modifying oral solid dosage forms is available through resources such as the approved product information and the CMI (Consumer Medicine Information) of the medicine. Compiled lists of medicines that should not be crushed or modified are available, such as the Handbook of Drug Administration via Enteral Feeding Tubes (White & Bradnam, 2007), the Australian Pharmaceutical Advisory Council (APAC) Guidelines for medication management in residential aged care facilities (Appendix F) (APAC, 2002), the Australian Medicines Handbook (AMH) Aged Care Companion (AMH, 2006) and electronic resources such as the list provided by the Institute for Safe Medication Practices (Mitchell, 2011). Many hospitals and aged care facilities have also developed their own lists. These lists, while providing a useful starting point, should not be relied upon as being all inclusive.

3.2 General guidelines for safe medicine use when modifying medicines

Pharmacists are able to play an important role in educating patients, carers and other health care professionals concerning the safe and effective administration of medicines. There are reported cases of fatalities that have occurred, not only in traditional care settings, for example where a patient was administered crushed controlled release tablets via a nasogatric tube (Schier, Howland et al., 2003), but also in the patients' home, where a patient chewed extended release diltiazem capsules, since they were too large to swallow (Ballard, 1996). This highlights the importance of patient counselling in a community setting, since pharmacists may be unaware of the medication administration techniques and practices adopted by patients at home.

The APAC guidelines for altering medicines (APAC, 2002) provide a six-step process, summarised in Figure 8 (Haywood & Glass, 2009), to ensure that patients receive the desired therapeutic response from their medicines.

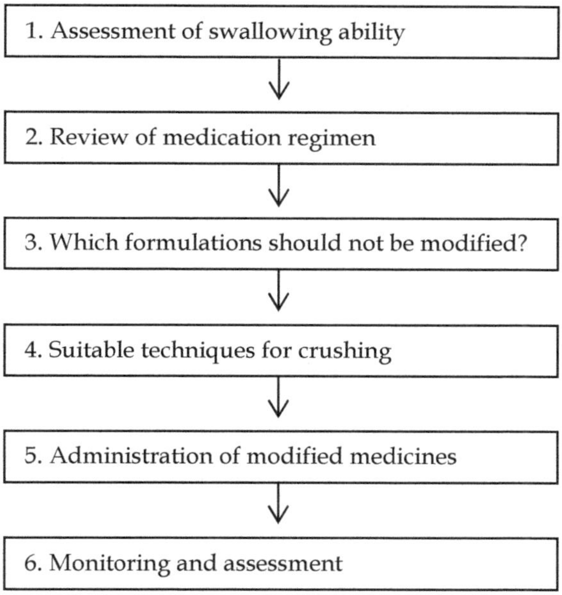

Fig. 8. Six-step process to ensure desired therapeutic response

3.2.1 Medication management, monitoring and assessment

There are many important considerations when administering modified medicines, either by the patient themselves at home, or with the assistance of carers or health care professionals in hospitals and aged care facilities.

A patient's swallowing ability should be carefully assessed since their ability to swallow may be a transient disability or may vary during the day. Changing the dosing time and renewed attempts to encourage taking unaltered dosage forms should be the first option, whenever possible. Instinctively, most patients will tip their head backwards when swallowing a tablet, but this actually narrows the oesophagus, making swallowing more difficult. Patients should tilt their head down rather than back as this will widen the oesophagus and may remedy difficulties in taking tablets. The use of liquid 'chasers' or swallowing tablets 'chased' by a few bites of a well chewed banana or soft food (e.g. bread) may be of assistance in some cases. Wherever possible patients should be upright, or as close as practically possible to upright when taking oral medicines. Patients who are unable to swallow may be able to take medicines that are not swallowed but are absorbed sublingually (e.g. glyceryl trinitrate tablets) or sucked (e.g. amphotericin lozenges) (Haywood & Glass, 2009).

Difficulties seen in swallowing oral medicines should provide a stimulus for a medication review. Various options should be considered prior to modifying a medicine, including: changing to a different formulation of the same medicine (e.g. oral liquids/drops, dispersible tablets or transdermal patches); seeking a therapeutic alternative in an appropriate dosage form; or stopping medicines that are no longer necessary. In cases where a different formulation or therapeutic alternative is not available, and a sustained-release product is prescribed, it may be necessary to administer an immediate-release preparation more often, for example immediate release verapamil tablets are usually given three times a day as opposed to once daily administration for the slow-release (SR) preparation, since these preparations can usually be crushed and also more cost-effective. Monitoring and assessing the therapeutic response is especially important when medicines are modified. Adverse effects or lack of expected effects should trigger a review of this practice (Haywood & Glass, 2009).

3.2.2 Techniques for crushing tablets

Appropriate equipment that is non porous and permits complete and reproducible recovery of powdered material should be used, such as a mortar and pestle or appropriate commercially available tablet crusher, to avoid dose inaccuracies. Equipment should be washed and dried between uses to prevent cross contamination. Cytotoxic medicines should have a dedicated set of equipment and special procedures must be adopted to minimise occupational exposure, for example a mortar and pestle enclosed in a transparent plastic bag with the operator wearing a mask and gloves (Haywood & Glass, 2009). Mixing powdered material with a small amount of food that the patient likes (e.g. jams, fruit purees) is sensible as it disguises unpleasant taste and aids in compliance. The crushed tablets or capsule contents should be given to the patient as soon as possible after crushing and mixing with any food or liquid, as this will minimise both the risk of medication degradation and, in an aged care setting, inadvertent administration to the wrong resident. Crushed tablets or contents of capsules should not be sprinkled onto meals where portions of the meal may be left uneaten, resulting in under dosing. Further practical guidelines to assist in safely modifying medicines are described by Haywood *et al* (Haywood & Glass, 2009).

3.2.3 Counseling and continuing education

It is important for pharmacists to develop and maintain a current list that is specific to their practice setting and is readily accessible to staff involved in patient counselling or administration of modified medicines. However, with the frequent addition of new medicines to the market and the challenge of maintaining an all-inclusive 'list', it is more important to educate patients, carers, staff and other health care professionals about safe medicine use and to provide them with a background concerning the safety and stability considerations of modifying medicines to ensure the quality use of medicines so that patients receive the best possible therapeutic outcome (Haywood & Glass, 2009).

4. Oral liquids in practice

The pharmacist, both in community and hospital pharmacy practice, is often challenged with the preparation of a liquid dosage form not available commercially for paediatric patients, those adults unable to swallow tablets or capsules, patients who must receive medications via nasogastric or gastrostomy tubes, and patients who require non-standard doses that are more easily and accurately measured by using a liquid formulation. It is common practice for these liquid dosage forms to be prepared from a commercially available solid dosage form such as a tablet/capsule. Although a number of parameters need to be considered in the formulation of a stable liquid dosage form, there is limited formulation and stability data available (Glass & Haywood, 2006).

4.1 Compounding pediatric mixtures

Children often require titratable individualised doses in milligrams per kilogram of body weight and most children under six years of age cannot swallow tablets (Nahata & Morosco, 2003). There are a limited number of suitable dosage forms commercially available for children due to the size of the market and resulting lack of financial viability for the pharmaceutical industry of these liquid dosage forms. In addition, there are complexities associated with the formulation of liquid dosage forms due to various physicochemical factors. The preparation of extemporaneous oral liquids, by pharmacists, recently reviewed by Nahata (Nahata & Allen, 2008) and Giam (Giam & McLachlan, 2008), therefore provides an appropriate solution to this problem. The following sections provide information on the preparation of safe, stable oral liquids for children and adults in those cases where suitable dosage forms are not available.

4.2 Physicochemical stability considerations in compounding oral liquids

Drug stability encompasses chemical, physical, microbiological, therapeutic and toxicological stability not only of the drug substance, but when taking account of the excipients, also the drug product. The extemporaneous preparation of oral liquids can be complex due to the addition of excipients to improve compliance or the stability of the final product. Further, if oral liquids are prepared from a commercially available solid dosage form such as a tablet or capsule, there may be potential interactions between the drug and the excipients (in the commercial product) in the prepared oral liquid. Such interactions may involve the vehicle, preservative, buffering agent, flavouring agent, wetting and suspending agent, viscosity enhancer or even the storage container. However, a review in 2006 of 83 oral liquids extemporaneously prepared by modifying an existing commercial dosage form revealed that only 7.2% of those compounded oral liquids exhibited stability concerns (Glass & Haywood, 2006). This review is a useful comprehensive summary of liquid dosage forms prepared from commercially available tablets or capsules and illustrates the low risk associated with these products, if cognisance is taken not only of the API but all those excipients present in the commercial dosage form, used in their preparation.

4.2.1 Active ingredient and excipients

In most practice settings, sourcing the API as a powdered raw material is not always practical and thus commercially available tablets and capsules are often used in compounding oral liquids. These solid dosage forms however contain many excipients, which although compatible in the solid state have the potential to interact both in solution or suspension, adversely affecting the potency of the API and thus impacting on the shelf-life of the compounded oral liquid. There are also certain tablets that should not be crushed and these are detailed in section 3 above. Excipients used in extemporaneous oral liquids commonly include suspending agents or viscosity enhancers, sweeteners and preservatives. They may also contain flavours, colours and buffers. It is not always necessary to colour a product. If a colouring agent is used, it should match the flavour (e.g. red for cherry) and should be used in minimal quantities to produce light-moderate density colours. The Therapeutics Goods Administration (TGA) has a recently updated list of colourings permitted in oral medicines (TGA, 2011). Mixtures may be required to be buffered to an optimum pH, or pH might need to be taken into account when using certain preservatives (e.g. benzoic acid) which require an acidic pH to be effective. Commercial liquid vehicles, that contain a combination of sweeteners, flavours, viscosity enhancers or suspending agents and preservatives are available. Some contain non-nutritive sweeteners and are therefore suitable for diabetic patients. Many stability studies are available in the literature that utilise these commercial liquid vehicles, thereby making them a convenient resource, especially since various practice settings may not hold a wide variety of excipients in stock (Haywood & Glass, 2010).

4.2.2 Designing an oral liquid formulation

Pharmacists should be encouraged to use the following process to assist in the compounding of safe, stable oral liquids (Figure 9) (Haywood & Glass, 2010). In the absence of existing commercially available dosage forms or therapeutic alternatives in a suitable dosage form, a compounded oral liquid is best prepared by searching for a suitable formula. It is important that the stability of these compounded oral liquids is evaluated by methods of analysis that are 'stability-indicating' (Prankerd, 2009). Examples of Journals containing stability-indicating methods of analysis include: *Journal of Pharmacy Practice and Research, International Journal of Pharmaceutical Compounding,* and the *Journal of American Health-System Pharmacists*. However, a suitable search engine will provide results from a wider range of journals. For example *Medline* (http://www.ncbi.nlm.nih.gov/pubmed) and *Google scholar* (http://scholar.google.com.au/schhp?hl=en&tab=ws) are free-access online search engines.

Useful texts include: *Allen's Compounded Formulations* (Allen, 2003), Paediatric Drug Formulations (Nahata, Pai et al., 2004), Stability of Compounded Formulations (Trissel, 2009), however it is important to check whether there is evidence for the stability of these formulations (Haywood & Glass, 2010).

Should no suitable formula be available in the literature, pharmacists may be required to design a formula from first principles. Designing an oral liquid, using sound scientific principles, is a lengthy process and would require careful consideration of a number of factors including: potential degradation of the active API by pathways such as oxidation, hydrolysis, photolysis or thermolysis; storage, preservation and packaging considerations and assigning a suitable shelf-life to the formulation; and interactions between excipients and the API, especially if tablets or capsules are used as the source. The manufacturer of the solid dosage form may be able to provide useful stability data for the API. Generally, a commercial liquid vehicle would be the preferred choice, since it would contain all the necessary excipients to formulate the oral liquid. It is also preferable to obtain the API in pure powder form as opposed to using a commercial product, since the excipients in the commercial product will add to the complexity of the final product. The final product should be packaged in a light-resistant container (since many colours and flavours are sensitive to light) with a child-resistant cap and be appropriately labelled (Haywood & Glass, 2010).

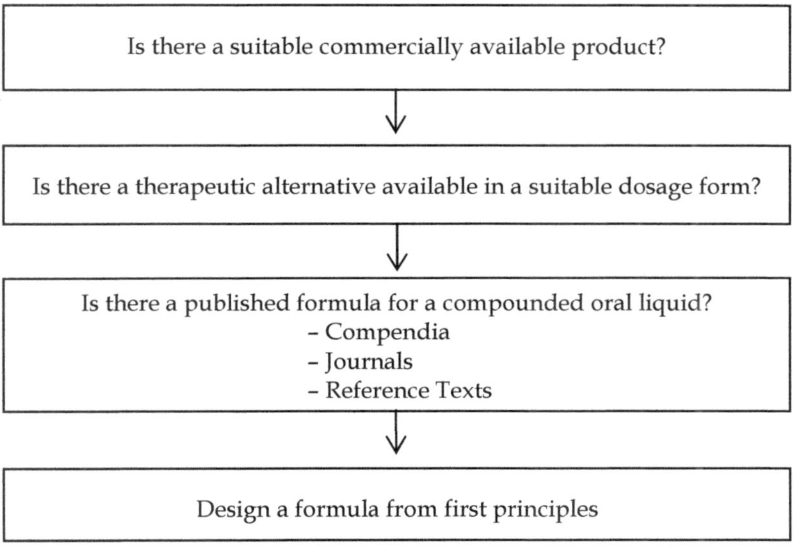

Fig. 9. Compounding safe, stable oral liquids.

4.2.3 Caution when modifying existing formulae

Because of the ability of excipients in tablets and capsules to interact with the API in the liquid dosage form, there are problems associated with using stability data for an oral liquid made from powdered raw material and in making the assumption that oral liquids prepared from commercial tablets or capsules will have comparable stability and thus shelf life. This was demonstrated in a study investigating the stability of an isoniazid (INH) liquid prepared using commercially available tablets and a British Pharmaceutical Codex (BPC) formula, where significant degradation of the INH (≥ 10 % after 3 days at both 4 and 25 °C) was shown whereas the control (using pure INH powder) retained the desired stability of > 90 % after 30 days, as specified in the BPC, under identical conditions (Haywood, Mangan et al., 2005). A replicate control formulation spiked with lactose (an excipient present in the commercial tablets), produced similar degradation profiles to that of the compounded oral liquid. INH is susceptible to hydrolysis and oxidation and is known to interact with reducing sugars (e.g. lactose), to form hydrazines (Figure 10). Although the BPC claimed 28 days stability for the extemporaneously prepared INH mixture, the use of INH powder, as opposed to INH tablets, was specified. This highlights the importance of considering not only the stability of the API but also the potential for interaction with excipients when modifying existing formulae.

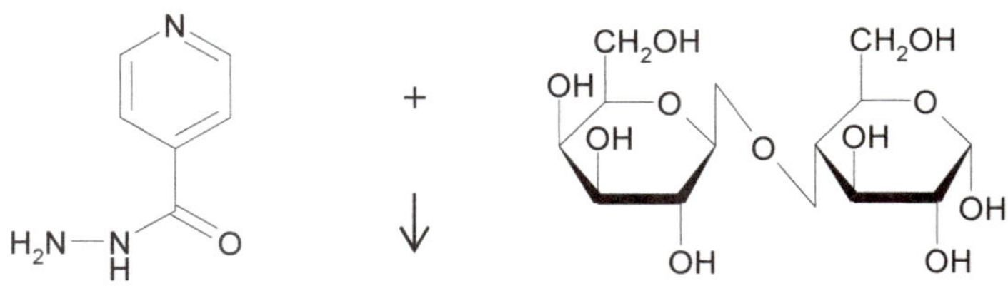

1-isonicotinoyl-2-lactosylhydrazine

Fig. 10. Degradation of isoniazid in the presence of lactose.

The stability of a drug substance in an extemporaneous oral liquid can be compromised by the addition of excipients. When considering the safety and efficacy of oral liquids prepared extemporaneously, it is important to consider not only the stability of the API but the entire formulation. Pharmacists are to be encouraged that by considering various factors such as drug stability, mechanisms and pathways of degradation, potential interactions with excipients in the tablets or capsules, and the availability of formulae (including methods and materials) for stable liquid dosage forms for oral administration in the literature, they are able to confidently dispense an oral liquid dosage form. However, clinical experience, including an assessment of bioavailability whenever possible, with extemporaneous liquid dosage forms for oral administration should be reported in the literature to further support health professionals in this important area of practice.

5. Repackaging and storage of medicines in developing countries

Solid dosage forms such as tablets are commonly procured in bulk and repackaged in developing countries in order control costs and therefore to ensure that governments are able to provide medicines to a large majority of the population. This repackaging process involving removal the drug from its primary packaging invalidates the stability guarantee of the manufacturer and often involves the transfer of the tablets to a small 'bank bag' which may be clear, opaque white, yellow, brown or red and manufactured from polyvinylchloride (Figure 11). Although the pharmaceutical industry is required to assure the stability of drugs under certain conditions of temperature light and humidity, these conditions may not be inclusive of in-use conditions. These in-use conditions may relate to accelerated temperature, light and humidity conditions, but may also simply be due to the fact that a request to store a drug product under refrigerated conditions cannot be adhered to in the a developing country, due to the patient simply not having access to a refrigerator.

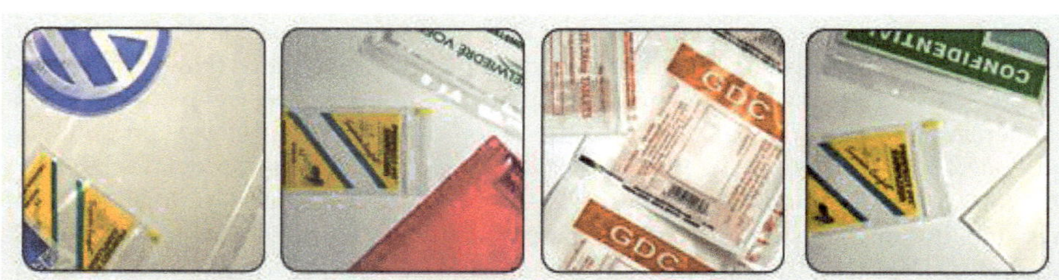

Fig. 11. Examples of Bank Bags (re-sealable polyethylene bags for dispensing tablets) - Stripform® Packaging.

5.1 In-use stability guidelines

Although the ICH Guideline on Photostability (ICH, 1996), describes a protocol for testing the stability to light of both the drug substance and drug product, it fails to provide a protocol for photostability testing under patient in-use conditions. Since this guideline was published in 1996 and has been implemented in the US, EU and Japan and the Pharmaceutical Industry has experience in these testing procedures, it has been suggested by Baertschi *et al* (Baertschi, Alsante et al., 2010), that the time has come to revise these guidelines to accommodate in-use conditions. Although the aim of photostability testing is that it should show that exposure to light does not result in an unacceptable change in both the drug substance and the drug product, by excluding any in-use testing the quality of the product to the end consumer, the patient could be compromised. Figure 12 represents a sequential testing which provides some recommendations as to those drug substances and products which are not included in the guideline such as topical agents e.g. creams, transdermal patches and intravenous preparations. There is however no reason why this should not be extended to the repackaging of oral solid dosage forms such as tablets and capsules and apply to all stability testing protocols, including in addition to light, appropriate in-use conditions of temperature and humidity. Lagrange has concurred with the findings by Baertschi *et al*, in a review on the current perspectives on the repackaging of solid oral dosage forms and stability implications for this process (Lagrange, 2010). Recognising the lack of availability of stability data, they are proposing that factors such as the barrier properties of the packaging materials, the hygroscopicity and light sensitivity of drug molecules and any pre-formulation stability data should be used to generate a list of drugs which should not be repackaged. Because of the advantages associated with the repackaging of drugs into compliance aids, they do recommend that it is the pharmacist who is required to assess the risk of this process.

STEP 1: Is the medicine stable when exposed to light unpacked – if YES to STEP 2

↓

STEP 2: Is the medicine stable when exposed to light in primary pack – if YES to Step 3

↓

STEP 3: Is the medicine stable when exposed to light in secondary pack – if YES to Step 4

↓

STEP 4: Do not transfer to a less protective pack in community/hopsital pharmacy practice

Fig. 12. Sequential photostability testing protocol

5.2 Risks and benefits associated with repackaging in developing countries

The obvious benefits associated with repackaging of medicines in developing countries are that of cost-reduction due to bulk purchases of drug products. However the risks associated with repackaging reported in the literature have to a large extent been centred on issues other than drug quality in terms of stability considerations and include (Leon Villar, Iranzo Fernandez et al., 2001; Kelly & Vaida, 2003): (i) medication errors in the repackaging of medicines; (ii) high cost of repackaging low-quantity prescriptions, which might lead to greater cost due to damage of the packaging by typing directly onto the label; and (iii) expiry dates may be extended beyond what is safe for the patient. Strategies may be implemented in order to avoid repackaging errors, including training pharmacy staff on safety issues, and storage and labelling of drug products in the pharmacy, also referred to as the design of a quality control line. The United States Pharmacopoeia (USP) standards on packaging of drug products, expiry dates and monitoring of temperature conditions have also been revised (Obeke, Bailey et al., 2000).

5.3 Repackaging of light-sensitive medicines

Three drug products (tablets) have been evaluated in terms of their stability to light when repackaged in various 'bank bags'. Frusemide and chlorpromazine are known to be light sensitive (Bundgaard, Nørgaard et al., 1988; Chagonda & Millership, 1989), while pyrazinamide, exhibits good stability and is frequently repackaged into 'bank bags' in developing countries for the treatment of patients with tuberculosis.

5.3.1 Frusemide and chlorpromazine

Frusemide tablets placed in a white opaque polyvinyl bag, the original blister (red) packing and in a white opaque securitainer were exposed to window-filtered sunlight under ambient temperature (25-40 °C) conditions for 10 days (Soneman, 1995). Both the appearance of the tablets and the drug content (using a validated HPLC method) was monitored. In all cases discoloration of the tablets, similar to that observed on repackaging of these tablets into a DAA of these tablets was noted. The percentage decrease in the frusemide content was 14.49%, 12.73% and 21.27% respectively, with further investigation revealing that temperature played a significant role in accelerating the photodegradation of the repackaged frusemide. Recommendations were that patients be counselled on the storage of repackaged frusemide tablets. Chlorpromazine similarly exposed to window filtered sunlight under ambient conditions in a white opaque polyvinyl bag, discolored, with only 76% of the drug remaining after the exposure period of 1 month (Jacobsen, 1997). For chlorpromazine it was recommended that in addition to counseling on storage of tablets to be protected from light, that tablets would be repackaged bi-monthly.

5.3.2 Pyrazinamide

Since pyrazinamide is commonly repackaged into 'bank bags', the effect of exposure to window-filtered sunlight for 3 days in both a 'yellow bag' commonly used in clinical practice and a 'brown bag' was investigated. For both bags tested, photodegradation of pyrazinamide was noted with the extent of degradation dependant both on the time of exposure and the 'type' of bag. Results indicated that pyrazinamide was most stable in the 'brown bag', with 95% of the drug remaining while in the 'yellow bag' 8% of the drug degraded. Although the protection offered by the 'brown bag' was superior, its use was limited in practice due to lack of aesthetic appeal and patient acceptance (Defferary, 1993).

5.4 Repackaging moisture-sensitive medicines

Some moisture sensitive drugs, because of their hygroscopicity absorb moisture resulting in tablets becoming unusable, due to loss of physical form. However there are instances where the absorption of moisture by drugs may cause a polymorphic transition and as result altered drug dissolution and bioavailability, which has the potential to seriously compromise the quality of the drug product.

5.4.1 Carbamazepine

The antiepileptic drug carbamazepine has the ability to lose its effectiveness when exposed to moisture due to a transformation from the anhydrous form to the dihydrate, which although thermodynamically stable is less soluble and thus less bioavailable (al-Zein, Riad et al., 1999). A study investigated the effect of repackaging of carbamazepine tablets into white opaque 'bank bags' stored under ambient (25°C; 60% RH), accelerated (40°C; 75% RH) and light (window-filtered sunlight) conditions for 30 days (Maharaj, 1997). Results from this study however confirmed both the chemical and physical stability of the repackaged carbamazepine tablets, under all storage conditions and thus the integrity of the 'bank bag'.

6. Conclusion

This chapter highlights that despite the rigor of the requirements for the pharmaceutical industry to provide quality medicines, there are a number of factors which have the ability to contribute to the patient receiving a medicine, which is not of an appropriate quality. This research confirms the role of health professionals especially pharmacists, because of their specialist knowledge in medicines of being able to contribute substantially to ensuring the quality of medicines for patients. It also raises the question as to whether we should be looking more closely at not only undertaking more research independently into in-use drug stability, but adding to the requirements for in-use stability testing to be undertaken on medicines by the pharmaceutical industry.

7. Acknowledgment

The authors wish to acknowledge James Cook University, Griffith University and Dr Sherryl Robertson, Mrs Martina Mylrea and Ms Victoria Llewelyn.

8. References

al-Zein, H., Riad, L. E. & Abd-Elbary, A. (1999). Effect of packaging and storage on the stability of carbamazepine tablets. *Drug Dev Ind Pharm* Vol.25, No.2, pp. 223-227

Allen, L. V. J. (2003). *Allen's compounded formulations: the complete US pharmacist collection,* Washington DC, American Pharmaceutical Association

AMH (2006). *Australian medicines handbook. Drug Choice Companion: Aged Care, 2nd ed,* ISBN 9780980579031, Adelaide, Australian Medicines Handbook Pty Ltd

Andrisano, V., Gotti, R., Leoni, A. & Cavrini, V. (1999). Photodegradation studies on Atenolol by liquid chromatography. *Journal of Pharmaceutical and Biomedical Analysis* Vol.21, No.4, pp. 851-857

Annemans, L., Wittrup-Jensen, K. & Bueno, H. (2010). A review of international pharmacoeconomic models assessing the use of aspirin in primary prevention. *Journal of Medical Economics* Vol.13, No.3, pp. 418-427

APAC (2002). *Australian Pharmaceutical Advisory Council. Guidelines for medication management in residential aged care facilities, 3rd ed,* ISBN 0-642-82113-5, Canberra, Commonwealth of Australia

Aryal, S. & Skalko-Basnet, N. (2008). Stability of amlodipine besylate and atenolol in multi-component tablets of mono-layer and bi-layer types. *Acta Pharmaceutica* Vol.58, No.3, pp. 299-308

Aulton, M. E. (2007). *Aulton's Pharmaceutics. The Design and Manufacture of Medicines,* ISBN 978-0-443-10108-3, Edinburgh, Churchill Livingstone

Australian_Government (2010). The Fifth Community Pharmacy Agreement between the Commonwealth Government and the Pharmacy Guild of Australia. 7 July 2011, Available from http://www.5cpa.com.au/iwov-resources/documents/The_Guild/PDFs/Other/Fifth%20Community%20Pharmacy%20Agreement.pdf

Baertschi, S. W., Alsante, K. M. & Tonnesen, H. H. (2010). A critical assessment of the ICH guideline on photostability testing of new drug substances and products (Q1B): Recommendation for revision. *Journal of Pharmaceutical Sciences* Vol.99, No.7, pp. 2934-2940

Ballard, D. (1996). Reminder of danger from chewing extended-release products. *American Journal of Health-System Pharmacy* Vol.53, No.16, pp. 1962, 1964

Bowen, L., Mangan, M., Haywood, A. & Glass, B. D. (2007). Stability of frusemide tablets repackaged in Dose Administration Aids. *Journal of Pharmacy Practice and Research* Vol.37, No.3, pp. 178-181

Bundgaard, H., Nørgaard, T. & Nielsen, N. M. (1988). Photodegradation and hydrolysis of furosemide and furosemide esters in aqueous solutions. *International Journal of Pharmaceutics* Vol.42, No.1, pp. 217-224

Chagonda, L. F. & Millership, J. S. (1989). The determination of chlorpromazine, related impurities and degradation products in pharmaceutical dosage forms. *Journal of Pharmaceutical and Biomedical Analysis* Vol.7, No.3, pp. 271-278

Chan, K., Swinden, J. & Donyai, P. (2007). Pilot study of the short-term physico-chemical stability of atenolol tablets stored in a multi-compartment compliance aid. *European Journal of Hospital Pharmacy Science* Vol.13, No.3, pp. 60-66

Church, C. & Smith, J. (2006). How stable are medicines moved from original packs into compliance aids? *The Pharmaceutical Journal* Vol.276, pp. 75-81

Defferary, L. A. (1993). A stability study of pyrazinamide solid dosage forms. Port Elizabeth, South Africa, University of Port Elizabeth. Honours Thesis (BPharm).

Giam, J. A. & McLachlan, A. J. (2008). Extemporaneous product use in paediatric patients: A systematic review. *International Journal of Pharmacy Practice* Vol.16, No.1, pp. 3-10

Glass, B., Mangan, M. & Haywood, A. (2009). Prochlorperazine tablets repackaged into dose administration aids: can the patient be assured of quality? *Journal of Clinical Pharmacy and Therapeutics* Vol.34, No.2, pp. 161-169

Glass, B. D. & Haywood, A. (2006). Stability considerations in liquid dosage forms extemporaneously prepared from commercially available products. *Journal of Pharmacy and Pharmaceutical Sciences* Vol.9, No.3, pp. 398-426

Glass, B. D., Haywood, A., Llewelyn, V. & Mangan, M. (2009). Compliance aids and medicine stability: new evidence of quality assurance. *Current Drug Safety* Vol.4, No.1, pp. 74-78

Haywood, A. & Glass, B. D. (2009). Dosage form modification for the elderly. *Australian Pharmacist* Vol.28, No.11, pp. 960-964

Haywood, A. & Glass, B. D. (2010). Paediatric mixtures. *Australian Pharmacist* Vol.29, No.4, pp. 316-330

Haywood, A., Llewelyn, V., Robertson, S., Mylrea, M. & Glass, B. (2011). Dose administration aids: Pharmacists' role in improving patient care. *Australasian Medical Journal* Vol.4, No.4, pp. 183-189

Haywood, A., Mangan, M. & Glass, B. D. (2006). Stability implications of repackaging paracetamol tablets into Dose Administration Aids. *Journal of Pharmacy Practice and Research* Vol.36, No.1, pp. 21-24

Haywood, A., Mangan, M., Grant, G. & Glass, B. D. (2005). Extemporaneous Isoniazid mixture: stability implications. *Journal of Pharmacy Practice and Research* Vol.35, No.3, pp. 181-182

ICH (1996). ICH Harmonised Tripartite Guideline. Stability Testing: Photostability testing of new drug substances and products Q1B: ICH Steering Committee, 7 July 2011, Available from http://www.ich.org/fileadmin/Public_Web_Site/ICH_Products/Guidelines/Quality/Q1B/Step4/Q1B_Guideline.pdf

ICH (2003). ICH Harmonised Tripartite Guideline. Stability Testing of New Drug Substances and Products Q1A(R2): ICH Steering Committee. 7 July 2011, Available from http://www.ich.org/fileadmin/Public_Web_Site/ICH_Products/Guidelines/Quality/Q1A_R2/Step4/Q1A_R2__Guideline.pdf

Jacobsen, L. J. (1997). A study into the effect of packaging on the stability of chlorpromazine tablets. Grahamstown, South Africa, Rhodes University. Honours Thesis (BPharm).

Kawamura, O., Easterling, C., Aslam, M., Rittmann, T., Hofmann, C. & Shaker, R. (2004). Laryngo-upper esophageal sphincter contractile reflex in humans deteriorates with age. *Gastroenterology* Vol.127, No.1, pp. 57-64

Kelly, K. & Vaida, A. J. (2003). Warning! - Repackaged substances may easily be confused. *Pharmacy Times* Vol.69, No.5, pp. 30-32

Kohara, T., Koyama, T., Fujimura, M., Tanaka, H., Maeda, J., Fujimoto, T., Yamamoto, I. & Arita, M. (2002). Y-931, a novel atypical antipsychotic drug, is less sensitive to oxidative phenomena. *Chemical and Pharmaceutical Bulletin (Tokyo)* Vol.50, No.6, pp. 818-821

Lagrange, F. (2010). Current perspectives on the repackaging and stability of solid oral doses. *Annales Pharmacetiques Francaises* Vol.68, No.6, pp. 332-358

Lee, J. K., Grace, K. A. & Taylor, A. J. (2006). Effect of a pharmacy care program on medication adherence and persistence, blood pressure, and low-density lipoprotein cholesterol: a randomized controlled trial. *The Journal of the American Medical Association* Vol.296, No.21, pp. 2563-2571

Leon Villar, J., Iranzo Fernandez, M. D., Ventura Lopez, M., Najera Perez, M. D. & Victorio Garcia, L. (2001). Línea de Control de Calidad en el reenvasado de medicamentos. Bases metodológicas. Diseño. Utilidad. *Farmacia Hospitalaria* Vol.25, No.1, pp. 31-37

Llewelyn, V. K., Mangan, M. F. & Glass, B. D. (2010). Stability of sodium valproate tablets repackaged into dose administration aids. *Journal of Pharmacy and Pharmacology* Vol.62, No.7, pp. 838-843

Maharaj, S. (1997). Carbamazepine tablets: An in-use stability evaluation. Grahamstown, South Africa, Rhodes University. Honours Thesis (BPharm).

Marriott, J. L. & Nation, R. L. (2002). Splitting tablets. *Australian Prescriber* Vol.25, No.6, pp. 133-135

Mitchell, J. F. (2011). Oral dosage forms that should not be crushed. Institute for Safe Medication Practices, 7 July 2011, Available from www.ismp.org/tools/donotcrush.pdf

Moore, D. E. & Tamat, S. R. (1980). Photosensitization by drugs: photolysis of some chlorine-containing drugs. *Journal of Pharmacy and Pharmacology* Vol.32, No.3, pp. 172-177

Morris, H. (2005). Administering drugs to patients with swallowing difficulties. *Nursing Times* Vol.101, No.39, pp. 28-30

Nahata, M. C. & Allen, L. V., Jr. (2008). Extemporaneous drug formulations. *Clinical Therapeutics* Vol.30, No.11, pp. 2112-2119

Nahata, M. C. & Morosco, R. S. (2003). Stability of tiagabine in two oral liquid vehicles. *American Journal of Health-System Pharmacy* Vol.60, No.1, pp. 75-77

Nahata, M. C., Pai, V. B. & Hipple, T. F. (2004). *Pediatric drug formulations. 5th ed.,* Cincinnati

Nejmeh, M. & Pilpel, N. (1978). The effect of the photodecomposition of chlorpromazine on lecithin monolayers. *Journal of Pharmacy and Pharmacology* Vol.30, No.12, pp. 748-753

Nunney, J. M. & Rayner, D. K. T. (2001). How are multi-compartment compliance aids used in primary care? *The Pharmaceutical Journal* Vol.267, pp. 784-789

Obeke, C. C., Bailey, L., Medwick, T. & Grady, L. T. (2000). Revised USP standards for product dating, packaging, and temperature monitoring. *American Journal of Health-System Pharmacy* Vol.57, No.15, pp. 1441-1445

Perks, S., Robertson, S., Haywood, A. & Glass, B. D. (2011). Clozapine Repackaged into Dose Administration Aids: A Common Practice in Australian Hospitals. *International Journal of Pharmacy Practice*pp. (in press)

Prankerd, R. J. (2009). Compounded products - Stability studies in Hospital Pharmacy departments. *Journal of Pharmacy Practice and Research* Vol.39, No.1, pp. 5-7

PSA (2007). Pharmaceutical Society of Australia. Guidelines and standards for pharmacists. Dose Administration Aids Service. 7 July 2011, Available from http://www.psa.org.au/site.php?id=2065

PSA (2010). *Professional Practice Standards. Version 4. Standard 7: Dose Administration Aids Service*, Pharmaceutical Society of Australia

RPSGB (2003). *Medicines, Ethics, and Practice: A Guide for Pharmacists, Volume 27,* ISSN 0955-4254, London, Royal Pharmaceutical Society of Great Britain

Schier, J. G., Howland, M. A., Hoffman, R. S. & Nelson, L. S. (2003). Fatality from administration of labetalol and crushed extended-release nifedipine. *The Annals of Pharmacotherapy* Vol.37, No.10, pp. 1420-1423

Simmons, D., Upjohn, M. & Gamble, G. D. (2000). Can medication packaging improve glycemic control and blood pressure in type 2 diabetes? Results from a randomized controlled trial. *Diabetes Care* Vol.23, No.2, pp. 153-156

Soneman, L. J. (1995). Evaluation of the relative light stability of furosemide when packed in blister packs, securitainers and approved bank bags. Grahamstown, South Africa, Rhodes University. Honours Thesis (BPharm).

TGA (2011). Therapeutic Goods Administration. Colourings Permitted in Medicines for Oral Use. 7 July 2011, Available from http://www.tga.gov.au/industry/cm-colourings-oral-use.htm

Trissel, L. A. (2009). *Trissel's Stability of Compounded Formulations. 4th ed,* Washington, DC, American Pharmacists Association

USP (2010). US Pharmacopeial Convention Inc. US Pharmacopeia 33 - National formulary 28 <681> Repackaging into single-unit containers and unit-dose containers for nonsterile solid and liquid dosage forms.

Walker, R. (1992). Stability of medicinal products in compliance devices. *The Pharmaceutical Journal* Vol.25, pp. 124-126

Ware, G. J., Holford, N. H., Davison, J. G. & Harris, R. G. (1994). Unit-of-issue medicine administration: time and cost-benefits in a geriatric unit. *Australian Journal of Hospital Pharmacy* Vol.24, No.4, pp. 329-332

White, R. & Bradnam, V. (2007). *Handbook of Drug Administration via Enteral Feeding Tubes,* ISBN 978-0-85369-928-6, London, Pharmaceutical Press

13

Quality Assurance in the Career of Nursing

Cecilia Latrach, Naldy Febré and Ingrid Demandes
Universidad Mayor
Chile

1. Introduction

Assessing the quality of university education has been presented as one of the main issues on the agenda of education reforms worldwide (Villanueva, 2004). Evaluation and accreditation processes as tools emerge to regulate the system of higher education from the perspective of the quality of educational services offered.

Creation and promotion of accreditation and evaluation mechanisms are important to ensure and promote the quality, comparability and transparency of the educational offer strengthening national and international recognition of educational systems and institutions (Mercosur, Estados asociados 2008).

Ensuring quality is a combination of planned and systematic actions that are necessary to provide the adequate reliability that a product or service meets the requirements given for quality, which should be supported in meeting the expectations of customers. Quality assurance is based on the implementation of a documented system of work, establishing clear, fixed and objective rules including all aspects related to the operational process. This process begins with the design, followed by planning, production, presentation, distribution, statistical techniques of control and staff training.

The complexity of the quality assurance system is that over the entire operating process a rigorous control should be kept over the correct application of the rules or technical specifications, methods and philosophies of quality

The quality assurance becomes important in higher education, because through this nursing schools provide training for highly skilled professionals who respond to country needs and protects the rights of citizens. It is here that according to UNESCO quality must be combined with the relevance and impact because it is not possible to conceive as a quality university the one that has no relevance to their social environment (Aguila, 2005).

In this chapter we develop the essential topics that should be considered by the various participants in the teaching-learning process of nursing professionals at university level.

2. Chapter objectives

1. To analyze the basic principles of the concept of quality assurance in university education in nursing.
2. Know the principles of quality assurance in the training of nursing professionals

3. Recognize the importance of the process of accreditation for quality education in the career of nursing.
4. Identify the standards and indicators to be considered for the management of quality assurance for nursing schools.

3. Initial definitions

Assurance: Specialized financial dictionaries define assurance as the "branch of accounting which aims to improve the quality of information to reduce business risk. Assurance extends beyond the financial information, seeks to verify the facts, processes and even qualitative information that relates to business practices and customer satisfaction.

Quality Assurance: It is a systematic, ongoing and continuous review, analysis and evaluation of the level of compliance with the standards set at local, national and international level.

Accreditation systems: The accreditation system is defined as a set of policies, strategies, processes and organizations whose main objective is to guarantee to society that higher education institutions are part of the national system that meets the highest standards of quality and the training of professionals. Accreditation is a witness of the State on the quality of a program or institution based on a prior assessment in which the institution, academic communities and Accreditation Agencies interfere.

Standards: These are conceptualized as the clear definition of a model, criterion, or rule of the minimum acceptable requirements for the operation of specific processes, to ensure quality in the provision of educational programs. The standards clearly state the expected behavior and desired processes and are used as guides to evaluate their performance and achieve continual improvement of services.

Indicators: Indicators are measures that quantify and assess the status of processes within an institution. They represent summary measures that capture relevant information on various attributes and dimensions of the evaluation processes and system performance.

Methods of evaluation: Defined as the methods for assessing processes by collecting objective information both quantitative and qualitative in a systematic way to inform any kind of decision. In the university context they can be defined by a variety of forms, types and evaluation procedures according to the different objectives, evaluation units, agents or consequences (Review and license states, regional accreditation agencies, self-study, Performance Indicators , state Testing for validation of professional qualifications, cyclical review programs (mandatory or voluntary), Specialized accreditation of professional programs, institutional assessment).

Higher/University Education: The education provided by universities, a social right, as such, is opposed to education understood as a privilege and it is understood as a "public good" that will benefit the community as a whole. It helps to train people in the values of freedom, social justice, solidarity and human rights, whose mission, task and results should be in the service of harmonious development of man and society, so first term, must respond and be accountable to the national community that surrounds and sustains it" (Ramirez, Ayarza et al, 1993). This leads to the evaluation of the work of institutions of higher education and the respective quality assurance, both the university as an institution and its academic programs.

4. Conditions in the quality assurance of universities worldwide

In the last decade, the culture of evaluation of university education has become a fundamental role in ensuring the quality of education at the international, regional, national and institutions from the voluntary self-assessments.

Quality assurance has been defined by many authors, however, to consider the terms of this chapter to that systematic, ongoing and continuous review, analysis and evaluation of the level of compliance and standards set at local, national and international level (Konrad, 2011).

Faced with the challenge of globalization, concern about the comparability and convergence among the various systems of higher education acquires special relevance for Educational Integration in Latin America. The strategy to facilitate the comparison of regional training programs is to establish objective processes of quality assurance in order to increase mutual trust among institutions of higher education (UNESCO, 2009).

Since the early 90's in response to the urgent need to develop strategies for quality assurance of educational provision/offer at the regional level, National Accreditation Agencies, Networks, Agencies and Regional Accreditation mechanisms emerged.

The traditional cycle of continuous improvement applied in different organizational systems is undoubtedly applicable as a strategy to improve the quality of the educational process at all levels.

4.1 The circle of Deming

The circle of Deming or PDCA (Plan, Do, Check, Act), is a strategy of continuous quality improvement and it is accomplished in four steps (Figure 1). They are then briefly assessed the development of the four phases of quality circle (Hernandez, Rodriguez, 2006).

4.1.1 Step 1, planning

At this step, we have the following activities:

a. First, it is the vision or goals, which is to establish the purpose of improvement.
b. Once set a goal, the person makes a description of his / her current situation, referring to all aspects and determines the areas that have problems or enhancements, making a selection of the most outstanding and greatest impact.
c. Determine the definition of a certain theory of solution in order to bring the variable to improve at a peak.
d. Finally, we define a work plan to carry out an implementation plan to test the theory of solution

4.1.2 Step 2, make

At this step, we carry out the planned work plan, establishing a tracking control to be sure of developing the program. To develop the implementation, there are tools such as Gantt chart or checklist of tasks performed, leading to check the progress of the process.

4.1.3 Step 3, checking

At this step the results are validated and a comparison with the planned is carried out.

4.1.4 Step 4, act

At this final step the cycle of quality concludes and it is checked if verification was successful and achieved the desired benefits, it is vital to systematize and make a documentation of the changes made to secure profits.

The quality circle becomes a process of continuous improvement that should be used systematically, becoming a permanent process of Planning, Doing, Checking and Acting.

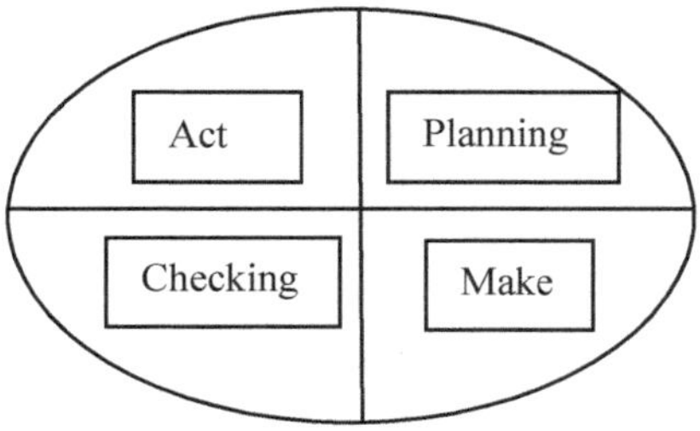

Source: Supperting empowerment with Deming´s PDSA cycle. Empowerment Organizations, MCB University Press. 1995.

Fig. 1. The deming circle

5. Quality in higher education for nursing schools

International practice recognizes various concepts of quality that lead to different types of evaluation and accreditation. Basically a distinction between: (i) quality as excellence as defined by the highest standards, (ii) quality and ability to meet the stated mission of the institution or program, (iii) quality of learning outcomes achieved by students; (iv) quality threshold as defined according to criteria and standards, (v) quality and continuous improvement where the center of gravity lies on the organizational learning, (vi) quality and value for money obtained, a concept used in situations where governments face fiscal constraints or seek to raise the efficiency of institutions or programs (Latrach, Soto et al, 2009).

The quality of university education is a philosophical concept; definitions vary and in some ways reflect different perspectives of the individual and society. In a democratic society, where there must be room for that many people to think differently, there is no single correct definition of quality, it is a relative concept that depends on the individual who uses it. To have a set of criteria from the perspective of different groups and no and not supporting a single definition of quality can provide a practical solution to a highly complex philosophical issue, not because it lacks an underlying theory, but because different groups have the right to bear different perspectives. This makes us understand that the concept of quality in education has many definitions, all related to different interpretations concerning the teaching-learning process. As it is obvious to conclude that it is not possible to choose one and take it for universally valid, since this type of definition does not exist, but it is necessary to refine the definition and we must align with the mission, vision and exit profile of the institution and the school of nursing without forgetting the following premises in relation to education delivered: 1) meets the expectations of the graduate, 2) meets the expectations of potential employers; 3) enables the practitioner to able to make an effective contribution to society (CNA, 2008).

As the quality of care, university quality has dimensions defined and evaluated in different systems of accreditation of university quality; these are relevance, effectiveness, availability of resources, efficiency, effectiveness and processes, generally evaluated by the following components: 1) characteristics of the organization, 2) teaching-learning process, 3) human Resources, 4) Infrastructure.

6. Accreditation system for schools of nursing

The accreditation process is a process that contributes significantly to the quality management of the university and the school nurse, whose installation should be gradual and constant over time. To display the management which we aim to some critical factors should be considered (Garcia, Cordero et al, 2008).

- •Installing a policy of institutional quality, which is an explicit definition of the core guidelines of the institution, incorporated into its mission and vision, which consequently leads to a cultural change.
- •Prioritization of quality issues over others, in order to reorganize institutional structures, leading to financial and human resources development of the plan to run.
- •The management of the institution should lead the establishment of a culture-based quality improvement processes. Involve all stakeholders in the establishment "Quality is not the responsibility of the heads, but everybody else's"
- •Development of a formal quality structure with a resolution of hours and duties.
- •The system incorporates incentives for prestige among the institutions and these generate competition to offer better benefits and services, which will directly benefit the users, who will be more satisfied with Institutions that present the highest quality standards.
- •When an entity develops process improvements aimed at achieving compliance with accreditation standards, this entity gets an increase in the probability that the customer is treated with full respect for their rights.

We will understand accreditation systems as the process by which an educational institution or program provides information about its operations and achievements to an external body which assesses and judges independently, this information is reflected in a public statement about the value or the quality of the program or institution. Accreditation is not permanent, but is granted for a period that can vary depending on years of the accreditation system selected, from which it can be renewed or withdrawn, based on the results obtained in the process of reassessment.

Consequently, the university accreditation is seen as a mechanism by which the educational community establishes and maintains self-regulation, and guarantees to the direct and indirect users of the services offered, their integrity, relevance and quality at levels that make it worthy the public trust and respect through their ability to create cultural, scientific and technological innovation and training of human resources (Latrach, Soto et al, 2009).

The evaluation and accreditation are related processes whose practice is intertwined, as shall be credited as a result of a process of evaluation and monitoring, however more than a diagnosis that leads to action by the institution itself. The accreditation is proof of credibility by the society and demanding public of educational services.

Accreditation systems for nursing schools generally have three stages (Figure 2): 1) Self-assessment process, 2) Process visit peer reviewer, 3) Final report of the accreditation. Accreditation is an improvement strategy, where the preparation for this is the beginning of the process. Processes are the unit of change, the main element "institutions do not change unless they change their processes.

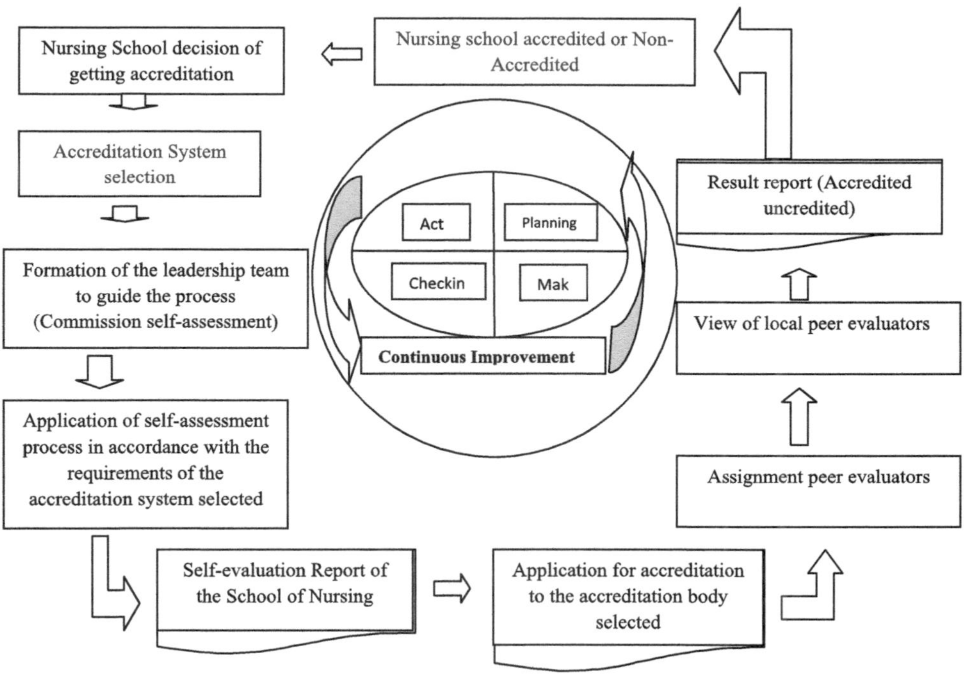

Fig. 2. A process of accreditation

7. Proposed indicators

The requirements of the different dimensions of quality applied to university education, not only must identify and understand the behavior of the variables that influence this process, but also determine the minimum level of quality and productivity necessary to ensure effective and efficient higher education, and secondly the continuity and development of the institution.

The level of quality is a concept that can be measured and evaluated by comparison with peers, perceptions of users and stakeholders, among others. The quality assessment should be objective and absolute, which requires the design of different control mechanisms by building standards and indicators that allow linking operation, resources and outcomes for activities, events, processes, organizational units and other components of the institution of higher education.

The quality of the educational process is evaluated through the so-called "indicators", defined as data that "suggest" how they develop different activities derived from the educational process, compared to how they should be done. Therefore, an indicator is a characteristic or variable that can be measured. The construction of an indicator means to account for a phenomenon in education, by absolute numbers, rates or more sophisticated indices. The indicators are classified into three indicators of structure, process and outcome.

The objectives of using standards and indicators for the management of the nursing schools are: 1) Identify problem situations or likely to be improved, 2) Incorporate improvement cycles to solve the identified problems, 3) Internal comparisons over time, 4) Comparisons with other institutions. These indicators will: a) Measure "how well executed", b) Develop evaluation criteria; **c)** Design a program of continuous improvement.

7.1 Attributes of good indicators in education

The quality of an indicator, although much depends on the quality of the data from which to build (components), also depends on the quality of information systems or information sources. You should also enjoy certain features, outlined below (Valenzuela, 2005).

Attributes of good indicators in education
Be useful: Gives answers and is designed for a specific purpose.
Validity: Measures what it intends to measure.
Reliability: The same results are reproduced if the measurement is repeated under similar conditions.
Specificity: Measures only the phenomenon being measured.
Sensitivity: Measures the changes in the phenomenon being measured.
Measurability: Based on data available or easy to obtain and easy to use. It is imperative to collect a limited amount, but feasible and valid rather than trying to complicate the indicators by systems impractical or complex parameters.
Relevance: Able to give clear answers to relevant issues embedded in health policies.
Cost Effectiveness: That the investment in time and other resources needed to construct the indicator is justified by its use and results
Integrity: Means that the required data are complete.
Internal Consistency: Refers that in the indicators, seen alone or in groups, the values are consistent and sensitive to change
Transparency: Refers to be easily understood and interpreted by users
Dissemination: To be accessible to users through periodicals
Dynamism: To update and correct as far as the environment changes. This may change in terms of the specific conditions described by indicators, data availability, scientific knowledge, or in levels of interest and needs of users.

Table 1. Attributes of good indicators in education

Once the indicators are established, they must be subjected to permanent monitoring of the quality and establish a mechanism to disseminate them, including the timing and frequency of compilation.

Structure indicators measure the quality of the characteristics of the framework in which services and state resources to provide, allow to assess how the academic unit and the University are organized and equipped, allowing to determine if resources are available and organized to facilitate the educational process. The evaluation of the structure involves:

- Material resources such as facilities, equipment and monetary budget.
- Human Resources: Number and qualifications of personnel
- Institutional or managerial aspects of management: Documentation relating to existing processes and organization

The evaluation of the structure indicators is easy, fast and objective because it encompasses a series of static characteristics and previously established, however, the most perfect structure does not guarantee the quality of management of the academic unit or university.

Process indicators assess and measure the quality of the process, defined as a set of intertwined actions with a purpose that lead to a result. These focus on how care is given, measured if all the steps of a process are done correctly, these indicators focus on the "how" the educational process is carried out, evaluate and measure whether all the steps of a given process are carried out as correct.

Outcome indicators assess the expected effect, that is to say, they allow an analysis of the results (cohort graduation rate) refer to the benefit achieved in students, providing opportunities to assess efficacy, effectiveness and efficiency of the program evaluated

Here, a proposal for some indicators of management structure, process and outcome to be considered by nursing schools to determine the quality of training will be evaluated.

INDICATOR NAME	STAGE OF THE PROCESS EVALUATED	TIPE INDICATOR
Admission Rate of Return	Sign up for students	Process
Admission Offered Rate	Sign up for Students	Process
Satisfaction of students, teachers, employers.	The whole process of training	Process
Rating support services	The whole process of training	Structure
Cohorts for performance	The whole process of training	Process
Retention cohorts	The whole process of training	Process
Cohort dropout rate	The whole process of training	Process
Repetition rate by cohort	The whole process of training	Process
Evolution for Teachers day academic hierarchies	The whole process of training	Structure
Teacher retention rate	The whole process of training	Structure
Cohort graduation rate (according to the actual number of years program)	Completion of the educational process	Result
Average time spent by cohort	Completion of the educational process	Result
Employability rate of graduates per year	Completion of the educational process	Result

Table 2. Indicators of management

8. Conclusions

Nursing is a profession and discipline deeply rooted humanist, whose essence is the care of people. Its foundations give support to a practice, profoundly humanist, which has evolved with scientific and technical progress. In keeping with its purpose to improve, exercises the power that tends to quality assurance systems in health services, seeking to satisfy the needs of society. Thus, quality has become an essential element of health services considering the fundamental requirement for the training of nursing professionals.

The accreditation of the School of Nursing is the result of a process of evaluation and systematic monitoring and voluntary compliance of university functions, which allows getting accurate and objective information on the quality of the academic unit evaluated. To certify the quality of trained human resources and the various educational processes taking place in it. It is the formal and public recognition given to a nursing school that has made significant progress in fulfilling its mission and stated goals, and meets an agreed set of criteria, indicators and standards of relevance and quality.

This activity is based on the constant search for excellence and represents the collective effort of all participants in the teaching-learning process to be accountable to themselves and society, on the appropriateness, relevance and quality of his being and institutional work.

9. References

Águila V. (2005). El concepto de calidad en la educación: clave para el logro de la competitividad institucional. *Revista Iberoamericana de Educación*, vol.35, 5, (marzo 2005), paginas 1-8, ISSN:1681-5653.

Comisión Nacional de Acreditación de pregrado. Chile (CNA). *Sistema Nacional de aseguramiento de la calidad de Educación Superior*. Extraído enero 2008, Fecha de acceso 10 de julio 2011, disponible desde http: www.cnachile.cl/portada/port_presgeneral.html

García G, Cordero R, Peralta A, Santamaría M & González J.(2008) Sistema de evaluación, acreditación y certificación de la educación superior en América latina y el Caribe- Problemática regional y plan integral de desarrollo. Aprobado en la LXXVI reunión del consejo ejecutivo de la UDUAL 25 y 26 de noviembre, Monterrey, Nuevo León, México, 2008. Disponible desde: http://www.udual.org/Evaluacion/SEACESALC/SEACESALC.pdf

Hernández S, Rodríguez. *Introducción a la Administración*. Ed Mcgraw-Hill Interamericana, 4° ed. 2006.

Konrad Fundación (2011).*Sistema Institucional de aseguramiento de la calidad SIAC*. Fecha de acceso 10 de julio 2011, disponible desde: http://www.konradlorenz.edu.co/images/stories/reglamentos/SISTEMA_INSTITUCIONAL_DE_ASEGURAMIENTO_DE_LA_CALIDAD_SIAC.pdf

Latrach C, Soto P, González I, Caballero E & Inalaf C (2009). Aseguramiento de la calidad en la formación de las enfermeras desde la perspectiva de los procesos de acreditación nacional. *Cienc. enferm.*, 15. 2. (Agosto 2009), paginas 79-94. ISSN: 0717-2075.

Mercosur y Estados Asociados (2008). *Acuerdo sobre la creación e implementación de un Sistema de Acreditación de Carreras Universitarias para el Reconocimiento Regional de la Calidad Académica de sus respectivas titulaciones en el MERCOSUR y Estados Asociados.* XXXVCMC-San Miguel de Tucumán, 30/VI/08. Fecha de acceso 10 de julio 2011, disponible desde: http://www.sice.oas.org/trade/mrcsrs/decisions/dec1708s.pdf

Ramírez S, Ayarza H, Char R, González L & Silva M (1993). *Calidad de la enseñanza universitaria, criterios de verificación.* Fecha de acceso 10 de julio 2011, disponible desde: http://200.45.54.151/novedades/documentos/gonzales_ayarza.pdf

UNESCO. (2009). Conferencia Mundial de Educación Superior 2009: Las Nuevas Dinámicas de la Educación Superior y de la Investigación para el Cambio Social y el Desarrollo. París, 5-8, Julio de 2009.

Valenzuela M, (2005). Indicadores de salud, características, uso, ejemplos. *Cienc. Trab.* Vol 7. 17, (julio-septiembre), páginas:118-122, 2005.

Villanueva, E. (2004). La acreditación en América Latina: el caso de Argentina en la RIACES y en el MERCOSUR. *Revista Iberoamericana de Educación,* vol.35, (mayo-agosto 2004), paginas 99-112, ISSN 1681-5653.

14

Patterns of Medical Errors: A Challenge for Quality Assurance in the Greek Health System

Athanassios Vozikis and Marina Riga
University of Piraeus, Athens
Greece

1. Introduction

A widely accepted definition of quality, as given by the Institute of Medicine is: "The degree to which health services for individuals and populations increase the likelihood of desired health outcomes and are consistent with current professional knowledge" (Lohr, 1990). According to other definition, "quality is conceptually complex and represents a synthesis of lessons, methods and acquired knowledge from a range of disciplines" (Dalrymple & Drew, 2000). So, we can realize that the subject of quality in healthcare organisations has been the object of numerous attempts at quick fixes. Thus, quality management in various types of heath services organizations is an important issue to improve the quality and safe patient care, promote quality patient and organizational outcomes and in general to improve health (Kelly, 2006).

For delivering quality in a heath care system, we should deal with six areas - dimensions of quality (World Health Organization [WHO], 2006). These dimensions are:

- *effective*, delivering health care that is adherent to an evidence base and results in improved health outcomes for individuals and communities, based on need;
- *efficient*, delivering health care in a manner which maximizes resource use and avoids waste;
- *accessible*, delivering health care that is timely, geographically reasonable, and provided in a setting where skills and resources are appropriate to medical need;
- *acceptable/patient-centred*, delivering health care which takes into account the preferences and aspirations of individual service users and the cultures of their communities;
- *equitable*, delivering health care which does not vary in quality because of personal characteristics such as gender, race, ethnicity, geographical location, or socioeconomic status;
- *safe*, delivering health care which minimizes risks and harm to service users.

Although error is inherent in all fields of human activity, it is however possible health professionals to learn from mistakes and prevent their reoccurrence and that health-care providers and organisations that have achieved a high level of safety have the capacity to acknowledge errors and learn from them.

Patients should participate in decisions about their health care, and recognising that those working in health-care systems should provide them with adequate and clear information about potential risks and their consequences, in order to obtain their informed consent to treatment.

Noting, also, the relevance of the World Health Organisation (WHO) "Health for All" targets for the European Region (target 2) and of its policy documents on improving health and quality of life and having regard to its Health Assembly Resolution 55.18 (2002) on "Quality of care: patient safety", which recognises the need to promote patient safety as a fundamental principle of all health systems.

Considering, thus, that patient safety is the underpinning philosophy of quality improvement and that all possible measures should therefore be taken to organise and promote patient-safety education and quality of health-care education (Council of Europe [C.E.], 2006).

For years, experts have recognized that medical errors exist and compromise health care quality, but the response to the November 30, 1999, release of the Institute of Medicine's (IOM) report, *"To Err is Human: Building a Safer Health System"*, brought medical errors to the forefront of public attention (Institute of Medicine [IOM], 1999). In March 2001, the second IOM report, *"Crossing the Quality Chasm: A New Health System for the 21st Century"*, was published (IOM, 2001). The 'chasm' report extends the findings of the 'error' report to other important dimensions of healthcare quality.

The reports concluded that the majority of these errors were the result of systemic problems rather than poor performance by individual providers, and outlined a four-pronged approach to prevent medical mistakes and improve patient safety. Much has been written worldwide about medical errors and improvements in their reporting and handling since then.

Recently, the Euro barometer survey, which was released by the European Commission (E.C., 2005) found that almost half of those surveyed said that hospital patients should be worried about being victims of medical errors.

In this research paper, we present various patterns of medical errors in Greece, after the analysis of 141 cases coming from administrative courts awards and Greek Ombudsman's reports for the years 2000 to 2007. We also present some of the current activities, as well as recommendations for additional activities to reduce errors through increased awareness of medical errors.

In the following sections 2 and 3, the definitions, classifications and epidemiology and root causes of *adverse events* and *medical errors* are given.

In sections 4 and 5, the measurement process and tools, as well as the underreporting factors of *medical errors* are presented.

In section 6, we present our research findings and finally, in section 7, we discuss our findings and we propose additional policies to reduce errors through increased awareness of medical errors.

2. Definitions, context and classifications

2.1 Definitions and context

The lack of standardized nomenclature and a universal taxonomy for adverse events and medical errors complicates the development of a response to the issues outlined in this paper. A number of definitions have been applied to medical errors and patient safety.

The World Health Organization (WHO) Collaborating Centres for International Drug Monitoring defines an adverse drug event as follows (WHO, 1984):

"Noxious and unintended and occurs at doses used in man for prophylaxis, diagnosis, therapy, or modification of physiologic functions."

In *To Err is Human*, the IOM (1999) adopted the following definitions:

"An error is defined as the failure of a planned action to be completed as intended or the use of a wrong plan to achieve an aim."

"An adverse event is defined as an injury caused by medical management rather than by the underlying disease or condition of the patient."

In an effort to thoroughly consider all of the relevant issues related to medical errors, the Quality Interagency Coordination Task Force (QuIC) expanded of the IOM definition, as follows (Quality Interagency Coordination Task Force, 2000):

"An error is defined as the failure of a planned action to be completed as intended or the use of a wrong plan to achieve an aim. Errors can include problems in practice, products, procedures, and systems."

2.2 Classifications

Most people believe that medical errors usually involve drugs, such as a patient getting the wrong prescription or dosage, or mishandled surgeries, such as amputation of the wrong limb. However, there are many types of medical errors. The following seven categories in Table 1 summarize types of medical errors that can occur (Lazarou et al., 1998) :

Medication Error	Such as a patient receiving the wrong drug.
Surgical Error	Such as amputating the wrong limb.
Diagnostic error	Such as misdiagnosis leading to an incorrect choice of therapy, failure to use an indicated diagnostic test, misinterpretation of test results, and failure to act on abnormal results.
Equipment failure	Such as defibrillators with dead batteries or intravenous pumps whose valves are easily dislodged or bumped, causing increased doses of medication over too short a period.
Infections	Such as nosocomial and post-surgical wound infections.
Blood transfusion-related injuries	Such as a patient receiving an incorrect blood type.
Misinterpretation of other medical orders	Such as failing to give a patient a salt-free meal, as ordered by a physician.

Table 1. Types of medical errors

There are many possible ways to categorize *medical errors*, but no universally accepted taxonomy. Classifications have included:

- Type of health care service provided (e.g., classification of medication errors by the National Coordinating Council for Medication Error Reporting and Prevention) (NCCMERP, 1998).
- Severity of the resulting injury (NQF, 2007) (e.g., sentinel events, defined as "any unexpected occurrence involving death or serious physical or psychological injury" by the Joint Commission on Accreditation of Healthcare Organizations (JCAHO, 2011)).
- Legal definition (e.g., errors resulting from negligence (IOM, 1999)).
- Type of setting (e.g., outpatient clinic, intensive care unit), and
- Type of individual involved (e.g., physician, nurse, patient).

Also, Leap (1993) proposed a classification of *medical errors'* types as presented in the Table 2:

Diagnostic	Error or delay in diagnosis
	Failure to employ indicated tests
	Use of outmoded tests or therapy
	Failure to act on results of monitoring or testing
Treatment	Error in the performance of an operation, procedure or test
	Error in the administering the treatment
	Error in the dose or method of using a drug
	Avoidable delay in treatment or in responding to a abnormal test
	Inappropriate (not indicated) care
Preventive	Failure to provide prophylactic treatment
	Inadequate monitoring or follow-up of treatment
Other	Failure of communication
	Equipment failure
	Other system failure

Table 2. Types of medical errors

Finally, the National Coordinating Council for Medication Error Reporting and Prevention (NCC MERP) Index for Categorizing Errors (National Coordinating Council for Medication Error Reporting and Prevention (NCC MERP), 1998), proposes the following categories of *Adverse Events* severity in the NCC MERP Index:

Category A	Circumstances or events that have the capacity to cause error
Category B	An error that did not reach the patient
Category C	An error that reached the patient but did not cause harm
Category D	An error that reached the patient and required monitoring or intervention to confirm that it resulted in no harm to the patient.
Category E	Temporary harm to the patient and required intervention
Category F	Temporary harm to the patient and required initial or prolonged hospitalization
Category G	Permanent patient harm
Category H	Intervention required to sustain life
Category I	Patient death

Table 3. Index for categorizing adverse events severity

Categories A, B, C and D describe medication errors that do not cause harm, while categories E, F, G, H, and I of the NCC MERP Index describe errors that do cause harm.

3. The epidemiology and the root causes of medical errors

3.1 The epidemiology of medical errors

It is clear that, although the United States provides some of the best health care in the world, the numbers of errors in health care are at unacceptably high levels. The Institute of Medicine's report (IOM, 1999) estimates that more than half of the adverse medical events occurring each year are due to preventable medical errors, causing the death of tens of thousands. The consequences of medical mistakes are often more severe than the consequences of mistakes in other industries—leading to death or disability rather than inconvenience on the part of consumers—underscoring the need for aggressive action in this area.

As shown in the following table 4, the estimated total number of iatrogenic deaths -that is, deaths induced inadvertently by a physician or surgeon or by medical treatment or diagnostic procedures- in the US annually is $783,936 (Barczak et al., 1997; Burger et al., 2003; Healthcare Cost and Utilization Project [HCUPnet]; IOM, 1999; Lazarou et al., 1998; Morbidity and Mortality Weekly Report [MMW], 2000; Null et al., 2007; Starfield, 2000a; Tunis & Gelband, 1994; Weinstein, 1998; Xakellis et al., 1995). It is evident that the American medical system is itself the leading cause of death and injury in the US (U.S. National Center for Health Statistics [USNCHS], 2003). By comparison, approximately 699,697 Americans died of heart in 2001, while 553,251 died of cancer.

Condition	Deaths	Cost
Adverse Drug Reactions	106,000	$12 billion
Medical error	98,000	$2 billion
Bedsores	115,000	$55 billion
Infection	88,000	$5 billion
Malnutrition	108,800	
Outpatients	199,000	$77 billion
Unnecessary Procedures	37,136	$122 billion
Surgery-Related	32,000	$9 billion
Total	**783,936**	**$282 billion**

Table 4. Estimated annual mortality and economic cost of medical intervention

European *medical errors* statistics are difficult to acquire. Unlike in the US, there is no official authority collecting data relative to *medical errors* occurrences. Nevertheless, existing figures indicate an increase in reported medical malpractice incidents in recent years.

A United Kingdom estimate of clinical risks by University College London (GeneralCologneRe, 2002) suggests that nowadays 3 - 4% of patients in the developed world are harmed during a hospital stay. For 70% of them the resulting adverse effect is short-lived, 16% endure permanent disabilities, while 14% subsequently die. The Kellog Foundation (National Coalition on Health Care and the Institute for Healthcare Improvement, 2000) found that Britain's medical malpractice death rate is comparable to that of the United States. Medical error is the third most frequent cause of death in the United Kingdom after cancer and heart disease, killing up to 40,000 people a year. The number of medical errors deaths therefore is four times greater than the number of deaths due to all other types of accidents. In Germany, 1999 estimates put the number of medical errors incidents at 400,000 per year (GeneralCologneRe, 2002).

In Ireland, one in every 100 patients is estimated to experience some form of medical error (GeneralCologneRe, 2002). In Greece, there are not any official *medical errors* statistics. Nevertheless, calculations indicate that about 20 to 30 patients die every day and other 200 are harmed because of preventable medical errors (Vozikis, A. & Riga, M., 2008a).

Rates in the later adverse event studies from UK, Denmark and New Zealand are remarkably similar, all being around 10%; US rates are much lower, with Australia seemingly much higher. The lower rates in the US might reflect better quality care, but could also reflect the narrower focus on negligent injury rather than the broader quality improvement focus of most other studies (Thomas et al, 2000a).

Although differences in adverse event rates between countries attract a great deal of media attention, much debate and occasional recrimination, the whole issue needs to be set in a broader context. Other attempts to compare health systems have produced a completely different picture.

3.2 The root causes of medical errors

According to a variety of sources, the root cause of medical errors is due to the complexity of today healthcare systems. The IOM (1999) emphasized that most medical errors are systems related and not attributable to individual negligence or misconduct. The key to reducing medical errors is to focus on improving the systems of delivering care and not to blame individuals. Health care professionals are simply human and, like everyone else, they make mistakes. The FDA reports that many patient deaths and injuries are associated with the use of FDA-regulated medical products within a complex and time-pressured health care system. Reducing the incidence of medical errors can save thousands of lives and billions of dollars. The Institute for Health Care Improvement has identified the leading cause of medical mistakes as the increasing complexity of health care (Griffin & Resar, 2009). His general recommendations were for more simplification and greater standardization, such as the use of bar codes to ensure that the right patient receives the right dose of the right medication.

Several issues have contributed the incidence of medical errors (Bates, 1998) including:

- Complexity of the health care system
- Reluctance of doctors to admit errors
- Lack of leadership
- Insurance reimbursement system that rewards errors since hospitals can still bill for additional services when patients are injured but often will not pay for practices that reduce those errors

This orientation of thought - that systems, not individuals, produce errors - has profound implications for caregivers and reporters. Medical leaders believe that focusing on systems is the best way to prevent errors. Assigning blame helps keep them hidden. A systems approach emboldens the health care policy officers to come forward with information needed to understand how mistakes occur.

Many doctors and nurses would like nothing better than for the media to stop skewering them individually and report on errors in the safer, neutral language of system failures. Indeed, if there is one overarching critique of quality stories from the medical profession, it is that the reporters look for victims and villains and blame caregivers too much. For journalists, however, reporting on system failures can be difficult; as such stories tend to be dry and antiseptic. But some journalists say the medical professionals may have a point. These reporters believe that the deepest stories include not only personal practices but also some sense of how these errors result from system failure. Reporters cannot let individuals off the hook, but the media should remember that no one acts alone. An error can usually be traced back to the system that sustains and directs the "perpetrator."

4. Identifying and measuring *medical errors* and *adverse events*

4.1 Framework for identifying medical errors

The overall goal of improved safety in health care is to reduce patient injury or harm, which underscores the importance of distinguishing between errors and harm. Although detection and analysis of errors is important in understanding failure-prone aspects of health care delivery systems and designing strategies to prevent and mitigate these failures, there is special value in quantifying actual harm.

Medical errors are failures in processes of care and, while they have the potential to be harmful, numerous reports have shown they are often not linked to the injury of the patient. Because events of harm are clear clinical outcomes, they are particularly likely to engage both clinicians and administrators in a thorough review of the system factors that led to the adverse event, with a clear focus on improving patient outcomes.

By concentrating on the events actually experienced by patients, a hospital can begin to foster a culture of safety that shifts from individual blame for errors to comprehensive system redesign that reduces patient suffering. To address the clear need to quantify adverse patient outcome, this paper focuses on the identification of harm or injury to the patient (Healthcare Cost and Utilization Project [HCUPnet], 2003-modified), as shown in Figure 1:

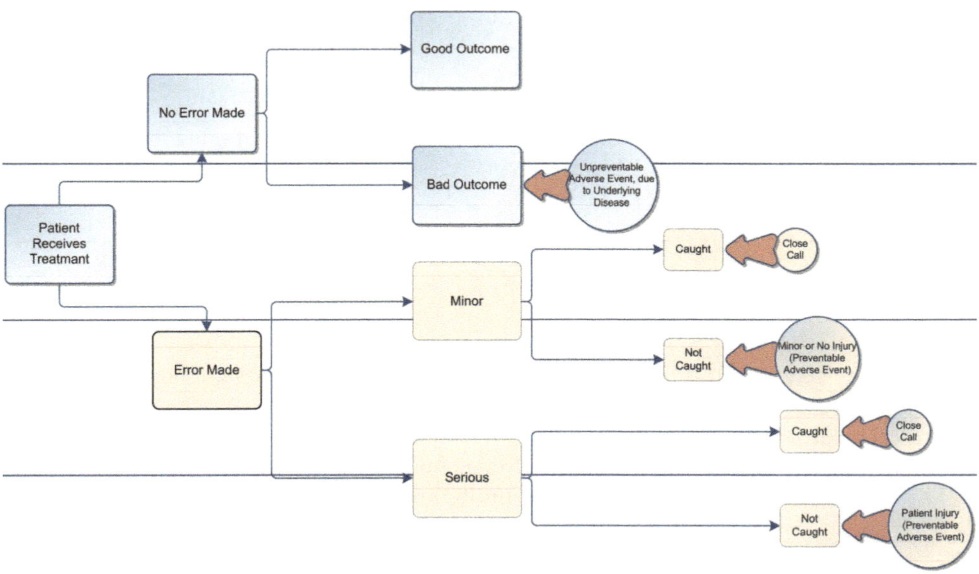

Fig. 1. Framework for identifying medical errors

4.2 Methods for studying and measuring medical errors

There are a number of methods of studying errors and adverse events, each of which has evolved over time and been adapted to different contexts. Each of the methods has particular strengths and advantages, and also weaknesses and limitations.

Table 5 illustrates a general framework to help select error and adverse event measurement methods (Thomas & Petersen, 2003) :

Study method	Advantages	Disadvantages
Morbidity and mortality conferences and autopsy	Can suggest contributory factors Familiar to healthcare providers	Hindsight bias Reporting bias Focused on diagnostic errors Infrequently used
Case analysis/root cause analysis	Can suggest contributory Factors Structured systems approach Includes recent data from interviews	Hindsight bias Tends to focus on severe events Insufficiently standardized in practice
Claims analysis	Provides multiple perspectives (patients, providers, lawyers)	Hindsight bias Reporting bias Non-standardized source of data
Error reporting systems	Provide multiple perspectives over time Can be a part of routine operations	Reporting bias Hindsight bias
Administrative data analysis	Uses readily available data Inexpensive	Might rely on incomplete and inaccurate data The data are divorced from clinical context
Record review/ chart review	Uses readily available data Commonly used	Judgments about adverse events not reliable Medical records are incomplete Hindsight bias
Review of electronic medical record	Inexpensive after initial investment Monitors in real time Integrates multiple data sources	Susceptible to programming and/or data entry errors Expensive to implement
Observation of patient care	Potentially accurate and precise Provides data otherwise unavailable Detects more active errors than other methods	Time consuming and expensive Difficult to train reliable observers Potential concerns about confidentiality Possible to be overwhelmed with information
Active clinical surveillance	Potentially accurate and precise for adverse events	Time consuming and expensive

Table 5. Advantages and disadvantages of methods used to study adverse events and medical errors

Although these methods can provide important and actionable information about systems, they also have weaknesses. They are incapable of providing error or adverse event rates because they are imprecise, primarily because of the various factors that influence whether an error or adverse event leads to a claim, incident report, or autopsy.

In the error reporting systems method, errors witnessed or committed by health care providers may be reported via structured data collection systems. Analysis of error reports may provide rich details about latent errors that lead to active errors and adverse events.

But error reporting systems alone cannot reliably measure incidence and prevalence rates of errors and adverse events because numerous factors may affect whether errors and adverse events are reported. Providers may not report errors because they are too busy, afraid of lawsuits, or worried about their reputation. High reporting rates may indicate an organizational culture committed to identifying and reducing errors and adverse events rather than a truly high rate. Despite these limitations, error reporting systems can identify errors and adverse events not found by other means, such as chart reviews, and can thereby be used in efforts to improve patient safety.

5. Underreporting of iatrogenic events

As few as 5% and no more than 20% of iatrogenic acts are ever reported (Barczak et al., 1997; Bates et al., 1995; Leape, 1994; Starfield, 2000a; Starfield, 2000b; Thomas et al., 2000b). A study conducted in two obstetrical units in the UK found that only about one-quarter of adverse incidents were ever reported, to protect staff, preserve reputations, or for fear of reprisals, including lawsuits (Bates et al., 1995).

An analysis by Wald and Shojania (2001) found that only 1.5% of all adverse events result in an incident report, and only 6% of adverse drug events are identified properly. The authors learned that the American College of Surgeons estimates that surgical incident reports routinely capture only 5-30% of adverse events. In one study, only 20% of surgical complications resulted in discussion at morbidity and mortality rounds (Vincent et al., 1999). From these studies, it appears that all the statistics gathered on medical errors may substantially underestimate the number of adverse drug and medical therapy incidents. They also suggest that our statistics concerning mortality resulting from medical errors may be in fact be conservative figures.

Standard medical pharmacology texts admit that relatively few doctors ever report adverse drug reactions to the FDA. The reasons range from not knowing such a reporting system exists to fear of being sued. Yet the public depends on this tremendously flawed system of voluntary reporting by doctors to know whether a drug or a medical intervention is harmful.

If hospitals admitted to the actual number of errors for which they are responsible, which is about 20 times what is reported, they would come under intense scrutiny (Vincent et al., 1999). Jerry Phillips, associate director of the FDA's Office of Post Marketing Drug Risk Assessment, confirms this number. "In the broader area of adverse drug reaction data, the 250,000 reports received annually probably represent only 5% of the actual reactions that occur."(Bates, 1998)

Dr. Jay Cohen, who has extensively researched adverse drug reactions, notes that because only 5% of adverse drug reactions are reported, there are in fact 5 million medication reactions each year (Dickinson, 2000).

6. Research findings

The aim of the present study is to present various patterns of medical errors in Greece. For the present research, an extensive search was carried out to find the relevant authorities and the organisations where the various stakeholders affected by the medical errors turn to. The material of our analysis consists from 141 cases coming from the administrative courts awards and Greek Obudsman's reports for the years 2000 to 2007.

For every case, we record the year of the recourse or the award publication, the legal status of the health care organization, the doctor's specialty, the type of medical error, the severity of the adverse event and the amount which the Administrative Court of First Instance imputed. The estimation of the financial cost is based on 31 lawsuits for which the Administrative Court of First Instance published awards during the period 2003-2007.

All the cases refer to Public or Not-for-Profit Health Care Organizations, because the administrative courts and the Greek Ombudsman have the authority to inquire cases only concerning Public and Not-for-Profit Organizations.

The assessment of patient safety should be carried out through both qualitative and quantitative methods. The qualitative methods (Institutionalization of Quality Assurance Project Report [QAP], 2001) map the various activities that exist in the routine delivery of services, for example using methods used in pathways analysis without, however, recommending one pathway as more appropriate than another. The purpose of the descriptive phase is to "map the genome of safety" in the delivery of care and services. The quantitative approach (C.E., 2006) uses indicators and epidemiological methods of analysis to systematically quantify distinct aspects of processes and their immediate outputs in relation to:

- adverse events;
- adverse events causing harm to patients;
- adverse events causing harm to providers; and
- for the risk of adverse events.

Surgeons and Obstetricians are the specialties most involved in medical errors as presented in the Table 6:

Specialty	Cases
General Surgeon	29
Obstetricians / Gynecologists	14
Orthopedic Surgeons	13
Ophthalmology Surgeon	11
Pathologists	10
Cardiac Surgeon	9

Clinical Microbiologist	8
Anaesthesiologist	7
Urologists	5
Gastroenterologist	3
Plastic Surgeon	3
Otolaryngologists (ENT surgeons)	2
All Other	18
n/a	9
Total	**141**

Table 6. Doctors' specialty

The most common medical errors are those referred to the category *"Error in performance of an operation, procedure or test"*, following with the *"Error or delay in diagnosis"*. The allocation of medical errors by error type (as defined in Table 2) is presented in the Figure 2:

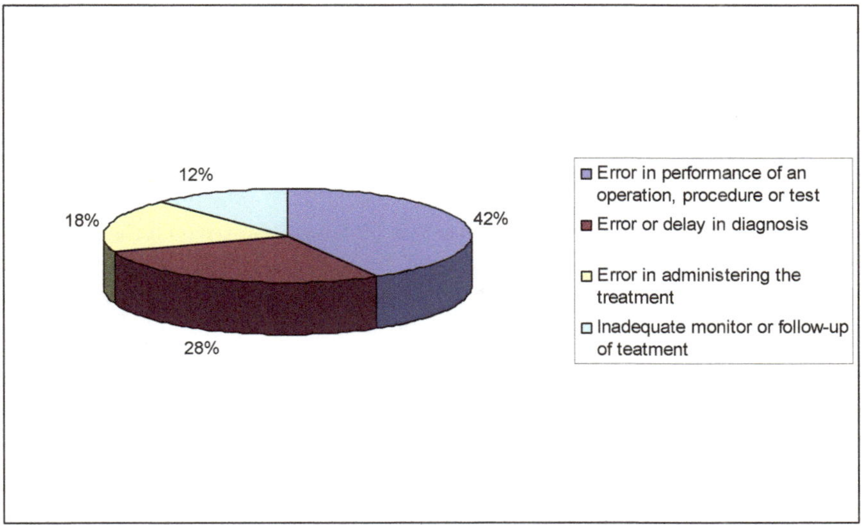

Fig. 2. Type of Medical error

The recorded medical errors caused various adverse events, with the most common the "**Category E**: *Temporary harm to the patient and required intervention & Category F: Temporary harm to the patient and required initial or prolonged hospitalization*" closely followed by the "**Category I**: *Patient death*".

Below, in the Figure 3, we present the allocation of medical errors by severity category, according to the National Coordinating Council for Medication Error Reporting and Prevention (NCC MERP) Index for Categorizing Errors (as defined in Table 3):

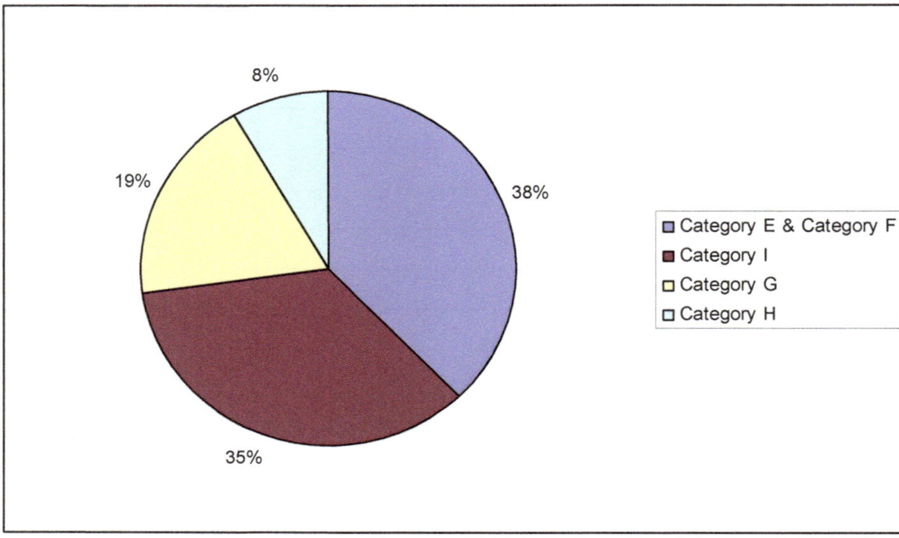

Fig. 3. Severity of medical error's adverse event

In Greece, contrary to other countries, the estimation of the total financial burden due to medical errors is very difficult due to the absence of an organized information system. Thus, the implementation of a system in order to identify, report and analyze medical errors and patient's adverse events following the international standards is crucial.

The estimation of the financial cost was based on 31 lawsuits for which the Administrative Court of First Instance published awards during the period 2003-2007. Our research pointed out that in Greece, the economic cost due to medical errors is worryingly high. In addition, the amount of mean and final compensation has been dramatically increased during the period 2003-2007 (Table 7 and Figure 4):

Year	Cases	Mean compensation	Total compensation
2003	3	136.972 €	410.916 €
2004	3	194.676 €	584.029 €
2005	5	50.972 €	254.860 €
2006	6	285.453 €	1.712.720 €
2007	2	375.000 €	750.000 €

Table 7. Mean and total compensation for years 2003-2007

The most injurious specialties are General Surgeons and Anaesthesiologists, while Anaesthesiologists have the higher mean compensation (Table 8 and Figures 5, 6 & 7):

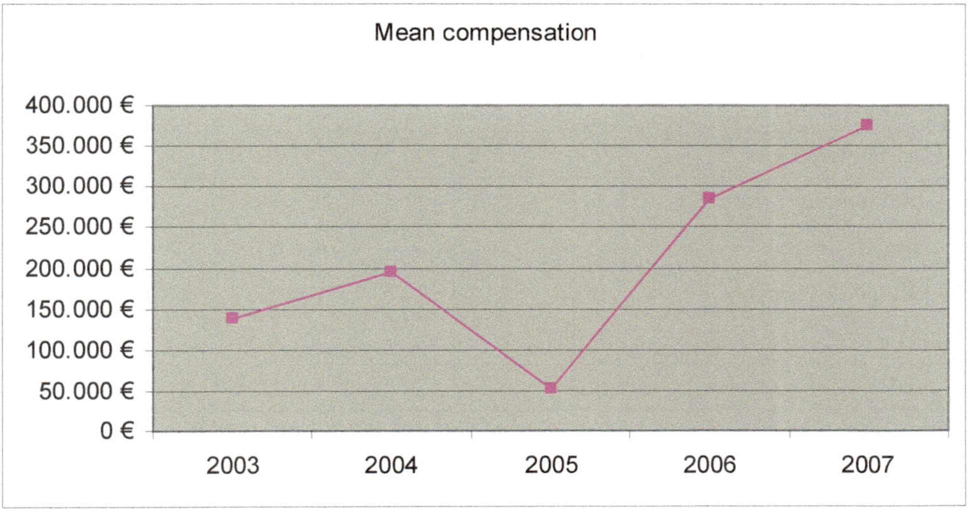

Fig. 4. Mean compensation for years 2003-2007

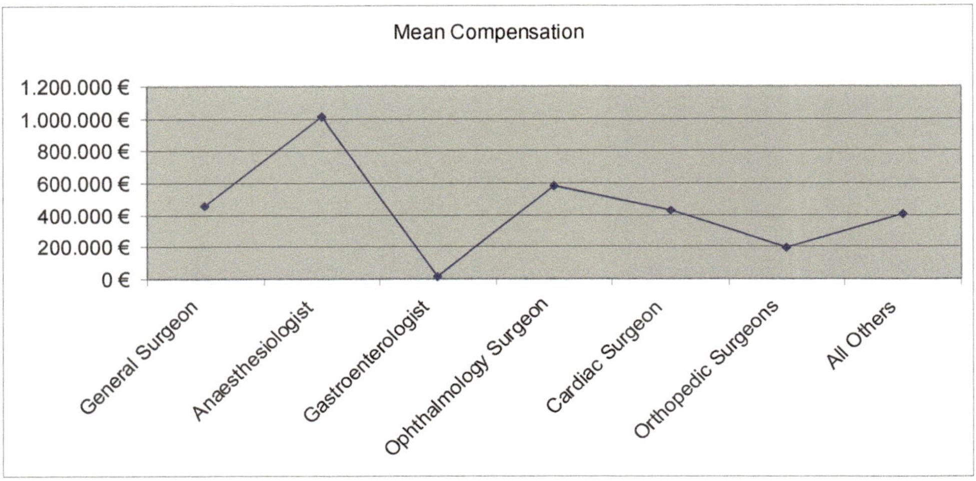

Fig. 5. Mean compensation for various specialties

Specialty	Cases	Mean compensation	Total compensation
General Surgeon	10	457.428 €	4.574.283 €
Anaesthesiologist	4	1.012.570 €	4.050.279 €
Gastroenterologist	2	17.500 €	35.000 €
Ophthalmology Surgeon	1	586.940 €	586.940 €
Cardiac Surgeon	2	426.838 €	853.675 €
Orthopedic Surgeons	3	196.950 €	590.850 €
All Others	9	407.333 €	3.665.995 €
Total	31	463.130 €	14.357.022 €

Table 8. Mean and total compensation for various specialties

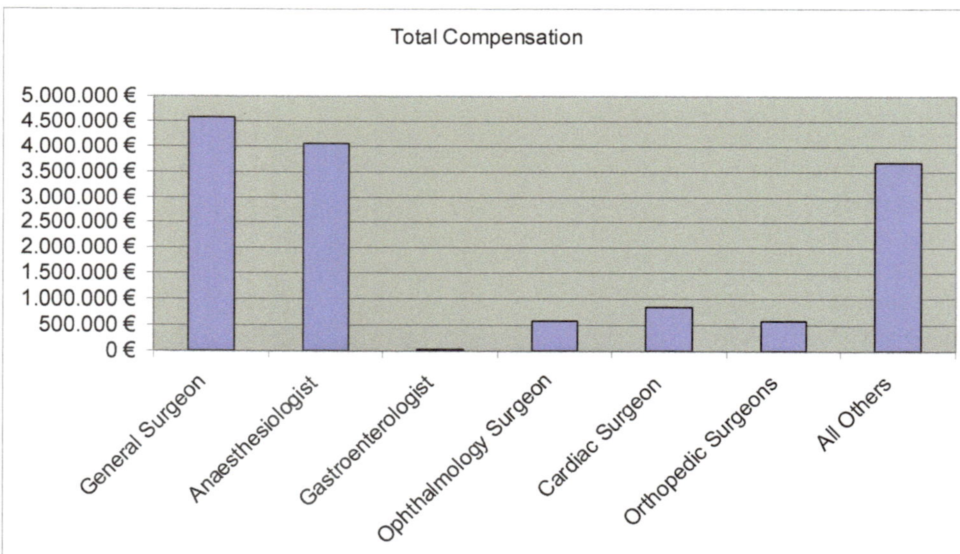

Fig. 6. Total compensation for various specialties

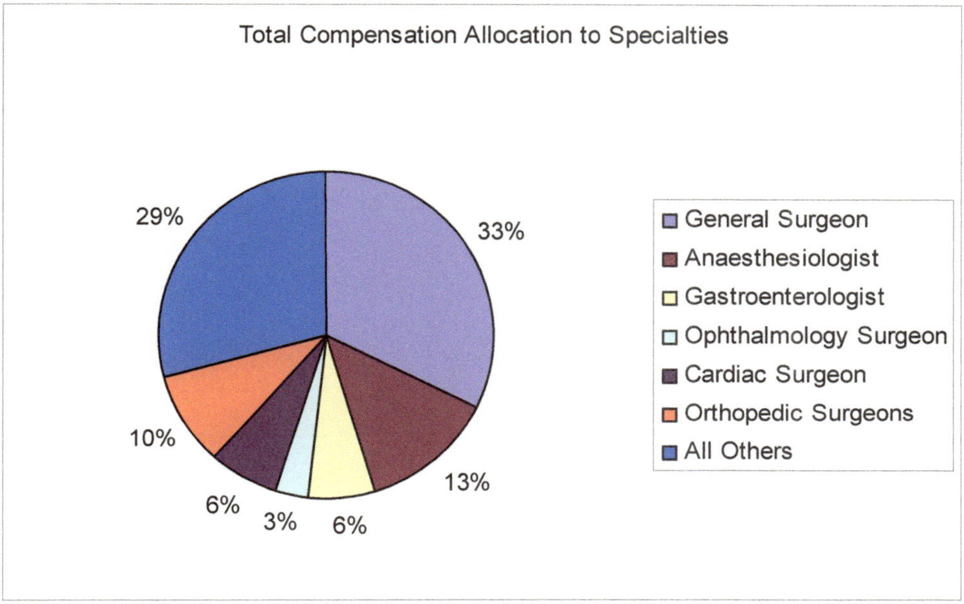

Fig. 7. Total compensation allocation to specialties

The highest mean compensation awarded to unknown severity adverse events and to the **Category H:** Intervention required to sustain life. It is remarkable that the awarded mean compensations for the **Category I:** Patient death is lower than those to adverse events categories with minor severity (Table 9 and Figure 8):

Severity	Cases	Mean Compensation	Total Compensation
Unknown	1	1.267.790 €	1.267.790 €
E	9	31.024 €	279.212 €
F	1	733.675 €	733.675 €
G	7	755.786 €	5.290.501 €
H	2	1.154.135 €	2.308.269 €
I	11	407.052 €	4.477.575 €
Total	31	463.130 €	14.357.022 €

Table 9. Mean and total compensation for adverse events severity categories.

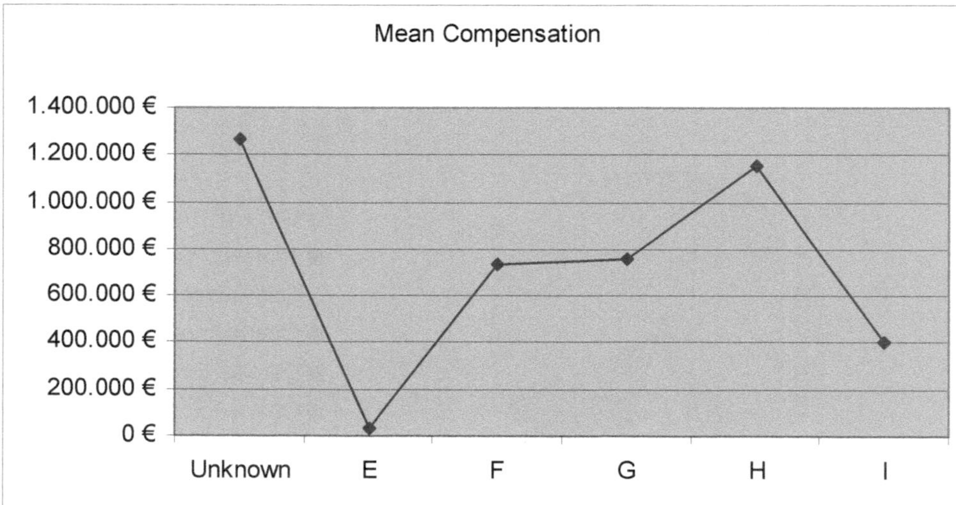

Fig. 8. Mean compensation for adverse events severity categories

7. Conclusions

Patient safety is extraordinarily important to the public, but the policy issues around adverse event detection and medical errors are questionable in practice. Within the Greek Health Care System today, most adverse events are being detected using spontaneous reporting, which identifies only a small number of adverse events. This is probably the major reason that problems with patient safety have been overlooked until recently.

Unfortunately, given the current structure of Greek Health Care System, there are strong incentives for Healthcare Organizations to turn a blind eye to medical errors and adverse events. In particular, serious, preventable adverse events typically should be reported to the Hospital Board and the Ministry of Health.

Unfortunately such events often lead to long lasting internal investigations from the hospital's management or end up in the press or in the courts, with adverse consequences for the doctor involved, the hospital or for the Health Care System as a whole.

Our research points out, that medical errors are a common phenomenon in Greek Health Care System (as in every Health Care System worldwide). Furthermore, they cause severe harm and substantial economic and psychological burden to the patients and to their relatives, professional medical liability to the doctors involved and a high economic burden to the Greek Health System and to the Greek Insurance Industry.

Though adverse events have negative connotations to many, our current system offers few incentives to healthcare organizations to look for them and possibly correct them aggressively.

It is clear that a systematic effort to understand and reduce medical errors will be the cornerstone of health care providers' professional responsibility in coming years, primarily due to the high costs associated with them.

Below, we present some of the current activities, as well as recommendations for additional activities and policy proposals to reduce medical errors:

- Building Public Awareness of Medical Errors
- Building Purchasers' Awareness of the Problem
- Working With Providers to Improve Patient Safety
- Using Decision-Support Systems and Information Technologies
- Using Standardized Procedures, Data Integration
- Checklists, and the Results of Human Factors Research

Specifically, Greek government should ensure that patient safety is the cornerstone of all relevant health policies, in particular policies to improve quality. For this reason, they due to develop a coherent and comprehensive patient-safety policy framework by promoting the development of a reporting system for patient-safety incidents in order to enhance patient safety, by reviewing the role of other existing data sources, such as patient complaints and compensation systems, clinical databases and monitoring systems as a complementary source of information on patient safety and also by producing regular reports on actions taken nationally to improve patient safety (C.E., 2006).

In developing patient-safety strategies, government should take a proactive, preventive and systematic attitude: to admit that errors happen, to identify and manage risk points in processes, to learn from errors and minimise their effects, to prevent further occurrences of patient-safety incidents and to encourage both patients and health-care personnel to report those patient-safety incidents they are confronted with. This could be achieved by proactive management and systematic design of safe structures and processes.

Patient safety should be recognised as the necessary foundation of quality health care, and should be based on a preventive attitude and systematic analysis and feedback from different reporting systems: patients' reports, complaints and claims as well as systematic reporting of incidents, including complications, by health-care personnel. The patient-safety strategy should become an integral component of the overall continuing quality-improvement programme. Investment in patient safety, as in quality improvement, should be considered as economically sound and good value for money (Institutionalization of Quality Assurance Project Report [QAP], 2001).

Support from the government to health professionals is crucial to make disclosure of the incident possible and to enable continuation of work in health care, where risks will always exist and adverse events happen (C.E., 2006).

Reducing the risk of error in health care will require a substantial and sustained effort at all levels of the health care system. It must become a priority goal wherever care is given—the doctor's office, the hospital, and the nursing home. That goal must be supported by the commitment of both human and financial resources. The Ministry of Health -and other regulators and accrediting bodies- must articulate the vision of safe care that they call upon others to work toward (National Coalition on Health Care and the Institute for Healthcare Improvement, 2000).

Of course, the key task for the future effectiveness of any medical errors' reduction strategy and policy will be to identify quality assurance practices that could respond effectively to system data. We must not forget, that even countries with a long history of error reporting, have not yet implemented comprehensive programs to correct problems once they are identified.

The primary objective of an incident reporting system (Vozikis & Riga, 2008b; Vozikis, 2009) is the enhancement of patient safety, by learning from adverse events and mistakes made. Reporting and collection of incident data is meaningful only if the data is analysed and evaluated and if feedback is given to the professionals involved in the incident, and to all others who could learn from the incident. Although the medical literature has focused primarily on medication- and procedure-related errors, there is little information on the potential benefits and hazards associated with the use of new medical technologies.

To sum up, patient safety deals with safe practices in a safe health care system where the health providers analyze the quality and safety indicators to prevent future adverse events.

8. Acknowledgement

This research has been co-financed by the European Union (European Social Fund - ESF) and Greek national funds through the Operational Program "Education and Lifelong Learning" of the National Strategic Reference Framework (NSRF) - Research Funding Program: Heracleitus II. Investing in knowledge society through the European Social Fund.

9. References

Barczak, C.A., Barnett, R.I., Childs, E.J., & Bosley, L.M. (1997). Fourth national pressure ulcer prevalence survey. Advances in Wound Care. Jul-Aug;10(4), pp. 18-26

Bates, D.W. (1998). Drugs and adverse drug reactions: how worried should we be? *JAMA*, Apr 15;279(15), pp. 1216-7

Bates, D.W., Cullen, D.J., Laird N. et al. (1995). Incidence of adverse drug events and potential adverse drug events. Implications for prevention. ADE Prevention Study Group. *JAMA* . Jul 5;274(1), pp. 29-34

Burger, S.G., Kayser-Jones, J., & Bell, J.P. (2000). Malnutrition and dehydration in nursing homes: key issues in prevention and treatment. *National Citizens' Coalition for Nursing Home Reform*

Council of Europe, Committee of Ministers. (2006). *Recommendation Rec7 of the Committee of Ministers to member states on management of patient safety and prevention of adverse events in health care,* 965th meeting of the Ministers' Deputies, Committee of Ministers on 24 May 2006

Dalrymple, J., & Drew, E. (2000). "Quality: On the Threshold or the Brink?" *Total Quality Management,* 11 (4-6), pp. 697-703

Dickinson, J.G. (2000). FDA seeks to double effort on confusing drug names. *Dickinson's FDA Review,* Mar;7(3), pp. 13-4

European Commission (2005). Special Eurobarometer

GeneralCologneRe (2002). Insurance Issues in Europe: Impending Changes in the European Health Care Sector and the Effect on Risk Management and Malpractice Insurance, December 2002

Griffin, F.A. & Resar, R.K. (2009). *IHI Global Trigger Tool for Measuring Adverse Events (Second Edition).* IHI Innovation Series white paper. Cambridge, MA:Institute for Healthcare Improvement

Healthcare Cost and Utilization Project. (2003). In: *Agency for Healthcare Research and Quality Rockville, MD,* Available from: http://www.ahrq.gov/data/hcup/hcupnet.htm

Institute of Medicine. (1999). *To Err is Human: Building a Safer Health System.* Washington, DC: National Academy Press

Institute of Medicine. (2001). *Crossing the Quality Chasm: A New Health System for the 21st Century.* Washington, DC: National Academy Press

Institutionalization of Quality Assurance Project Report. (2001). Published for the U.S. Agency for International Development (USAID) by QAP, Bethesda, Maryland, U.S.A.

Joint Commission on Accreditation of Healthcare Organizations (JCAHO). (2011) Sentinel Event Policy and Procedures, Available from http://www.jointcommission.org/Sentinel_Event_Policy_and_Procedures/

Kelly, D. (Ed.). (2006). *Applying quality management in healthcare: a systems approach,* Health Administration Press, AUPHA Press, 2nd ed., ISBN-13 978-1-56793-260-7, Washington, DC

Lazarou, J., Pomeranz, BH., & Corey, PN. (1998). Incidence of adverse drug reactions in hospitalized patients: a meta-analysis of prospective studies. *JAMA,* Apr 15;279(15), pp. (1200-5)

Leape, L.L. (1993). Preventing Medical Injury. *QRB Quality. Review Bulletin.,* 19 5, pp. 144-149

Leape, L.L. (1994). Error in Medicine. *JAMA,* 21;272, pp. 1851-1857

Lohr, K. N. (Ed.). (1990). *Medicare: A Strategy for Quality Assurance.* Washington, DC: National Academies Press

Morbidity and Mortality Weekly Report. (2000). *Fourth Decennial International Conference on Nosocomial and Healthcare-Associated Infections,* Vol. 49, No. 7, p.138

National Coalition on Health Care and the Institute for Healthcare Improvement (2000). Reducing Medical Errors and Improving Patient Safety Success Stories From the Front Lines of Medicine, Monograph, in Accelerating Change Today (A.C.T.) foe America's Health

National Coordinating Council on Medication Error Reporting and Prevention. (1998). Taxonomy of Medication Errors

NQF. (2007). *Serious Reportable Events in Healthcare – 2006*, ISBN 1-933875-08-9, Washington D.C.

Null, G., Dean, C., Feldman, M., Rasio, D., & Smith, D. (2007). *Death by Medicine*, Life Extension Foundation

Quality Interagency Coordination Task Force. (2000). Report to the President, February, Doing What Counts for Patient Safety: Federal Actions to Reduce Medical Errors and Their Impact, Accessed September 7, 2003, Available from: http://www.quic.gov/report/mederr4.htm#evidence

Starfield, B. (2000a). Is US health really the best in the world? *JAMA* , Jul 26;284(4), pp. 483-5

Starfield, B. (2000b). Deficiencies in US medical care. *JAMA*, Nov 1;284(17), pp. 2184-5

Thomas , EJ & Petersen LA. (2003). Measuring Errors and Adverse Events in Health Care, *Journal of General Internal Medicine 18* (1), pp.61–67.

Thomas, E.J., Studdert, D.M., Burstin, H.R., Orav, E.J., Zeena, T., Williams, E.J., et al. (2000b). Incidence and types of adverse events and negligent care in Utah and Colorado. *Medical Care,* 38(3), pp. 261–271

Thomas, E.J., Studdert, D.M., Runciman, W.B., Webb, R.K., Sexton, E.J., Wilson, R.M., et al. (2000a). A comparison of iatrogenic injury studies in Australia and the USA. I: Context, methods, casemix, population, patient and hospital characteristics. *International Journal for Quality in Health Care,* 12, pp. 371-8

Tunis, S.R. & Gelband, H. (1994). Health care technology in the United States. *Health Policy,* Oct-Dec;30(1-3), pp. 335-96

U.S. National Center for Health Statistics. (2003). National Vital Statistics Report, vol. 51, no. 5, March 14

Vincent, C., Stanhope, N. & Crowley-Murphy, M. (1999). Reasons for not reporting adverse incidents: an empirical study. *Journal of Evaluation in Clinical Practice*. Feb;5(1), pp. 13-21

Vozikis, A. & Riga, M. (2008a). Medical Errors in Greece: The Economic Perspective through the Awards of Administrative Courts, *Journal of Society, Economy and Health Vol. ,2, pp.* 423-437

Vozikis, A. & Riga, M. (2008b). M.E.R.I.S: Information System for the identification, collection, analysis and reporting of Medical Errors, *Proceedings of 2nd International Scientific Conference eRA-2,* Greece-Athens, September 2007, pp. 423-437

Vozikis, A. (2009). Information management of medical errors in Greece: The MERIS proposal. *International Journal of Information Management,* 29, pp. 15–26

Wald, H. & Shojania, K.G. (2001). Incident reporting. chap 4. Evidence Report/Technology Assessment, In: *Making Health Care Safer: A Critical Analysis of Patient Safety Practices. Rockville, MD: Agency for Healthcare Research and Quality,* Shojania, K.G., Duncan, B.W., McDonald, K.M., et al, eds., No. 43. AHRQ publication 01-E058

Weinstein, R.A. (1998). Nosocomial Infection Update. *Emerg Infect Dis,* 4(3), pp. 416-20

World Health Organization. (1984). Publication DEM/NC/84.153(E)

World Health Organization. (2006). *Quality of care. A process for making strategic choices in health systems.* WHO Library Cataloguing-in-Publication Data, ISBN 92 4 156324 9, 978 92 4 156324 6, WHO Press, Switzerland

Xakellis, G.C., Frantz, R., & Lewis, A. (1995). Cost of pressure ulcer prevention in long-term care. *Am Geriatr Soc,* 43(5) pp. 496-501

15

Critical Success Factors for Quality Assurance in Healthcare Organizations

Víctor Reyes-Alcázar, Antonio Torres-Olivera,
Diego Núñez-García and Antonio Almuedo-Paz
Andalusian Agency for Health Care Quality
Spain

1. Introduction

In recent years in the health services the need has been stressed to use management tools which highlight the central role of professionals and which support the implementation of a clinical leadership based on greater autonomy and decision-making in management. The purpose is to allow alignment with the ultimate goals of health systems, which are to strive for a service with a high degree of effectiveness, to respond to the needs and expectations of citizens, and to have medium and long term sustainability.

There are many conceptual approaches in the field of clinical management: its realization in defined organizational spaces (Clinical Units); synergy with other management tools such as the process or clinical pathway approach and the definition of competences; the use of clinical guidelines or the use of scientific evidence as guarantees of clinical effectiveness; the new relationship between professionals and patients where more proactive models promote the use of patient decision aids that encourage participation in shared decision-making, where core dimensions of healthcare quality such as continuity and safety are reasonably assured.

Clinical management could be defined as the ability of health professionals to manage the resources they use in their clinical practice efficiently and effectively (Torres-Olivera & Reyes-Alcázar, 2011). It establishes the effective involvement of professionals in achieving the objectives of the healthcare organization and is associated with greater decision-making capacity and autonomy. It also establishes greater commitment to the citizen (taking into account their needs and expectations) and to the healthcare organization by promoting sustainability in the short, medium and long term. There are many initiatives aimed at boosting clinical management in the public health sector as a strategy to ensure greater effectiveness and sustainability of the health services. This chapter seeks to analyze possible critical success factors to be considered when starting this type of project.

2. Critical Success Factors

A Critical Success Factor (CSF) is a particular feature of the organization's internal or external environment that is important for it to achieve its objective. A factor is critical if its fulfilment is absolutely necessary to achieve these objectives, thus requiring action from the sectors involved. Critical success factors that can determine the proper development of clinical management can be summarized as follows:

2.1 Patient-centred care

From a public service perspective, the focus of healthcare is the citizen, so any approach to intervention should be based on their needs and expectations. To achieve this, these needs and expectations must be explored and understood. Their fulfilment must be seen as a fundamental dimension in the quality of service provided. In the last 40 years a large body of literature "*has been advocating a patient-centred approach in healthcare*" (Mead and Bower, 2000, 1087). Since then, different definitions of the concept have emerged. Byrne and Long (1976) proposed a definition that emphasized a style of medical consultation that uses the patient's knowledge and experience to guide the interaction between themselves and the healthcare professional. From a clinical management point of view rather than healthcare, Laine and Davidoff (1996) understood patient-centred care as more "*congruent with, and responsive to patients' wants, needs and preferences.*" This entails: the inclusion of a biopsychosocial perspective in the treatment of disease; the patient being conceived as "*a personal experience rather than an object*" suffering from an illness (Mead & Bower, 2000: 1089); sharing decisions and responsibilities in a more cooperative manner; assuming an inter-subjective character in the doctor-patient relationship, strengthening its human dimension. This way of approaching healthcare attempted to oppose the conventional "*biomedical model*" (Friedson, 1970) of doctor-patient relationship. In that conventional model the relationship is reduced to identifying the clinical signs and symptoms of a disease and the subsequent prescription of a standardized treatment.

At present the clinical setting is much more open. The traditional reliance of service organizations on changes in their environment was not applicable to health services imbued with technical self-importance, the guarding of information and one-sided communication processes with patients. The paradigm shift aimed at making the citizen the real objective of clinical practice, has opened a new scenario in which the units must build a new relationship with the patient in which new skills and abilities should form part of the action plans of professional teams.

From the clinical management standpoint, the model of patient-centred care carries certain elements of responsibility for the patient. In this sense, it is assumed that the information must be accurate and shared, rights respected and participation guaranteed. These elements define perfectly the changes of direction that a clinical unit or department must take in order to develop a clinical management plan, and which basically involve:

- Knowing the users and potential users of the health services. Knowing who they are and how many they may be, but above all, knowing their needs and expectations. A clinical unit or department must build its services on that foundation.
- Knowing the level of satisfaction among patients who have used a clinical unit or department and using this information to detect new areas for improvement and development.
- Providing adequate and accurate information, promoting opportunities for participation and shared decision-making, properly handling procedures for informed

consent and patient decision aids. To achieve this goal it is essential to improve the communication skills of the professionals in the clinical unit.

- Transparency, to bring the results of the clinical unit or department to the attention of society, opening channels of communication that support the transmission of information and knowledge, and encouraging participation through social networks, patient forums, etc.

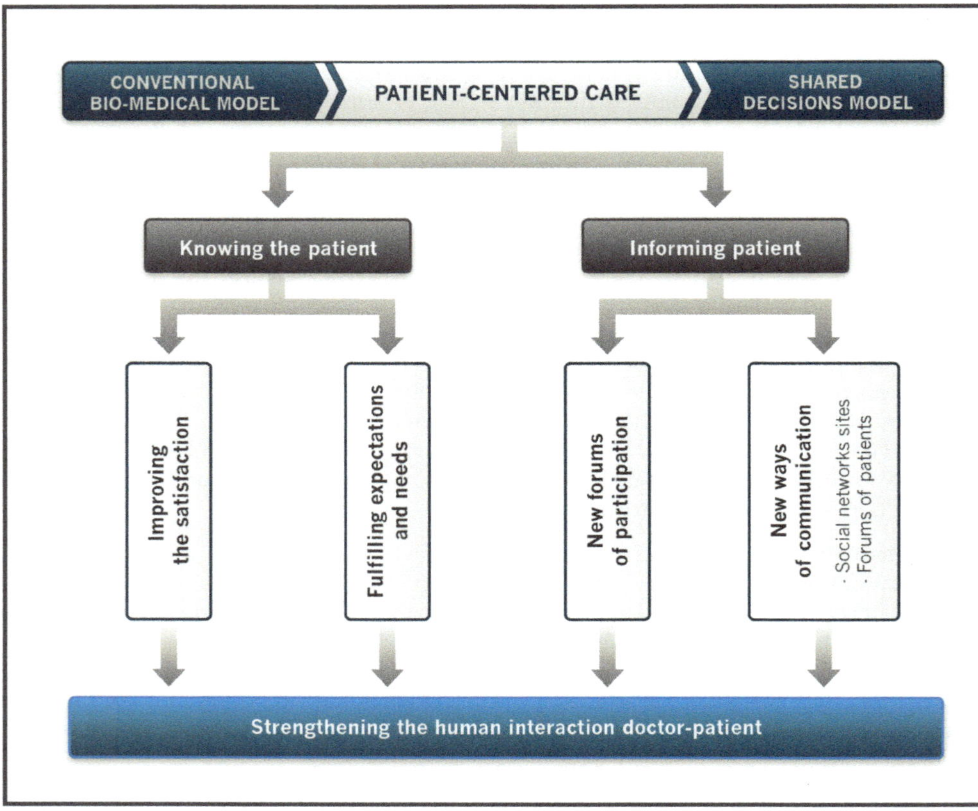

Fig. 1. Patient-centered care

2.2 Leadership

Clinical leadership is central to the strategy of clinical management. Studies, such as those of Cunningham and Kitson (2000) and West et al. (2004), have demonstrated the importance of clinical leadership in improving the quality of care and how the use of evidence-based medicine has resulted in a more patient-centred practice. The importance of the role played by the clinical leadership is acknowledged by the government agencies responsible for ensuring the quality of care, such as the National Health Service of Scotland being given responsibility for "*driving service improvement and the effective management of teams to provide excellence in patient/client care*" (NHS Scotland 2004: 4). Also, other organizations such as the

Joint Commission speak of "*effective leadership*" that determines the strength and consistency of performance in areas such as planning, management, coordination, provision and improvement of health services (Joint Commission Resources, 2009).

However, this leadership is not an isolated factor and requires other support. As stated by West et al. (2004), other support mechanisms and staff must be deployed and promoted for clinical leadership to be successful in achieving greater quality of care. Among these mechanisms, those associated with the empowerment paradigm (Abdelrazek et al., 2010) are highlighted. There is broad academic consensus on the importance of empowerment in work environments. On the one hand, there is a macro perspective on the contextual agents and socio-structural conditions that facilitate empowerment. On the other, the micro perspective focuses on the feelings of employees about their role within the organization, and has been called "*psychological empowerment*" (Knol & van Linge 2009). Both aspects of empowerment, structural and psychological, provide clinical leadership with performance tools that improve both individual work and the work context in which they are located.

Therefore, professional leadership must be encouraged in all its aspects, making the stakeholders of the clinicians the objectives of the organization, striving for commitment to them and providing effective leadership tools from the management structures.

This process necessarily involves the need to define the map of the competencies, attitudinal and/or knowledge and abilities, that healthcare professionals have to incorporate. In other words, it is a matter of cataloguing best practices to guide professionals to achieve their own development goals. It establishes a set of competencies that helps determine the gap between the actual competencies that a particular professional may have and those required by their competencies map. An appropriate training programmes can then be established to reduce any gap. Different international organizations, such as the International Union for Health Promotion and Education (IUHPE), have encouraged and agreed standards for career development within the healthcare sphere, helping to establish a common international professional development framework (Shilton, 2009).

In this regard, the launch of an ambitious professional development project by the Andalusian Agency for Healthcare Quality (Spain) is highlighted. A tool for the Management of Individual Development Plans has been designed and implemented which allows the competence gaps to be identified among the professionals of the clinical units. This is achieved through a process of self-assessment based on identifying levels of achievement in relation to a catalogue of best practices for the job. Training or professional development programmes are established according to the identified gap.
(http://www.juntadeandalucia.es/agenciadecalidadsanitaria/formacionsalud/)

It is also important to clearly define the figure of the manager or director of the clinical unit, establishing their competency map and their assigned functions. Interaction with the managers of other healthcare centres and political structures of the healthcare system is stressed as one of the core attributes that define clinical leadership, according to Christian and Norman (1998). They also play the role of manager of all the resources (human, financial, equipment and materials) that make up the Clinical Unit.

On this topic, it should be noted that the Observatory for Quality of Health Training-Spain, using the tool for the Management of Individual Development Plans, has defined the competency map of the managers or directors of the Clinical Units of the Andalusian Public

Health System. This map consists of 26 specific practices and 106 requirements with different levels of achievement (from the essential to the strategic) in 12 key competencies: [1] Attitude of continuous learning and improvement [2], Scientific and technical capacity, [3] Capacity for decision-making, [4] Communication [5] Management and Planning, [6] Promotion of professional development, [7] Management of the quality of clinical safety, [8] Efficient management of resources, [9] Innovation and Leadership, [10] Promotion of research and teaching, [11] Guidance to citizens [12] Results-orientation. (http://www.juntadeandalucia.es/agenciadecalidadsanitaria/)

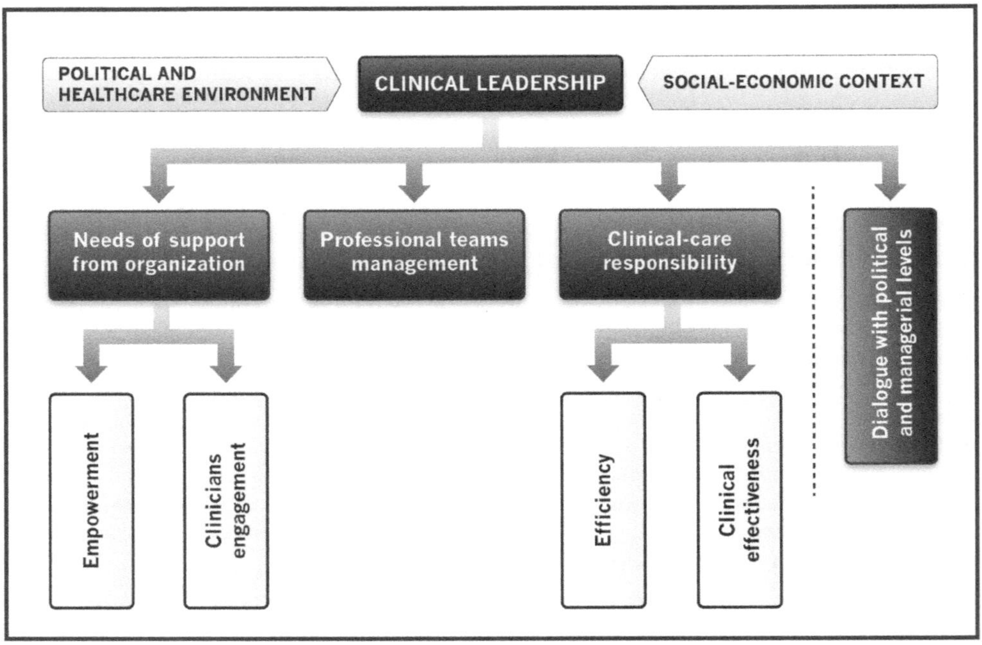

Fig. 2. Leadership

2.3 Teamwork

A patient-centred approach to healthcare involves a multidisciplinary care process built around a team with common goals and developed in an integrated organizational model. The importance and effectiveness of teamwork in healthcare processes have been widely studied in recent decades. In fact, some studies have empirically validated their effectiveness through the study of their structures and dynamics (Mickan, 2005). Mickan (2005) indicates two large blocks of benefits of effective teamwork: collective and individual benefits. Among the collective benefits may be found those affecting the organization, such as reducing the time and costs of hospitalization and greater accessibility for patients, and those affecting the team such as a more efficient use of healthcare services or improved communications. The individual benefits affect the patients by improving satisfaction, greater acceptance of treatment and improved health outcomes, and the professionals, who experience improved satisfaction and well-being at work.

In the view of this chapter's authors, teamwork will play an important role in the effectiveness of the results obtained in clinical practice. In other words, the capacity of health professionals in a particular clinical unit to integrate will be directly related to their clinical effectiveness and their decision-making ability. This will therefore determine the final results of the unit. For example, the Strategic Plan 2010-2013 of the Hospital Clinic of Barcelona is based on teamwork as an essential element in the organization of healthcare units (Castells, 2011).

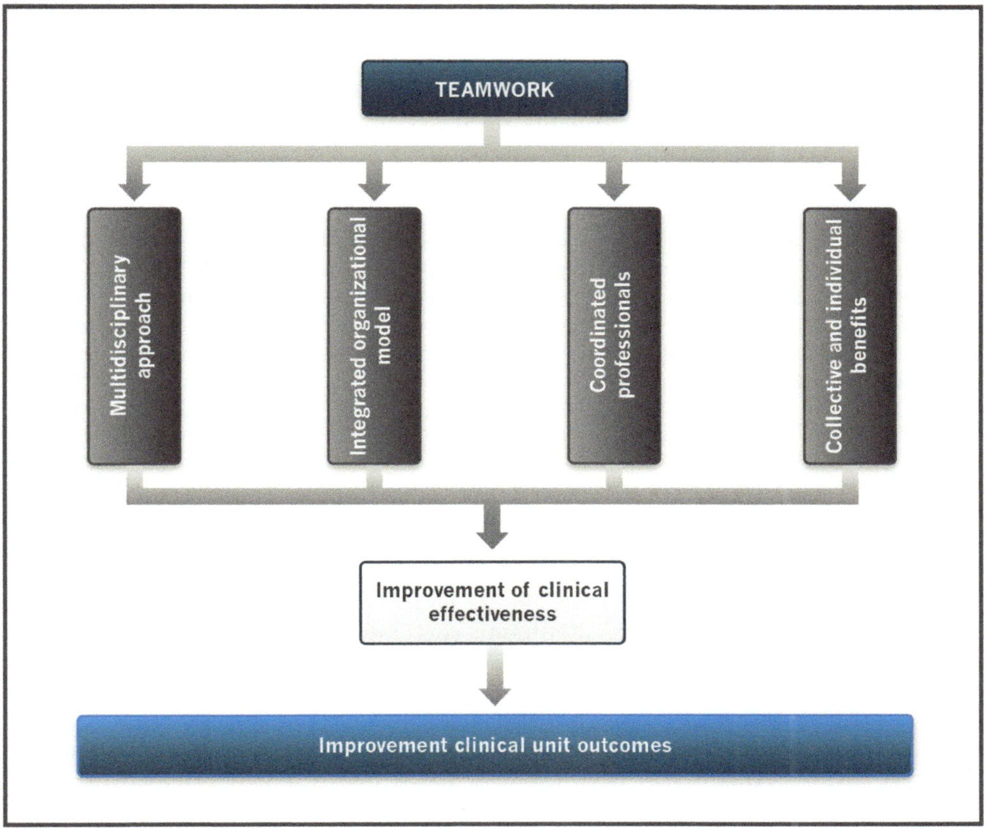

Fig. 3. Teamwork

Teamwork encourages a sense of belonging and differentiation as a motivating element. The team must know what the objectives of the unit are, and what their individual contribution to them is, regardless of the professional group and level to which they belong. Similarly, they should know what their results are, both individually and collectively, and actively participate in the improvement processes that are instigated. The working model must transcend the limits of the clinical unit, with continuity of care being a dimension that the healthcare unit must take into account in order to improve clinical outcomes through consensus building between different levels of professionals involved in patient care.

2.4 Autonomy and responsibility

The ability to make decisions on resources in clinical practice is the core concept of clinical management. That capacity has to be based on the commitments established with the organization through procedural agreements that define the playing field between healthcare units (departments) and Corporate Management. This is probably one of the most difficult success factors to develop in practice because it incorporates a different culture, which clashes with the values, or rather with the roles traditionally assigned to the different actors in the system. There is widespread academic agreement on the need to maintain a certain degree of autonomy for healthcare professionals in order to improve not only their ability to cope with everyday challenges, but also as a fundamental aspect of their personal performance and motivation (Harrison and Dowswell, 2002). The demands of autonomy and responsibility for the clinical units stem from those benefits that have been empirically demonstrated in the professional field (Schulz et al. 1991, Akre et al. 1997, Kapur et al. 1999). The main focus of these analyses has been on the ability of professionals and clinical units to determine or establish their own clinical strategies and the assessment of their own performance. However, demands for autonomy and "*decentralization*" of decision-making in clinical practice have not been without developmental difficulties. As in any other field of work, autonomy in clinical practice has faced problems that derive from the ongoing "*bureaucratization*" of the processes and "*mechanization*" of the tasks of professionals, with their decisions subject to ever increasing controls. (McKinlay & Arches, 1985).

There is little tradition of managers delegating tasks and responsibilities to healthcare professionals, even though it is the clinical decisions that have a more direct impact on resource consumption. Furthermore, healthcare professionals are not inclined to take responsibilities which they believe are outside their healthcare role. To this can be added a traditional distrust between the two camps. The nature of this distrust is rooted in the culture clash between the medical corpus and those that Alford (1975) termed "*corporate rationalists*", these are managers and healthcare policy-makers that need not be medically qualified (Camprubí, 2011). For Alford, the corporate rationalists are confronted by structures that represent the interests of the "*professional monopolists*", i.e., the healthcare professionals, and they stress rational planning and efficiency over the decisions of healthcare experts (Alford, 1975, Lewis, 2006). Along the same line, Lewis states "*...the power of the medical profession in the health policy arena, by analysing which actors are perceived as influential, and how influence is structured in health policy*" (Lewis (2006). Despite the claims of other studies, Lewis empirically demonstrates the increasing capacity of healthcare professionals to influence health policy, and therefore they have greater responsibility in decision-making.

It is therefore essential to incorporate new management styles where the prime role of managers is to facilitate healthcare processes and to support the operation of clinical units. On the other hand, professionals must be introduced to, and trained in, the various competencies that encourage their participation and involvement in decision-making on the use of resources. Finally, all this must be articulated through management or grant agreements that clearly establish the commitments made by both sides, linking the achievement of individual objectives, and those of the clinical unit, to an incentives model. Although there is broad consensus on this, when it comes to it being put into practice, healthcare organizations face difficulties caused largely by the, not always peaceful, coexistence of strong, pre-existing corporate interests or divisional structures that detract from the patient-centred vision and which make decision-making difficult at the clinical unit level. At other times, it is the managers of the healthcare organizations themselves who are reluctant to redirect their own role to allow real decentralization of management processes, incorporating and engaging clinical decision makers. It is probably necessary to introduce different rules which enable some professional and management roles to be redefined, or to create less hierarchical structures that more obey effectiveness and efficiency criteria, that allow effective integration of resources to give a more cohesive response to the needs and expectations of the citizens.

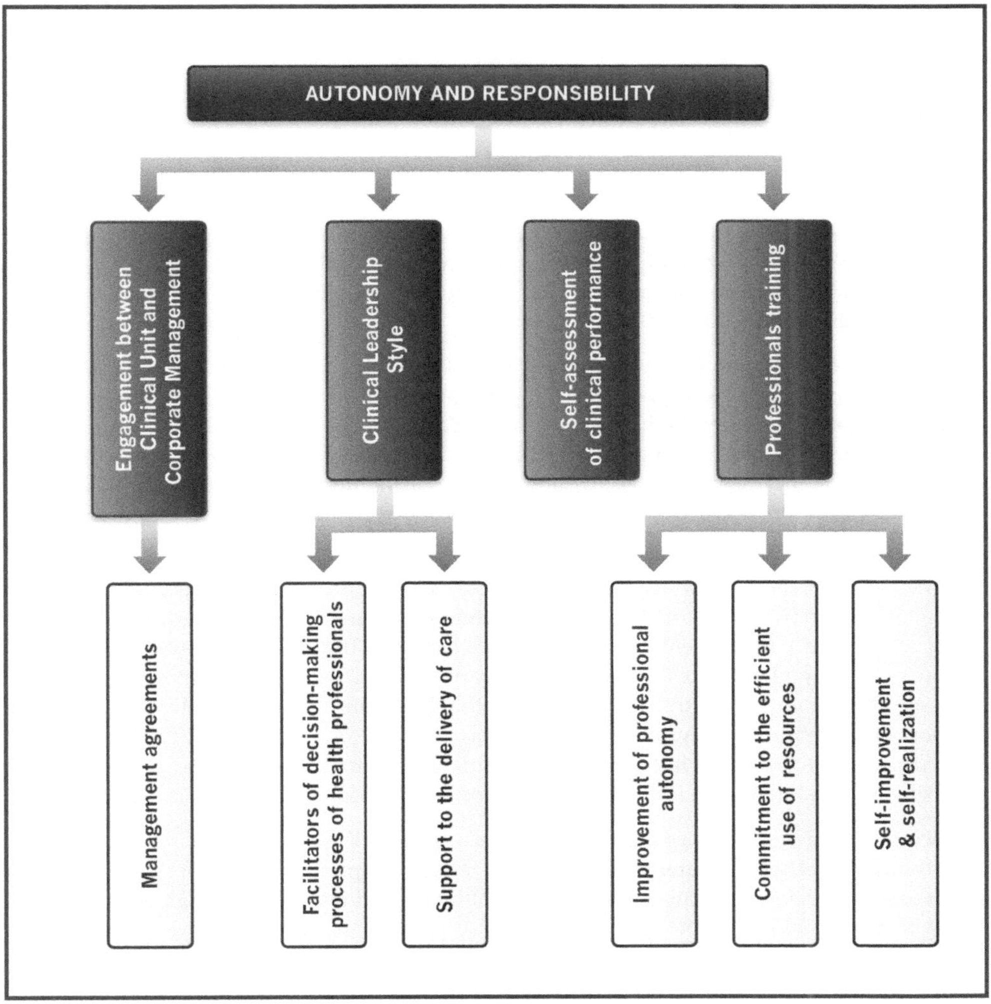

Fig. 4. Autonomy and responsibility

2.5 Care organized from an integrated view of processes

In today's information society the citizen plays a major role, in such a way that their expectations have a direct impact on the modes of action of any service delivery organization. Factors such as user-perceived quality have become essential elements that bring legitimacy to any organization providing services, and thus become an important value to be analyzed by any healthcare organization (Torres Olivera, 2003).

This scenario requires an analysis of what healthcare organizations are doing, how they are doing it and what level of response is given to the needs and expectations of the user of healthcare services. However, this may not be sufficient unless it is done from a comprehensive view of healthcare processes. For instance, it is difficult to refer to concepts such as total quality, or a comprehensive view of quality, if some issues on continuity of care and appropriateness of healthcare processes have not been previously resolved.

The possibility of receiving comprehensive and continuous healthcare currently faces a number of difficulties, largely caused by how the health services are organized. Departmental segmentation is possibly due more to the interests of managers and professionals than the needs of users. Organizations can be unnecessarily complex, and often "*super-specialization*" does not properly address the diversity or multisystemic nature of the problems that health services face. If inadequate mechanisms for coordination between different levels of care and a poor patient-centred tradition are added, the idea of comprehensiveness that encompasses the concept of total quality may seem a little beyond reach (Torres Olivera, 2003 2004).

On the other hand, if the expectations of citizens regarding what they demand from healthcare organizations, especially if they are public, are explored in depth, it is seen that they express a desire for higher quality healthcare in terms of accessibility, effectiveness , safety and information. Satisfaction must be measured, but it is increasingly conditioned by elements of comprehensiveness. In fact, poor coordination between levels of care has a negative effect on accessibility to services, and the effectiveness and safety of clinical performance.

Evolving from "*how much*" is it done to "*how*" is it done, is undoubtedly an important qualitative step in healthcare organizations, and is related to a different vision of healthcare that assumes a commitment to quality and which requires a comprehensive view of the processes taking place in their services. The integrated approach to these care processes should pursue the following objectives: [1] Ensure continuity of care through a continuous and shared vision of healthcare [2] Adapt the functional structure of the services to the needs and demands of citizens, [3] Link professional effort to the final objective, the patient outcome, sharing responsibilities, [4] Put the resources (costs) in the right place (greater benefit).

Process management is organized through a number of key elements such as: [1] user-centred approach, [2] greater involvement of healthcare professionals, [3] support from the best scientific evidence available, and [4] use of integrated information systems.

Process management attempts to ensure continuity of care at all times, and is therefore intended to seek a unique and coordinated provision of services, avoiding fragmentation of care at multiple levels (Consejería de Salud - Regional Ministry of Health, 2001). Therefore, process management is understood as the set of elements that are linked sequentially to meet the health needs of citizens. It provides a vision that goes beyond inter-level coordination and attempts to delve into the meaning of "*Care Continuity*" (Torres Olivera, 2001). This approach requires healthcare organizations to order the actions that are performed in different areas, by different professionals at different times. The processes are thus approached horizontally, involving all levels of care and all the professionals involved in them.

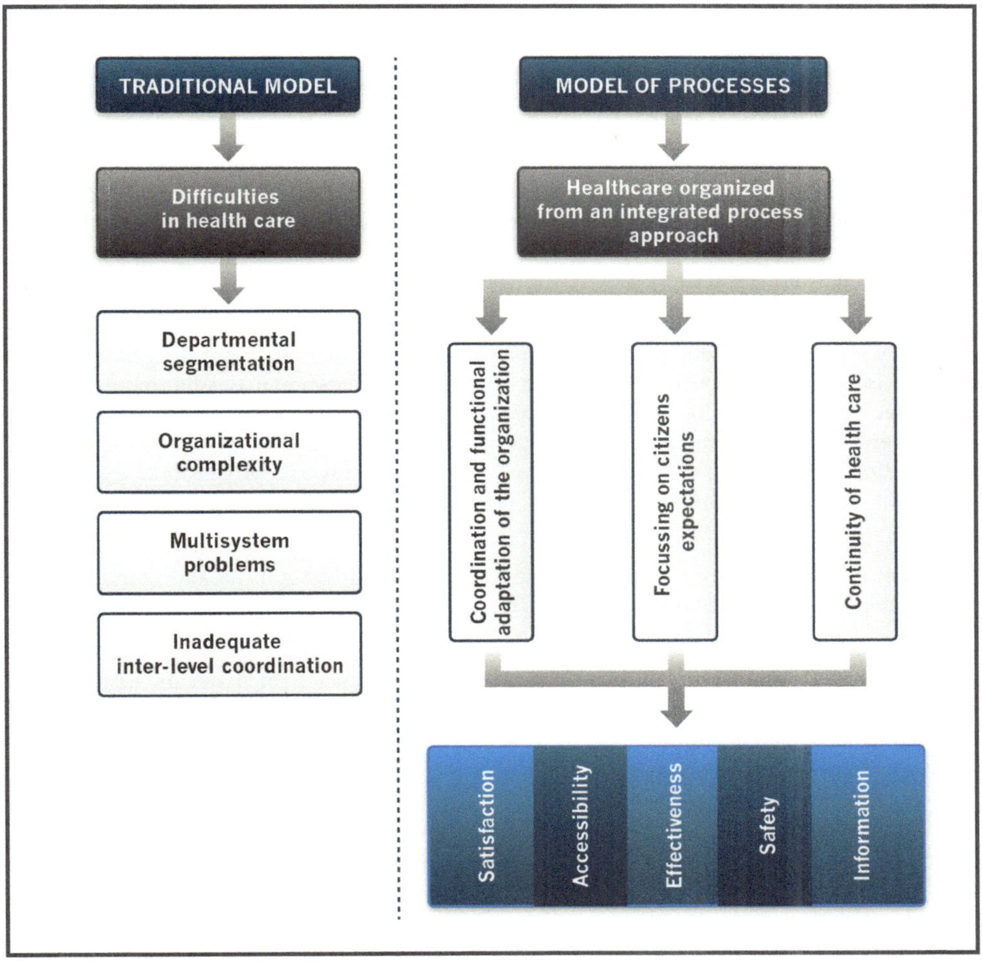

Fig. 5. Care organized from an integrated view of processes

Integrated process management is an approach to patient healthcare that seeks to coordinate resources across the health system (Fernández, 2003). This means an integrated approach to preventive actions and health promotion, the use of clinical practice guidelines, appropriate criteria in the management of resource and support to diagnosis, the appropriate use of drugs and the evaluation of results.

This means trying to analyze working methods in the light of the best available evidence, proposing elements for improvement so that added value is given to recipients of the healthcare process, i.e. the citizens and patients. Process mapping a clinical unit or department will determine its portfolio of services and its quality characteristics, as well as establish roadmaps for patients in the healthcare setting and establish the competencies that professionals have to develop (Mora Martínez, 2002). The approach to healthcare with a focus on processes and the definition of roadmaps for patients allows the identification of critical elements related to information or communication with the patient, or the introduction of important aspects related to healthcare quality such as patient safety.

2.6 Professional competencies

The term *Competence* in its current sense is due to David McClelland, a psychologist at Harvard University, an expert in motivation theory, who published an article in 1973 which caused a radical shift at the time. The article suggested going beyond the traditional evaluation methods in human resources management to focus on directly searching for those behaviours that were shared by those who were excellent at their job within a specific culture, and which differentiated them from the rest (McClelland, 1973).

From this perspective, the determination of professional competences, and the generation of tools that manage them, becomes a key to continuing professional development (Reyes-Alcázar et al., 2011). Competence management not only encourages professional development but also innovation in learning models and processes of exchange and dissemination of knowledge. It also directs an organization towards professional excellence, cooperative work and the development of models of professional recognition and incentives.

The map of competencies of healthcare professionals should be set as a Gold Standard that contains the knowledge, skills and attitudes desirable for developing excellent clinical practice.

A typical competency map consists of a set of key competencies for a specific job and the best practices that a professional should develop for optimum performance in that job. The map should be related to the desired results and the individual and organizational objectives. Consequently, competence refers to a person's ability to efficiently undertake a specific job. Thus, competence has to meet certain criteria: [1] Consider the context, i.e., the real world scenarios where these competencies have to be developed. [2] Identify the desired results in terms of level of development, achievement or mastery of tasks or functions. [3] Associate the level of development of each task or function with the performance, requirement or evidence criteria. [4] Involve the areas of responsibility of the professional.

By comparison with the competence map, the healthcare unit professionals can carry out a process of self-assessment to establish the competence gaps, and produce individual professional development plans that will determine the training and learning processes that are needed for their development.

One of the most important projects currently under development in the field of competence management is that undertaken in Andalusia (Spain) by the Andalusian Agency for

Healthcare Quality for the Public Health Service since 2006. This accreditation model is intended to recognize excellence of professionals through a self-assessment system that identifies the competences that a particular professional should have, the good practices that should be present in each job and the evidence that must be provided to demonstrate those competences.

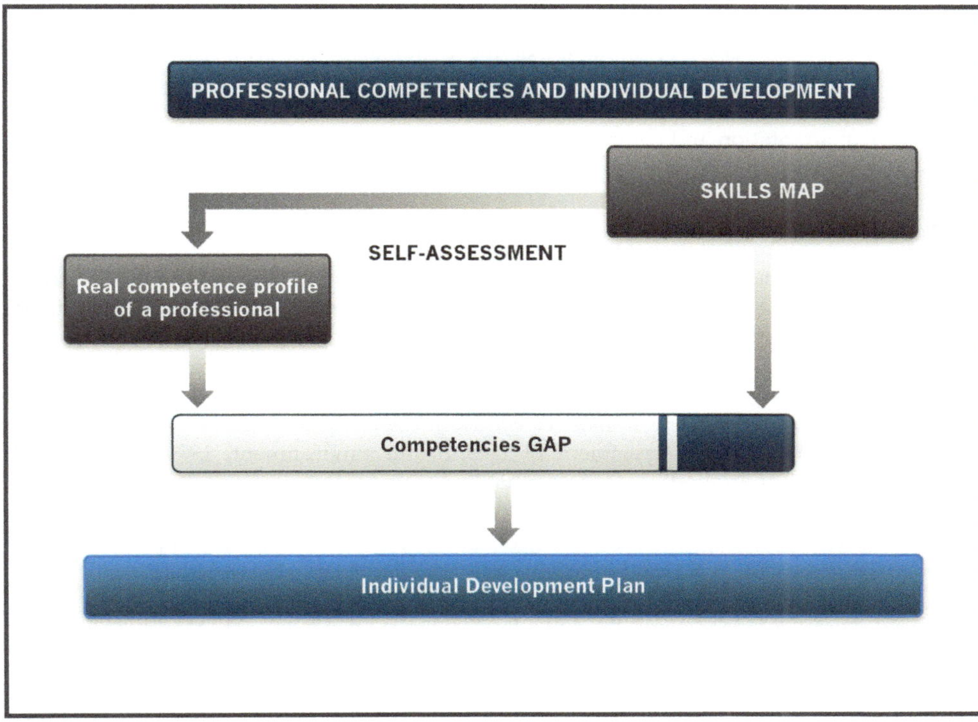

Fig. 6. Professional competencies

In the Competence Management Model of the Andalusian Public Health System, the *Accreditation* is defined as a recognition, explicit and public, of meeting the necessary requirements for the provision of quality care, as well as the beginning of a line of continuous improvement by a professional. (Almuedo-Paz et al, 2011).

The Competence Management Model of the Andalusian Public Health System has defined a total of 70 Competence Manuals, aimed at all professional healthcare groups. Specifically, 50 Competence Manuals have been published for the medical specialties, 9 for nursing specialties, 3 for the pharmacy disciplines and another 8 for healthcare specialties.

These Competence Manuals are structured into 5 blocks and 10 criteria, each of which contains a set of Competencies. In turn, each Competence is expressed through a set of observable and measurable behaviours called "*Best Practices*". The fulfilment of Best Practice is demonstrated through the contribution of Evidence by each professional, from their actual practice. This Evidence is an objective tool for assessing their competence.

Also in Spain, the Catalan Institute of Health (ICS) began Project COM-VA© in 2005, a healthcare and management competence assessment for hospital nursing professionals. (Juve-Udine, 2007). In this model, the competencies encompassed the individual qualities indicative of effective performance. These attributes include knowledge, abilities and attitudes that enable the professional to make the right decisions in each case. Moreover, the term competence also means the sphere of responsibility or professional area in which the law gives the professional the right to make autonomous decisions. The healthcare competencies of the COM-VA© model define 6 domains: 1) caring for the patient; 2) assessing, diagnosing and addressing changing clinical situations; 3) helping the patient to keep to their treatment; 4) helping to ensure safety and the healthcare process; 5) facilitating the process of adaptation and coping, 6) teamwork and adapting to a changing environment (Juve-Udine, 2009).

2.7 Results-orientation

Results-oriented concerns in clinical management are such that some authors have stated that "*healthcare management is inconceivable without a measurement of results*" (Marín-León, 2011: 90). Healthcare units have to establish mechanisms to measure their results, as much in terms of healthcare outcomes as in management and user satisfaction. An evaluation programme is defined as "*the systematic collection of data related to a programmes activities and outcomes that results in decisions to improve efficiency, effectiveness, or adequacy*" (Keller et al., 2002; Washington County Department of Health and Environment, 1999; Patton, 1997; Berkowitz, 1995). The evaluation process must be accompanied by a prior planning process that delineates the compliance objectives, both those that are strategic to the unit and those operational and care objectives of the processes. This will allow the objectives in the Procedural Agreements made with the Corporate Management to be assessed.

In this sense, healthcare units must have a Balanced Scorecard to collect information in summary form, identifying areas for improvement and allowing the unit to establish benchmarking processes. The information systems of the centre (hospital, primary care, etc.) must support the collection of information necessary to enable the unit to analyze its results and propose actions for improvement.

According to Keller et al. (2002), an integrated evaluation programme and results-oriented planning consists of two fundamental elements in its design. First, it involves the selection of aims, which will clarify the objectives to be achieved. Generally, the objectives are difficult to measure and should be broken down into measurable indicators, according to criteria of continuity in time and external comparison. Second, it involves designing a specific strategy to enable the achievement of those aims and objectives identified. From the standpoint of healthcare units, the selection of an appropriate strategy will help improve quality in terms of efficiency and effectiveness, while the identification of aims will improve quality in terms of adaptation.

Indeed, the development of a Balanced Scorecard helps to translate the strategy of an organization into measurable objectives. In a sense, the healthcare unit is put to the test to see if it is capable of monitoring what it does. The Balanced Scorecard developed by Kaplan and Norton in 1992 has gained enormous popularity in recent times. It was the result of a year-long study that arose out of a general notion that as knowledge became a basis for competition, conventional financial measures were becoming obsolete (Kaplan and Norton, 1992 en Keller et al., 2010). The term balanced reflects the balanced consideration given to long- and short-term objectives, financial and nonfinancial measures, leading and lagging indicators, and external and internal performance perspectives (Keller et al., 2010). The BSC is a system of causal relationships among composite indicators (Key performance areas), that integrate large amounts of information (Key performance Indicators) into an easily understood single metric (Lovaglio, 2011). This basic outline of corporate and business management has also been exported to the healthcare setting. The work conducted by Baker and Pink (1995) is highlighted among the first in this field. In that study, the authors proposed a balanced scorecard model for hospitals with which they hoped to obtain a balanced view between the areas for improvement identified by the organization and those identified by patients. As Lovaglio (2011) confirms, the basic principles of a balanced scorecard model for the health sector have been adequately agreed in the scientific literature (Chow et al. 1998, Zelman et al. 1999), and have also been widely applied in public healthcare systems and organizations (Inamdar, Kaplan and Bower, 2002; Northcott and France, 2005). This reinforces the certainty of its feasibility and applicability for any healthcare unit. The indicators of a clinical unit must include those of an operational nature that assess processes and tasks from the standpoint of efficiency, and those of a comprehensive nature that allow the monitoring of fulfilment of the strategic goals previously agreed by the organization. Recently, studies have proliferated which investigate the factors affecting the variability of healthcare. Variability attributed to medical practice itself is, from the standpoint of clinical management, an inefficient use of resources because resources are allocated to services of dubious effectiveness (Bernal-Delgado, 2008; Peiró et al., 2009). In other words, setting strategic goals also means avoiding variability in clinical practice and moves the model closer to evidence-based practice.

The existence of both strategic and operational level indicators in clinical units facilitates designing a more accurate results orientation, and thus management structures can align the Mission, Vision and Values of the unit with the desired results. In general, it can be theorized that a Mission of a clinical unit or department would be to efficiently and effectively manage the resources used in its normal clinical practice. Its Vision would be to achieve a continuity of care model to meet the expectations of patients and professionals. Finally, the Values that underpin the daily work would be, at least, transparency, integrity, cooperation, and scientific rigor.

In short, promoting results orientation within healthcare units provides four basic results focused on quality improvement: 1) A set of healthcare indicators that can be subjected to longitudinal comparison over time; 2) Another set of indicators of the satisfaction with the care given to users/patients; 3) A map of areas that need improvement; 4) The possibility of establishing standardized processes of comparison with the best units, or in other words, monitoring performance using the benchmarking criteria.

2.8 Capacity of self-assessment and external assessment

The concept of Continuous Quality Improvement (CQI) should permeate the entire development of clinical management. Initially designed for the business sphere, the concept of CQI has been gradually incorporated into healthcare models (Chovil, 2010; Hyrkäs & Lehti, 2003; LeBrasseur, Whissell, & Ojha, 2002; Shortell, Bennett, & Byck, 1998). Its adoption by a healthcare unit "*enables it to be proactive rather than reactive by relying on a continuous evaluation of processes and outcomes*" (Chovil, 2010: 22).

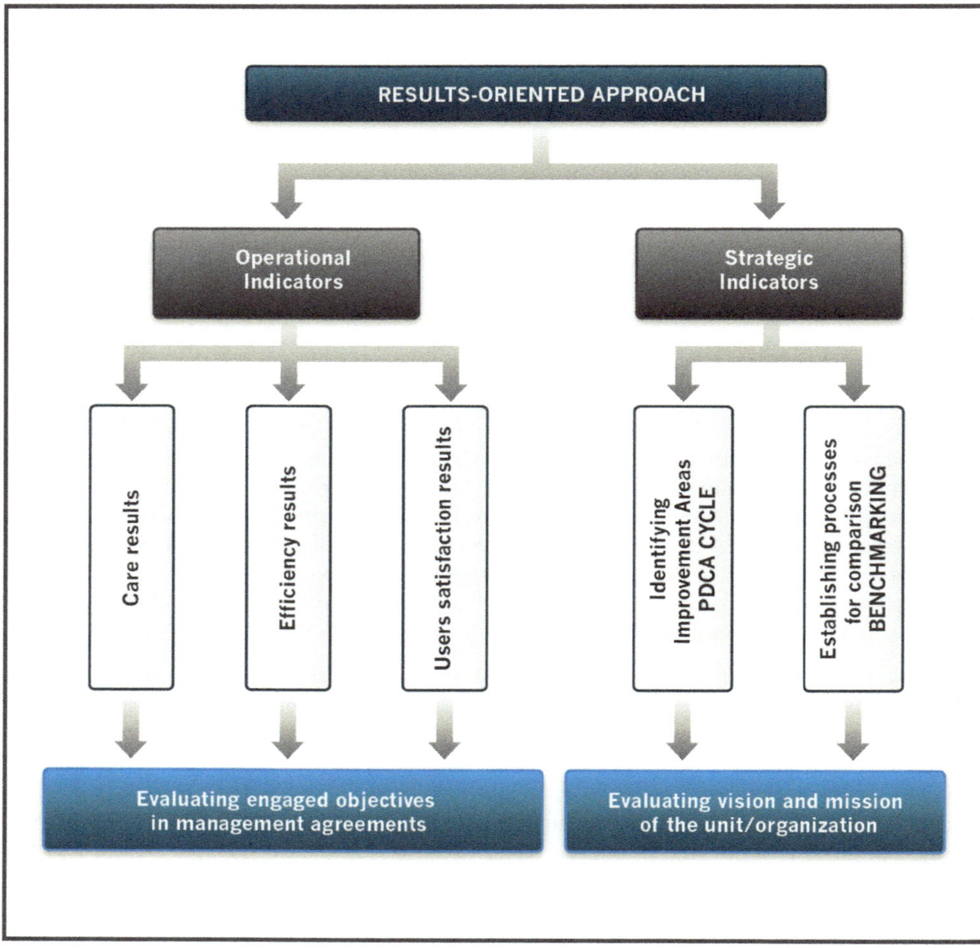

Fig. 7. Results-orientation

One of the most widely used methods to achieve this has been the process of self-assessment by professionals and healthcare units. Studies such as that of Hyrkäs & Lehti (2003) have demonstrated the favourable impact of self-assessments in healthcare units on the satisfaction perceived by patients. However, self-assessment cannot be the only means of monitoring the performance of healthcare units. This self-assessment process must be endorsed on a regular basis through a process of external assessment to objectively establish achievements and examine the quality of healthcare provided by the healthcare unit.

As noted above, the external assessment of performance of a healthcare unit can be given by indicators of satisfaction of the users/patients. In fact this is a process of external assessment of results focused on the evaluation by end users. However, external assessments do not address all aspects that must be evaluated to identify the greatest possible number of weaknesses, areas for improvement and subsequent corrective actions. All these dimensions, much more specific regarding the efficiency of processes and reflections on the potential of healthcare units, must be determined by agents specialized in the field of healthcare assessment. In recent years, the commitment to healthcare quality has encouraged the emergence of agencies, organizations and international agreements responsible for the assessment and accreditation of healthcare professionals, centres and units.

Some of the most important international, national and regional organizations and agencies dedicated to the accreditation of healthcare units and centres are the following.

Firstly, The International Society for Quality in Health Care (ISQua) which is a not-for-profit organization that emerged in 1986. The International Journal for Quality in Health Care is published on behalf of ISQua. Also, ISQua organizes an international conference which provides a forum and meeting place for agencies specialized in accreditation and external assessment, organizations dedicated to healthcare and all types of stakeholders.

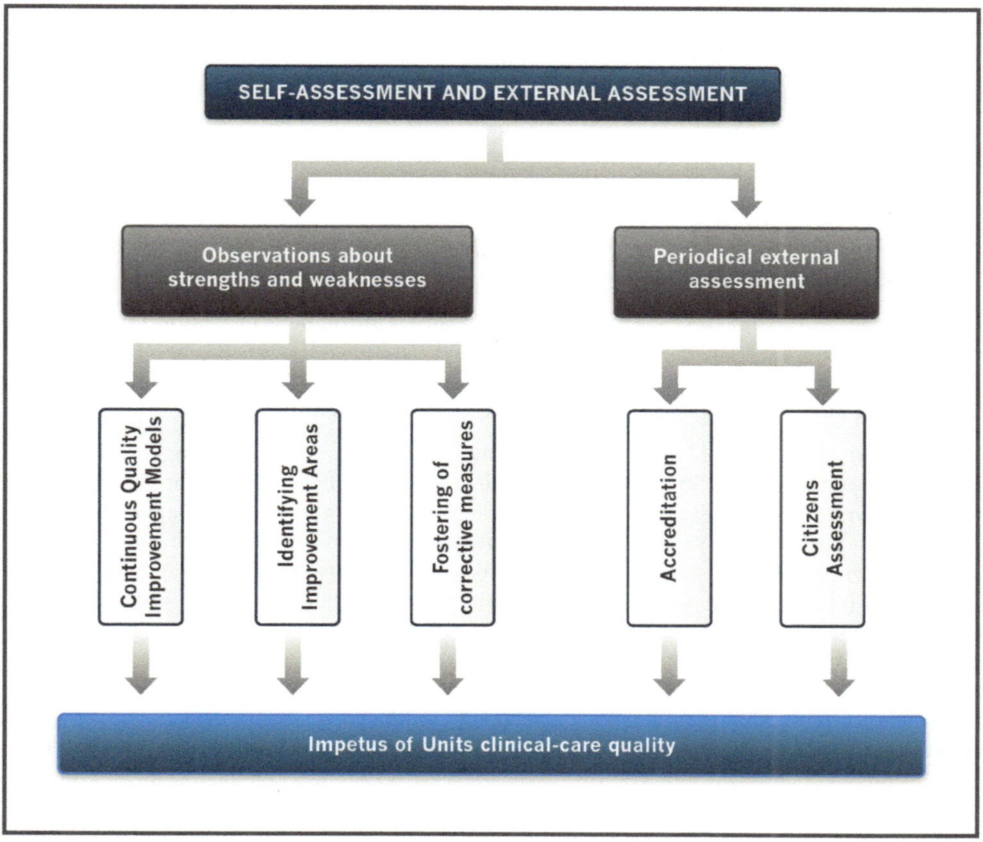

Fig. 8. Capacity of self assessment and external assessment.

The Joint Commission of the USA is highlighted among the accreditation and external assessment organizations in the Anglo-Saxon sphere. This independent not-for-profit organization was originally founded in 1917 and re-founded in 1951 with the purpose of hospital accreditation. The stated mission of the organization is "*To continuously improve health care for the public, in collaboration with other stakeholders, by evaluating health care organizations and inspiring them to excel in providing safe and effective care of the highest quality and value*" (Joint Commission, 2011).

The following are also highlighted in the Anglo-Saxon sphere: the Australian Council on Healthcare Standards (ACHS), Australian General Practice Accreditation Limited (AGPAL), Australian Quality Improvement Council (QIC), Accreditation Canada, Agency for Healthcare Research and Quality (AHRQ) of the USA, Institute for Healthcare Improvement (IHI) of the USA, National Committee for Quality Assurance (NCQA) of the USA and Quality Health New Zealand.

Among the accreditation and external assessment organizations in the European sphere are highlighted: the French National Authority for Health (HAS), Netherlands Institute for Accreditation in Healthcare (NIAZ), Health Information and Quality Authority (HIQA) of Ireland, Healthcare Accreditation and Quality Unit (CHKS-HAQU) of the United Kingdom, and the Andalusian Agency for Health Care Quality (ACSA) in Spain, among others.

In Latin America are highlighted: the Technical Institute for the Accreditation of Health Care Establishments (ITAES) of Argentina, Consortium for Brazilian Accreditation (CBA), Superintendence of Health of Chile, and the Colombian Institute for Technical Standards and Accreditation (ICONTEC).

Finally, in Asia and Africa are highlighted: Malaysian Society for Quality in Health (MSQH), Health Service Accreditation of Southern Africa (COHSASA) and the Taiwan Joint Commission on Hospital Accreditation (TJCHA).

In short, the external assessment of healthcare units through accreditation means the certification of compliance with predetermined quality standards by an external organization. This accreditation ensures the commitment of the healthcare unit to the continuous quality improvement developed in its process of self-assessment.

As discussed above, there are a significant number of quality accrediting or certifying agencies or organizations in different countries, some public but most of them private entities. Their approaches are different even though their terms of reference (standards) are very similar. The differences lie primarily in the field of accreditation, [a] in how the self-assessment processes are managed, [b] in whether the assessment processes are systemic or fragmented [c] in the possibilities of exploiting the information that results from the certification process. In this sense a model focused on continuous improvement in the context of a clinical unit is a model best suited to promote cultural change in healthcare organizations. This is because they take a different approach to clinical practice compared with models oriented to the global accreditation of health centres -which may improve some aspects of management and organization of the centre but have little impact on actual clinical practice- (Cutler, 2000). The continual use of external assessment results to promote the identification of areas for improvement in ongoing self-assessment processes is an important way of promoting real changes in the way of doing things at the level of clinical units and departments (Schrijvers 2003).

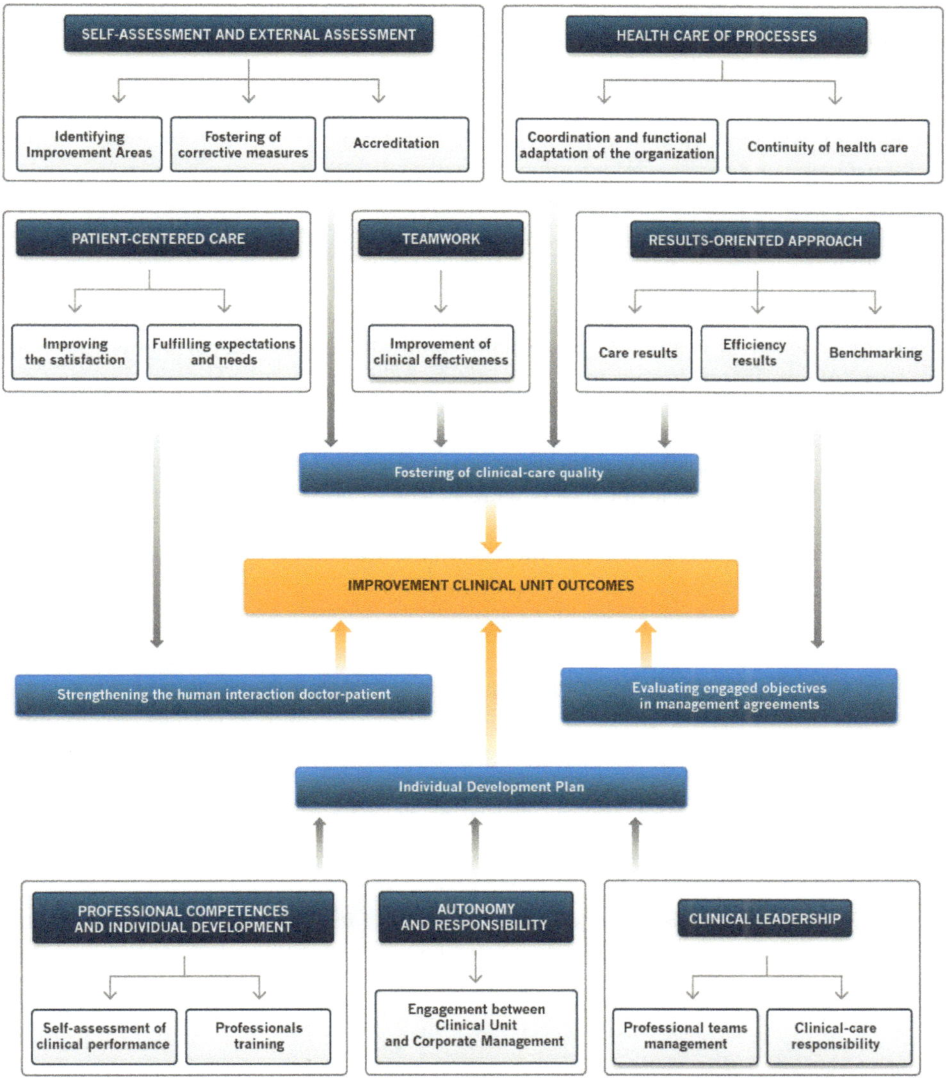

Fig. 9. Critical Success Factors for quality assurance in healthcare organizations.

3. Conclusion

Clinical management encourages the capacity of self-organization and professional autonomy, it stimulates accountability in the management of the resources used in clinical practice, and it instils the culture of continuous improvement and results orientation. It thereby facilitates an organizational design which is more adaptable to the needs of professionals and citizens.

Critical success factors that can determine the proper development of clinical management can be summarized in the following key ideas:

1. Orient health services to citizens and patients, exploring their needs and expectations, promoting their participation by providing accurate and quality information. To ensure the achievement of this critical success factor it is essential to measure patient satisfaction and use this information to improve healthcare.
2. Form multidisciplinary teams which provide integrated responses and share common goals.
3. Establish procedural agreements with corporate management which clearly set out the commitments of both parties, agreements that signify a greater capacity for decision-making and greater accountability on the part of the clinical units and departments.
4. Define the set of healthcare processes of clinical units and departments that address their quality characteristics and critical safety points.
5. Determine professional competence maps for members of healthcare units, departments and organizations, which allow individual development plans and specific training plans to be established.
6. Set indicators so that healthcare outcomes, satisfaction, efficient resource management and benchmarking can be measured.
7. Promote self-assessment and submit to periodic accreditation processes and/or external quality assessment.

By taking into account the critical success factors outlined above, a management plan for a clinical department or unit can be approached with a guarantee of success. Essentially it involves: Optimizing the existing knowledge in the organization and putting it at the service of citizens. Increasing the quality in healthcare processes through further development of competence of the professionals involved in them. And finally, using a different vision of the organization to orientate it to processes, by decentralizing decision-making and encouraging greater involvement of professionals in corporate objectives.

4. Acknowledgment

We would like to acknowledge the research's support team of the Andalusian Agency for Health Care Quality. This includes our research's technicians, Manuel Herrera-Usagre and Ángela Aguilar-de-Leyva, PhD, as well as the contributions Habibullah Rodríguez Contreras, graphic designer.

5. References

Abdelrazek F., Skytt B., Aly M., El-sabour M.A., Ibrahim N. & Engstro¨ M. M.(2010). Leadership and management skills of first-line managers of elderly care and their work environment. *Journal of Nursing Management,* vol.18, pp.736–745

Akre, V., Falkum, E., Hoftvedt, B.O. and Aasland, O.G. (1997). The communication atmosphere between physician colleagues: competitive perfectionism or supportive dialogue? A Norwegian study. *Social Science and Medicine*, vol.44, no.4, pp.519–26

Alford, R.R. (1975). *Health Care Politics*, University of Chicago Press, Chicago

Almuedo-Paz A, Brea-Rivero P, Buiza-Camacho B, Rojas-De-Mora Figueroa A, Torres-Olivera A. (2011). Skills Accreditation as an impetus for continuous professional development. *Revista de Calidad Asistencial*, Vol.26, no.4, (forthcoming)

Bernal-Delgado, E. (2008). How can we improve effectiveness (quality) by reducing the gap between research and action? 2008 SES-PAS Report. *Gaceta Sanitaria*. vol.22, no.1, (April 2008) pp.19-26

Byrne, P. & Long, B. (1976). *Doctors Talking to Patients*, HMSO, London:

Camprubí, J. (2011). Una nueva forma de trabajar que mejora las decisions: Gestión Clínica de la Efectividad, *Una nueva forma de trabajar... mirando al futuro III Congreso de Gestión Clínica*, pp. 79-85, ISBN 978-84-694-2735-4, Madrid, Spain, Oct 18-19, 2010.

Chovil, N. (2009). One Small Step at a Time: Implementing Continuous Quality Improvement in Child and Youth Mental Health Services. *Child and Youth Services*, vol.31, no.1, pp.21-34.

Chow, C.W., Ganulin, D., Teknika, O., Haddad, K., Williamson, J. (1998). The balanced scorecard: a potent tool for energizing and focusing healthcare organization management. *Journal of Healthcare Management*. Vol.43, no.3, pp.263 - 280

Christian, S.L., Norman, I.J. (1998). Clinical leadership in nursing development units. *Journal of advanced nursing*, vol.27, no.1, pp.108-116

Cunningham, G. & Kitson, A. (2000). An evaluation of the RCN Clinical Leadership Development Programme: Part 2. *Nursing Standard*, vol. 15, no. 13-15, pp. 34-40.

Cutler, D. McClellan, M. Newhouse, J. (2000). How does managed care do it? *Rand Journal Economist* vol.31, pp.526-48

Castells, A. (2011). Gestión Hospitales basada en resultados: Hospital Clinic de Barcelona, *Una nueva forma de trabajar... mirando al futuro* III Congreso de Gestión Clínica, pp. 105-109, ISBN 978-84-694-2735-4, Madrid, Spain, Oct 18-19, 2010

Deem, J.W., Barnes, B., Segal, S. & Preziosi, R. (2010). The Relationship of Organizational Culture to Balanced Scorecard Effectiveness. *S.A.M.Advanced Management Journal*, vol. 75, no. 4, pp. 31-39

Fernández Fernández, I., Fernández de la Mota, E., Sanz Amores, R. (2003). Gestión por procesos asistenciales: aplicación a un sistema sanitario público. *Cuadernos de Gestión* vol.9 pp.19-37

Friedson, E. (1960). Client control and medical practice. *American Journal of Sociology*, vol.65, pp.374-382.

Consejería de Salud (2001). *Guía de diseño y mejora continua de procesos asistenciales: calidad por sistema*. Available from: http://www.juntadeandalucia.es/salud ISBN 84-8486-024-8

Harrison, S. & Dowswell, G. (2002). "Autonomy and bureaucratic accountability in primary care: what English general practitioners say [Autonomie et responsabilité bureaucratique en soin primaire : ce que disent les praticiens généralistes Anglais]. *Sociology of health & illness*, vol. 24, no. 2, pp. 208-226

Hyrkas, K. & Lehti, K. (2003). Continuous quality improvement through team supervision supported by continuous self-monitoring of work and systematic patient feedback. *Journal of nursing management* vo.11, no.3, pp.177-88

Inamdar, N., Kaplan, R.S., Bower, M. (2002). Applying the balanced scorecard in healthcare provider organizations. *Journal of Healthcare Management*, vol.47, pp.179 - 195

Joint Commission Resources. (March 2009). *The New Standards: Section 3. Leadership*. 18.7.2011. available from http://www.jcrinc.com/The-New-Standards/The-New-Standards-Section-2-Improving-Organ/The-New-Standards-Section-3-Leadershi/

Joint Commission. Resources. (July 2011). *2011 Accreditation Process Guide for Hospitals.* 18.7.2011. available from http://www.jcrinc.com. ISBN: 978-1-59940-400-4

Juvé-Udina ,M.E., Farrero, S., Matud, C., Monterde, D., Fierro, G., Marsal, R. (2007). ¿Cómo definen los profesionales de enfermería hospitalarios sus competencias asistenciales? *Nursing* vol.25 no.7 pp.50-61

Juvé-Udina, M.E., Farrero Muñoz, S., Matud Calvo, C., Rius Ferrus, L., Monterde Prat, D., Cruz Llauna, R., Artigas Lage, M. (2009). Pesos competenciales asociados a las diferentes áreas de cuidados en el ámbito hospitalario. *Nursing* vol. 27, no.7, pp.54-58

Kaplan, R. S., & Norton, D. P. (1992). The Balanced Scorecard: Measures that drive performance. *The Harvard Business Review*, vol.70, no.1, pp. 71-80

Kapur, N., Appleton, K. & Neal, R.D. (1999). Sources of job satisfaction and psychological distress in GPs and medical house officers. *Family Practice*, vol. 16, no. 6, pp. 600–601

Knol, J. & van Linge, R. (2009). Innovative behaviour: the effect of structural and psychological empowerment on nurses. *Journal of Advanced Nursing*, vol. 65, no.2, pp.359–370

Laine, C. & Davidoff, F. (1996). Patient-centered medicine: a professional evolution. *Journal of the American Medical Association*, vol. 275, pp. 152-156.

LeBrasseur, R., Whissell, R. & Ojha, A. (2002). Organizational learning, transformational leadership and implementation of continuous quality improvement in Canadian hospitals. *Australian Journal of Management*, vol.27, no.2, pp.141–62

Lewis, J.M. (2006). Being Around and Knowing the Players: Networks of Influence in Health Policy. *Social science & medicine*, vol. 62, no. 9, pp. 2125-2136

Linda, O.K., Schaffer, M.A., Lia-Hoagberg, B. & Strohschein, S. (2002). Assessment, program planning, and evaluation in population-based public health practice. *Journal of Public Health Management and Practice*, vol. 8, no. 5, pp. 30-43

Lovaglio, P.G. (2011). Model building and estimation strategies for implementing the Balanced Scorecard in Health sector. *Quality and Quantity*, vol. 45, no. 1, pp. 199-212

Marín-León, I. (2011). Gestión de Hospitales basada en resultados, *Una nueva forma de trabajar… mirando al futuro* III Congreso de Gestión Clínica, pp. 89-94, ISBN 978-84-694-2735-4, Madrid, Spain, Oct 18-19, 2010.

McClelland, D. (1973). Testing for Competence Rather Than for Intelligence. *American Psychologist*, vol.28, no.1, pp.1-14

McKinlay, J.B. & Arches, J. (1985). Towards the Proletarianization of Physicians. *International Journal of Health Services*, vol. 15, no. 2, pp. 161-195

Mead, N. & Bower, P. (2000). Patient-Centredness: A Conceptual Framework and Review of the Empirical Literature. *Social Science and Medicine*, vol. 51, no. 7, pp. 1087-1110 ISSN: 0277-9536

Mickan, S.M. (2005). Evaluating the effectiveness of health care teams. *Australian Health Review*, vol. 29, no. 2, pp. 211-217

Mickan, S.S. & Rodger, S.S. (2000). Characteristics of effective teams: a literature review. *Australian Health Review*, vol. 23, no. 3, pp. 201-208

Millward, L.J. & Bryan, K. (2005). Clinical leadership in health care: a position statement. *International journal of health care quality assurance*, vol. 18, no. 2, pp. R13-R13-R25. ISSN: 09526862

Mora Martínez, J.R., Ferrer Arnedo, C., Ramos Quirós, E. (2002). Gestión clínica de procesos: mapa de procesos de enfermería en centros de salud. *Revista de Administración Sanitaria* vol. 6, pp.135-59

National Health Service Scotland (2004) *Leadership Development Framework for Discussion.* NHS, Scotland.

Northcott, D. & France, N. (2005). The balanced scorecard in New Zealand health sector performance management: dissemination to diffusion. *Australian Accounting Review*, vol.15, no.37, pp.34 - 46

Patton, M.Q. (1997) *Utilization-Focused Evaluation.* Thousand Oaks, Sage Publications, California, USA

Peiró, S., Meneu, R., & Bernal-Delgado, E. (2009). Effectiveness, variation and inequalities. Hysterectomies and prostatectomies due to neoplasm in Spain (2002-2004). *Revista Española de Salud Pública*, vol. 83, no. 1, pp. 109-121

Reyes-Alcázar, V., Salvador-Sanz, A., & Segura, C. (2011). Mapa de competencias y buenas prácticas de los profesionales de una UGC del corazón. In: *Gestión Clínica en Unidades del Corazón,* Torres-Olivera, A. and V. Reyes-Alcázar, (Ed.), 93-118, Sociedad Española de Calidad Asistencial, ISBN 978-84-615-0671-2, Madrid, Spain

Schrijvers, G., Oudendijk, N. & ,de Vries, P. (2003). In search of the quickest way to diseminate health care innovations. *International Journal of Integrated Care,* Vol. 3, pp.1-22, ISSN 1568-4156

Schulz, R.I., Girard, C., Harrison, S. & Sims, A.C.P. (1991). Perceived autonomy and work satisfaction amongst psychiatrists. *Journal of Management in Medicine*, Vol. 5, no. 2, pp. 54–65

Shilton, T. (2009). Health promotion competencies: providing a road map for health promotion to assume a prominent role in global health. *Global Health Promotion*, vol. 16, no. 2, pp. 42-46, 61, 75

Shortell, S. M., Bennett, C. L., & Byck, G. R. (1998). Assessing the impact of continuous quality improvement on clinical practice. *The Millbank Quarterly* vol. 76, 593–624

Torres Olivera, A., Fernández, E., Paneque, P., Carretero, R., & Garijo A. (2004). La Gestión de la calidad asistencial en Andalucía. *Revista de Calidad Asistencial*, vol.19, no.3, pp. 105-112

Torres Olivera, A., Lledó, R. (2001) La coordinación entre niveles asistenciales. *Revista de Calidad Asistencial* vol. 16, pp.232-234

Torres Olivera, A. (2003). Desarrollo de los Procesos Asistenciales Integrados en Andalucía. *Cuadernos de Gestión,* vol.9, pp.127-134

Torres Olivera, A. (2003). La gestión por Procesos Asistenciales Integrados: una estrategia necesaria. *Atención Primaria,* vol.31, pp.561-563

Torres Olivera, A., Reyes-Alcázar, V. (2011). *Gestión Clínica en Unidades del Corazón,* Sociedad Española de Calidad Asistencial, ISBN 978-84-615-0671-2, Madrid, Spain.

Washington County Department of Health and Environment. (1999). *Community Health Service Plan, 1992–1996 and 1996– 2000.* Department of Health and Environment. Stillwater, MN: Washington County, USA

West, B., Lyon, M.H., McBain, M. & Gass, J. (2004). Evaluation of a clinical leadership initiative. *Nursing Standard,* vol. 19, no. 5, pp. 33-41

Zelman, W.N., Blazer, D., Gower, J.M., Bumgarner, P.O., Cancilla, L.M. (1999). Issues for academic health centers to consider before implementing a balanced-scorecard effort. *Academic Medicine,* vol.74, no.12, pp.1269 - 1277

16

Quality Improvement Through Visualization of Software and Systems

Peter Liggesmeyer, Henning Barthel, Achim Ebert, Jens Heidrich, Patric Keller, Yi Yang and Axel Wickenkamp
Fraunhofer IESE / University of Kaiserslautern
Germany

1. Introduction

Many organizations still lack support for obtaining control over their system development processes and for determining the performance of their processes and the quality of the produced products. Systematic support for detecting and reacting to critical process and product states in order to achieve planned goals is often missing. As systems and software become bigger and more complex, classic approaches reach their limits, due to the difficulty of extracting relevant information from a large volume of measures. Here, suitable visualization and virtual reality solutions can offer a clear advantage by representing the relevant information in a more easily recognizable form. However, many resulting visualizations are still hard to understand, even for experts. This opens the door for researching modern, human-centered approaches that provide the user with visualization and interaction models for visually analyzing and understanding the underlying complex data. This chapter focuses on two main topics: system visualization and software visualization.

Software visualization continues a very successful direction of software engineering, namely the topic of "software measuring/software analysis". Software visualization in general does more than visualizing the product "software" itself. Visualization mechanisms are also typically applied to software development processes and to support the tracking of project progress (e.g., as a central component in project dashboards).

System visualization aims at a better understanding of system properties, e.g., safety. These properties are usually influenced by mechanical components, microelectronics, and software.

2. Software Visualization

The term "Software Visualization" refers to a broad range of visualization mechanisms that can be applied to many different issues relating to engineering-style software development. For instance, it may be used to visualize and analyze software development processes, different artifacts created as part of the development process (such as designs or code), or even aspects of the project itself (like the communication of the team or the information flow between development locations). This section focuses on providing an overview of the variety of different visualization techniques that may be used. First, the visualization of product and process/project characteristics will be discussed. After that, some mechanisms for dealing with complexity and nesting issues in general will be highlighted.

2.1 Related work

Depending on the focus of a software development organization (e.g., safety-critical software or commercial off-the-shelf software), different characteristics of software products such as security, reliability, or maintainability may be important. ISO/IEC 25000 (see (ISO, 2011)) gives an extensive overview of different product characteristics and some recommendations on how to quantify and measure different attributes of the software. There are lots of measurement tools on the market for analyzing static and dynamic aspects of a software product. Examples are Metrics Eclipse Plugin (Metrics, 2011), CodePro Studio (Instantiations, 2011), JDepend (JDepend, 2011), or McCabe IQ (McCabe, 2011). Each of these tools has different capabilities in terms of the metrics supported and the visualization capabilities offered. A detailed description of such tools can be found in (Rech, 2005).

Typically standard visualization mechanisms are used for project control, such as simple line, bar, and pie charts, spider plots, tables and matrices, simple graphs (networks), Gantt charts, or box plots. The type of visualization used depends on the level of abstraction and the viewpoint for which the data is visualized. Different kinds of exploration capabilities are provided for project control, such as drill-down mechanisms aimed at getting a more detailed view on the data, or aggregation mechanisms aimed at getting an overview containing multiple, abstract pieces of information. Moreover, visualization techniques may provide basic interaction mechanisms for browsing the data, such as scrolling data along a time axis in order to browse the history and development of indicators over time. More advanced visualizations such as visualization metaphors are seldom used within project control frameworks. However, they can be found in special-purpose tools, e.g., visualizing software systems using 3D models of landscapes and urban structures (Balzer, 2004).

In software visualization, graph drawing techniques are used to describe and comprehend the complex structures of software systems, development processes, and software quality characteristics. The most popular graph drawing techniques used are node-link diagrams. The standard algorithm was proposed by Sugiyama (Sugiyama, 1981). Due to the inherent complexity of solving the sub-problems using this algorithm, many heuristics have been proposed, e.g., by Gansner et al. (Gansner, 1993) and Eiglsperger et al. (Eiglsperger, 2005). A detailed description of graph visualization techniques using node-link diagrams can be found in (Herman, 2000), (Battista, 1999), and (Kaufmann, 2001). Since the space requirements of node-link diagrams grows rapidly with the size of a graph, space-filling techniques like Treemaps (Johnson, 1991), Sunbursts (Stasko, 2000), or Icicle Plots (Kruskal, 1983) have been applied to realize a space-constrained visualization of large hierarchical data sets. However, in contrast to node-link diagrams, it might be more difficult to understand hierarchical relations using these space-filling techniques. Node-link diagrams might be combined with space-filling techniques to visualize both hierarchical and non-hierarchical relations within one view. Fekete (Fekete, 2003) uses a Treemap to visualize the hierarchy, while Bézier curves are used to show additional non-hierarchical relations connecting two Treemap regions. See (Fekete, 03) for more details on combining node-link diagrams with space-filling visualization techniques.

2.2 Software product characteristics

Fine-grained measurement of software product characteristics tends to result in a huge amount of measurement data. With simple high-level charting and diagram visualizations (Fig. 1), only condensed views on the measurement data can be provided.

Fig. 1. High-level charting and diagram visualization techniques.

In order to provide a goal-oriented visual exploration and analysis of product characteristics, the structural information about the system needs to be utilized, preserving the user's mental map of the system architecture. Hence graph-based techniques and especially UML diagrams are often used as a base visualization technique, with relevant attributes of the software system being analyzed mapped onto the structural visualization using a set of additional graphical elements such as text, geometric shapes, and icons (Fig. 2). If product characteristics are to be analyzed on the source code level, image-based techniques combined with traditional brushing and shading mechanisms may be used. By utilizing appropriate scaling and interaction techniques, this approach allows the visualization of several levels of details of the source code within one picture: Some areas may show the original source code, while others may visualize several lines of code as a shaded region using color coding and shading to map measurement data onto the visual item (Fig. 3).

To get a deeper insight into the quality of a software product, it is not sufficient to use these visualization techniques with just simple interaction techniques only. Rather, the specification and selection of visualization techniques has to be tightly integrated with the specification and selection of metrics used to collect the measurement data. With such tight integration, the user may apply tailored metrics and visualization metaphors on different level of details when visually exploring the product characteristics.

At Fraunhofer IESE we have realized this approach with our scalable measurement tool *M-System*. We store an abstract syntax tree (AST) representation of the source code in a relational database. This database is then used to define metrics and collect measurement data by issuing appropriate SQL statements. The same SQL frontend is further used to filter the measurement data and to select and tailor the visualization technique for use on different levels of detail. Fig. 4 and Fig. 5 show example visualizations generated with the *M-System*, used to analyze the product characteristics of a software system on different levels of detail. The user starts with a 3D-Treemap overview visualization showing a condensed view of the product characteristics. The 3D-Treemap technique maps both the basic code structure as well as the quality data (e.g., component size and complexity measures) to cubes using different graphical properties (position, size, height, and color).

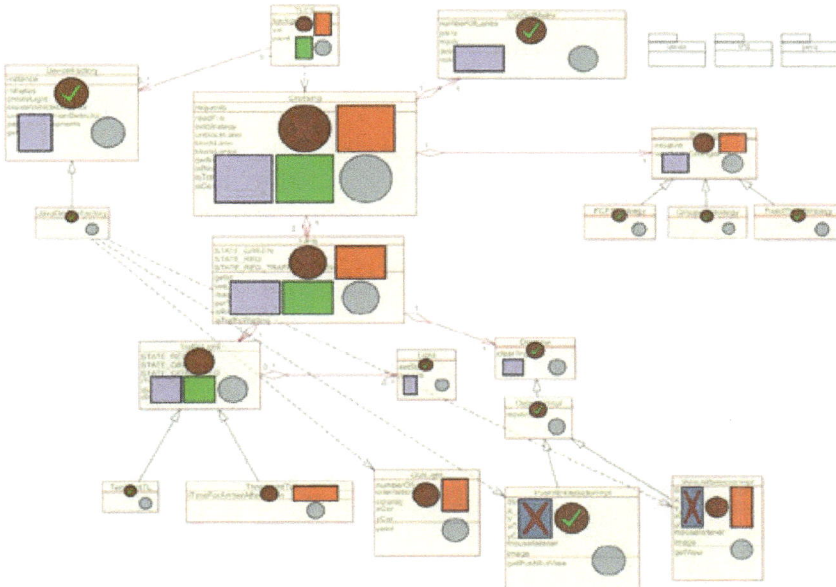

Fig. 2. Graph-based techniques (Termeer, 2005) using UML diagrams.

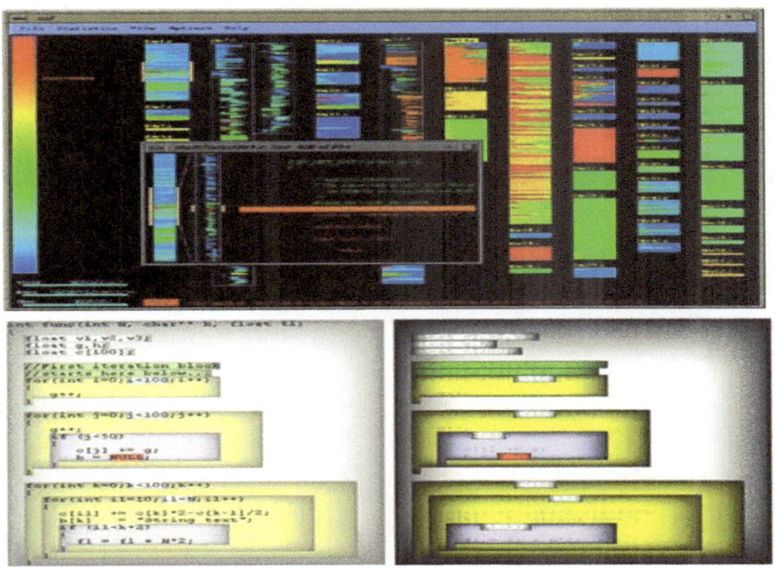

Fig. 3. Image-based techniques (Ball, 1996) and brushing (Lommerse, 2005) on the source code level.

Critical components (e.g., a red cube in the 3D-Treemap) might then be further analyzed by selecting such components within the visualization and adapting the measurement data and the visualization technique to be used in the subsequent analysis process. For example, in Fig. 5, we use a node-link graph visualization to highlight critical components. We take the same measures, but simply adjust the visualization to color code the data according to a given threshold value. This simplifies the identification of critical parts of the component being analyzed.

Using the SQL frontend to define metrics and collect measurement as well as to filter the measurement data and to select and tailor the visualization technique has proven to be a very flexible and powerful approach. Nevertheless, making this frontend more user-friendly as well as integrating further visualization techniques into the *M-System* is ongoing work at Fraunhofer IESE.

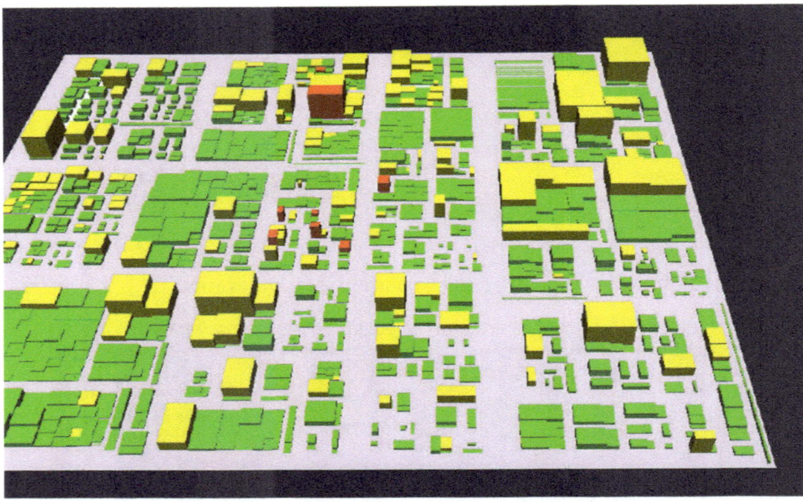

Fig. 4. 3D-Treemap visualization of software quality on different levels of detail using Fraunhofer IESE's M-System.

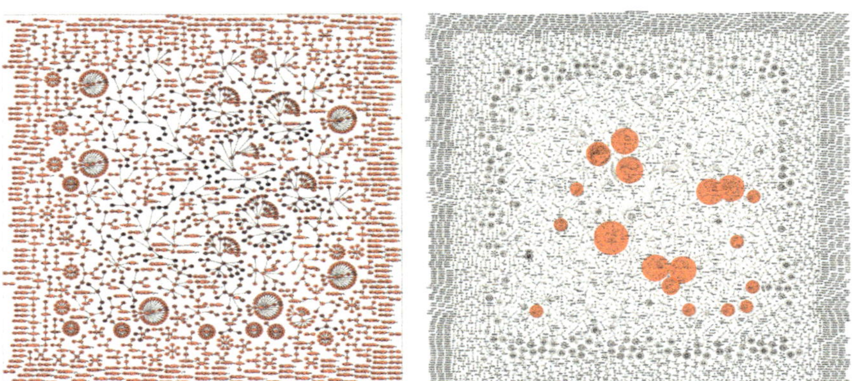

Fig. 5. Node-link graph visualization for highlighting critical components.

2.3 Software process and project characteristics

Visualization techniques may also be used for analyzing complex software development processes. For instance, the German V-Modell®XT has more than 1000 entities with different relationships among them. Analyzing the commonalities between the different versions of such a process and identifying inconsistencies and bottlenecks is a non-trivial task that requires advanced visualization techniques. Fig. 6 presents two analysis views for getting an overview of the evolution of the V-Modell®XT based on the DeltaP approach developed at Fraunhofer IESE (Soto, 2008). The left-hand side demonstrates the increase of entities (activities, artifacts, roles, etc.) over different versions of the V-Modell®XT. The vertical lines indicate official releases of an internal version. From version 1 to version 600, more than 200 entities were added. The right-hand side shows the number of changes in different modules of the V-Modell®XT over time. The size of the circle reflects the number of changes (additions, deletions, modifications) of a certain module. In order to obtain a simple footprint of the evolution, the changes over time are mapped to a bar code. What can be observed from the chart is, e.g., that the major work before release 1.0 was completed at the end of 2005, but it took more than one month with close to zero changes before the actual release took place. A second observation is, e.g., that a huge number of changes were postponed after release 1.1. As soon as the main development branch was released, some major changes were implemented into release 1.2.

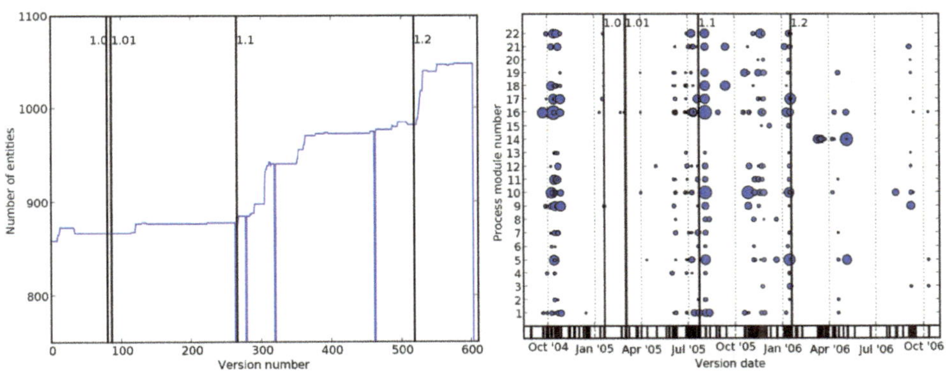

Fig. 6. Visual analysis of the V-Modell®XT evolution (Soto, 2008).

Another very common area for software visualization is project monitoring and control. Many different aspects of a development project have to be visualized on different levels of abstraction, depending on the goals and characteristics of the project. This typically includes the schedule of the project as well as effort/cost aspects, but also quality-related measures, such as the defect density, the requirements already implemented, or the test coverage. One challenge for the visualization of this data is how to show trends for identifying potentially critical project states in time and how to initiate corresponding countermeasures. Another challenge is how to provide visualization mechanisms for condensing and aggregating the data (e.g., making use of corresponding quality models). Fig. 7 presents a Software Project Control Center (SPCC) prototype created at Fraunhofer IESE. The Specula approach is able to dynamically create an SPCC interface from project control building blocks for data collection, data processing, and data visualization. The screenshot on the right shows on overview of available control charts and displays the selected Gantt charts for monitoring the project's schedule and progress. The screenshot on the left side aggregates all data into a simple matrix structure for analyzing the trend with respect to core project measurement objects and properties.

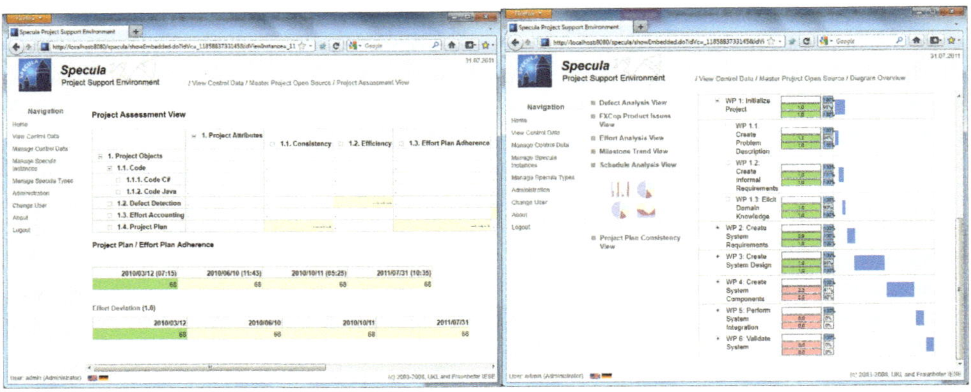

Fig. 7. Visualizing software project control data (Heidrich, 2010).

2.4 Complexity and nesting

A very general underlying problem of software visualization is how to deal with highly complex structures and nesting. Typically, graph representations are used for describing these structures, and graph visualization techniques are applied to improve the comprehension of software products, projects, and processes or to facilitate decision support for software architects and project managers.

However, using off-the-shelf graph visualization tools may be inadequate in the software engineering domain because they often do not reflect the user's context and role-specific visualization and interaction requirements in such distinct application scenarios as Quality Measurement, Architectural Analysis, Reengineering, Process Management, Process/Product Evolution, Risk Management, or Security and Safety. Rather, specially engineered tool support is required, providing domain-centered visualization and interaction techniques that support the interactive visual exploration of highly complex data structures within each application scenario.

At Fraunhofer IESE we align our tool support according to this strategy (see sections 2.2, 2.3, 3.1 and 3.2 targeted at relevant application scenarios). Based on the user's need, we combine traditional visualization techniques with suitable interaction and navigation techniques on different levels of detail. The user might start with a 2D or 3D visualization providing an overview of the whole data set, which acts as the starting point for the visual exploration (see Fig. 8). However, the user can also use traditional node-link diagrams or a combination of node-link diagrams and space-filling techniques if he is used to that (see Fig. 9).

The user might then decide to further investigate parts of the dataset using more appropriate visualization and interaction techniques. For example, if the user performs an architectural analysis and needs to know more details about the relations between components of a subsystem, he might select the subsystem in the overview visualization and switch to a more suitable visualization like that in Fig. 10.

Practical examples from industry indicate that it is reasonable to realize domain-specific, user-centered visual exploration tools in the software engineering domain that are tailored to the user's context and role-specific visualization and interaction requirements. Realizing this tool support is ongoing work and an interesting research topic at Fraunhofer IESE.

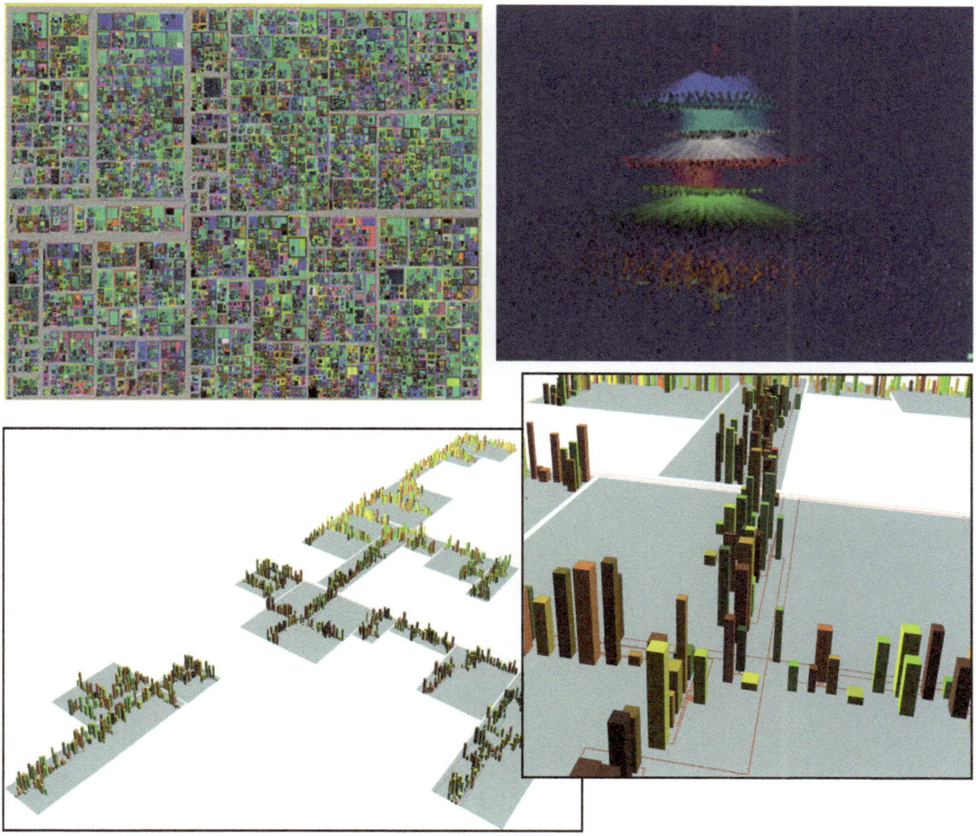

Fig. 8. Exemplary overview visualizations that might be used as a starting point for visual exploration.

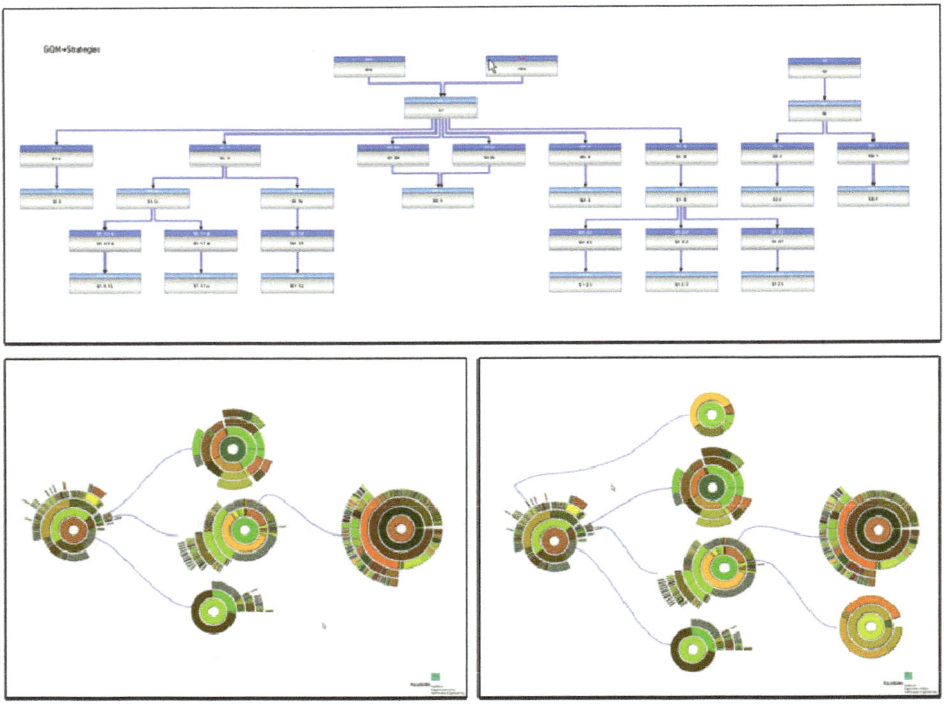

Fig. 9. Traditional node-link diagram visualization (top) and a combination with space-filling techniques (bottom) using sunbursts to visualize hierarchy nodes and Bezier-Curves to visualize links.

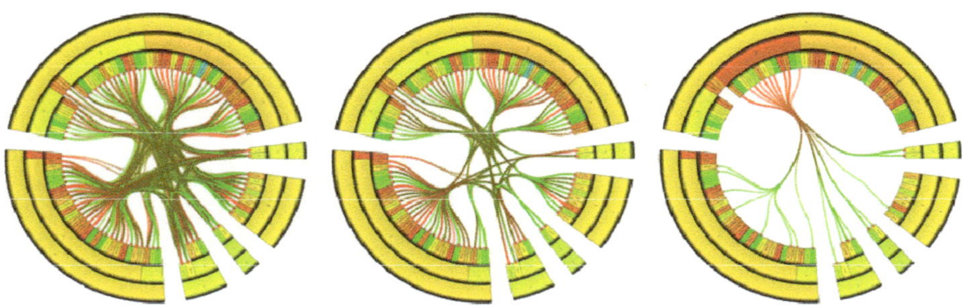

Fig. 10. Combined space-filling and node-link visualization. Edge bundling (Holten, 2006) is used to visualize non-hierarchical relations, based on the multivariate information taken from the original dataset.

3. System visualization

The importance of system visualization increases due to growing size, critical properties (e.g., safety, security), and new application domains (e.g., ubiquitous systems / ambient systems), which are characterized by complex behavior (e.g., dynamic aspects). Many safety-critical embedded systems need to be certified by certification authorities before being released for public use. This requires techniques that are capable of displaying detailed information about the safety properties of large systems and that can, at the same time, prevent critical information overload.

3.1 Safety visualization

Safety is a quality characteristic that is very important in many application domains, e.g., rail systems, avionics, or medical systems. According to the IEC 61508 standard, safety is the absence of unacceptable risks.

3.1.1 Safety analysis

The analysis of safety-critical systems is organized to follow certain rules. The corresponding techniques and methods tend to analyze the causes leading to critical system conditions that violate the specified safety objectives. One technique frequently used in this context is the so-called Fault Tree Analysis (FTA). It describes the relationship between low-level failures, e.g., failures on the component level, and system failures. Fault Trees (FT) use logical gates to model these cause-effect relationships. Low-level failures (Basic Events (BEs)) are represented as leaves of the FT, whereas the root represents a safety-critical system failure (Top Event (TE)). Besides being a qualitative analysis, FTA allows quantifying the probability of occurrence of the TE if the probabilities of the corresponding BEs are known.

A Minimal Cut Set (MCS) is defined as a minimal set of BEs whose simultaneous occurrences cause the TE. Traditional approaches for analyzing MCSs primarily use text-based ((ESSaREL, 2011), (RELIA, 2011), (Fig. 11 (a)) or table-based ((RelexArchitect, 2011), (SYNC, 2011), (Fig. 11 (b)) representations. The information listed there usually includes an ID for identifying the individual MCSs and the BEs contained together with their corresponding failure probabilities. In many cases, table-based representations rely on regular interaction techniques such as filtering and sorting. However, the possibilities of these manipulation techniques are limited. They neither provide appropriate graphical assistance facilitating better understanding nor do they support the analysis of correlations between BE and MCS within large data sets. This hampers efficient handling and consequently might lead to situations in which information intended to contribute to safety improvements may become lost. To reflect the way the effects of the current MCS propagate along the FT (Fig. 11 (c)), some approaches ((ESSaREL, 2011), (ISOGRAPH, 2011)) represent MCS via highlighted paths between BEs and the TE.

The idea of modeling cause-effect relationships was extended to so-called Component Fault Trees (CFT). CFTs permit encapsulating sub-trees that correspond to technical components and thus provide the possibility to deal with different levels of abstraction.

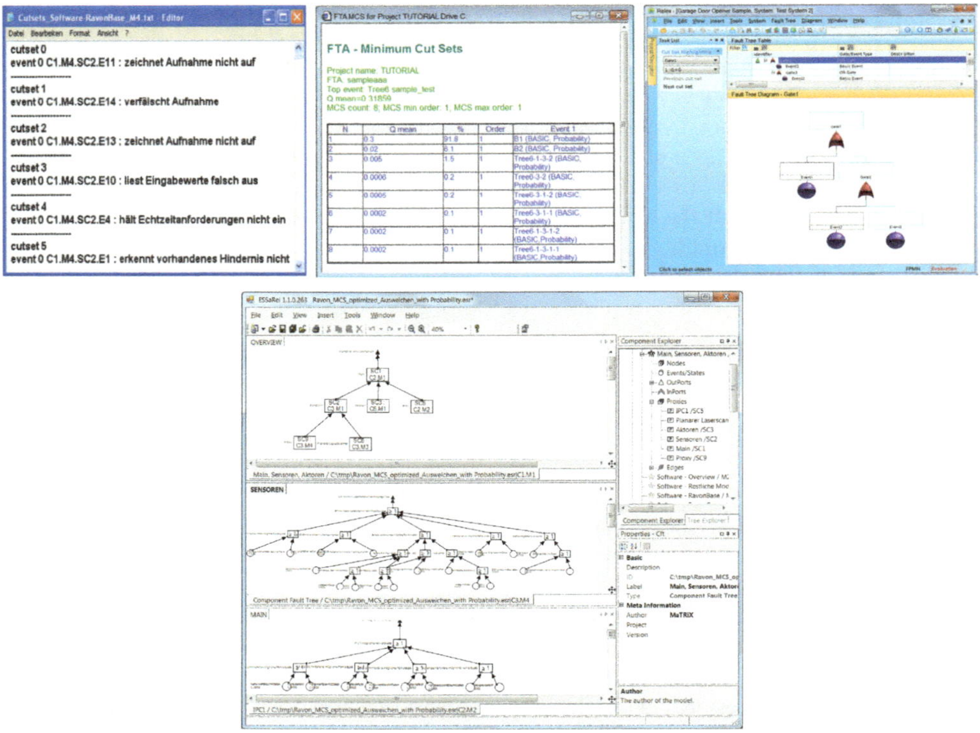

Fig. 11.Common methods for representing MCSs: (a) Plain text (ESSaREL, 2011). (b) Table format (ALD, 2011). (c) Associated paths highlighting (RelexArchitect, 2011). (d) Component Fault Tree (ESSaREL, 2011).

The tool ESSaREL (ESSaREL, 2011) supports the use of CFTs (Fig. 11 (d)). Additional MCS information in the form of plain text is provided in separate views (Fig. 11 (a)).

3.1.2 ViSSaAn

A new visualization tool called ViSSaAn (Visual Support for Safety Analysis) developed at the University of Kaiserslautern (Yang, 2011) uses a matrix-based visualization to efficiently represent the information associated with MCSs of FTs and CFTs, respectively.

The matrix assigns the MCSs to the rows, whereas the failure probabilities of the individual MCSs, the MCS IDs, the orders of the MCSs (Fig. 12, area 1), and the BEs (Fig. 12, area 2) are assigned to the columns. The tool classifies the MCSs into three different safety categories according to their failure probabilities: high, moderate, and acceptable; the thresholds for this classification can be set according to the specific needs of the analysis process. The color scheme applied for visualizing these relations uses red to indicate high, yellow to indicate moderate, and green for acceptable probabilities. In this way the user is able to perceive the criticalities of the individual MCSs very intuitively. The three-level categorization concept is also applied in the same way for characterizing and visually highlighting the BEs.

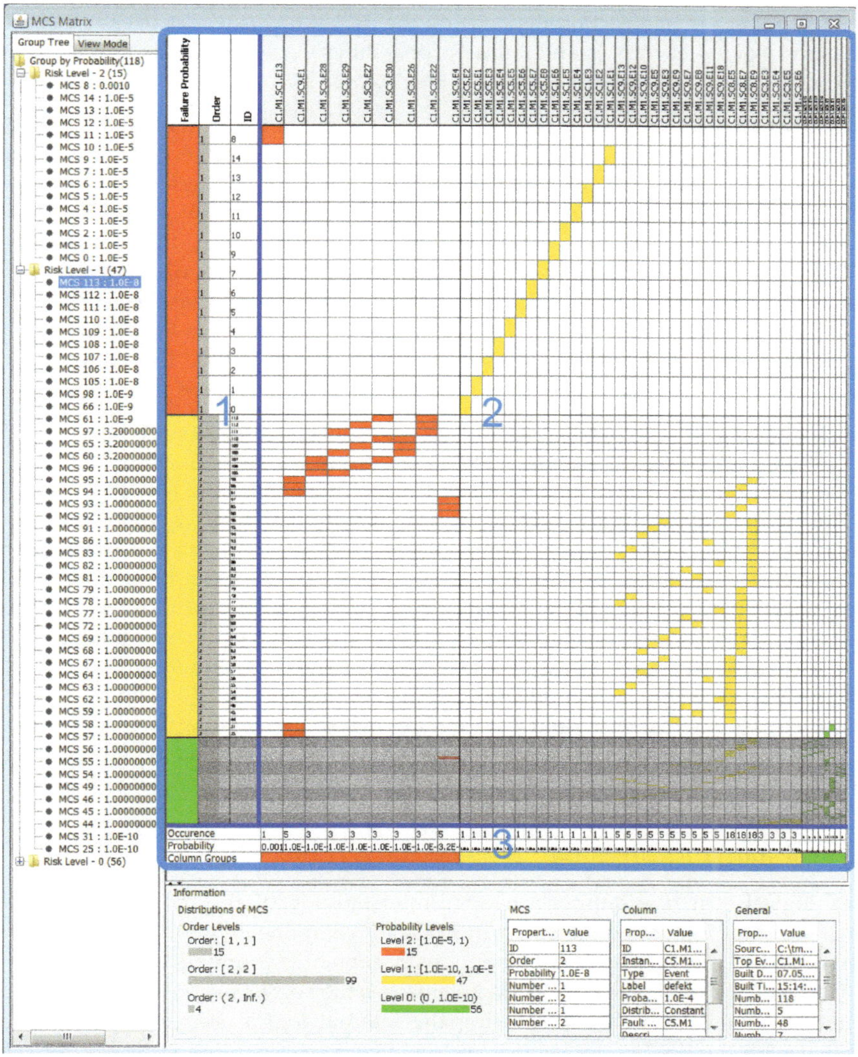

Fig. 12. Overview of ViSSaAn: matrix view (center), navigation tree (left), and information panel (bottom). In matrix view, area 1: properties of MCSs; area 2: containing relationships between MCSs (rows) and BEs (columns); area 3: properties of BEs and indicators of groups for BEs.

Furthermore, ViSSaAn uses translucent bar charts to represent the orders of the MCS. The order in this regard is defined by the number of BEs contained in an MCS. In general, this means: the smaller the order, the more critical the MCS.

Whenever a BE belongs to an MCS, the matrix indicates this dependency by coloring the cell at the location where the BE column intersects with the MCS row (see Fig. 12, area 2). The cell color used at the intersection depends on the BE's safety category. In addition, ViSSaAn displays the exact occurrences and failure probabilities of the BEs at the bottom of matrix to support an extended quantitative analysis (Fig. 12, area 3).

In order to quickly identify essential dependencies within a large number of MCSs, ViSSaAn provides a range of sorting and grouping functionalities. The *sorting function,* for instance, allows sorting the MCSs according to their failure probabilities in descending and ascending order. The resulting groups of MCSs having high, moderate, and acceptable probabilities separating the matrix vertically into three distinct segments with different colors (Fig. 12, area 1 shows the three MCS groupings in red, yellow, and green). The sorting function automatically groups the MCSs according to their criticalities, so that the user can efficiently delimit the number of MCSs to be examined. The sorting function is also available for the columns in order to intuitively identify and delimit important BEs. The grouping in case of BEs is illustrated by the last row of the matrix view in Fig. 12 (area 3).

To satisfy the need for providing a suitable overview when addressing large-scale MCS data sets, ViSSaAn additionally provides flexible scaling interaction functionalities like *uniform scaling* and *scaling by groups*. *Uniform scaling* allows narrowing the row heights in order to depict as many MCSs as possible to exploit the limited display space. However, one problem immanent to this scaling approach concerns the reduction of the space used to depict other MCS-related information. To address this issue, ViSSaAn uses a scaling approach following the degree-of-interest (DOI) concept referred to as *scaling by groups*. It preserves important MCS information, while simultaneously maintaining a suitable overview (Fig.12, area 1). The heights of the rows are adapted to the safety criticality of its corresponding MCS: the more critical the MCSs, the larger the row heights. This way, the users can focus on the important information while maintaining the overview as their context. Both scaling approaches can be applied to rows as well as to columns in order to reduce information overhead by simultaneously maintaining all details necessary for an efficient analysis.

3.1.2.1 Embedded CFT structures

The FT logical structure supports better understanding of the influences of MCSs in a system. ViSSaAn integrates the structures of CFT components into the matrix view using focus+context techniques in combination with semantic zooming concepts. Whenever a color-filled cell is double-clicked, ViSSaAn enlarges the corresponding cell area and displays an embedded view (see Fig.13) showing the internal logical structure of the component associated with the BE. The leaf nodes are again colored according to their safety criticality levels. The color-filled nodes represent the BE corresponding to the selected column, whereas the nodes having a thick border represent all other BEs contained by the MCSs of the current row. Those leaf nodes or BEs not directly related to the MCS are marked with a thin border. Integrating such embedded logical structures allows ViSSaAn to provide additional information about how the BEs of an MCS influence the component of a CFT along the internal structure. This allows the users to focus on the detailed internal structures while maintaining the overall context. ViSSaAn also provides interactions for the internal structure inside the embedded view, such as smooth zooming and panning, which allow adapting the view to the current visualization needs. A more detailed description of the ViSSaAn tool and its possibilities is provided in (Yang, 2011).

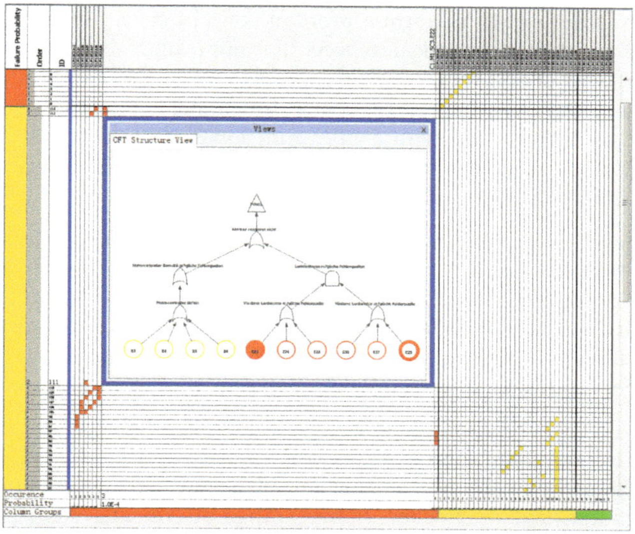

Fig. 13. Logical structure of the CFT component embedded in the matrix view. The color-filled node represents the BE in the current column. The nodes with a thick border are contained by the MCS in the current row. BEs not related to the current MCS are depicted with a thin border.

3.2 Virtual Reality

Virtual Reality (VR) is considered a technology for providing virtual mappings of real-world objects, which are allowed to be explored and manipulated by users. It combines 3D visualization techniques with human-centered interaction techniques. Depending on the display technology, VR is classified into four main categories: immersive (e.g., head-mounted displays), semi-immersive (e.g., car driving simulator), projected (e.g., CAVE - Cave Automated Virtual Environment, 3D Walls etc.), and desktop (e.g., monitor), each with its own advantages and disadvantages.

3.2.1 Current approaches In VR

Regarding the usage of VR technology, several fields in science and engineering have greatly benefitted from its application. This includes, among other things, that VR contributes to a better understanding of real-world systems and processes. VR can emphasize which information is important for the user and simultaneously allows blanking out other kinds of information. It also supports the exploration of complex structures, processes, and relations (Kreylos, 2003).

In addition, VR facilitates the process of developing large systems and environmental structures by providing means for efficient design and analysis. In this context, VR is being adopted in domains like automotive (e.g., the design of cars), aeronautics (e.g., in the development of airplanes and satellites), as well as in the construction industry to provide a first glance at final products, e.g., in the form of virtual prototypes. Following up on this idea makes it possible, e.g., to perform a proof of concept even without any real system at hand. When we consider economical aspects, the application of VR techniques provides great benefits in terms of cost reduction and time savings during development. Particularly, in scenarios where the existence of expensive functional models is required, it may no longer be necessary to come up with a fully functional real-world system. The article by Purschke et al. (Purschke, 1998) gives an example of how VR techniques have recently been used with respect to developing vehicle systems at Volkswagen.

Another application area uses VR for the purpose of training and operator qualification. Virtual training systems are built with the intention to reduce expenses for education and training and to provide a safe way of preparing critical operations and procedures that otherwise would most likely endanger persons or equipment. This is possible since VR systems are, by their very nature, safe, controllable, and economical. VR applications were developed e.g., for education (Oliveria, 2007), preparation of medical surgeries (Heng, 2004), and others.

In addition, VR is preferably used in cases where safety-critical systems have to be tested with regard to dangerous situations. This may become necessary within the context of common certification procedures concerning a range of critical system properties. Not all situations can be covered by real-world testing - be it due to time or money restrictions or simply because of safety reasons. In (Barret, 2010), the authors introduce a virtual system allowing the design and exploration of safety-critical scenarios in the context of electrical engineering.

3.2.2 VR and safety

In the last few decades, the number of electronic and electronically programmable devices deployed within embedded systems has risen dramatically. This not only had effects on the size and complexity of those systems but also on their quality. Especially safety is a quality characteristic strongly influenced by this evolution.

One possible step towards "safer" systems is to take advantage of extended means provided by Virtual Reality (VR) technologies. VR is increasingly becoming a powerful tool for tackling the challenges found in quality assurance of embedded systems. It allows incorporating knowledge/information from different sources, ranging from physical domains like object shapes and materials to more imaginary quantities like system quality and usability, and allows presenting it in a way intuitively accessible for human perception. In addition, special interaction techniques facilitate the handling of the provided representations. The idea of supporting the analysis of complex embedded systems by using VR technology is pursued by the project ViERforES (ViERforES, 2011) and its successor ViERforES 2 (ViERforES, 2011). These projects especially aim at improving the quality of complex embedded systems by providing methods and techniques for virtually accessing and evaluating particular system characteristics such as safety, security, and reliability.

Regarding the analysis and evaluation of safety-related properties, the goal is to identify critical system parts, relating them to possible system failures, and providing quantitative measures concerning the probability of occurrence of hazardous situations. A system failure in this sense is an undesired deviation of the actual behavior from the specified one. Performing such an analysis necessitates the abstraction of certain system properties to yield a view that represents a reduced image of the system. To accomplish this, most common techniques rely on graph- or text-based approaches (see section on safety analysis). These allow capturing specific safety-related aspects of the system under investigation. However, in general this entails the loss of context information, which may also indirectly contribute to system failures and hence is important with regard to the completeness of the analysis.

When considering embedded systems, the failing components at the most abstract stage can be hardware (HW) or software (SW). By nature, HW and SW components have a great variety of characteristics, modalities, and working conditions, and are generally not representable by a single model. For example, installation prerequisites, exposure to environmental interferences, as well as mutual interference between HW components are additional properties not reflected by the models used for safety analysis.

Let's assume that the safety of a control unit exposed to internal/external forces such as extreme heat, shock, and/or radiation has to be evaluated. Here, only a safety analysis is capable of making assumptions about the failure probability of the controller and how it may contribute to a system crash. However, it does not consider the ways the controller correlates with the aforementioned physical factors, i.e., it cannot answer the question whether or not there is an actual dependency between the way the controller is installed and the system's safety. Deriving cross-related knowledge from other models or system views is not possible from the safety models or is only provided up to a certain limit. To ascertain such extended coherences, the incorporation of knowledge/information from additional sources is required. Collecting all relevant information and representing them in a suitable way is necessary for the hidden relations to become apparent.

The above observations hold for HW as well as for SW, which has a variety of essential characteristics that indirectly contribute to safety. For example, important issues for assessing the criticality of a SW component can be its complexity (e.g., lines of code, number of dependencies between components, etc.), its state space situation, or its dependencies between workload and the system response time in critical scenarios. Again this represents context information not always considered in standard safety models.

Another important aspect concerns the distribution of the components of an embedded system. Generally, the sort of systems that belong to this category feature the property of being distributed. This means that communication takes place, which might also be affected either directly by intended attacks or indirectly by physical interferences. The connection between the vulnerability of communication channels in general and the way they affect the safety/security criticality are explored best if there is a combined virtual sight.

In the ViERforES 2 project, VR is used to combine safety-related information with such kind of "hidden" context information. This provides the chance to visually integrate safety-critical relationships with the corresponding static and dynamic physical dependencies of the investigated system. The related paper (Al-Zokari, 2010) proposes a way to link static safety-related information in the form of Minimal Cut Sets (MCS) (see section 3.1) with information about the physical system structure. A direct visual connection between the critical components and the ways they cause the system to fail is established. A new metaphor representing all possible constellations of cause-effect relationships allows the effective identification of most critical failure causes. The linkage between the MCS safety indicators and the physical context information in the form of a geometric representation of the system components is based upon highlighting those components that contribute to a selected failure scenario. The tool allows to intuitively trace the causes of safety-critical system failures back to the components responsible for their origination. This way the users (safety engineers and system engineers) can explore relations that could help, for example, to efficiently identify critical system regions. Moreover, it provides a common view for system and safety engineers to support inter-connected collaboration.

The ongoing research could be seen as a starting point regarding further development aimed at the integration of multiple models from complex embedded systems. The possibilities of application of the VR technology are various and provide great potential concerning future research.

Virtual Reality (VR) provides promising possibilities for visually assessing the characteristics of large embedded systems. Analyzing functional and non-functional aspects generally necessitates uncovering hidden properties and relations between system parts that heavily affect overall system functioning and system quality. Particularly, properties related to software are not always visible at first glance. The research within the ViERforES (Virtual and Augmented Reality for Maximum Safety and Reliability of Embedded Systems) project (ViERforES, 2011) concentrates on capturing and visualizing qualitative aspects of safety-critical systems to support system analysis and improvement. The process of tracing back undesired system characteristics to their roots constitutes the first step in identifying critical parts of a system. Generally, this produces huge amounts of data that are strongly correlated and not accessible without the usage of the proper context information. The adoption of VR techniques allows the user to look at every aspect of the system in detail. Furthermore, it facilitates the exploration of hidden properties and supports the collaboration of domain experts such as system and safety engineers. It associates safety-related information with virtual models of the examined systems and provides interaction techniques to support the process of identifying safety-critical elements.

4. Conclusion

The visualization of software and systems is becoming increasingly important for many organizations. Well-known visualization and interaction techniques are applied to obtain better insight into and control over system development processes and the quality of produced products. Even though these methods have proven to be advantageous, more domain-specific, user-centered approaches are necessary to support users in recognizing and finding relevant information more easily. Based on practical examples from industry and the experience of the authors, this chapter presented an overview of such goal-oriented, domain-specific visual exploration tools in the software engineering domain.

We showed that tightly coupling the specification and selection of visualization techniques with the specification and selection of metrics used to collect measurement data better supports the visual recognition of software product, process, and project characteristics and their interrelations.

In addition, we presented visualization techniques aimed at a better understanding of system properties such as safety and security in highly complex, dynamic embedded systems.

More sophisticated human-centered visual exploration techniques are needed to be able to analyze such system properties, which generally necessitates uncovering hidden properties and relations between system parts that heavily affect overall system functioning and system quality. The presented work allows interactively exploring information associated with the results of a safety analysis to maintain an overview of critical elements and relations. Approaches like the matrix-based representation presented here facilitate the understanding of the structural composition of failure causes and, in particular, allow to efficiently determine weak points of safety-critical systems.

5. References

ALD RAM Commander. (2011) http://www.aldservice.com.

Al-Zokari, Y., Khan, T., Schneider, D., Zeckzer, D., Hagen, H. (2010). *CaKES: Cake Metaphor for Analyzing Safety Issues of Embedded Systems.* Dagstuhl Follow-Ups.

Ball, T., Eick, S. (1996). *Software visualization in the large.* IEEE Computer, pp. 33–43.

Balzer, M., Noack, A., Deussen, O., Lewerentz, C. (2004). Software Landscapes: Visualizing the Structure of Large Software Systems. In: Proceedings of the Joint Eurographics - IEEE TCVG Symposium on Visualization (VisSym), pp. 261-266.

Barret, M., Blackledge, J. Eugene, E. (2010). *Using Virtual Reality to Enhance Electrical Safety and Design in the Built Environment.* Dublin: Dublin Institute of Technology.

Battista, G. D., Eades, P., Tamassia, R., Tollis, I. G. (1999). *Graph Drawing: Algorithms for the Visualization of Graphs.* Prentice Hall, Upper Saddle River, NJ, USA.

Eiglsperger, M., Siebenhaller, M., Kaufmann, M. (2005). *An efficient implementation of Sugiyama's algorithm for layered graph drawing.* InGraph Drawing (GD 04), vol. 3383 of Lecture Notes in Computer Science, Springer, pp. 155-166.

ESSaREL. (2011). http://www.essarel.de.Fekete, J.-D., Wang, D., Dang, N., Aris, A., Plaisant, C. (2003). *Overlaying Graph Links on Treemaps.* In Proc. InfoVis'03, Poster Compendium, IEEE Press.

Gansner, E. R., Koutsos, E., North, S. C., Vo, K. P. (1993). *A Technique for Drawing Directed Graphs.* IEEE Trans. Softw. Eng. 19, 3, pp. 214-230.

Heidrich, J. Münch, J. (2010). *Goal-oriented customization of software cockpits.* Journal of Software Maintenance and Evolution: Research and Practice, Volume 22 Issue 5.

Heng, P.A., Cheng, C.Y., Wong, T.T., Xu, Y., Chui, Y.P., Chan, K.M., Tso, S.K. (2008). *A virtual-reality training system for knee arthoscopic surgery.* IEEE Transaction on Information Technology in Biomedicine 8 (2), pp. 217-227.

Herman, I., Melancon, G., Marshall, M. S. (2000). *Graph Visualization and Navigation in Information Visualization: A Survey.* IEEE Transactions on Visualization and Computer Graphics 6 (1), pp. 24–43.

Holten, D. (2006). *Hierarchical Edge Bundles: Visualization of Adjacency Relations in Hierarchical Data.* IEEE Transactions on Visualization and Computer Graphics (Proc. of INFOVIS'06) 12 (5), pp. 741–748.

Instantiations CodePro Studio. (2011). http://www.instantiations.com

ISOGRAPH FaultTree+. (2011). http://www.isograph-software.com.

ISO – International Organization for Standardization. (2011). http://www.iso.org

JDepend. (2011). http://clarkware.com

Johnson, B., Shneiderman, B. (1991). *Tree-maps: A spacefilling approach to the visualization of hierarchical information structures*. Proceedings of Visualization '91, (San Diego, California), pp. 284–291.

Kaufmann, M., Wagner, D. (2001). *Drawing Graphs, Methods and Models*. Lecture Notes in Computer Science, Springer Verlag.

Kreylos, O., Bethel, E.W., Ligocki, T.J., Hamann, B. (2003). *Virtual-Reality Based Interactive Exploration of Multiresolution Data*. In: G. Farin, H. Hagen and Bernd Hamann: Hierarchical Approximation and Geometrical Methods for Scientific Visualization. Springer Verlag, pp. 205-224.

Kruskal, J. B., Landwehr, J. M. (1983): *Icicle Plots: Better Displays for Hierarchical Clustering*. The American Statistician 37 (2), pp. 162–168.

Lommerse, G., Nossin, F., Voinea, L., Telea, A. (2005). *The Visual Code Navigator: An Interactive Toolset for Source Code Investigation*. Proceedings of the 2005 IEEE Symposium on Information Visualization, IEEE Computer Society.

McCabe IQ. (2011). http://www.mccabe.com

Metrics Eclipse Plugin. http://sourceforge.net/projects/metrics

Oliveria, D.M., Cao, S.C., Hermida, X.F., Rodríguez, F.M. (2007). *Virtual Reality System for Industrial Training*. IEEE International Symposium on Industrial Electronics, pp. 1715-1720.

Purschke, F., Schulze, M., Zimmermann, P. (1998). *Virtual Reality - New Methods for Improving and Accelerating the Development Process in Vehicle Styling and Design*. Computer Graphics International.

RelexArchitect. (2011). http://www.relexsoftware.co.uk.

Rech, J., Weber, S. (2005). *Werkzeuge zur Ermittlung von Software-Produktmetriken und Qualitätsdefekten- Studie zu Software-Messwerkzeugen*. Fraunhofer IESE.

RELIA BlockSim. http://www.reliasoft.com/BlockSim.

Stasko, J., Zhang, E. (2000). *Focus+context display and navigation techniques for enhancing radial, space-filling hierarchy visualizations*. In INFOVIS '00: Proceedings of the IEEE Symposium on Information Visualization 2000, pp. 57-68.

Soto, M., Ocampo, A., Münch, J. (2008). *The Secret Life of a Process Description: A Look into the Evolution of a Large Process Model*. International Conference on Software Process (ICSP) - Proceedings. Berlin: Springer, pp. 257-268.

Sotograph. http://www.software-tomography.de

Sugiyama, K., Tagawa, S., Toda, M. (1981). *Methods for Visual Understanding of Hierarchical System Structures*. IEEE Transactions on Systems, Man, and Cybernetics SMC-11(2), pp. 109-125.

SYNC DPL-Faulttrees. (2011). http://www.syncopationsoftware.com/faulttree.html.

Termeer, M., Lange, C. F., Telea, A., Chaudron, M. R. (2005). *Visual Exploration of Combined Architectural and Metric Information*. VISSOFT '05 Proceedings of the 3rd IEEE International Workshop on Visualizing Software for Understanding and Analysis, IEEE Computer Society.

ViERforES Project. (2011). http://www.vierfores.de.

Yang, Y., Zeckzer, D., Liggesmeyer, P., Hagen, H. (2011). *ViSSaAn: Visual Support for Safety Analysis.* Dagstuhl Follow-Ups, Vol. 2.

17

The ACSA Accreditation Model: Self-Assessment as a Quality Improvement Tool

Antonio Almuedo-Paz, Diego Núñez-García,
Víctor Reyes-Alcázar and Antonio Torres-Olivera
Andalusian Agency for Health Care Quality
Spain

1. Introduction

Excellence in healthcare is based on the concept of "continual improvement". It must start from the premise that everything we do is always capable of being improved (Pomey et al, 2005). It is not an isolated action, but rather a culture and a dynamic concept that must backbone the entire organization without impediment. This "know how" will allow the reinforcement of strengths and the reduction of those weaknesses (Lorenzo, 2004) found when making a previous consideration of how everyday practice is performed and how it can be improved.

The quality and safety of care is found in all European policy agendas, which currently show differences in the quality of services provided (Lombart et al, 2009). Consequently strategies are suggested to improve quality and safety in different health systems.

Healthcare institutions have a social responsibility to respond to health-related needs with the provision of important services, that are equitable, cost effective and of high quality. Similarly, health professionals have the responsibility to remain competent to attain standards of excellence, generate and disseminate knowledge and commit to defending the interests and welfare of patients, responding to the health demands of society (Perez & Oteo, 2006).

The above is generally assumed by everyone involved in improving health services, such as health institutions, scientific societies, professional associations and the professionals themselves. Consistent with this, we are living at a time of expansion of continuing education activities (Pardell, 2005), that are accompanied by a significant investment in resources. Similarly, significant efforts are being made to promote the accreditation and quality of healthcare institutions (Pomey et al, 2005).

However, what methodology should be followed? The better this "know how" is carried out the closer results will be to excellence, and it must be done by measuring results and performance (Laing, Hohh & Winkelman, 2004) and taking into account the meanings of effectiveness and efficiency. In this context, the concept of accreditation is used as a tool to drive the continual improvement of care quality. The detection of areas for improvement is highlighted as one of the main strengths of the process (Aranaz et al, 2003). This accreditation process may include a serie of essential interventions to facilitate and encourage professionals and organizations to assess themselves (Claveria, 2004), participating in identifying and prioritizing problems and developing areas for improvement. It is essential that this reflection involves the largest number of professionals, including managers (Gutt et al, 2006), something that can enrich the process. Performance is compared with expectations and objectives, thereby identifying opportunities for improvement. Many accreditation models use self-assessment as a reflection tool that allows the identification of the strengths and areas for improvement of the organization (Viswanathan & Aslmon, 2006, Giraud, 2001, Greenfield & Braithwaite, 2008).

2. Self-assessment in professional development

Among the methodologies for the evaluation of competence, the portfolio is characterized by developing a self-assessment phase. This reflection exercise is one of the strengths of the portfolio (Arnau-Figueras & Martinez-Carretero, 2007; Prados, 2005). That is why the self-assessment phase of the process is the main generator of benefits for the continuing professional development (CPD) of those professionals who undertake the assessment process (Almuedo-Paz et al, 2011).

There are several experiences of the use of the portfolio in postgraduate education and its role in CPD support (Tochel et al, 2009). Similarly, in the health professions self-evaluation is used to induce continual improvement in professional work (Casey & Egan, 2010; Cowan, Wilson-Barnett, & Norman, 2007; O'Neill & Kurtz, 2001). In the health field, this type of competence evaluation based on self-assessment is most widespread at the resident stage (Caverzagie, Shea & Kogan, 2008; Pasquina, Kelly & Hawkins, 2003; Staccini & Rouger, 2008).

2.1 Self-assessment in health institutions

With regard to models of self-assessment of quality, the EFQM (European Foundation for Quality Management) Excellence Model is the most widely accepted with regard to self-assessment (Del Rio et al, 2006, Brun et al, 2004 ; Pariente et al, 2004; Martínez-Pillado, 2008; Editorial Committee, 2004), allowing total quality management to be applied in the health sector to obtain better results (Ugalde, 2003). However, the complexity of adapting it to primary and specialized care decreases its applicability (Arcelay et al, 1998). Importantly, among the models for the accreditation of quality specific to the health sector, there is an important commitment to self-assessment as a cornerstone of the accreditation process, such as in the Australian Council on Healthacare Standards (Greenfield & Braithwaite, 2008), Haute Autorité de Santé (Fortes, Mattos & Baptista, 2011) and Accreditation Canada (McCurdy et al, 2009).

As the result of a need for a model more suited to the healthcare reality, already demanded in public health organizations (Guix Oliver, 2005), and favoured by further development of specific continual improvement tools that enable management of healthcare quality (Torres-Olivera et al, 2004), the ACSA accreditation model was born (Almazán-González, 2006). In this model, professionals and institutions that initiate the process, at the first stage of self-assessment, have reference standards that address aspects of quality related to the citizen, professional, healthcare processes, support elements and results. The working methodology is to reflect on what is done today to reach those standards (objectifiable facts or positive evidence), how it can be demonstrated, what results are obtained and how they can be improved (where work needs to be done to strengthen or achieve compliance), that is, areas for improvement, a cornerstone in the process of continual improvement. The greater number of professionals that work on accreditation projects, the more enriching are the improvement cycles that are designed. In this model, professionals can participate as assessors (reflecting on the implementation and improvement of standards) and as agents for improvement (they are responsible for implementing the identified improvements).

This model attempts to identify areas or activities where attention must be focussed to achieve higher quality levels in institutions throughout the whole Health System, whilst promoting the development of a culture of quality and transparency based on continual improvement.

3. ACSA model of accreditation

The Andalusian Agency for Healthcare Quality was founded in 2002 as part of the strategy promoted by the Andalusian Regional Ministry of Health, to improve and ensure quality in the healthcare provided to citizens within the Andalusian Public Health System (SSPA).

Quality plans of the Andalusian Regional Ministry of Health arise from a firm commitment to excellence, innovation and professional development. One of its strategic processes is focused on "ensuring the management of quality in the health services."

These strategic lines include the development of the accreditation programmes of the Agency for Healthcare Quality, based on the continual improvement of quality and the progress and development of professionals.

At this time, the Agency has established and implemented its own unique accreditation model, has designed a series of accreditation programmes as tools for the continual improvement and safety at the service of the professionals, units and organizations, and has established a methodology that facilitate their application in practice and maximizes results.

3.1 The accreditation model

Accreditation is seen as the process that observes and recognizes how the healthcare provided to citizens responds to our quality model, always with the aim of encouraging and promoting continual improvement of our institutions, professionals, training, etc.

From this perspective, the accreditation model of the Andalusian Health System boasts a number of typical characteristics:

- It is consistent with the plans and management tools for continual improvement in the SSPA: clinical management, process management, competence management and knowledge management.
- The standards present in the different programmes reference the health regulatory framework of Andalusia, the strategic elements of the SSPA, the recommendations on best practice, safety elements, the needs and expectations of citizens, the results of satisfaction surveys, etc.

- It addresses quality from an integrated approach, through a series of accreditation programmes targeting different elements involved in healthcare: health centers and clinical management units, continuing education, professional competences and web pages.
- It has a progressive nature, identifying different stages or degrees of progress towards excellence. Going beyond an isolated benchmark or a recognition obtained at a given time, accreditation is a dynamic, continual and evolving process that reflects and reveals not only the situation at the time of its implementation but also, above all, the potential for development and improvement to grow in quality.

The accreditation model of the Andalusian Health System provides a common framework for all accreditation programmes within it.

3.2 The accreditation programmes

The Agency has accreditation programmes in several areas:

- Programmes of accreditation of centres and units (hospitals, primary care and hospital clinical units, haemodialysis units, clinical laboratories, diagnostic imaging units and blood transfusion centers).
- Programmes of accreditation of professional competences for 70 different health disciplines.
- Accreditation Programmes for continuing education (activities, programmes and centers for continuing education).
- Programmes of accreditation of healthcare web sites.

Accreditation programmes share the same structure and contemplate, from each of their perspectives, the same key areas of quality management. They are grouped into five blocks, five dimensions around which are grouped the content of the various programmes. These blocks are related to:

- The citizen
- Integrated health care
- Professionals
- Areas of support
- Efficiency and results

Each programme is developed with the participation of the professionals in the Health System through technical advisory committees.

3.3 Methodology

Our accreditation model grants a role to self-assessment in all the programmes. Continual improvement is based on the immense potential possessed by the people and organizations. In the self-assessment phase, different groups or professionals identify their current position, determine their aspirations and plan actions to achieve them.

Self-assessment creates a space for consensus and shared improvement in which the different actors are involved (professionals, managers and citizens).

For example, in the process of accreditation of centres and units, the professionals analyze the standards and their purpose and reflect on:

- What is done? and How can it be demonstrated?, from where those positive evidences arise.
- What are the results? How can they be improved?, Questions that lead to areas for improvement.

Similarly, in the process of accreditation of competences, during the self-assessment the professional examines the competences and good practices related to the achievement of outstanding results in their work, and provides evidence and proof of the presence of these good practices in their daily performance.

The accreditation model starts with reference standards with which the centres, professionals, units, etc. move towards improving services to citizens and the implementation of management tools to improve quality.

From our view of quality, the standards are a continually evolving system which the citizen contributes to by incorporating their needs and expectations. Its definition, review and continual updating give dynamism to the processes of accreditation.

Accreditation also means an explicit and public recognition that the requirements needed to develop quality care have been fulfilled, and that a line of continual improvement has been undertaken. As a tool, and not an end in itself, accreditation promotes and encourages processes of improvement and evaluation within the health system.

The road proposed for continual improvement involves the entire organization, from the highest management to the entire group of people who work there, all of whom must be firmly committed to this process.

Additionally, given that accreditation has been regarded as a dynamic process, it should not be understood as the end of a road, but as an opportunity to establish new and alternative paths to improve quality.

External evaluation is another of the common elements of the accreditation process, for both centres and professional competences. At this stage, teams of the Agency's surveyors observe and recognize the evidence presented (documents, by observation, in interviews)that are associated with different elements of quality and safety, and identify the level of accreditation obtained, the strengths, the potential and the areas for improvement.

Each of the phases, and especially self-assessment, are based on a series of Web-based applications called ME_jora developed by the Andalusian Agency for Healthcare Quality, which allows each accreditation process to be conducted securely and with the support of Agency professionals, and also enables the dissemination and exchange of knowledge and the elements of quality identified in them.

4. Accreditation programme for healthcare units

As in the other accreditation programmes, the standards for the accreditation of clinical units are divided into three groups:

- The standards of group I relate to the vested rights of citizens, the aspects related to the safety of citizens and professionals, the ethical principles that should be contemplated in all healthcare policies and those priority items for the SSPA.
- In group II the standards include those elements associated with the further development of the organization (information systems, new technologies and redesigning of organizational spaces).
- The group III standards relate to showing that the clinical unit generates innovation and development for society in general.

Group I includes those standards that are considered mandatory and therefore must necessarily be present and stabilized to achieve any level of accreditation.

The following table summarizes the distribution of programme standards for clinical management units by type of standard:

Standard Type	Definition
Group I	Standards that provide for the vested rights of citizens, issues related to the safety of citizens and professionals, the ethical principles that should be contemplated in all activities of the clinical management units and those priority items for the Andalusian Public Health System
Group II	Standards governing elements associated with the further development of the organization (information systems, new technologies and redesigning of organizational spaces)
Group III	Includes those standards that demonstrate that the clinical management unit generates innovation and development oriented to society in general

Table 1. Distribution of programme standards for accreditation of clinical unit.

4.1 Results of accreditation

The accreditation model articulates progression in different grades, each more complex and demanding than before, thereby facilitating continual improvement. Accreditation levels are advanced, optimum and excellent.

However, as long as the system is constantly evolving and seeking ways to improve, it would be wrong to understand the final grade as a final or last stage. Rather (and as a result of improvements that will occur due to new technologies, new services, new forms of organization and new demands from the citizen user and professional user), the standards established for the various grades will be updated periodically. For example, what can be seen today as far for any system, with continual improvement, may be excellence tomorrow.

Ultimately, the accreditation model for the Andalusian Health System is a useful methodological tool that allows checking of the extent to which activities are carried out according to quality standards, and the light of external evaluations provides public and express recognition of those institutions and professionals who comply and demonstrate such.

The result of the accreditation process can be:

Pending the stabilization of obligatory standards: situation is maintained until the plans for improving the clinical unit meet the obligatory standards of group I. Achieving these will qualify for some level of accreditation.

Accreditation - Advanced: accreditation obtained by achieving over 70% compliance with group I standards (all those considered obligatory are included within this percentage).

Accreditation - Optimal: is reached when there is 100% compliance with group I standards and more than 70% compliance with group II standards.

Accreditation - Excellent: the level of excellence is achieved when there is 100% compliance with group I and group II standards, and more than 70% compliance with group III standards.

After obtaining an accreditation rating of advanced or optimal, clinical management unit, after at least a year, may voluntarily choose to attempt accreditation in successive ratings.

4.2 Structure of the standards manual

The manual of standards of clinical management units is divided into five blocks and eleven criteria, which are described below:

I. The person, central to the Health System	The person as an individual asset
	Accessibility and continuity of care
	Clinical information
II. Organization of person-centred activity	Management of integrated healthcare plans and processes
	Health promotion in the community
	Clinical unit management
III. Professionals	Clinical unit professionals
IV. Support Processes	Structure, equipment and suppliers
	Information systems and technology
	Continual improvement
V. Results	Clinical unit results

Table 2. Structure of accreditation programmes for clinical units

I. The person, central to the health system

This block represents 24.77% of all standards in the manual, referring to the rights, expectations and participation of users, proffesionals and units petitioning of the clinical unit; to the elements related to privacy and accessibility to available resources; to the relationship between professionals and healthcare and to interdisciplinary actions linked with the use and safekeeping of the clinical and personal information of the user. There is differentiation into three criteria:

- The person as an individual asset
- Accessibility and continuity of care
- Clinical information

II. Organization of person-centred activity

This block constitutes 22.94% of the standards contained in the manual. They primarily concern issues related to:

- Management of integrated healthcare plans and processes
- Health promotion in the community
- Management of the clinical unit

III. Professionals

These account for 11.01% of all standards and cover from the induction of professionals to the adequacy of professional resources for care, while facilitating the updating of their competences and professional development and enhancing research work in the clinical unit.

IV. Support processes

This block accounts for 30.28% of the standards in the manual, and is dedicated to the management of the structure of the centre and its facilities, supply processes, equipment, safety measures and functionality for users and professionals to achieve the proposed objectives. It analyzes the areas relating to new technological advances in the field of information technology, the protection of personal data and strategies for managing risks and specific quality plans. It is differentiated into the following three criteria:

- Structure, equipment and suppliers
- Information Systems and Technology
- Continual Improvement

V. Results

Finally, there is a set of standards which make up 11.01% of the contents of the manual and which reflect the results obtained by the clinical unit in terms of activity, efficiency, accessibility, satisfaction and scientific- technical quality.

4.3 Phases of the accreditation programme

PHASE 1. Preparation. Application for accreditation and introductory visit

The head of the unit requests the Andalusian Agency for Healthcare Quality to start the accreditation process by completing an application in the ME_jora C program, available on the website of the Andalusian Agency for Healthcare Quality at http://www. juntadeandalucia.es / agenciadecalidadsanitaria.

This application results in joint planning of the whole process of accreditation between the unit and the Agency.

Subsequently, the Andalusian Agency Healthcare Quality appoints a project manager, and the unit names an internal accreditation process manager to facilitate the development of the process and communication with the Agency.

Finally, by agreement with the unit, a visit is planned to present the accreditation process.

PHASE 2. Internal focus: self-assessment

Self-assessment is conceived as the permanent testing of the areas for improvement in the organization. The standards manual is taken as a reference for this assessment. During this phase, the professionals in the unit conduct an exercise of reflection, and must observe the strengths (i.e. positive evidence) and in turn identify areas for improvement.

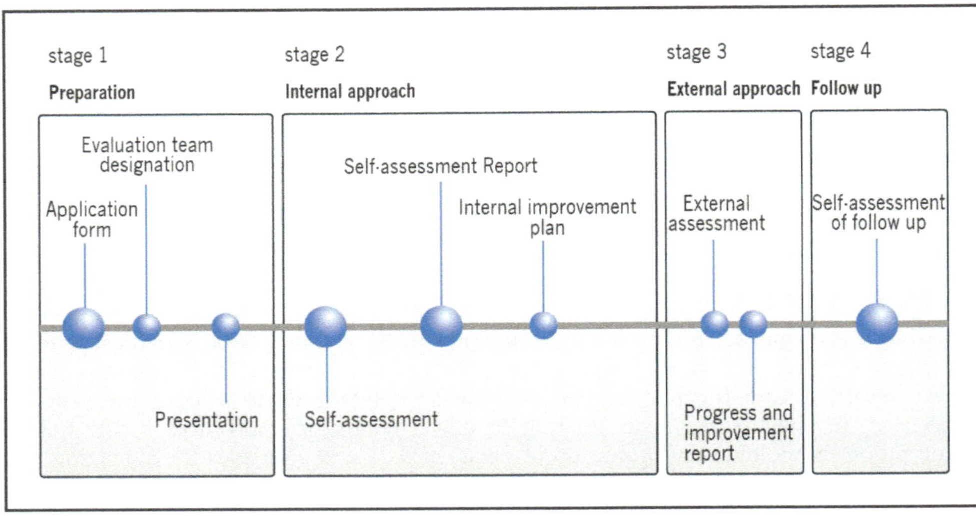

Fig. 1. Phases of the accreditation program.

Self-assessment has the following objectives:

- To provide the unit with a way towards continual improvement and towards accreditation through:
  - To identify strengths, in order to maintain, and even improve them. To identify areas for improvement in order to work on them and turn them into strong points.
  - The expansion of information on the purpose, scope of the standard and the provision of examples of good practice.
- To enable periodic self-assessment, both within and outside of the accreditation cycle, to assess progress on an ongoing basis.
- To promote learning and knowledge management in the health system.

In the design of self-evaluation, a qualitative approach has been chosen to determine the level of compliance using the Deming PDCA cycle.

Thus, by following the PDCA (Plan - Do - Check - Act) method for each of the standards, we develop a continual improvement cycle. This prevents the compliance with a standard remaining as a static fact, or point, associated with the moment of assessment. By using the PDCA cycle of continual improvement it is intended that the organization review its approach to compliance with standards. It must PLAN ahead, DO what is required, CHECK the effectiveness of the standard and ACT to improve implementation and development, thus ensuring the consolidation and stabilization of the standard over time.

In turn, stabilization of a standard does not only involve the compliance with it, it also involves the mobilization of the unit in a process that will ensure future compliance.

The unit will therefore review which phase of the PDCA cycle (Plan, Do, Check, Act) each standard is at, according to the steps described in the table below:

PHASE	PRECIS	STEP
PRIOR	Preliminary improvement profile	The influence of the standard on the organization has been determined prior to starting the accreditation process
PLANNING	Specify objective and information system	Indicators have been defined that identify the attainment of the standard
	Plan	The actions needed to achieve the standard have been defined
	Define functions	The responsibilities and human resources needed to meet the standard have been identified and assigned
	Communicate	All those involved in the initial process to reach the standard have been informed of the plans
	Proportion resources	All the resources (materials, training, etc.) needed to achieve the standard are defined and assigned
COMPLIANCE	Complies	The purpose of the standard is achieved in accordance with its influence and with the defined indicators.
EVALUATION	Appraise	Non-conformities have been identified in the results
ADAPTATION	Correct and enhance	Actions have been undertaken to eliminate the observed non-conformities

Table 3. Phases of the improvement cycles

The standards manual is based on a software application that:

- Provides accessibility to the accreditation process with secure access via user profiles from any post or workplace in the healthcare unit.
- Acts as a document manager for all the information generated in the accreditation process.

The self-assessment phase consists in turn of the following phases:

1. The management team sets the objectives and action plan for the self-assessment of the clinical management unit.

How the self-assessment will be deployed, and the format for internal and external communications, etc, will be determined in the planning.

A manager is designated in the unit for the accreditation process.

It is advisable to appoint a manager for the accreditation process in the unit, who will assume the role of key person in organizing and coordinating the whole process and communicate with the Andalusian Agency for Healthcare Quality. Their main functions will be to:

- Set the schedule to be followed and ensure compliance.
- Participate in the selection of assessors, giving them support and help in training.
- Establish and conduct meetings of the assessors.

2. The assessors are selected and assigned to certain standards.

A group of assessors is designated in this phase, depending on the size of the unit. It is recommended that these assessors be multidisciplinary, as this will enhance learning and creation of organizational knowledge. Given the dynamic nature of both standards and the process itself, this team should not only be for a one-time-only self-assessment exercise, but rather should continue to work on identified areas for improvement and on the recommendations of the external evaluation, and would regularly update the self-assessment. To promote teamwork and make it efficient, it is also advisable to divide the standards among professionals who will participate in the self-assessment.

The assessors are trained in handling the IT application.

The unit manager of the accreditation process will provide the assessors with:

- The complete standards manual and the list of standards upon which each will perform self-assessment.
- Access to software where groups can register positive evidence, the areas for improvement and the degree of compliance according to the PDCA cycle.

The manager for the accreditation process of the Andalusian Agency for Healthcare Quality will be responsible for providing sufficient training to ensure the management of the software and the monitoring of the process.

3. The self-assessment files are completed.

Coordinated by the internal manager of the accreditation process, the working group reviews the standards and completes the forms in the software application available on the Web. The review of standards entails reflection on whether the standard is met, in which case the positive evidence supporting it must be described. If there was no evidence to demonstrate compliance with the standard, the self-assessor must describe the areas for improvement that the clinical management unit should address in order to comply and stabilize the standard. The software allows the attachment of files to the positive evidence and improvement, for thus acting as a document manager.

4. The self-assessment group shares the findings (positive evidence and areas for improvement) and finalizes the self-assessment.
5. The results from the assessment are then pooled, and some of the responses are clarified, and the information shared and completed.
6. Priority is given to the positive evidence and areas for improvement. After sharing the results obtained by the different groups, the areas for improvement are approached globally, searching for common lines of action. The software allows the prioritization, planning and allocation of managers of the areas of improvement.
7. The improvement plans are developed and implemented. The software allows the description of actions for each area of improvement, which along with the ability to plan and designate managers, makes it an easy management system for continual improvement in the unit.
8. The process of self-assessment Is evaluated and improved. Finally, the process of self-assessment is contemplated as a learning formula, to introduce improvements and thus prepare the successive self-assessments. The software has a results module that facilitates and supports the planning, monitoring and achievement of the actions arising from the management of improvement made in the self-assessment phase.

PHASE 3. External approach. Evaluation visit

Once the self-assessment phase is completed, a visit will be made by the team of surveyors from the Andalusian Agency for Healthcare Quality, who will study the self-assessment and perform an external evaluation. This visit will planned with the agreement of the management team of the clinical unit.

Thus, throughout this phase, the evaluation team of the Andalusian Agency for Healthcare Quality is responsible for verifying compliance with the standards based on positive evidence and areas for improvement provided by the healthcare unit during the self-assessment and on other significant evidence that will be collected during the visit and that will be in the form of documents, interviews, and direct observation.

PHASE 4. Reports

After the external evaluation visit, the evaluation team of the Andalusian Agency for Healthcare Quality prepares a progress report which specifies the degree of compliance with standards and recommendations.

This report is sent from the Andalusian Agency for Healthcare Quality to the management team of the unit.

PHASE 5. Monitoring and collaboration between the clinical unit and the Andalusian Agency for Healthcare Quality

A specific module in the ME_jora C application has been defined and implemented in order to carry out accreditation project monitoring over the five year validity period of accreditation.

The objectives of the follow-up phase are:

- Consolidate the results obtained by the stabilization of standards compliance over five years.
- Maintain and increase the momentum of improvement through the implementation of identified areas for improvement, along with the opportunity to continue identifying new areas for improvement.

After obtaining accreditation, the unit has the self-assessment monitoring sheet available, so that at two and four years from the date of accreditation a follow-up evaluation is carried out consisting of the following four sections:

- Analysis of the prior considerations about structural and organizational changes that could have been produced in the clinical unit and that could affect the scope of accreditation.
- Positioning and analysis of compliance with the mandatory standards, in order to ensure maintenance of compliance over time.
- Update of the areas for improvement identified in the self-assessment phase yet to be fulfilled.
- Update indicators of activity and healthcare processes over the last two years.

5. Professional competences accreditation programme

5.1 Conceptual framework

The professional competences accreditation programme of the SSPA has been designed to recognize the achievements of professionals in real and daily practice, and as a tool to promote professional development and continual improvement.

The accreditation programme is based on the methodological and conceptual framework of management by competences, as a comprehensive model for configuring the processes of selection, performance appraisal, training management, promotion and incentives.

The concept of competence refers to a capacity or stable personal characteristic causally related to desirable outcomes in an organization.

A key element of management by competences is identifying these capacities as measurable elements. The most coherent acceptance of the competencies approach is that which considers them as a set of observable behaviours which are measurable in a reliable and valid way, and causally related to good or excellent performance.

In the healthcare system, and in its measurement, competence is defined as the ability of health professionals to integrate and apply knowledge, skills and attitudes associated with the "good practices" of their profession to resolve the situations that arise.

This conceptual approach focuses on what the professional does. For the professional to develop good practices, i.e., the observable behaviours associated with a competence (to do), it is necessary for all five components of the competence to be present: to know it (knowledge), know-how to do it (skills), know how to be (attitudes), want to do it (motivation) and the ability to do it (professional competence and means) (Fig. 2).

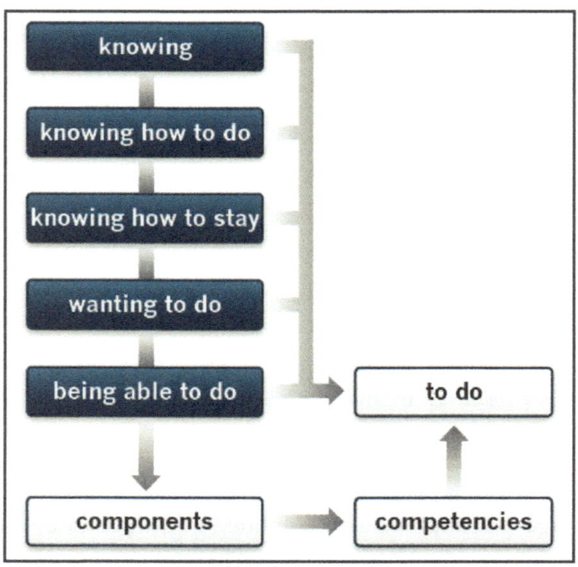

Fig. 2. Components of the competence.

The set of competences required by a professional in a job is their "competence map". It identifies competencies and the best practices (observable behaviours) associated with them, as well as the evidence (or verification criteria for determining the presence of good practices) and tests (measurement and evaluation tools that determine compliance of the evidence of good practice integrated in a professional competence).

As described above, most of the tests to determine compliance with the evidence included in the competence accreditation programme are based on "what the professional does" (in real situations, in their results, and so on.), as an ideal way to recognize and accredit professional competences.

The accreditation of professional competences is conceived as a process that systematically observes and recognizes the proximity between the competences that a professional really has and those defined in the competence map (Almuedo, 2006).

The manuals for the accreditation of health professionals have been developed with the participation of over 600 professionals and representatives from 70 scientific societies, which have formed technical advisory committees, one for each discipline or specialty, each of which has developed its specific manual of competencies.

Each of these technical committees has identified the competencies that a particular professional must possess, as well as the good practices that should be present in the performance of their work.

In all the manuals, the professional competences are grouped around 5 blocks and 10 criteria, which address the quality model of the Andalusian Public Health System (Table 4)

Block I	**The citizen**	1. Orientation to the Citizen (satisfaction, participation and rights)
Block II	**Integrated Health Care**	2. Health Promotion, Prevention and Community Care
		3. Attention to the individual and the family
		4. Management by Integrated Healthcare Processes
Block III	**The Practitioner**	5. Teamwork & Inter-professional Relationships
		6. Attitude of Progress and Professional Development
		7. Commitment to Teaching
		8. Commitment to Research
Block IV	**Efficiency**	9. Efficient use of resources
Block V	**Results**	10. Results oriented professional development

Table 4. Structure of the professional competences accreditation manuals

Each competence is associated with a number of good practices, and each good practice must be backed up by evidence and tests that the professional must provide to demonstrate that they do indeed possess such competences.

5.2 Levels of accreditation

Accreditation means getting explicit and public recognition of compliance with the requirements to provide quality care and the beginning of a line of continual improvement by a professional. Thus, accreditation is not an end in itself, but a dynamic, continual and evolving process, which provides professionals with the opportunity to establish development options to grow in quality.

When a professional is competent in a particular area of their professional performance, they present a series of observable and measurable behaviours, which verify the presence of such competence: this set of behaviours constitute their good practices, which can be observed and measured through evidence and tests (Brea-Rivero et al, 2001).

The evidence used to verify the existence of good practice has been classified by level of complexity and can be of various types:

- Essential evidence (which is essential for professional compliance)
- Evidence Group I (indicating that the professional progresses towards maturity),
- Evidence Group II (consolidating the professional's maturity)
- Evidence Group III (which makes the professional a benchmark for other professionals in the system).

The tests are instruments or objective elements of measurement or evaluation which determine the fulfilment of the evidence associated with each good practice of a professional competence.

The tests to be provided are primarily:

- Non-attendance tests:
  - The self-audit is a review that the professional makes of sample health records of patients seen over a certain period of time.
  - The reports consist of conducting a brief summary of a health history, in which the professional shows what his performance was in certain situations. Also, the reports can be "reflection" and / or "clinical practice".
  - Certificates are documents accrediting the performance of a particular activity.
- Attendance tests simulating a clinical situation. These tests are called by the Agency for Healthcare Quality on determined dates, which the professional can attend or not, as these tests are not mandatory).
- Non-attendance tests subject to call (also called "contextualized cases"). They can consist of using the Internet to make a critical reading of a scientific or medical journal based on the evidence.

Evaluation methods are mainly focused on the highest level of Miller's pyramid (Miller, 1990).

With regard to how much evidence and tests a professional has to provide, it is important to note that the evidence contained in the manual of competencies corresponds to your professional group, depending on the level of accreditation intended to demonstrate or achieve, the quantity and percentage of evidence required is different.

Based on these percentages of evidence obtained (essential, group I, II and III), the result may be accredited in any of the following grades (Table 5):

- Advanced grade
- Expert grade.
- Excellent grade.

The accreditation will have a term of five years. After this period, the accreditation ceases to have effect, unless the process of reaccreditation began before expiry.

	ADVANCED	EXPERT	EXCELLENT
GROUP I	70%	70%	60%
GROUP II		70%	70%
GROUP III			80%

Table 5. Evidence required to achieve different levels of accreditation.

5.3 Accreditation process

The process of accreditation of competences is a voluntary process by which the professional systematically reviews their own practice, demonstrating a level of competence, which they already had, or have achieved during the accreditation process. Thus, the competence accreditation programme attempts to ensure the presence and / or acquisition of new competences and a determined level of their development throughout professional life.

Accreditation is a dynamic process that entails a periodic evaluation every five years to verify the presence or acquisition of new skills and their level of development (certification and recertification).

The accreditation process consists of three phases:

- Phase 1: The Application.
- Phase 2: The self-assessment.
- Phase 3: The recognition and certification.

Step 1. Request

The accreditation of professional competences begins with a formal request via the Web, which contains the information needed to correctly identify the professional and their accreditation choice.

Access to the professional competences accreditation programme, is made through the website of the Andalusian Agency for Healthcare Quality, in the ME_jora P application, which is designed to facilitate the process of professional accreditation.

http://www.juntadeandalucia.es/agenciadecalidadsanitaria/acsa_profesionales/

Once the application is accepted, the professional has access to all information relating to their accreditation process, thereby enhancing the autonomy of the professional to manage their accreditation path with transparency throughout the process.

From this moment, the Andalusian Agency for Healthcare Quality will supply credentials to access the accreditation programme and the manual of competencies for your professional group, assigning it a professional surveyor of the Agency (tutor guide) who will accompany the whole process, either through personal meetings, telephone contact, or electronic communication through the Web. Moreover, the ME_jora P application includes a video with the information needed to provide accreditation.

PHASE 2: Self-assessment

This phase is the most important for the professional. It consists mainly in collecting and provide evidence from their own practice, real and daily (depending on the contents of the manual of competencies concerned), which demonstrate good practice in professional performance, allowing a certain level of competence to be proven, which was previously possessed, or that has been achieved during the accreditation process. Therefore a portfolio type methodology is applied to assessing competence, which is the greatest benefit of continuing professional development.

Since the self-assessment can last indefinitely, depending on the professional, the evidence has a period of validity, outside of which it expires.

The Agency has developed a Web-based software application (ME_jora P), which allows the professional to provide the necessary evidence to attain accreditation, as well as accessing the content of the corresponding manual of competencies, examples and references to consult in relation to good practice, it facilitate the development of self-assessment, customizes the process and establishes permanent contact with the Agency to resolve concerns.

PHASE 3: Recognition and certification

Once the professional has completed their self-evaluation, the Agency for Healthcare Quality reviews the evidence provided, by expert professionals in each discipline or specialty, and depending on them, issue a report of results and the corresponding certification of the results of the evaluation, according to the criteria and standards established at the level of development that the professional has reached: advanced, expert or excellent.

In the report of results, the Agency for Quality provides the professional with vision on the percentage of evidence provided, and the level of compliance of such evidence after the assessment phase, identifying the level of professional development in each of the competences contained in the specific manual.

In addition, the practitioner may request a review of the report of the results of process of assessing the level of professional competence.

5.4 Support tools

To facilitate the professional accreditation process, the Agency offers several support tools:

- Manual of competencies: in digital or paper format, containing the competencies and best practices that have been defined for each professional group or specialty. In addition, the manual contains all the evidence and proof to be provided to advance the accreditation.
- Tutor guide: when the professional requests starting their competences accreditation process, they are assigned an Agency professional to accompany them throughout the process. The tutor-guide will be in constant contact, either by personal meetings, telephone contact, or electronic communication through the Web.
- ME_jora P: this Web-based software application has been designed to especially facilitate the self-assessment phase. ME_jora P allows the professional to have an updated version of their manual of competencies, it provide the necessary evidence relating to their good practices, answer doubts with your tutor guide and check the status of accreditation at any stage.

6. Results

6.1 Results of the programme of accreditation centres and units

The number of accreditation projects has increased very significantly from 2003 to the present, from the 14 self-assessments initiated in 2003, to the 790 in 2011. During this period, 285 healthcare units and centres have been accredited and there have been 410 primary survey visits and 93 additional visits in those projects requiring a second visit for the verification of certain aspects that were improved after the first. (Table 6)

In the years 2009, coinciding with the release of the first general document of accreditation of services (rules governing the certification process), those files that were in the phase of self-assessment for over a year began to be closed.

Our model includes 2 follow up visits at 2 and 4 years from the first survey visit, for which a monitoring self-assessment is necessary in the 3 months preceding the visit itself. Up to the first quarter of 2011, 323 monitoring self-assessments had begun.

	2003	2004	2005	2006	2007	2008	2009	2010	2011
Applications received	14	51	60	47	81	50	196	172	119
Ongoing self-assessments	10	30	57	66	126	130	203	194	172
Initiated surveys	4	31	32	38	22	47	62	142	78
Visits made	4	34	42	55	36	57	97	146	78
Certifications granted	-	12	28	32	27	20	51	68	75
Monitoring self-assessments started	1	13	28	46	23	32	57	86	61
Certifications due	-	-	-	-	-	3	26	23	18
Suspended Certificates	-	-	-	-	-	-	-	4	11
Certificates withdrawn	-	-	-	-	-	-	-	-	4

2011: 1st Quarter and 2nd Quarter

Table 6. Results of accreditation projects.

As can be seen in Table 7, the average times for self-assessment of the different clinical units are variable. Note that the haemodialysis units took the most time to formalize the self-assessment phase (15 months), and emergency centres took the least time to complete this phase (6 months). The primary care clinics and hospital clinical units took very similar times to complete the self-assessment phase: 11 and 12 months respectively.

Approximately 54 areas have been identified for improvement by process during the self-assessment phase. The primary care units had more areas of improvement identified and haemodialysis units the least. There is a difference between the areas for improvement identified by the primary care units and those of hospitals, specifically 100 per project in the first case and 78 in the second, despite the time spent on the self-assessment phase being very similar.

Of the 30,497 areas identified for improvement by centres and units, 60% of them were planned and implemented during the self-assessment. It can be seen that the primary care units (66%) had the most implemented areas for improvement in relation to those identified, and the emergency centres (2%) and clinical laboratories (40%) the least. In the identification and implementation of these areas for improvement, during the self-assessment phase 4893 assessors participated (professionals of the clinical centres and units who reflected on the standards compliance). Most assessors per project were found in the hospitals, followed by the primary care units, and least in the haemodialysis units.

	HOS	E	PU	HA	HM	L	DI	TC	Total
Indicators (process completed)									
Av. Time for self-assessment (months)	12	6	11	12	15	8	10	8	10
Av. Areas for Improvement by process	98	30	100	78	10	26	65	25	54
Overall Results (all processes)									
No. Centres /Units	**29**	**9**	**215**	**174**	**26**	**23**	**9**	**8**	**544**
Improvement areas identified	2.521	30	16.349	9.883	265	836	480	133	30.497
Areas of improvement achieved	1.385 (55%)	2 (7%)	10.732 (66%)	5.450 (55%)	154 (58%)	335 (40%)	220 (46%)	87 (65%)	18.365 (60%)
Self Assessments	424	88	2.472	1.598	52	117	94	48	4.893
Number of Queries	1.587	50	8.315	7.182	482	687	256	215	19.466

HOS (Hospitals). E (Emergency). PU (Primary Units). HA (Hospital area). HM (Haemodialysis). L (Labs). DI (Diagnostic imaging). TC (Transfusion Centre)

Table 7. Description of accreditation projects by clinical sphere.

Table 8 presents the results obtained in the accreditation process in relation to the number of areas for improvement identified in each of the blocks (dimensions of quality assessed). In clinical units, both primary care and hospital, the largest number of areas for improvement were identified in relation to the standards that address the evaluation of service being provided to people within the health system (privacy, accessibility, continuity of care, ethics, information, etc.), while in hospitals and clinical laboratories the greater number of areas for improvement were identified in the support process block (equipment, emergency plans, information systems, infection control, appropriate use of medication, etc.).

Figure 3 shows the classification of areas for improvement and the rate of implementation during the self-assessment.

Sphere	The citizen	Organization of patient-centred activity	Professionals	Support Processes	Results
Hospital	825	312	168	1,152	63
Emergency	11	2	8	9	23
Clinical Laboratory	267	107	61	342	59
Diagnostic Imaging	189	68	45	156	22
Blood Transfusion	37	16	11	66	3
Primary Care Units	6,400	3,304	1,358	4,646	641
Hospital units	4,164	1,855	803	2,715	346
Haemodialysis	122	53	27	52	11

Table 8. Areas for improvement identified in the various dimensions of quality discussed.

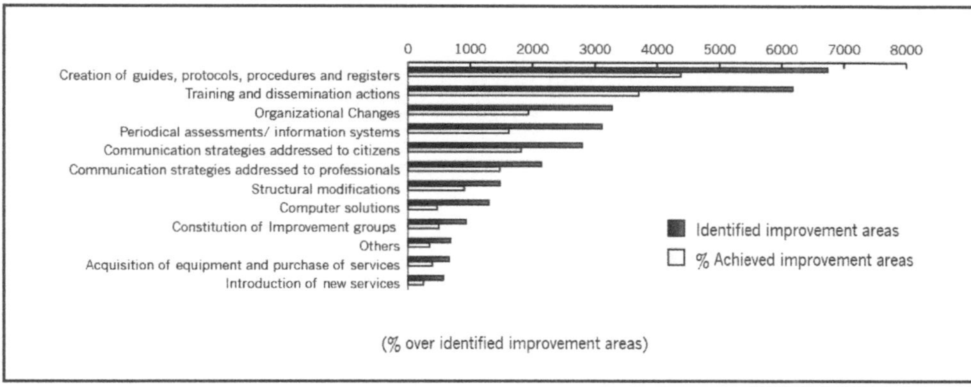

Fig. 3. Classification of areas for improvement.

6.2 Results of the programme of accreditation of professional competences

Following a pilot exercise in 2005, version 1 of the manuals of competencies were offered to the professionals in 2006. In January 2008, and based on experience gained since the beginning of the programme, version II was born, which included various improvements in the wording of the evidence, in the software application that supports the accreditation process and other actions aimed at facilitating the issuance of certificates for the workplaces of the professionals. Similarly, the average amount of evidence decreased through the manuals of competencies and, therefore, the evidence to be provided by professionals, fell from 74 on average in version I to 67 in version II.

Currently there are 70 version II manuals of competencies available to many other professional disciplines. (Figure 4).

Fig. 4. Accreditation manuals for professional competences.

Since the programme began, the number of professional accreditation processes has grown steadily (Fig. 5). In June 2011, more than 16,000 accreditation processes had been initiated, representing over 40% of medically qualified staff of the SSPA.

Of these, 3430 professionals had achieved some level of accreditation: 1,328 at advanced level, 1430 at expert level and 672 at excellent level.

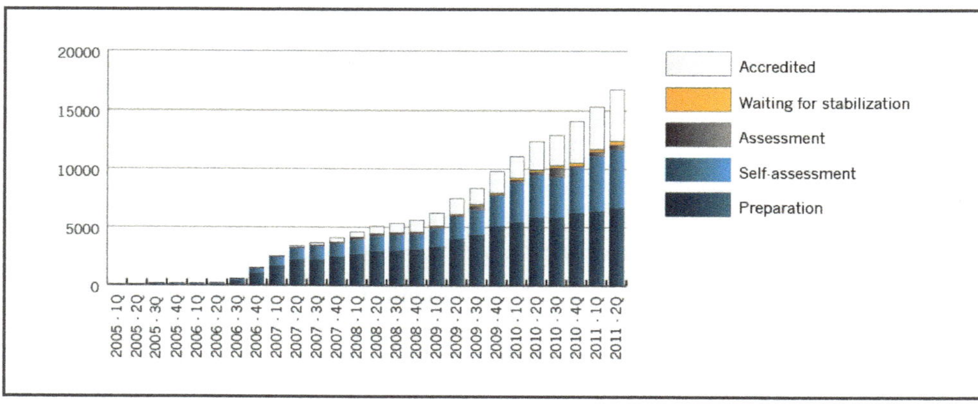

Fig. 5. Evolution of the accreditation processes.

The manual of competencies with the highest number of self-evaluation processes in absolute values are:

- Nursing care (hospitalization and special care).
- Primary care nursing.

- Family Medicine primary care.
- Nursing care hospital.

The procedures for these four maps together account for 42% of the ongoing processes.

As the demand for accreditation, hospital professionals account for 61% and professional primary care 24%, the rest belong to professionals and emergency blood transfusion centres.

In the matter of gender, more applications for accreditation are received from women than men in proportions of 56% - 44%, with considerable heterogeneity in response to the competence map.

The competency profile through accredited professionals to date, and grouped by criteria, is shown in Figure 6:

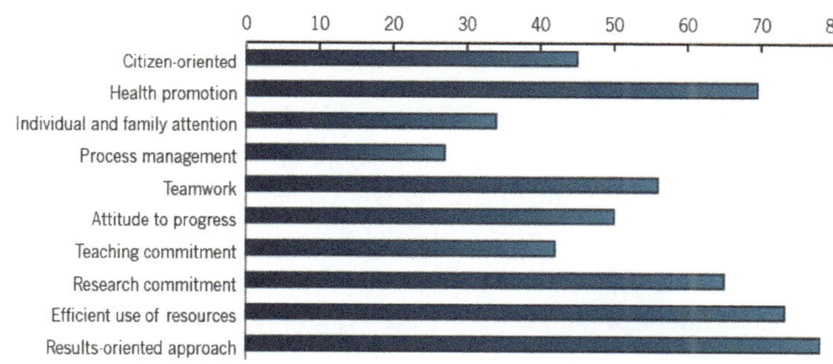

Fig. 6. Competency profile through accredited professionals.

Accredited professionals have invested an average time of 150 days on the self-assessment of their practices with version II of the manuals.

Professionals who have closed their accreditation process, over 36% have decided to restart the process in order to achieve a higher level of competence, so in fact they are already providing evidence and testing.

As a support to the continual improvement of the accreditation programme, the questionnaire of perceived satisfaction has been clearly defined. The validation of this questionnaire was carried out by the Andalusian School of Public Health, ensuring its content validity by using the following methodology:

- Logical validity in terms of the population in the study, validity of the method of application and the structure of the tool, translation and adaptation (semantic equivalence, conceptual and cultural) of items and scales found in the literature (which may improve the content of the tool) and format of the tool's presentation.
- Revision (trial judges) was performed by experts in survey research methodology, research methodology in health services, quality and accreditation of centres, professional development organization and management of health services, to issue an assessment and critical assessment of the measuring tool.

The questionnaire and instructions were sent to 100% of the population under evaluation after each call. This was done via a message on the personal accreditation page, eliminating selection bias. The online nature of the accreditation process familiarises the professional with the use of technology, limiting the effect of the technological barrier for non-response. Responses were received, at the same preserving anonymity, and distinguishing between professionals who had used the accreditation process using different versions of the manuals.

The questionnaire includes several dimensions that are valued by different questions and statements using a Likert scale with response elements at levels of 0 to 10.

In this study, the assessment of the following elements were analyzed:

- Overall satisfaction with the accreditation process.
- Usefulness of the accreditation process in the "self-study" and "reflection of the practice."
- Usefulness of the accreditation process in the "maintenance and improvement of competences."
- Usefulness of the accreditation process in "maintaining and improving results."

Of the professionals who have completed this, the response rate has been 62.08%. It verified that the scales of the questionnaire items were adequate, had good discriminatory power and showed no such bias as "floor and ceiling", ie, they present acceptable proportions of upper and lower ends of its distribution.

The overall assessment obtained by the accreditation programme is 8.01 out of 10. On the other hand, analyzing the perceived usefulness in continuing professional development, we identified the following results:

- Usefulness of the "self-study and reflection on practice": 8.1 out of 10
- Usefulness on "maintaining and improving their competences": 8 out of 10
- Usefulness on "maintaining and improving their results": 8 out of 10

7. Conclusions

1. Healthcare organizations are making a major effort to promote quality in the care they provide, but a disturbing question has arisen, especially within the environment of the economic crisis we live in, Are you using the right tools to increase quality ? There are posts warning against complacency with current methodologies that are proven ineffective and urging areas of self-assessment and reflection to be sought by professionals, followed by external validation.

2. The ACSA accreditation model offers healthcare organizations and professionals self-assessment and reflection, along with external validation by qualified professionals. One of the conceptual pillars of this model is considered to be the ultimate aim in continual quality improvement.

3. The Andalusian Agency for Healthcare Quality has established and implemented its own unique accreditation model to improve and ensure quality in the healthcare provided to citizens within the Andalusian Public Health System. To do that, the Agency, has designed a

series of accreditation programmes as tools for the continual improvement and safety at the service of the professionals, units and organizations, and has established a methodology that facilitates their application in practice and maximizes results.

4. Accreditation programmes share the same structure and contemplate, from each of their perspectives, the same key areas of quality management: The citizen, Integrated health care, Professionals, Areas of support and Efficiency & results.

5. For the Andalusian Agency for Healthcare Quality, Accreditation means an explicit and public recognition that the requirements needed to develop quality care have been fulfilled, and that a line of continual improvement has been undertaken. As a tool, and not an end in itself, accreditation promotes and encourages processes of improvement and evaluation within the health system.

6. Each of the phases, and especially self-assessment, are based on a series of Web-based applications called ME_jora developed by the Andalusian Agency for Healthcare Quality, which allows each accreditation process to be conducted securely and with the support of Agency professionals, and also enables the dissemination and exchange of knowledge and the elements of quality identified in them.

7. The return obtained through the questionnaire of perceived quality is encouraging; our results are consistent with experiences using self-assessment in other areas.

8. Acknowledgment

We would like to acknowledge the research's support team of the Andalusian Agency for Health Care Quality. This includes our research's technicians, Manuel Herrera-Usagre and Ángela Aguilar-de-Leyva, PhD, as well as the contributions Habibullah Rodríguez Contreras, graphic designer.

9. References

Almazán González, S. (2006). Modelo de acreditación andaluz. Agencia andaluza de calidad sanitaria. *Revista de Administración Sanitaria.* Vol.4, no.2, pp.189-194

Almuedo-Paz, A. (2006). Programa de acreditación de competencias profesionales en el Sistema Sanitario Público Andaluz. Cir Andal, vol.17, pp.160-162

Almuedo-Paz, A., Brea-Rivero, P., Buiza-Camacho, B., Rojas-De-Mora Figueroa, A., Torres-Olivera, A. (2011). Skills Accreditation as an impetus for continuous professional development. *Revista de Calidad Asistencial,* Vol.26, no.4, (forthcoming)

Aranaz, J.M., Leutscher, E., Gea, M.T. & Vitaller, J. (2003). El aseguramiento y la acreditación sanitaria. Qué opinan sobre la acreditación los directivos sanitarios españoles. *Revista de Calidad Asistencial,* Vol.18, no.2, pp.107-114

Arcelay, A., Lorenzo, S., Bacigalupe, M., Mira, J.J., Rodríguez-Marín, J., Sitges, E. (1998). La aplicación del modelo europeo de gestión de la calidad total al sector sanitario: ventajas y limitaciones. *Revista de Calidad Asistencial,* Vol.13, pp.92-97

Arnau-Figueras, J., Martínez-Carretero, J.M. (2007). *Comparativa de instrumentos de evaluación de la competencia. Evaluación de la competencia clínica: Análisis comparativo de dos instrumentos (ECOE versus Portafolio).* Plan de Calidad para el Sistema Nacional de Salud. Ministerio de Sanidad y Consumo. Agència d'Avaluació de Tecnologia i Recerca Mèdiques de Cataluña, Informes de Evaluación de Tecnologías Sanitarias, nº. 2006/14, Barcelona

Brea-Rivero, P., Almuedo-Paz, A., Rojas-de-Mora Figueroa A., Torres-Olivera, A. (2011). Buenas prácticas que promueven la seguridad. *Revista ROL Enfermería,* vol.345, no.7-8, pp.525-530

Brun, A., García-Aísa, J.C. (2004). Sistema de calidad del Servicio Aragonés de Salud. *Revista de Calidad Asistencial,* Vol.19, no.3 pp.131-120

Casey, D.C., Egan, D. (2010) The use of professional portfolios and profiles for career enhancement. *British Journal of Community Nursing,* vol.15, no.11, pp.547-552

Caverzagie, K.J., Shea, J.A., Kogan, J.R. (2008). Resident identification of learning objectives after performing self-assessment based upon the ACGME core competencies. *Journal of General Internal Medicine,* vol.23, no.7, pp.1024-1027

Clavería, A. (2004). El papel de la Agencia de Calidad en el Sistema Nacional de Salud. De la teoría a la práctica. *Revista de Calidad Asistencial,* vol.19, no.3 pp.205-210

Comité editorial. (2004). Una cartografía de la gestión de calidad en España. *Revista de Calidad Asistencial* vol.19, pp.103

Cowan, D.T., Wilson-Barnett, J., Norman, I.J. (2007). A European survey of general nurses' self assessment of competence. *Nurse Education Today,* vol.27, no.5, pp.452-458

Del Río, M.P. Feliu, J., Roch, I., Sánchez, A., Ordóñez, A., García-Caballero, J., González-Barón, M. (2006). Aplicación Del Modelo Europeo De Calidad En Oncología. *Revista Clínica Española,* vol.206, no.3, pp.129-136

Fortes, M.T., Mattos, R.A., Baptista, T.W. (2011). Accreditation or accreditations? A comparative study about accreditation in France, United Kingdom and Cataluña. *Revista da Associação Médica Brasileira,* vol.57, no.2, pp.239-246

Giraud, A. (2001). Accreditation and the quality movement in France. *Quality in Health Care,* vol.10, pp.111-116

Greenfield, D. & Braithwaite, J. (2008). Health sector accreditation research: a systematic review. *International journal for Quality in Health Care,* vol.20, no.3, pp.172-183

Guix Oliver, J. (2005). Calidad en salud pública. *Gaceta Sanitaria.* vol.19, no.4, pp.325-332

Gut, R., Knudsen, J.L., Freil, M., Jensen, A.J. (2006). Clinical leaders' and staff's assessment of the current quality related to the patient pathways and expectations of and knowledge about accreditation abaseline evaluation in Copenhagen County. *UGeskr Laeger,* vol.168, no.46, pp.3993-3997

Laing, A., Hohh, G., Winkelman, D. (2004). Healthcare and the information revolution: reconfiguring the health-care service encounter. *Health Services Management Research,* vol.17, pp.188-199

Lombarts, M.J.M.H., Rupp, I., Vallejo, P., Suñol, R., Klazinga, N.S. (2009). Application of quality improvement strategies in 389 European hospitals: results of the MARQuIS project. *Quality & safety in health care,* vol.18, pp.28-37

Lorenzo, S. (2004). La Gestión de la Calidad: de los modelos a la excelencia. *Revista de Gestión Clínica y Sanitaria,* vol.6, no.4, pp.131-132

Martínez-Pillado, M., Méndez, D., Trabado, J.L. Martín Casañas, F., Ramos, S., Coello, C., Afonso-Rodríguez, J.J. (2008). Método de implantación de una acción de mejora en el Servicio de Ginecología del Hospital Universitario Nuestra Sra. de Candelaria. *Revista de Administración Sanitaria Siglo XXI*, vol.6, no.4, pp.691-698

McCurdy, B.M., Duggan, L., Howlett, S., Clark, B.G. (2009). A comparison of medical physics training and education programs--Canada and Australia. *Australasian Physical and Engineering Sciences*. Vol.32, no.4, pp.251-260

Miller, G.E. (1990). The assessment of clinical skills/competence/performance. *Academic Medicine*, vol.65, no. suppl. 9, pp.563-567

O'Neale, M., Kurtz, S. (2001). Certification: perspectives on competency assurance. *Seminars in perioperative nursing*, vol.10, no.2, pp.88-92

Pardell, H. (2005). La acreditación de la formación médica continuada y la mejora de la calidad. Situación actual en España y en el ámbito internacional. *Revista de Calidad Asistencial*, vol.20, no.3, pp.171-177

Pariente, E., Ramos, C., Dierssen, T., Esteban, J., Amaya, A., Rodríguez de Lope, C., Soler, G. (2004). Gestión de la calidad en el Servicio Cántabro de Salud. *Revista de Calidad Asistencial*, vol.19, no.3, pp.121-127

Pasquina, P.F., Kelly, S., Hawkins, R.E. (2003). Assessing clinical competence in physical medicine & rehabilitation residency programs. *American Journal of Physical Medicine & Rehabilitation*, vol.82, no.6, pp.473-478

Pérez, E., Oteo, L.A. (2006). *Función directiva y recursos humanos en Sanidad*. Ediciones Díaz de Santos. Madrid, Spain.

Pomey, M.P., Lemieux-Charles, L., Champagne, F., Angus, D., Shabah, A., Contandriopoulos, A.P. (2010). Does accreditation stimulate change? A study of the impact of the accreditation process on Canadian healthcare organizations. *Implementation Science*, vol.5, p.31

Pomey, M.P., François, P., Contandriopoulos, A.P., Tosh, A., Bertrand, D. (2005). Paradoxes of French accreditation. *Quality & safety in health care*, vol.14, no.1, pp.51-55

Prados, J.A. (2005). Preguntas sobre la evaluación de las competencias y alguna respuesta: portfolio. *Atención Primaria*, vol.36, no.3, pp.3-4

Staccini, P., Rouger, P. (2008). Modeling and using a web-based and tutored portfolio to support certification of professional competence in transfusion medicine. *AMIA Annual Symposium Proceedings*, pp.697-701

Tochel, C., Haig, A., Hesketh, A., Cadzow, A., Beggs, K., Colthart, I., Peacock, H. (2009). The effectiveness of portfolios for post-graduate assessment and education: BEME Guide No 12. *Medical Teacher*, vol.31, no.4, pp.299-318

Torres Olivera, A., Fernández, E., Paneque, P., Carretero, R., & Garijo A. (2004). La Gestión de la calidad asistencial en Andalucía. *Revista de Calidad Asistencial*, vol.19, no.3, pp. 105-112

Ugalde, M. (2003). Excelencia en la Gestión. ¿Existen resultados o es solo un acto de fe? *Revista de Calidad Asistencial*, vol.18, no.6. pp.338-41

Viswanathan, H.N., Aslmon, J.V. (2006). Accrediting Organizations and Quality Improvement. *American Journal of Managed Care*, vol.6, no.10, pp.1117-1130

18

Automatic Maintenance Routes Based on the Quality Assurance Information

Vesa Hasu and Heikki Koivo
Aalto University, Department of Automation and Systems Technology
Finland

1. Introduction

Maintenance and repairing of widely distributed and large scale networks of sensors and actuators can easily become overwhelming - physically and financially. This applies especially, if the units are located so that performing the maintenance operations is neither easily accessible nor cheap. In order to make effective and efficient maintenance decisions, we need methodologies for describing the condition of the units. For that purpose, we propose using general quality concept descriptions. This methodology links linguistic terms and numerical values into easily understandable concepts on the network performance.

The market interest in large-scale networks has risen with the ability to utilize wireless communication nodes. The arranging of the maintenance of such large-scale networks becomes tedious due to sheer size of the problem. In this work, we assume a network, which is distributed spatially in wide area. That is, the movement of maintenance personnel takes much time and physically this means that repairing every occurring fault is too time time-consuming. Economically this means that the distances are large enough to cause significant costs. Statistically "widely distributed" means that measurement stations next to each other are not exactly valid reference measurements for reliability estimation. In overall, any automated help is welcome to making automated maintenance decisions.

In our approach, we take the essential thing is to express the operation of a device in a simple numerical form of one index and analyze device maintenance needs based on the numerical value. In general we consider this index as reliability. In here, the reliability refers specifically to the *unit* performance reliability. That is, this reliability refers to a larger concept than an occurrence fault failure. This is because, for example, in the measurement networks an exact detection of faults is impossible or fuzzyfied.

In Isermann (2006) reliability is defined as the "ability of a system to perform a required function under stated conditions, within a given scope, during a given period of time" and availability is defined as "probability that a system or equipment will operate satisfactorily and effectively at any period of time." These definitions apply at some level in this work too, but since the errors are not exactly traceable, we aim to use performance indices to describe these concepts instead of *e.g.* failure probabilities or failure densities. Hence, in our work, the proposed concepts of reliability and availability are described by indices varying between 0 and 1. We emphasize that these indices are not probabilities; they are measures of performance in the index framework.

The reliability in a spatially distributed system is considered *e.g.* by Sun *et al.* (2005), which approaches the reliability through redundant measurements and voting procedures. In here the redundancy of measurements is not assumed, only through the spatial dependency of the measurement field.

De Souza *et. al.* (2005) explore monitoring of power networks with a fitness function methodology, which include the critical system measurement weighting. In the context of logistics systems, spatially distributed systems are considered by Wang *et al.* (2006), who describe accurate mathematical definitions of reliability.

In here, more emphasis is laid on describing the overall ideas in measurement networks and going through mathematical details by a case example on weather station networks. This chapter is based on the performance indices to the measurements networks presented by Hasu and Koivo (2007, 2009) but we consider new and improved methods for route solving. Additionally, we emphasize that the automatic maintenance route approach can be utilized in actuator networks. In this case, the performance indices used for industrial plant level by Hölttä and Koivo (2009, 2011) are interesting.

This chapter studies application of the quality assurance knowledge in the automatic maintenance route planning. The original framework is aimed at sensor network maintenance but the basic ideas of route planning are transferable also into actuator networks. The suggested approach has three essential stages: the basic quality assurance, the evaluation of performance indices and the maintenance route planning.

At first, quality assurance techniques determine whether sensors and/or processes work correctly or suffer from suspicious behavior. Second, specially designed performance indices are applied to describe the observation history. The key is to describe the reliability in one index as well as possible. The strengths of the performance index approach include that it enables forgiving bad functioning periods. If the problems are gone, maintenance actions for that component are futile and they result in unnecessary costs. The second main strength of performance indices is the ability to give a quick overlook on the performance without a need to see the whole history with its QA information. The third stage of our approach is the automatic planning of maintenance routes based on performance indices. Since this route-planning problem is close to the travelling salesman problem, routes are solved using either heuristic or evolutionary computing methods using somewhat similar ideas. This chapter demonstrates the approach using surface weather stations.

The maintenance route-planning problem (MRPP) discussed in this paper is a version of a multiple TSP with profits, in some cases named as a multivehicle routing problem with profits (Feillet *et al.* 2005). MRPP has similarities also to a team orienteering problem (Chao *et al.* 1996a). However, we derive cost functions and constraints from a different point of view, and therefore the problem has a unique nature. The main differences are time-limited routes and multi-objective nature, which are included in MRPP. We provide solutions MRPP problem using heuristic and ant colony optimization (ACO) algorithms. The heuristic algorithm has similarities to the heuristic solution to the team orienteering problem presented in Chao *et al.* (1996b). Other solving algorithms are based on evolutionary computing methods. ACO is considered based on the vehicle routing approach (see Bell and McMullen 2004 and Yu *et al.* 2009). Additionally, a genetic algorithm (GA) for MRPP based on the ordinary GA solutions for TSP is presented in Hasu and Koivo (2009). Since we have found that ACO is more efficient for MRPP, we do focus on this solution in Section 3.

2. From quality information to the maintenance variables

Quality assurance systems process huge amounts of data and form information on the state of the process, the quality of which is under interest. From the maintenance perspective, the first issue to consider is how large the maintenance need of a sensor or actuator is. The reliability is considered through other concepts. In measurement networks, we can consider the reliability through availability and accuracy of observations. Availability refers to the ratio of available observations and the expected number of observations. Accuracy refers to the estimated accuracy of the measurement. Reliability is the combined effect of the availability and accuracy. That is, if either availability or accuracy is compromised, the measurement is not reliable. In probabilistic terms, the reliability l can be seen as the intersection of the availability and accuracy. That is, a completely reliable measurement is both available and accurate all the time. All described concepts are measured by indices varying from 0 and 1 with larger values corresponding to the better performance.

Hölttä (2009) has stated requirements for the plant performance indices, which are applicable for the maintenance variables in this chapter. First, no additional measurements should be required for the determination of indices. Second, setting up the indices should happen automatically. Third, the indices should be easy to understand. Fourth, the high level information should indicate the user easily towards the problematic point. The above concepts of reliability, availability and accuracy suit the demands very well.

The maintenance strategy of a sensor and actuator network should always be made according to the needs of the network. That is, the acceptable inaccuracy in operation should not affect the performance indices. In the end, the operational condition depends on the defined reliability, and therefore the reliability concept should include all of the essential elements to be reflected in the maintenance and nothing else.

The linguistic concepts depend on the mathematical definitions, which must be made according to the case. For example, different measurement networks have different requirements in spatial and temporal resolutions, as well as the measured fields have different time constants and spatial dependencies. For the spatial resolution, one factor influencing the mathematical definitions of performance indices is how acceptable is the estimation of missing measurements based on the neighbor measurement stations. A related issue is the ability to estimate the missing measurements. This could be rephrased as the ratio between the average station spacing and the desired spatial accuracy. In the temporal resolution, the essential issue is the difference between the system's time constant defining the dynamical behavior of the system and the sampling interval. This is related to the question of criticality of occasional missing measurement values.

2.1 Performance indices for sensor networks

The maintenance of a measurement network must be considered through the need for maintenance in each measurement. In here, the measurement *reliability* is considered through two other concepts: *availability* and *accuracy*. Availability refers to the ratio of measurements being available. Accuracy refers to the estimated accuracy of the measurement. The essence of this subsection is presented in Hasu and Koivo (2007).

Availability s_i expresses the availability ratio of the measurement i. If the measurement i has not ever been missing from the data bank, $s_i = 1$. If the measurement i has always been missing, $s_i = 0$. The measurement might be missing, for example, due to communication or measurement device errors.

The accuracy t_i describes the estimated accuracy of the measurement i. The accuracy takes into account all known defects in the measurement accuracy. An accurate measurement has t_i equal to one and a definitely inaccurate measurement has t_i close to zero. Hence the input of the accuracy can be *e.g.* decision variables of the fault detection. As in the availability, the accuracy should put emphasis on the most recent events. Additionally, if different types of errors, *e.g.* noise or drift, can be detected, all of these should be integrated into t.

Reliability is the combined effect of the availability and accuracy. That is, if either availability or accuracy is compromised, the measurement is not reliable. In probabilistic terms the reliability l can be seen as the intersection of the availability and accuracy. That is, a completely reliable measurement is both available and accurate all the time.

Since the measurements might run for long periods, it is advisable to use recursive techniques in order to have more emphasis on more recent events and reduce the computational requirements. Basically, this recursion increases and decreases the availability depending whether or not the measurement is available.

In Hasu and Koivo (2007), the availability was considered through the recursion

$$s_i(k) = \begin{cases} \lambda \cdot s_i(k-1) + (1-\lambda), & \text{if } x_i(k) \notin M, \\ \lambda \cdot s_i(k-1), & \text{if } x_i(k) \in M \text{ and } \exists j \in S \text{ s.t. } x_j(k) \notin M, \\ s_i(k-1), & \text{if } x_j(k) \in M \ \forall j \in S, \end{cases} \tag{1}$$

where λ is the forgetting factor of recursive methods, $\lambda \in (0, 1)$, M is the set of missing observations and S is the set of measurement stations in the system. In here λ serves also as a penalization factor - (1) shows that smaller λ causes a larger decrease in s when the measurement is missing - and larger increase if the measurement is available. Hence the small forgetting factor results in a smaller dependency to s further back in history. Note also that (1) does not penalize a single measurement station if all simultaneous observations are missing in the system. This is since these types of problems often refer to communication or database problems, which are not related to measurement stations.

The update on the accuracy index can depend very much on application. The simplest way is to utilize the existing quality control tests for the observation in similar fashion to (1). That is, the accuracy can be updated through a recursion

$$t_i(k) = \begin{cases} \mu \cdot t_i(k-1) + (1-\mu), & \text{if } x_i(k) \notin E, \\ \mu \cdot t_i(k-1), & \text{if } x_i(k) \in E, \end{cases} \tag{2}$$

where μ is the forgetting factor for accuracy, $\mu \in (0, 1)$, and E is the set of erroneous observations.

However, if the applied quality assurance techniques follow the measurement quality closer, their statistics can be utilized in the accuracy computation. Hasu and Koivo (2007) present an example, in which the filter residual $r_i(k)$ and its respective standard deviation $\sigma_{i,r}$ update the accuracy. Mathematically this is essentially represented by

$$t_i(k) = \left(\tfrac{\lambda_1}{2} + \tfrac{\lambda_2}{2}\right) \cdot t_i(k-1) + \left(1 - \tfrac{\lambda_1}{2}\right)\left(1 - f(r_i(k), \sigma_i(k)\right) + \left(1 - \tfrac{\lambda_2}{2}\right)\left(1 - g(\sigma_i(k), \sigma_i^{\min}(k)\right), \quad (3)$$

where λ_1 and λ_2 are forgetting factors for relative and absolute inaccuracies of the measurement, and

$$f(r_i(k), \sigma_i(k)) = \begin{cases} 0, & r_i(k) < 3\sigma_i(k), \\ 2 \cdot P\left(r_i(k)/\sigma_i(k) - 4\right), & r_i(k) \geq 3\sigma_i(k), \end{cases} \quad (4)$$

$$g(\sigma_i(k), \sigma_i^{\min}(k)) = P\left(3\sigma_i(k)/\sigma_i^{\min}(k) - 9\right), \quad (5)$$

$$P(x) = \left(1 + e^{-x}\right)^{-1}, \quad (6)$$

where $\sigma_i^{\min}$ is the minimum residual standard deviation in the measurement network, and P is the sigmoid function. Even though the update (3)–(6) looks a bit complicated, it includes only simple computations with variables.

As mentioned, the reliability is considered as the intersection of availability and accuracy. Hence, as in probability theory, we define the reliability as the product of these two:

$$l_i(k) = s_i(k)t_i(k). \quad (7)$$

Note that this definition enables the easy use of additional performance indices as the availability and accuracy in the determination of reliability. For example, other accuracy variables can be included in the form of $t = t^1t^2\ldots t^n$, where the superscripts indicate the numbers of accuracy subindices. In (7), we could use a variety of methods to integrate different performance indices into higher level indices (see Hölttä 2009 or Hasu and Koivo 2009). The other possibilities include taking minimum, maximum, arithmetic mean, harmonic mean or any other type of central tendency measure.

Hasu and Koivo (2007) present a numerical example of the performance indices for a weather sensor network. Figure 1 shows the availability and accuracy indices measured using the updates (1) and (3), respectively. Based on Figure 1, three stations can be described as follows. Kaukas station temperature measurement has poor availability and accuracy. Luhtaanmäki and Huhtionmäki stations suffer from a compromised temperature observation quality, and Luhtaanmäki has also occasional missing observations. The corresponding reliability values from these stations are drawn in Figure 2, where also a reference station of Kantele is included.

Figure 3 shows the observations from a reference station of Kantele, which does not suffer from any problems, as well as the Kaukas and Huhtionmäki stations presented in Figure 1. Comparing the Huhtionmäki observations with the Kantele observations shows how the

Kantele measurement suffers from noise in the five-minute sampling, and therefore the accuracy is compromised. Figure 3 shows that problems of Kaukas Station relate to a massive number of missing observations and large noise in the daytime observations.

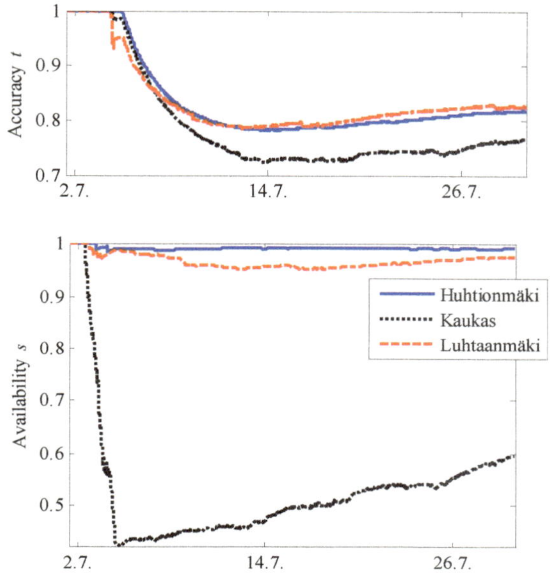

Fig. 1. Availability *s* and accuracy *t* from three temperature measurement stations in the Southern Finland.

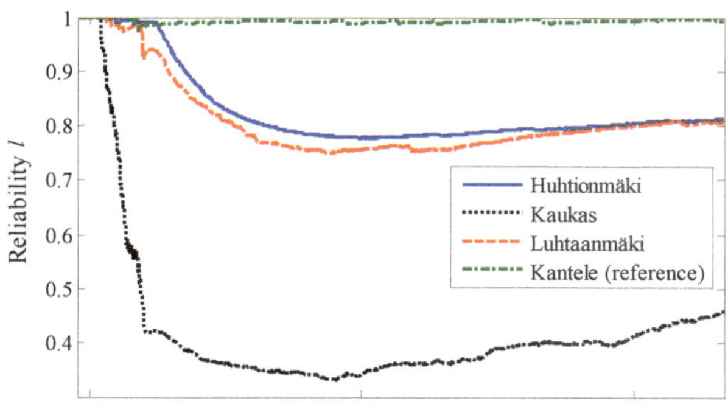

Fig. 2. The reliabilities of the above stations with a reference station Kantele.

In overall, the above numerical example exemplifies how the measurement quality can be described easily using a few index. First, the reliability index shows the overall quality. Second, the accuracy and availability describe the sources of potential problems. If desired, the hierarchy on the accuracy side could be further constructed to identify the nature of the problems - thus helping to choose the correct types of maintenance actions.

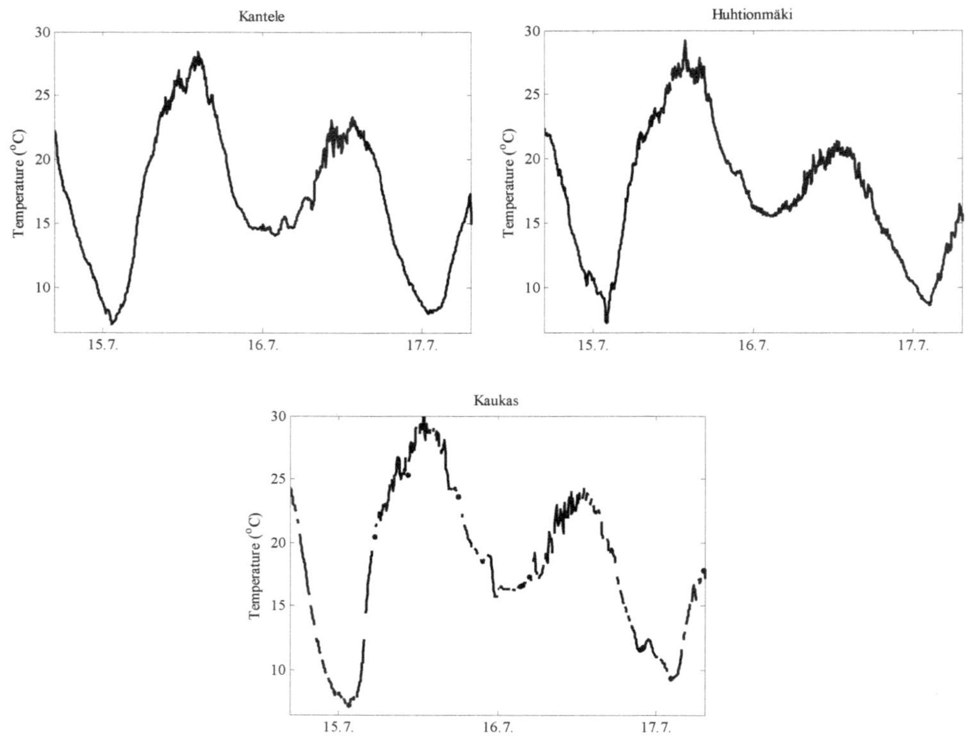

Fig. 3. Observations from Kantele (top left), Huhtionmäki (top right) and Kaukas (bottom) stations.

2.2 Performance indices for actuator networks

In the actuator networks, the actuator performance relates to the control performance of the (sub-)system. Hölttä and Koivo (2011) presents several performance indices for the determination of the control performance. Some of the performance indices are more application-related but in the following, we have collected a few generic performance indices for actuator performance.

The first actuator performance index, the duration of setpoint error, originates from the work of Jämsä-Jounela *et al.* (2003). This index follows the control loop performance by detecting the time of permanent error. That is, this index indicates if the control loop suffers from a permanent error. The duration index is updated using

$$I(k) = \lambda I(k-1) + (1-\lambda)\cdot\begin{cases}\text{sgn}(e), \text{ if } |e| > e_{\text{lim}}, \\ 0, \text{ if } |e| \leq e_{\text{lim}}, \end{cases} \tag{8}$$

where e is the tracking error between the setpoint and system output, e_{lim} represents the boundary of the acceptable region and sgn stands for the signum function. Since our goal is to have positive indices, for which zero and one stand for poor and good performance, repectively, and we are not interested in the sign of the setpoint error, the final index is

$$\eta(k) = P\left(\tfrac{k}{25}\left(\eta_{50} - |I(k)|\right)\right). \tag{9}$$

That is, the absolute value of I is scaled using the sigmoid function (6) using constants k and η_{50}, which describe respectively the scale and location (midpoint) of the sigmoid. The scaling constants depend on the application. Additionally, Hölttä and Koivo (2011) present a variation of (8) for the magnitude of the set point error, using essentially the following

$$I(k) = \lambda I(k-1) + (1-\lambda)\cdot\begin{cases}\operatorname{sgn}(e)\left(\operatorname{sgn}(e)e - e_{\lim}\right)^2, \text{ if } |e| > e_{\lim}, \\ 0, \text{ if } |e| \le e_{\lim}.\end{cases} \tag{10}$$

Again, the final index is obtained using (9). Obviously, this index describes the temporally weighted average for the setpoint error.

More detailed control performance indices can be formed. Hölttä (2009) presents a version of sluggish control index originally introduced by Hägglund (1999). This index grows if the control loop follows the control signal slowly. The index is defined as

$$I(k) = \lambda I(k-1) + (1-\lambda)\cdot \operatorname{sgn}\left(\Delta u \Delta y\right), \tag{11}$$

where Δu and Δy are the control and output increments, respectively. The values close to one indicate a sluggish control loop. Due to possible negative values, Hölttä uses the scaling similar to (9) in order to form the final performance index.

A numeric simulation exemplifies the strength of this approach. We simulate a system described in Figure 4, including a plant with transfer function 1/(s2 + 2s + 2) and a PI-controller. The pulse-like reference is drawn as the black line in Figure 5. The controller is applied with two tunings, from which the first one ($K = 2$, $T_i = 2/3$) is much slower than the second one ($K = 1/2$, $T_i = 1/2$), to which the controller is switched at $t = 50$.

The system output is plotted in blue in the top of Figure 5. The first tuning reaches the reference slowly compared with the second tuning. This is visible also in the performance index for sluggish control, which is included in the bottom of Figure 5. The permanent error index behaves similarly. It, however, does not reach as low values due to the change of reference. The performance indices are combined into a loop index (the black dotted line at the bottom of Figure 5) using multiplication.

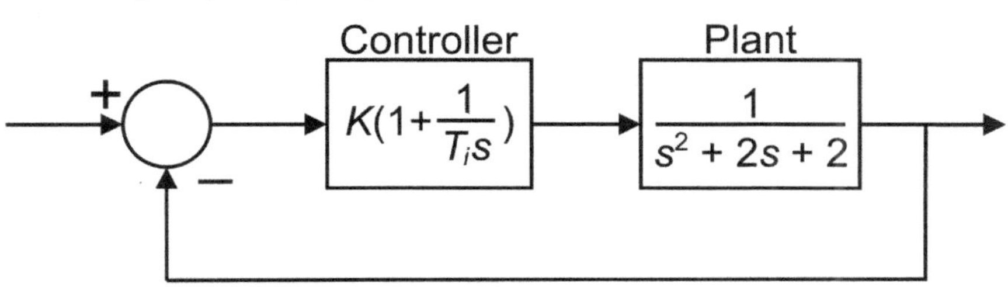

Fig. 4. The simulated control loop.

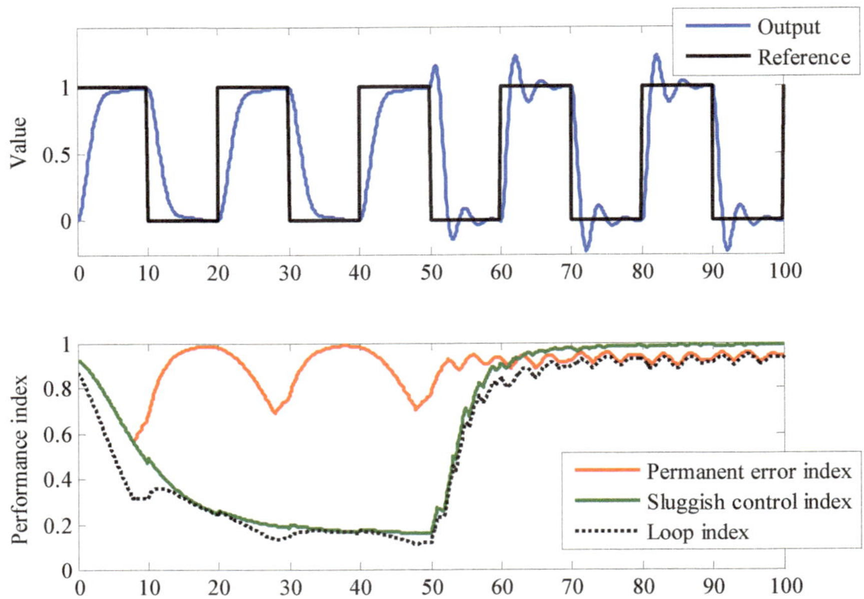

Fig. 5. The simulated output (top) and performance indices (bottom).

2.3 Performance Indices using fuzzy inference

In addition to the recursive techniques, the performance indices can be calculated using other methods - as well as the requirements for the indices mentioned at the beginning of Section 2 are met. One interesting option is to use fuzzy inference. Hasu and Koivo (2009) have presented one version how to apply fuzzy inference for the measurement accuracy and availability performance indices.

The idea of the fuzzy system is to replace the equations (1)-(6), which are applied in the index computation. Hence the inputs to the fuzzy inference include the previous index value. The other input to the accuracy is the QA information about the erroneousness of the observation and for the availability the other input is the period from the previous observation (as a multiple of sampling time). The applied fuzzy system is based on the Mamdani inference with triangular membership functions and multiplicative intersection, implication and aggregation. The membership functions for the accuracy and availability are presented in Figure 6.

The fuzzy rules for the accuracy and availability are collected into Table 1. The resulting fuzzy interference surfaces are included in Figure 7. Note that for the accuracy we have used only "good/erroneous observation" classification for the measurement quality - this corresponds to the accuracy index presented by (2).

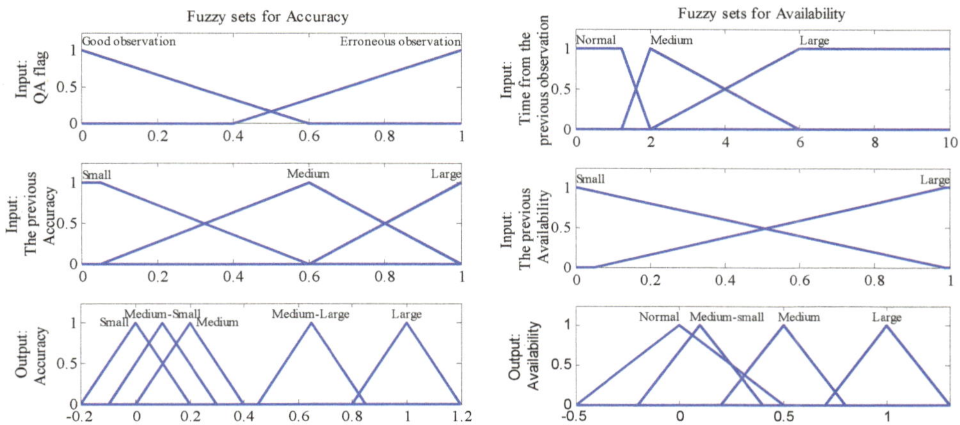

Fig. 6. The fuzzy sets for the accuracy and availability indices.

Accuracy			Availability		
Input: QA indication for the observation	Input: The previous accuracy	Output: Accuracy	Input: Time from the previous observation	Input: The previous availability	Output: Availability
Good	Large	Large	Normal	Large	Large
Good	Medium	Medium-large	Normal	Small	Medium-small
Good	Small	Medium-small	Medium	Large	Medium
Erroneous	Large	Medium	Medium	Small	Medium-small
Erroneous	Medium	Medium-small	Large	Large	Small
Erroneous	Small	Small	Large	Small	Small

Table 1. Fuzzy inference rules.

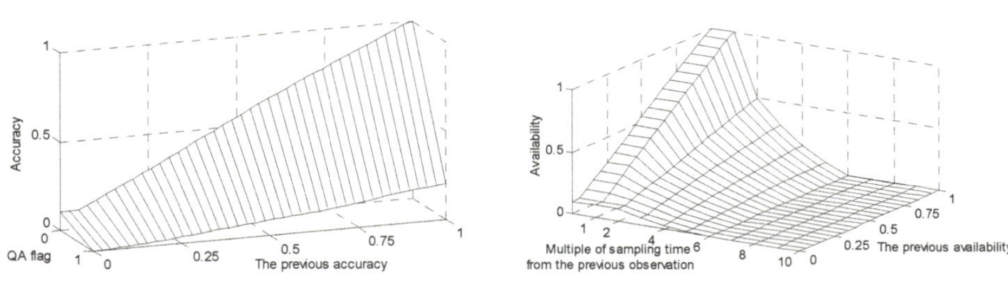

Fig. 7. The fuzzy inference surface for accuracy and availability indices.

3. Automatic maintenance route planning

3.1 Definition of maintenance route planning problem

The maintenance route planning aims for the maximum improvement of sensor network performance with minimal costs. Since services are nowadays often outsourced, an automated tendering of subcontractors must occur in the planning stage. Planning limitations are the number of available technicians and the lengths of working days. The route planning must take into account the distances between the stations as well as the maintenance times in each station. Costs include starting, repair, and travelling costs, which may be subcontractor dependent. Additional information to take into account is the importance of stations.

MRPP is a multiobjective optimization problem. On one hand, we want to maximize the network improvement by sending technicians to stations with poor performance, *i.e.* small performance indices. On the other hand, we like to keep the maintenance costs as low as possible while keeping the sensor network performance still in an acceptable level. In practice, the multiobjective problem means that MRPP will have a Pareto-optimal set of optimal solutions (*e.g.* [7]) and therefore the choice of the best route combination is up to one's preference. In this work, we deal with the problem by presenting so-called Pareto-boundary found by the algorithm.

The mathematical definition of MRPP starts from a definition of a route matrix **R**, $[R_{i,j}]_{i,j=1}^{n,m}$, where n is the maximum route length plus two and m is the number of available technicians. By definition, the first entry of each column corresponds to the current location, *i.e.* the starting point, of the technician. We use station index zero to refer to the depots of technicians. The rest of column j is for the indices of stations in the order of technician j's route. Since the routes are usually shorter than the maximum route length, the column is adjoint with a suitable amount of zeros.

MRPP consists of a multiobjective function to be maximized, working day constraints and minimum gain requirements. In mathematical terms, MRPP maximization is

$$\max_{\mathbf{R}\in\Re}\mathbf{f}(\mathbf{R}) = \left[\sum_{j=1}^{m}\sum_{i=2}^{n-1}W(R_{i,j})\big(1-l(R_{i,j})\big), -\sum_{j=1}^{m}\sum_{i=2}^{n}c_j d(R_{i,j},R_{i-1,j}) - \sum_{j=1}^{m}\sum_{i=2}^{n-1}C(R_{i,j})\right], \tag{12}$$

where the first entry is the network performance gain and the second entry is the opposite number of the cost. $\Re$ is the space of possible route matrices. $W(R_{i,j})$ and $l(R_{i,j})$ are the importance weight and reliability index, respectively. On the cost side of (12), c_j is the hourly cost of technician j, $d(R_{i,j},R_{i-1,j})$ is the temporal distance between the ith and the previous stations in the technician j's route and $C(R_{i,j})$ is the maintenance cost of station i by technician j.

By definition, the depots do not have repair costs, *i.e.* $C(0) = 0$, and distance from a place to itself is zero, *i.e.* $d(i,i) = 0$. Note that (12) assumes that the starting point is already repaired, and hence solving (12) enables the updates of routes during the working day.

The working day constraint for the technician j in MRPP is

$$\left(\sum_{i=1}^{n} d(R_{i,j}, R_{i-1,j}) + \sum_{i=2}^{n-1} t(R_{i,j})\right) \leq T_{\max,j}, \tag{13}$$

where $t(R_{i,j})$ is the maintenance time of station i of technician j and $T_{\max,j}$ is the length of working day of technician j.

Additional constraints for MRPP describe the required performance decrease of a station before maintenance is considered and the desired minimum gain from a route. These constraints remove unnecessary routes and maintenance actions, and they restrict the solution space of the optimization problem. If we define a total route gain of technician to be $G_j = \sum_{i=2}^{n-1} W(R_{i,j})\left(1 - l(R_{i,j})\right)$, the minimum station gain and the minimum route gain constraints are equal to

$$W(R_{i,j})\left(1 - l(R_{i,j})\right) \geq g_{\min}, \forall R_{i,j} \setminus \{0\}, \tag{14}$$

and

$$G_j \geq G_{\min} \vee G_j = 0, \tag{15}$$

respectively, where $g_{\min}$ is the minimum maintenance gain of a station and $G_{\min}$ is the minimum acceptable route gain.

3.2 Solving MRPP: A heuristic solution

The following heuristic solution of MRPP is a greedy type algorithm introduced originally in Hasu and Koivo (2009). The algorithm adds to routes a station, which maintenance improves the network performance the most in respect to the additional cost. This is measured with a ratio between the increased network performance and the increased cost while keeping constraints (13) and (14) in mind. After the addition of a station, the heuristic checks possible station swaps between routes in order to reduce the cost described in the second entry of (12). The station swaps from one route to another are somewhat similar to orienteering problem heuristic in Chao *et al.* (1996b).

The ratio of the network performance increase and cost is

$$\frac{W(k)(1 - l(k))}{\min\limits_{i,j}\left(c_j \Delta t(i,j) + C(k)\right)} \tag{16}$$

where $\Delta t(i,j)$ refers to the temporal increase of technician j's route if station k is added into ith station in that route. The time addition Δt includes both travelling and maintenance time. In here, "available" refers to a station, the addition of which satisfies (2), (3), and which is not already included in routes.

The heuristic algorithm for a subset of technicians follows the outlines described in Table 2. The route additions are greedy and the station modifications between technicians make sure that route costs do not grow unnecessarily large.

In order to make the tendering of subcontractors and to choose the best number of technicians, the heuristic algorithm determines the routes for all possible subsets of technicians and chooses the most advantageous route ensemble. For example, if two subcontractors with one technician are available for the maintenance, the heuristic determines separate routes for each of them separately (i.e. **R** is a column vector) and one route combination with both technicians (i.e. **R** is a two-column matrix). The algorithm chooses the route matrix, which maximizes the first entry of f in (12). If a route does not satisfy (15), the algorithm discards the whole solution in the corresponding route matrix.

```
WHILE ∃ station index k ∉ R
    FIND the maximum ratio (16) over k ∉ R satisfying (13) and (14), and over Δt(i,j) and ADD index k
    into ith slot of the route of technician j
    IF ∄ k satisfying the above conditions
        END WHILE
    END IF
    IF moving any station from technician j to other technicians would reduce the total costs
        MOVE the selected station from technician j to an another technician
    END IF
    IF swapping a station from technician j to any other technician's station would reduce the costs
        SWAP the selected stations
    END IF
END WHILE
```

Table 2. Heuristic maintenance route determination.

3.3 Solving MRPP: Ant colony optimization

ACO is a widely used method for solving combinatorial optimization problems, to which group includes also MRPP. ACO is inspired by the way ant colonies find their food in nature. Basically, ants leave *pheromone* to mark good routes to find food and the following ants are likely to choose the route parts with pheromone. In ACO, the routes ranked as good solutions to the problem receive pheromone, and therefore those parts are more likely used also in the future.

As mentioned in Section 1, MRPP has similarities to the vehicle routing problem and therefore our ACO algorithm for MRPP is based on a few vehicle routing algorithms. The bi-objectivity of the gain function **f** defined in (12) is taken care of using a similar approach to Schilde *et al.* (2009), where two different pheromones are used for the different objectives. That is, the pheromone additions are done with the similar all-for-the-few-best fashion. We collect the Pareto-optimal solutions to so-called Pareto-optimal set, in which no solution dominates completely other (*i.e.* none of the solutions is better than other in respect to the *both* gain function entries). The multiple starting depots of technicians are handled as in Yu *et al.* (2011): the first and last locations in routes are set to each technician's depot.

In ACO for MRPP, an ant population of q groups of m ants (m is the number of technicians) wander through to the possible destinations without violating the MRPP constraints (13) and (14) for each group. That is, during the route construction phase, m ants start from their respective depots in random order to build route $\mathbf{R}_n$, $n = 1,\ldots,q$, to a random station one station a time.

The main algorithm for ACO solving MRPP is presented in Table 3. In the following, we describe the algorithm steps more closely.

```
COMPUTE heuristic information matrices η_jik1 and η_jik2.
INITIALIZE the pheromone matrices τ_jik1 and τ_jik2.
WHILE iteration stopping condition is not met
    CREATE q groups of m ants with randomized starting order if there are more than one depots.
    FOR q groups
      WHILE ∃ stations that are feasible to add to current group, i.e. Ω(R) ≠ ∅.
        FOR ant 1 to ant m.
            SELECT randomly a new station to add to the route using weighting based on the
            pheromone and heuristic information.
            UPDATE the pheromones using the evaporation: τ_jika ← (1 − ρ) τ_jika + ρτ0.
        END FOR
    END WHILE
    APPLY improvement strategies: route permutation, switching and removing stations, and the
    replacing a route from the Pareto-solutions.
    END FOR
    UPDATE the pheromones with the global update according to the best route combinations based
    on the separate gain functions and the equally weighted gain.
    UPDATE the set of Pareto-solutions.
END WHILE
```

Table 3. ACO maintenance route determination.

Before ACO algorithm starts to increase route lengths, the heuristic information and initial pheromone levels must be determined. The heuristic information is applied in the route formation for the transition probabilities from one station to another – representing the natural attractiveness of ant j moving from the current station i to the station k. This attractiveness depends on the gain function, which is maximized. Since we are dealing with a bi-objective optimization (12), we have separate heuristic information for both of the goals of **f**. The heuristic information are calculated using

$$\eta_{jik1} = \frac{1}{c_j \Delta t(i,j) + C(k)}, \quad \eta_{jik2} = W(k)(1 - l(k)). \tag{17}$$

The first heuristic information is the inverse of the additional cost and the second is related to the gain corresponding to maintaining the station k. Note how these compare with the numerator and denominator of the heuristic method's decision variable (16). In addition to the heuristic information, the ACO pheromone levels must be initialized. Initially, the pheromones of technician j moving from station i to station k in respect to the objective a are set to value τ_0, *i.e.* $\tau_{jika} = \tau_0$.

In here, the next random station is chosen from the set of stations satisfying constraints (13) and (14) with probabilities given by

$$p(v_{jik}) = \begin{cases} \dfrac{\tau_{jik1}\eta_{jik1}^2 p_1 + \tau_{jik2}\eta_{jik2}^2 p_2}{\sum_{h \in \Omega(x)} \left(\tau_{jih1}\eta_{jih1}^2 p_1 + \tau_{jih2}\eta_{jih2}^2 p_2 \right)}, & \text{if } h \in \Omega(\mathbf{R}), \\ 0, & \text{otherwise,} \end{cases} \tag{18}$$

where $\Omega(\mathbf{R})$ is the index set of stations, which are available for a visit *i.e.* not visited by any of the m ants (technicians) and addition of these stations does not violate (13) or (14), and p_a is the weight of ath goal in (12) for the current group of m ants. If the station is added to a route of any ant in grouped in **R**, the station is removed from $\Omega(\mathbf{R})$.

Every time a new station k is added to route **R**, its pheromone is evaporated in order to reach the improved diversification of solutions. This means that for other groups of ants, unvisited stations become more attractive. The *pheromone evaporation* is done by

$$\tau_{jika} \leftarrow (1 - \rho)\,\tau_{jika} + \rho\tau_0, \tag{19}$$

where ρ is a forgetting factor called evaporation constant and a is 1 or 2. Since the attractiveness of moving between stations i and k does not depend on the direction of movement, we make also an additional update $\tau_{jkia} = \tau_{jika}$ for symmetry.

After a group of m ants has finalized its routes, *i.e.* no more stations can be added to routes, *route improvement strategies* are applied to reach better solutions. If the strategies improve the solution at least in Pareto sense (*i.e.* value at least one objective of **f** is improved), the solution is added to the population having originally q solutions. Literature knows a variety of such strategies (*e.g.* Yu *et al.* 2011, Schilde *et al.* 2009, Bell and McMullen 2004).

In our solution, we use three techniques, from which two first ones are based on the literature (*e.g.* Schilde *et al.* 2009) and the last one is developed especially for this problem. First, we *permutate the stations* in the route of each technician to find if the stations can be arranged to lower cost order. If the route has many stations, we limit the permutations to 50. Second, for a fraction of groups, we either *switch* one station from route to another *or* completely *remove* one station from the route. The first option tries to minimize the total costs and the second option finds possible Pareto-optimal solutions. Third, we replace a single technician's route with a route from the set of Pareto-optimal solutions and see if the new route combination is better at least in Pareto sense. This way we increase the number of route combinations using known good routes and improve the Pareto set of solutions.

In order to emphasize the good solutions, after each round of making all q sets of routes, we make a global update for the pheromones. The updates are made as in Schilde *et al.* (2009); the pheromones of the best solution in respect to one of the entries in gain function (12) is increased using $\tau_{jika} \leftarrow \tau_{jika} + \tau_0$, and for the second best solution the update is $\tau_{jika} \leftarrow \tau_{jika} + \tau_0/2$, where a is the number of the gain function entry.

In addition to emphasize the multi-objectivity, we make also a compromise update between the gain function entries for the routes, which reach the maximum value when both gain function entries are weighted equally. These pheromone updates are done by $\tau_{jika} \leftarrow \tau_{jika} + \tau_0/2$, and for the second best solution the update is $\tau_{jika} \leftarrow \tau_{jika} + \tau_0/4$, where a is 1 *and* 2. Also for the global update, we do the symmetry update $\tau_{jkia} = \tau_{jika}$.

After each round, the solutions are compared with the Pareto-optimal set of solutions. In the Pareto-optimal set is included all routes, which do not have completely dominating solutions, *i.e.* no other solution has better values in both entries of the gain function **f**.

3.4 Maintenance routing examples

The following simulated case examples demonstrate the maintenance routing. In all simulations, the working day length is limited to 7.5 hours, and the travel times between stations are calculated based on 60 km/h speed. The MRPP parameters G_{min} and g_{min} are set to 2 and 0.5, respectively. In all cases, the heuristic algorithm routes are included to the first iteration round of ACO as an initialization help for the pheromone. The maintenance takes 0.5 hours per station in the smaller examples and 1 hour per station in the large example. In our examples we use the initial ACO pheromone value $\tau_0 = 1$.

Figure 8 shows a small simulated example of maintenance route formation, in which the routes of heuristic and ACO algorithms are included. The ACO algorithm was used with 75 ant groups for 50 iterations. In the left part of Figure 8, we have one technician starting from the depot (black diamond) and in the right part we have two technicians starting from different depots. The heuristic algorithm routes reach the total gains of 8.9 and 13.2 and the total costs of 12.5 and 24.0, whereas the ACO algorithm routes have the total gains 8.9 and 13.5 and the total costs 12.4 and 23.1. The ACO routes are chosen from the Pareto set of solutions based on the closeness to the heuristic solution.

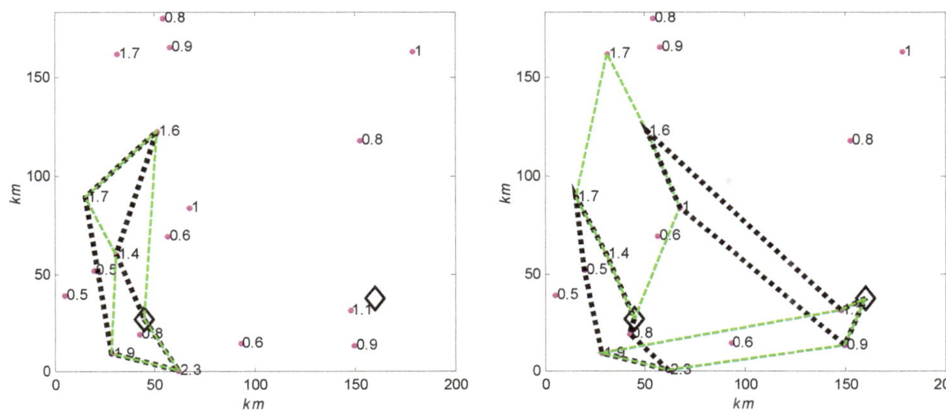

Fig. 8. Maintenance route examples for two depots. Magenta dots are the stations to be maintained and the numbers next to them present the maintenance gain available. Black dotted lines present the heuristic algorithm routes and the green dashed lines are the ACO routes.

Figure 9 shows an example with larger number of maintenance technicians and stations requiring maintenance. In here, two depots have both two technicians and one depot has a technician for the maintenance of the network. The ACO algorithm is run with 100 ant groups of five for 50 iterations.

For the routes in Figure 9, the heuristic algorithm routes the total cost and gain are 53.9 and 27.0, respectively, whereas the presented ACO route solution has the cost of 53.4 and the gain of 28.0.

The experiences on these types of large problems with many technicians having relatively short routes is that the improvement strategy of picking good routes from the Pareto-optimal solutions is very important in improving the overall optimization results.

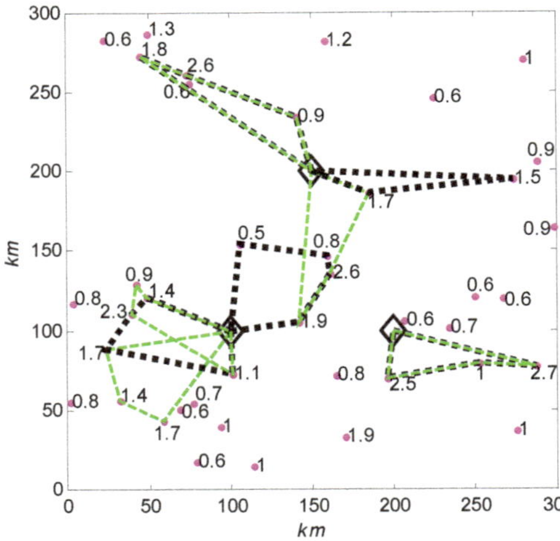

Fig. 9. A large maintenance routes example with three depots and five technicians. Magenta dots are the stations to be maintained and the numbers next to them present the maintenance gain available. Black dotted lines present the heuristic algorithm routes and the green dashed lines are the ACO routes.

4. Final remarks

The maintenance is an essential part in the continuum of quality assurance. Automated maintenance planning can offer a valuable tool for extensive and widely distributed networks, for which the maintenance managing would otherwise be tedious. The approach presented in here has three main steps. First, the performance indices are extracted from the essential QA information. As demonstrated in Section 2, the indices are applicable in sensor and actuator networks. Second, the performance indices are transformed into gains, which measure how much maintenance actions can improve the network performance. Third, the maintenance routes are solved using the methods in Section 3.

Solving the MRPP presented as in Section 3 offers several properties needed in practical systems; such as the optimization of network performance with given resources, tendering the outsourced maintenance services on case-by-case-basis, and the ability to update the maintenance strategy if the situation has changed in the network. The multiobjective nature of the problem leads to the Pareto-optimal set of solutions and the best solution must be determined depending on the relation between the improved performance and costs in the network.

5. Acknowledgments

This work was funded by the Finnish Fund for Automation, Finnish Funding Agency for Technology and Innovation (Tekes) and Vaisala Oyj.

6. References

J.E. Bell, and P.R. McMullen, 2004: "Ant colony optimization techniques for the vehicle routing problem", Ad. Eng. Informatics, vol. 18, no. 1, pp. 41–48.

I.M. Chao, B.L. Golden, and E.A. Wasil, 1996a: "The Team Orienteering Problem", Europ. J. Operational Res., vol. 88, no. 3, pp. 464–474.

I.M. Chao, B.L. Golden, and E.A. Wasil, 1996b: "A Fast and Effective Heuristic for the Team Orienteering Problem", Europ. J. Operational Res., vol. 88, no. 3, pp. 475–489.

D. Feillet, P. Dejax, and M. Gendreau, 2005: "Traveling Salesman Problems with Profits: An Overview", Transportation Science, vol. 39, no. 2, pp. 188–205.

T. Hägglund, 1999: "Automatic detection of sluggish control loops", Control Engineering Practice, vol. 7, no. 12, pp. 1505–1511.

V. Hasu, and H. Koivo, 2007: "Maintenance variables for spatially distributed measurement networks," Proc. IEEE Int. Conf. Systems, Man, and Cybernetics, Montreal, Canada, IEEE, pp. 1327-1332.

V. Hasu, and H. Koivo, 2009: "Automatic maintenance route planning of large-scale sensor networks," Proc. IEEE Int. Conf. Computat. Intellig. for Meas. Syst. and Appl., Hong Kong, China, IEEE, pp. 18–23.

V. Hölttä, 2009: "Plant Performance Evaluation in Complex Industrial Applications", Doctoral Thesis, Helsinki University of Technology, 123 p.

V. Hölttä, and H. Koivo, 2009: "Quality index framework for plant-wide performance evaluation", J. Process Control, vol. 19, no. 7, pp. 1143–1148.

V. Hölttä, and H. Koivo, 2011: "Performance metrics for web-forming processes", J. Process Control, vol. 21, pp. 885-892.

R. Isermann, 2006: "Fault-Diagnosis Systems - An Introduction from Fault Detection to Fault Tolerance", Springer-Verlag.

S.-L. Jämsä-Jounela, R. Poikonen, N. Vatanski, A. Rantala, 2003: "Evaluation of control performance: methods, monitoring tool and applications in a flotation plant", Minerals Engineering, vol. 16, pp. 1069–1074.

C.M.H. Kuijpers, R.H. Murga, I. Inza, and S. Dizdarevic, 1999 "Genetic algorithms for the travelling salesman problem: A review of representations and operators", Artificial Intelligence Review, vol. 13, no. 2, pp. 129–170.

M. Schilde, K.F. Doerner, R.F. Hartl, G. Kiechle, 2009: "Metaheuristics for the bi-objective orienteering problem", Swarm Intell., vol. 3, no. 3, pp. 179–201.

J.C.S. de Souza, M.B. Do Coutto Filho, M.T. Schilling, C. de Capdeville, 2005: "Optimal Metering Systems for Monitoring Power Networks under Multiple Topological Scenarios", IEEE Trans. Power Systems, vol. 20, no. 4., pp. 1700–1708.

T. Sun, L.-J. Chen, C.-C. Han, and M. Gerla, 2005: "Reliable Sensor Networks for Planet Exploration", Proc. IEEE Conf. on Networking, Sensing, and Control ICNSC 2005, pp. 816–821.

N. Wang, J.-C. Lu, and P. Kvam, 2006: "Reliability Modeling in Spatially Distributed Logistics Systems", IEEE Trans. Reliability, vol. 55, no. 3, pp. 525–534.

B. Yu, Z.-Z. Yang, and B. Yao, 2009: "An improved ant colony optimization for vehicle routing problem", Eur. J. Operat. Res., vol. 196, no 1, pp. 171–176.

B. Yu, Z.-Z. Yang, and J.-X. Xie, 2011: "A parallel improved ant colony optimization for multi-depot vehicle routing problem", J. Operational Research Society, vol. 62, no. 1, pp. 183–188.

19

Improving Quality Assurance in Automation Systems Development Projects

Dietmar Winkler and Stefan Biffl
Christian Doppler Laboratory
"Software Engineering Integration for Flexible Automation Systems" (CDL-Flex),
Institute of Software Technology and Interactive Systems,
Vienna University of Technology,
Austria

1. Introduction

Development processes of large-scale automation systems, e.g., power plants and manufacturing systems involve engineers from various disciplines, e.g., mechanical, electrical, and software engineers, who have to collaborate to enable the construction of high-quality systems (Biffl *et al.*, 2009a). Engineers in individual disciplines apply domain-specific tools, methods, and data models, which are typically not seamlessly linked to each other. For instance, electrical engineers use circuit diagrams and technical data sheets to model the electrical behaviour of the systems, process engineers focus on process workflows for the instrumentation of the system, and software engineers use software models to develop and test control applications of the system (Hametner *et al.*, 2011).

Because of the heterogeneity of individual disciplines and the missing links between them, project management (e.g., project observation and control) and quality assurance (QA) activities across disciplines become even more difficult. Nevertheless, a comprehensive view on the project, frequent synchronization of systems engineering artefacts between disciplines, and QA activities are success-critical factors for developing large-scale automation systems.

Observations at our industry partner, a hydro power plant systems integrator, showed that these overlapping project activities (i.e., project management and QA) are currently not supported sufficiently (Sunindyo *et al.*, 2010)(Winkler *et al.*, 2011). In typical industry projects in a distributed and heterogeneous environment synchronization between disciplines and QA activities across disciplines are conducted manually and require high effort by experts, who have to overcome media breaks between the outcomes of different tools and data models. In addition, we observed strong limitations of QA activities which leave important and critical defects unidentified. To support systems development activities in heterogeneous environments for project management (PM) and QA, we identified three main challenges:

- *Project and Process Management*. Heterogeneity of tools and data models requires time-consuming activities to assess the current project state across all involved disciplines. Because of a lack of tool support experts have to collect and analyze data manually. Therefore, the current project state based on real data and facts derived from manual project analysis is available very infrequently and on request only. Nevertheless, continuous analysis of engineering projects and the availability of project status reports are key requirements of project managers to (a) enable a comprehensive view on the overall project state(s) and (b) control the course of events based on the analysis results more effectively and efficiently (Moser *et al.*, 2011).
- *Change Management*. Decoupled disciplines and workflows make engineering processes and change management processes more difficult, in particular, if heterogeneous disciplines are involved. For instance, changing a hardware sensor (e.g., an oil pressure sensor) from a digital to an analogue device (executed by the electrical engineer) affects process engineers (required changes in hardware wiring), and software engineers (required change of software variables according to value ranges and data types). Therefore, a second key requirement is to improve collaboration and interaction between engineers (coming from various disciplines) with respect to propagating critical changes to affected disciplines in a controlled way within a short time interval (Winkler *et al.*, 2011).
- *Quality Assurance*. Typically engineers apply isolated QA approaches recommended by standards and industry best practices to assess and improve product quality with a focus on their individual application domain. For instance, electrical engineers apply simulation approaches of wiring and electrical signals (Sage *et al.*, 2009) and software engineers conduct reviews (Sommerville, 2007), inspections (Laitenberger *et al.*, 2000), and testing approaches (Meyers *et al.*, 2004) to identify defects in the artefacts efficiently and effectively. Isolated QA methods focus on an individual discipline and are well-established. Nevertheless, we observed strong limitations regarding QA activities across disciplines and tool borders. New mechanisms are required to support QA across disciplines. Therefore, the third key challenge focuses on enabling and supporting QA in heterogeneous engineering environments across disciplines and domain borders.

Common to all three challenges/requirements is the need to linking heterogeneous environments to support synchronization and QA across disciplines and tool borders. Figure 1 illustrates these challenges on the semantic level. Three basic roles (see Figure 1; positions 1a - 1c), i.e., electrical, process, and software engineers work within their disciplines using specific tools and methods including best-practice QA approaches. Nevertheless, there is a strong need to synchronize artefacts and disciplines (represented by the overlapping areas in Figure 1), which could address specific risks and quality issues. Observations at our industry partner confirmed that QA activities with focus on the overlapping areas of two or more (heterogeneous) disciplines are not sufficiently addressed yet (see Figure 1; position 2) (Biffl *et al.*, 2011).

Common practices for synchronizing different disciplines focus on these overlapping areas, where experts have to discuss and exchange data to bridge these technical and semantic gaps manually (Biffl *et al.*, 2009b). Therefore, we see the need to support this synchronization process by providing inspection and testing approaches with focus on these overlapping areas to improve QA activities by means of increasing product quality and decreasing effort and error-proneness (caused by the manual synchronization process). Note that the goal of synchronizing individual disciplines is to focus on the overlapping areas, leaving discipline-specific data within their assigned tools. In this chapter we present an approach for identifying these common concepts as foundation for addressing these overlapping areas and show benefits for QA activities, i.e., defect detection across disciplines and project observation and control.

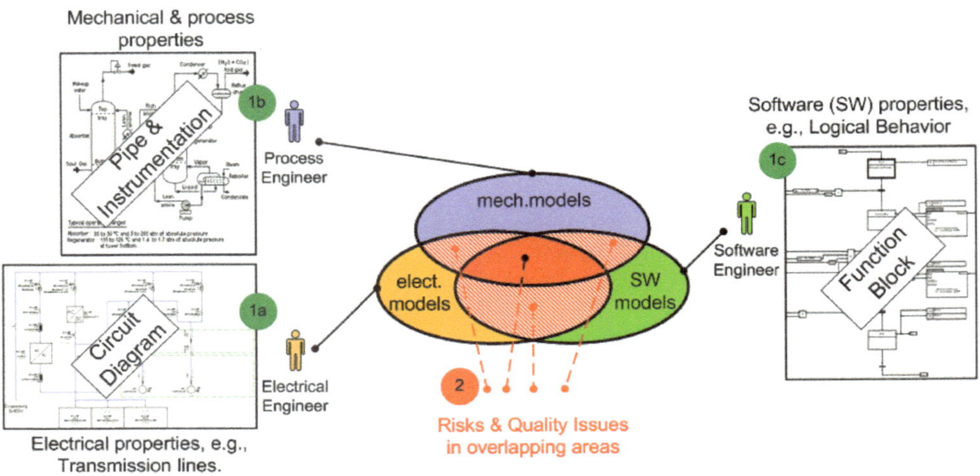

Fig. 1. Risks and Quality Issues in overlapping areas in heterogeneous engineering environments (Biffl *et al.*, 2011).

The reminder of this chapter is structured as follows: Section 2 provides an overview on the related work and Section 3 highlights the research issues. Section 4 describes the basic concepts of the Automation Service Bus (ASB) and Section 5 presents a pilot application for improving QA aspects based on the ASB. Finally, Section 6 summarizes, concludes and identifies future work.

2. Related work

This section summarizes related work of automation systems development processes and software QA as lessons learned from business IT software development for application in large-scale automation systems engineering projects.

2.1 Automation Systems development processes

Automations Systems (AS), such as power plants and industrial automation systems for manufacturing purposes, include distributed software components to control systems behavior (Biffl *et al.*, 2009b). Increasing complexity of software products require well-defined processes and methods for software and systems construction and verification and validation. Various software and systems engineering processes support engineers by providing sequences of steps for project planning and control, e.g., GAMP (Gamp, 2008), W-Modell (Baker *et al.*, 2008), eXtreme Programming (Beck *et al.*, 2004), Scrum (Schwaber, 2004), and V-Modell XT[1]. Nevertheless, process standards focus on the organizational structure of software and systems engineering projects with limitations on method support, tooling and synchronizing various and heterogeneous disciplines.

Observations at our industry partner showed a basic sequential engineering process in Automation Systems Engineering (ASE) development projects (see Figure 2 for details). The observed system development process includes a set of sequential steps including isolated (discipline-specific) QA activities conducted by experts or groups of experts in the individual domain. Because of the sequential process structure, changes from late phases of development (e.g., during test and/or commissioning) can have a major impact on previous phases of the project and can lead to project delays in case of critical changes. Note that these effects are common to sequential and waterfall-like development processes in homogeneous engineering environments (Sommerville, 2007).

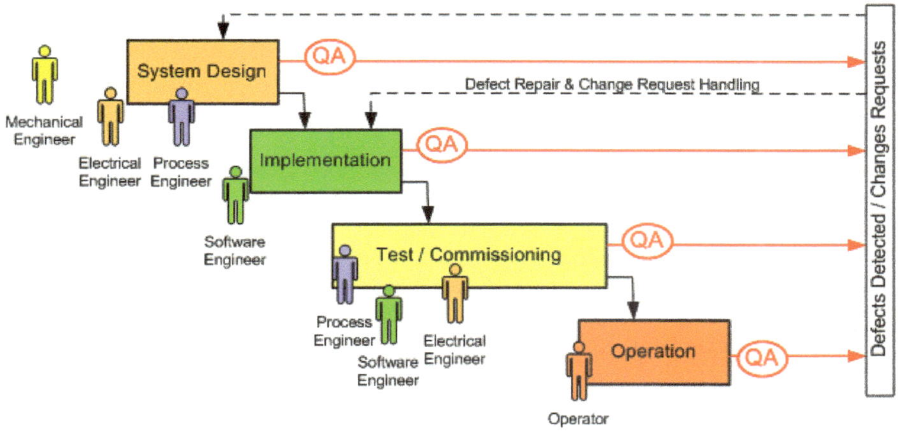

Fig. 2. Sequential Engineering Process with isolated Quality Assurance (QA) Activities.

Considering AS development projects, where experts work distributed in heterogeneous environments, effects of late changes are more critical, more risky, and error prone. Engineers from individual disciplines work concurrently during the development project. Therefore, frequent synchronization between these disciplines is a success-critical issue during development and change request handling (see Figure 3a). For instance, a wrong alarm indicator of an oil pressure sensor in the control center (identified during test and/or commissioning) might affect software engineers (because data handling could have been implemented incorrectly), electrical engineers (wrong alarm sensor type used or incorrectly wired) or the process engineer (incorrect sensor planned). This kind of defects might remain undetached if not tested appropriately. If such a defect is uncovered, there is a need for analyzing the defect and the origin of the defect (across all involved disciplines). A weak link between different disciplines will hinder efficient analysis of defects across disciplines

[1] For a description of the V-Modell XT see http://www.v-modell-xt.de

and could lead to quality problems and project delays. Please note that this analysis steps are typically conducted by experts who are familiar with at least two involved disciplines. Figure 3a presents a basic synchronization step applicable in every phase of the sequential process workflow. Technical integration of tools and semantic integration of data models could help supporting synchronization across disciplines and tool borders (see Figure 3b).

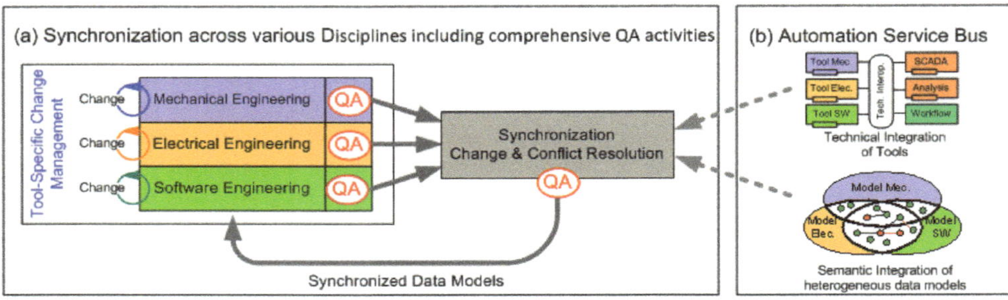

Fig. 3. Synchronization of heterogeneous disciplines with QA (Winkler *et al.*, 2011).

In current industry projects this synchronization step is conducted manually by experts. Expert knowledge is embodied in domain-specific standards, terminologies, people, processes, methods, models, and tools (Moser *et al.*, 2010b). Note that these standards typically do not support technical and semantic integration of tools and data models across disciplines. Assuming that technical and semantic gaps between different engineering experts lead to a lack of QA of artefacts and inefficient change management approaches (Schäfer *et al.*, 2007), a major challenge is to bridge the gap between heterogeneous disciplines on a technical and semantic level to enable efficient change management, quality assurance, and data collection for project monitoring and control during development, commissioning, and maintenance.

Technical and semantic integration of tools and data models across disciplines enables frequent synchronization and data exchange, supports efficient change management processes, and enables more effective and efficient QA. In addition, processes across disciplines and tools borders become observable and - as a consequence - enable effective and efficient project management (PM), project monitoring, and control. Figure 3b illustrates the basic contribution of the ASB approach for technical integration of tools and semantic integration of data models to support PM and QA more effectively and efficiently.

2.2 Quality Assurance aspects for Automation Systems development

Quality Assurance (QA) - embedded within isolated disciplines - is supported by appropriate methods and tools (Schulmeyer, 2008). Nevertheless, a key challenge is to conduct QA activities across domain and tool borders. Based on *Software Engineering Best Practices*[2] (Schatten *et al.*, 2010) specific methods from business IT software development, e.g., inspection and testing, are promising approaches for application in AS development projects.

---

[2] See http://bpse.ifs.tuwien.ac.at for additional material related to the book in English language.

2.2.1 Reviews and software inspection

Reviews and (more formal) software inspections are common and well-established QA methods in business IT software engineering to discover candidate issues and defects systematically. Depending on the configuration of the inspection process (Laitenberger *et al.*, 2000), software inspections are applicable to various types of artefacts, e.g., written text documents, models, drawings, and code. In addition, inspections are applicable in all stages of software and systems development. See (Kollanus *et al.*, 2007) for an overview on software inspection research and (Winkler, 2008) for an empirical evaluation of selected inspection variants.

Individual inspectors form an inspection team and apply a defined sequence of steps *(inspection process)* to identify defects (a) individually and (b) in teams (Biffl *et al.*, 2003). Based on individual inspection results, an aggregated team defect list is generated in an inspection team meeting based on interaction and discussions. Depending on the project scope and the problem domain, *reading techniques* support inspectors and inspection teams to focus on a certain type of defects and defect classes. Basically, reading techniques are structured approaches for reading the document under investigation systematically (Basili, 1997)(Biffl, 2001). Example reading techniques are *checklist-based reading* (inspectors apply a pre-defined and/or customized checklist), *perspective-based reading* (based on different perspectives and disciplines), and *usage-based reading* (use cases and scenarios represent the guidelines for defect detection and drive the inspection process)(Winkler, 2008). Assuming that different perspectives will lead to different defects and defect classes, perspective-based reading techniques focus on defect detection from different viewpoints on the artefact under inspection. For instance, the tester view might lead to defects regarding testability of requirements (e.g., based on the completeness of use cases), the developers might focus on a fully specified design and architecture (e.g., ability to implement requirements), and the user view might focus on end-user requirements (e.g., software solutions that must be usable in the customer domain). Usage-based reading focuses on business cases (typically described as use cases) and the value contribution of the software solution within the business domain. Therefore, this reading technique approach focuses on the most important use cases with the most valuable outcome of the software solution (e.g., based on prioritized use cases).

Analyzing the state-of-the practice at our industry partner, we identified software inspection as a candidate method for improving product quality in ASE projects. Experts analyzed overlapping areas during the synchronization process phase to identify defects in a rather unsystematic and informal way. Software inspections techniques and reading techniques (e.g., perspective based reading) can help engineers in better focusing on a certain type of defects coming from various disciplines.

2.2.2 Software and systems testing

Traditional software testing approaches focus on executing a program with the intent to identify defects (Kaner *et al.*, 1999). Nevertheless, the availability of executable code and test information (e.g., test case specification, test data, and test environments) are pre-conditions for applying software testing techniques (Meyers *et al.*, 2004). Traditional testing approaches - aligned with some V-Model approach - focus on different levels of detail within the overall system (see Figure 4): (a) detecting defects on *component level* (e.g., applying unit tests), (b) integration tests to verify and validate the design and the architecture on *architectural level*, and (c) systems and acceptance test with focus on customer requirements *(system level).*

Lessons learned from testing business IT software products showed the applicability of prominent basic testing techniques, i.e., Black-Box and White-Box testing techniques (Sommerville, 2007), as promising testing approaches on different levels of AS development projects. The component level focuses on testing and simulation of individual components located at isolated disciplines. Integration testing of components - aligned with the architecture - can be seen as testing across disciplines and domain borders, and acceptance testing seems to be similar to system testing and commissioning at the customers site. Figure 4 illustrates the different levels of testing AS project artefacts. Note that test cases and test scenarios can be defined early, following the Test-Driven (Beck *et al.*, 2004) or Test-First (Winkler *et al.*, 2009) approach based on agile software development, another Best-Practice learned from Business IT software development.

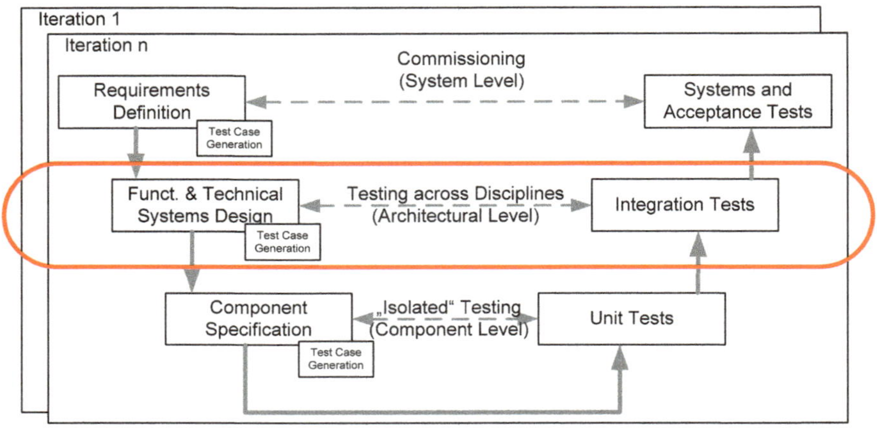

Fig. 4. Test levels in automation systems development according to the W-model (Baker *et al.*, 2008)(Winkler *et al.*, 2009).

In context of AS Black-Box can refer to the *'interfaces'* between various disciplines, e.g., wired connections at a control unit or interface to a software visualization component. Testing these interfaces refers to some kind of *'Black-Box Testing'*. The commissioning phase (comparable to system tests at the customer site), including all hardware and software components of the power plant or manufacturing system, is one of the most critical phases related to QA in the AS domain. Isolated subsystems are launched step by step with real hardware and software. Our observations at industry projects showed that defects - found during this phase - have to be detected manually by analyzing paper work (e.g., drawings) and hardware/software components. Therefore, the commissioning phase requires a very high effort by experts.

The ASB concept aims at supporting QA by enabling testing across domain borders, i.e., testing the overall system *from (hardware) sensor to (software) variables*, comparable to an integration test - well-known from testing business IT software products.

3. Research questions

Heterogeneous engineering environments suffer from weak or missing links between individual disciplines, e.g., mechanical, electrical, and software engineers, on technical and semantic level. Missing links between disciplines hinder efficient collaboration between engineers and makes PM (project observation and control), change management, and comprehensive QA more difficult, risky, and error-prone. Based on observations at our industry partners and related work, we derived the following set of research questions to improve collaboration, project and change management, and QA in heterogeneous environments across disciplines and engineering domains.

Research Question 1 (RQ1). *How can we link various disciplines on a technical and semantic level to enable efficient data exchange in heterogeneous ASE environments?* Efficient data exchange is a pre-condition for effective and efficient PM, change management, and QA. Figure 1 presented the need for collaboration regarding the overlapping areas of individual disciplines, where experts have to synchronize data (from various disciplines) manually. The first research question focuses on eliciting the common concepts, i.e., data represented in the common and overlapping areas, of related disciplines.

Research Question 2 (RQ2). *How can we support QA across disciplines in heterogeneous environments?* Quality assurance aspects can focus on identifying defects in engineering artefacts and overlapping areas of different artefacts coming from various disciplines in a heterogeneous environment. Observations at industry projects showed that these QA activities require a high manual effort provided by experts. We expect a significant improvement (in terms of reducing effort and increasing quality by means of identifying defects more effective and efficient) of QA performance. Therefore, the second research question focuses on providing mechanism to support defect detection in these overlapping areas.

Research Question 3 (RQ3). *How can we support project and quality managers in collecting and analyzing project data (from heterogeneous sources) more effective and efficient to enable continuous project monitoring and control?* Observations at industry projects revealed a high manual effort for collecting and analyzing data from different sources. Because of this high effort (conducted by experts) the project state is captured less frequent and hinder efficient and effective PM. The third research question focuses on providing a 'window to engineering data' across disciplines and domain borders to provide engineering project data tool-supported, frequently and fast.

4. Automation Service Bus for automation systems engineering projects

This section describes the basic concept of the Automation Service Bus (ASB) as a foundation for enabling tool-supported QA activities in heterogeneous environments across disciplines and tool borders (Biffl *et al.*, 2011).

Isolated disciplines apply individual data models and tools with limitations regarding collaboration and interaction (see Figure 1). In industry projects experts bridge the gap between these data models from heterogeneous sources manually. To overcome high effort for manual synchronization and improve the quality of manual activities the Automation

Service Bus (ASB) provides an tool-supported approach to link heterogeneous sources (e.g., data models, data formats, and tools) for PM and QA (Biffl *et al.*, 2009b). In contrast to existing solutions, e.g., Comos PT[3] or EPlan Engineering Center[4], the ASB concept provides a flexible and light-weight infrastructure based on the Enterprise Service Bus (Chappell *et al.*, 2004) concept.

'Flexibility' refers to the concept to respond to changed environments (e.g., introduction of new/modified tools and data models) easily and 'light-weight' refers to reducing the effort for synchronizing different disciplines by focusing on a subset of data (common concepts) similar for all related disciplines. Note that data synchronization can be limited to these common data without considering additional domain-specific data (not relevant for synchronization). These common concepts are represented by a *virtual common data model (VCDM)* (Biffl *et al.*, 2011).

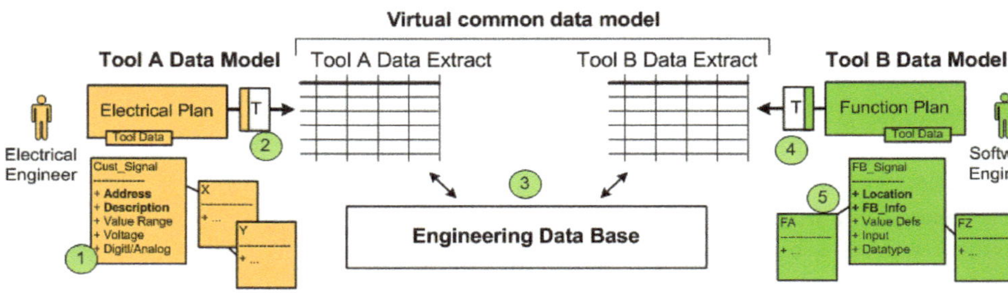

Fig. 5. Schematic overview on the virtual common data model (Biffl *et al.*, 2011).

Basically the VCDM aims at bridging this gap between heterogeneous sources. Figure 5 illustrates the VCDM and the relationship of two different tools from two different disciplines by example. Electrical engineers use tools for designing an electrical plan using defined tool data and attributes located within the (isolated) tool domain (1). On the other hand side, a software engineer (5) use specific tools for designing function plans, a common representation for the development of control applications. Both experts have to agree on a common language (the VCDM) to exchange data efficiently. In the AS domain, e.g., power plants, we observed signals as common concepts between different domains (Winkler *et al.*, 2011). For instance, a signal is represented as a voltage level of an electrical device (by the electrical engineer) and as a software variable (by the software engineer). Additional common information, e.g., hardware addresses and signal description, is used by both disciplines. The agreement of the VCDM results in a mapping table for translation purposes. Note that a transformation is required to map the VCDM to the individual tool data models (see (2) for the electrical plan transformer and (4) for the software model transformer). Finally, signal lists are passed from individual tools via transformer to the engineering data base (EDB) (3) holding the common data based on the VCDM. Therefore, signals and related

[3] Comos PT: http://www.automation.siemens.com

[4] EPlan Engineering Center: http://www.eplan.de/

common signal-specific information are used as foundation for efficient data exchange in AS development projects at our industry partner.

Note that this concept (a) enables interaction between related disciplines (and tools) via the VCDM and the EDB and (b) represents the foundation for PM and QA in the overlapping areas in heterogeneous environments.

5. Pilot application for Quality Assurance support in ASE Projects

This section provides a critical use case, i.e., signal change management (Winkler *et al.*, 2011), and highlights the results of a prototype implementation at our industry partner, a large hydro power plant systems integrator.

5.1 Use case: Signal change/deletion management with warning

The analysis of typical projects at our industry partner showed that an overall number of around 40,000 signal engineering objects are spread across several disciplines in a distributed and heterogeneous environment. Up to now manual synchronization processes were executed infrequently; as a consequence continuous PM and comprehensive QA become difficult and challenging.

Therefore, there is a strong need for synchronizing individual (discipline specific) signal lists more frequently to enable (a) efficient PM and (b) efficient QA across these disciplines. The VCDM enables tool-supported synchronization, efficient interaction, and data exchange between these disciplines. We identified signal change and signal deletion management as critical tasks within an engineering project in the AS domain and designed a basic workflow for signal handling (Moser *et al.*, 2010a). Figure 6a illustrates the process approach for change management and synchronizing of signal lists implemented at our industry partner. Signal deletion (i.e., an engineer removes a signal from the local tool data base) represents a critical issue because if the signal will be removed automatically, the entire signal and related automation aspects will be removed in the local data base of the participating engineer after updating the local tool data. Therefore, we implemented tool-supported notification regarding the removed signal to enable transparency of the deletion process. Figure 6b presents the extended change management process for handling deleted signals.

In detail, both processes are implemented as follows:

- *Signal Change Management (Figure 6a).* An electrical engineer modifies an existing electrical plan (1), e.g., changing the sensor type from a digital to an analogue sensor type. The local and isolated tool data base holds the modified data. The second step includes the submission (check-in) of the changed signals via the transformer (2) to the Engineering Data Base (EDB). Based on the available information in the EDB and the newly submitted signal list, a difference analysis (3) is conducted automatically (including separation of new and modified signals). The second engineer (in this example the software engineer) applies a check-out process (4) and synchronizes with his own local tool data base. If conducted frequently this concept enables a consistent data base available for all participating engineers.

- *Signal Deletion Handling with Warning (Figure 6b).* Locally removed signals are critical for collaboration in heterogeneous environments if signal deletion is propagated across the ASB without appropriate notification of related engineers. Note that the removed signal will disappear in the local data base after a check-out by the corresponding engineer. Note that a signal is considered to be 'removed' if the signal is stored in the EDB but the signal is missing in the new/modified signal list during the check-in process. To overcome this issue, we extended the change management process (Figure 6a) by adding tool-supported notification (Figure 6b). Similar to the signal change handling process, an electrical engineer conducts a check-in process (1) and passes the signal list via Transformer (2) and a difference analysis step (3) to the EDB. The removed signal is identified[5] (4) and results in an engineering ticket (5) including related information and contact information, which are passed to related engineers who are affected by the change/removed signals. Note that information about the involved engineers is stored in a project configuration. While handling the personalized notification the related engineers can respond to the engineering ticket and check-out (synchronize) the modifications if necessary.

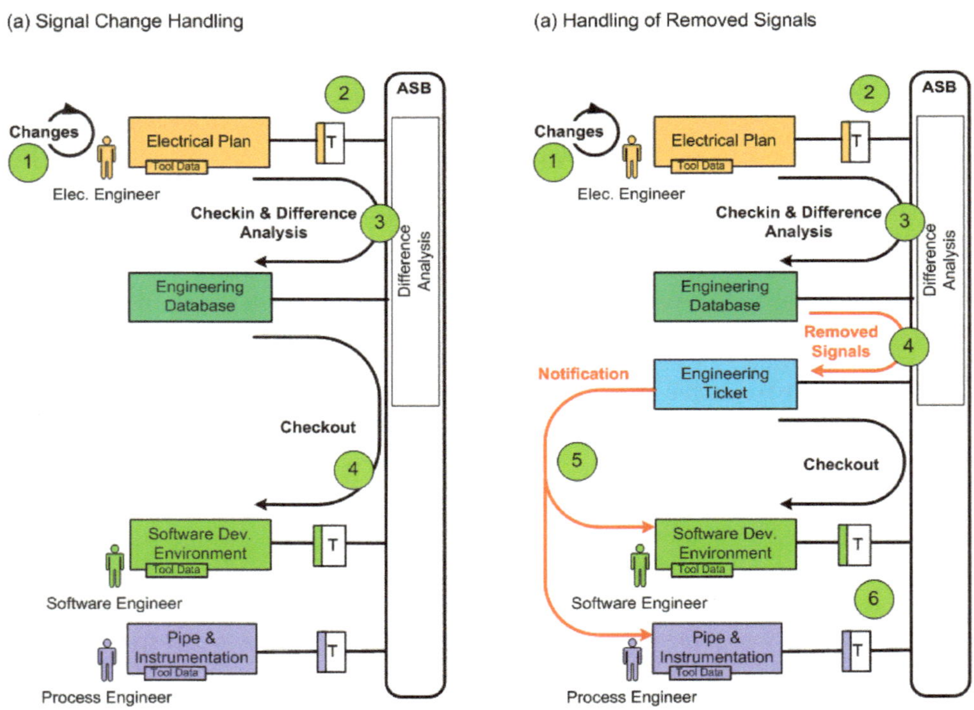

Fig. 6. Concept: Signal change (6a) and signal deletion with warning (6b).

[5] Note that the current scope of the check-in process, e.g., one or more engineering objects, is a mandatory pre-condition for assessing whether a signal has been removed or not.

Based on the VCDM and the signal change/signal deletion management process synchronization and data exchange between various disciplines and data models can be executed with tool support. Nevertheless, it remains open, how this concept can support QA across disciplines.

5.2 Quality Assurance in Automation Systems Engineering projects

In common ASE projects, QA activities, e.g., reviewing signal lists and/or following signal paths across several engineering documents (e.g., electrical plan, P&ID, software models) are conducted manually by experts. Because of a high number of signals (up to 40,000 signals in a common hydro power plant), these tasks are time consuming and include limitations regarding completeness and product quality. Therefore, experts use some supporting tools, e.g., macro solutions to support difference analyses. Nevertheless, these temporary solutions are created and used by individual engineers as small supporting tools. Because of these limitations it is hard to maintain and/or reuse these tools. For instance, changed environments (e.g., different projects or changed project settings) will typically require modification of the supporting tools. In addition knowledge transfer to other engineers, who are not familiar with these tools, becomes difficult.

Focused inspection. A major goal of the ASB concept is to enable experts/engineers focusing on relevant and critical signals during a check-in process. Therefore, an ASB solution should be able to provide only relevant changes to the experts (for discussion and conflict solving) to decrease synchronization effort significantly. Figure 7 illustrates the focus of the QA in context of synchronizing relevant signals across disciplines.

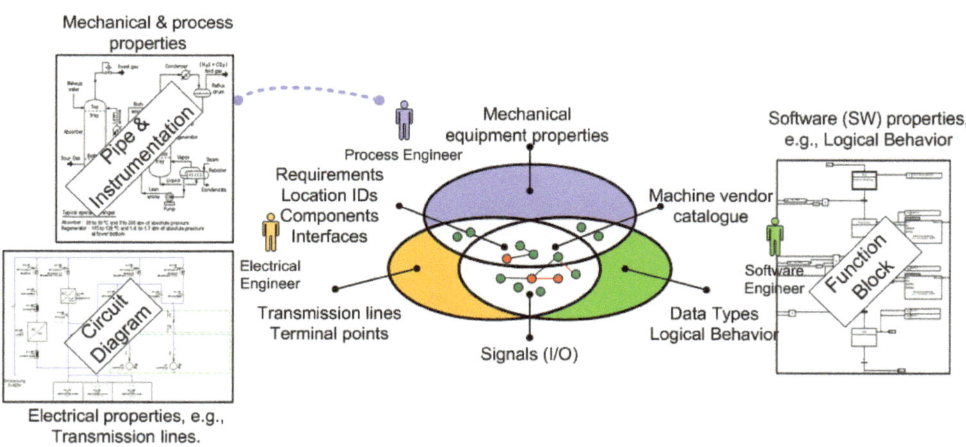

Fig. 7. Defect detection across disciplines (Biffl *et al*, 2011).

QA of signal lists includes two aspects: (a) local QA activities conducted by individual disciplines and tools limited to their application domain (not considered by the ASB approach) and (b) QA across disciplines in the overlapping areas, where experts have to synchronize signals and collaborate. Figure 7 illustrates the main aspects: the green marked dots represent unchanged and agreed signals; the red marked dots represent changes/conflicts/removed signals which can result in major issues in related disciplines. Expert discussions have to focus on these critical changes to identify missing, wrong, or inconsistent elements (signals) or relationships. In addition the difference analysis highlights conflicts coming from changes in more than one discipline.

Figure 8 presents the results of the difference analysis after a check-in conducted by an electrical engineer. Changes are highlighted by providing the old and the modified value. Experts can use this difference check for synchronizing changes across disciplines and either accept or reject the change. Note that additional lists with focus on newly introduced signals, unchanged signals, and removed signals are presented to focus on a defined set of changes. In addition we introduced a list of 'invalid' signals to present, whether the new signal list does not confirm to given guidelines regarding data formats and structure.

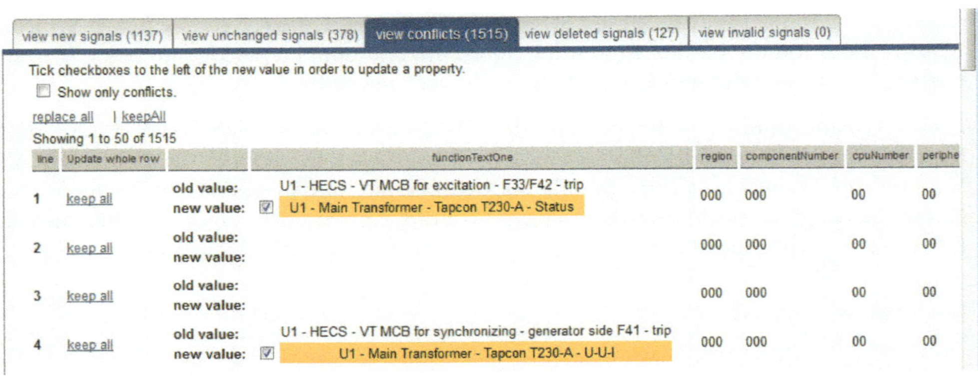

Fig. 8. Screenshot: Highlighted changes at a check-in sequence.

The main advantage is that experts can focus on the changes and conduct a 'focused' inspection from different perspectives, either from the point of view of an electrical engineer, process engineer, or software engineer (as illustrated in Figure 1). Because of this tool-supported approach for data exchange, the synchronization process will be conducted more frequently and can enable the construction of 'no-surprise' systems products.

Note that future work will include a more detailed presentation of signals, including checks for consistency to enable advanced QA activities (a) within one signal and (b) across a set of signals. For instance, a check for consistency within one signal can identify missing information, e.g., a missing hardware address or incompatibility of two or more information sets; a check for consistency can find duplicate addresses or inconsistencies between similar signals within one component. Nevertheless, the difference analysis and presentation of defects will support experts in solving conflicts more effective (completeness) and efficient (within a shorter time interval).

Integration Testing. A second critical QA approach focuses on an overall consideration of signals and their connections across disciplines - comparable to (software) integration tests -

which could hardly be found by (focused) inspection. For instance, a sensor is connected to a switchboard *(System Interface)* and connected to a Software Variable *(Software Interface)* used in a control application *(Software 'Behaviour')*.

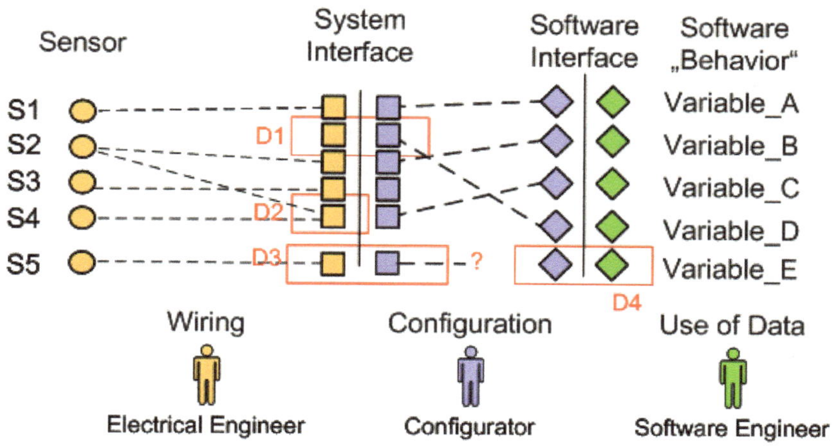

Fig. 9. Automation-supported testing across disciplines (Biffl *et al*, 2011)

Figure 9 illustrates the related disciplines and interfaces and their connection points (Biffl *et al.*, 2011). Sample candidate risks and quality issues are: (D1) Sensor data used by a software variable but without connection to a hardware sensor; (D2) multiple sensors are connected to one software variable; (D3) correctly wired sensor but no link to a software variable; (D4) Software variable not connected to a sensor data and sensors.

These types of defects across disciplines could not be found easily during signal check-in. In addition a manual inspection whether a sensor is wired and connected to the desired variable is a complex and time-consuming activity. Therefore, tracing of signals across disciplines is a valuable approach for supporting experts in their field in better identifying defects. Based on the VCDM (where all signals are available) defined queries focus on a certain class of defects, e.g., whether all sensors are wired to software variables, and deliver a result set of signals across disciplines where this assumption is not given. Experts can focus on the designated signal traces to check whether the traces (and the connection between sensor, switchboard, and software variable) are correct. See (Biffl *et al.*, 2011) for a detailed description and prototype application of the 'End-to-End-Test'.

5.3 Project management support

Observing and monitoring the current project state is a key requirement in ASE projects. Because of the heterogeneity of disciplines and the involvement of various stakeholders in deriving the current project state becomes challenging. Our observations at industry partners showed that distributed project data have to be collected and analyzed manually including a high effort. Therefore, the project state is captured infrequently and on request.

Based on a VCDM the ASB approach enables tool-supported data collection, analysis and visualization by providing project data to relevant stakeholder, e.g., to engineers or to the project manager. We developed an engineering cockpit (Moser *et al.*, 2011) to support PM and QA in ASE projects. Figure 10 presents a prototype of an engineering cockpit to observe changes and the impact of changes in an automation system engineering projects. Major components of this cockpit are (a) role specific views on engineering data and activities, (b) team awareness, and (c) status data regarding the project based on signals as common concepts. Figure 10 presents a snapshot of an engineering project at the industry partner. The data presentation section focuses on the project progress, i.e., the number of signals and the corresponding state of the signal, (e.g., in work, released, changed) per project phase and over time. Selecting defined data sets lead to a drill-down, i.e., a more detailed view on the engineering data within the selected scope. The example shows a selected set of components and the already implemented signals as wells as the expected number of signals for completing the components. Note that the engineering cockpit enables the presentation of data (signals) across disciplines and tool borders and enables a comprehensive view on the engineering project.

See (Moser *et al.*, 2011) for a more detailed description of the engineering cockpit prototype. Based on the prototype implementation of the engineering cockpit in an ongoing research project, we will also address additional information and activities with respect to provide a central starting point for PM and QA in AS development projects.

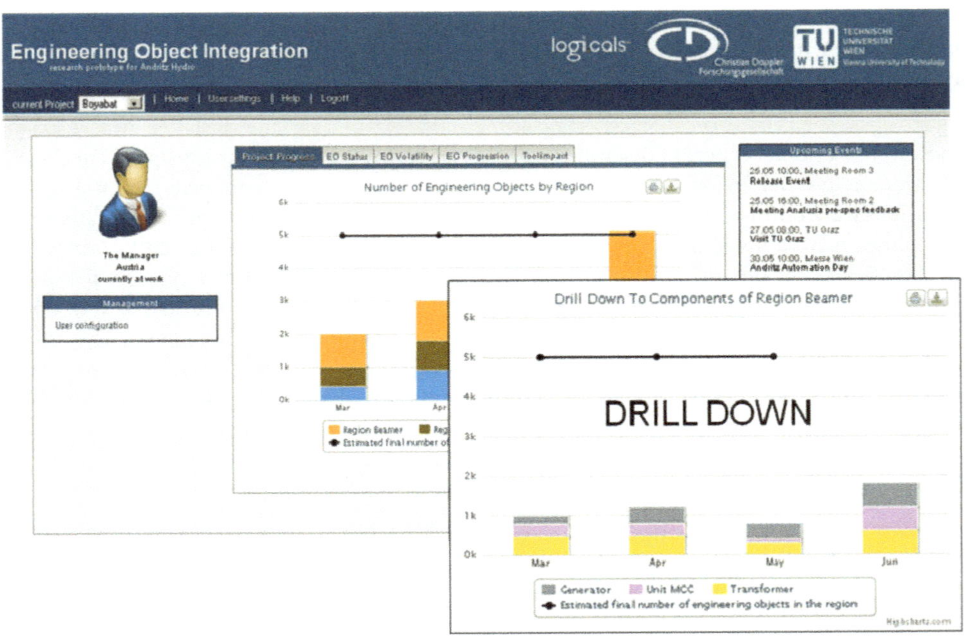

Fig. 10. Project observation with the engineering cockpit (Moser *et al.*, 2011).

6. Conclusion

This chapter presented a snapshot of an ongoing research project (CDL-Flex[6]) summarizing QA aspects for automation system development projects.

6.1 Summary and lessons learned

The development of large-scale AS, e.g., manufacturing systems and power plants, involve various disciplines, e.g., electrical, mechanical, and software engineers, who have to collaborate to construct high quality systems. Collaboration between disciplines and tool borders is a success-critical issue in AS development projects. Typically, disciplines are isolated using individual tools (technical heterogeneity), data models (semantic heterogeneity), and adapted development processes (process heterogeneity). These different levels of heterogeneity make collaboration more difficult and more risky. In addition distributed and heterogeneous data models hinder effective and efficient QA and PM. Efficient synchronization, collaboration, and QA mechanisms are required to support systems development projects. Up to now synchronization and QA across disciplines requires domain experts (familiar in at least two related disciplines), who perform these overlapping tasks manually. These manual activities require a high effort and include defects, which can be avoided by tool-supported mechanisms.

This chapter introduced to the Automation Service Bus (ASB), a flexible middleware platform with focus on the technical integration of tools and the semantic integration of data models in heterogeneous engineering environments (Biffl *et al.*, 2009). Technical and semantic integration is the foundation for bridging the gap between disciplines in the AS project and enable effective and efficient PM (project observation and control) and QA across disciplines and tools borders.

Research Question 1 (RQ1). *Enabling efficient data exchange in heterogeneous ASE environments.* To overcome the semantic gap between heterogeneous data models and data we introduced the Virtual Common Data Model (VCDM). The VCDM aims at bridging the gap between heterogeneous data models by (a) identifying common concepts (e.g., signals) and (b) map isolated data sources to a common Engineering Data Base (EDB) as a central repository for exchanging data between various sources efficiently. The VCDM is the foundation for synchronizing data from various sources efficiently.

Based on our observation and discussions with our industry partner we developed the signal change / signal deletion handling process (Winkler *et al.*, 2011). This process approach - embedded as a central use case for AS development projects - enables systematic data exchange between various stakeholders (across a centralized EDB) based on common concepts (signals), leaving additional discipline-specific data within the specialized domain. Therefore, this ASB approach is considered as a light-weight approach for handling heterogeneous engineering environments.

Research Question 2 (RQ2). *Support for QA across disciplines in heterogeneous environments.* Isolated disciplines and processes include individual QA activities with focus on defect detection within one (isolated) discipline. Nevertheless, QA across disciplines is still

[6] See CDL-Flex website http://cdl.ifs.tuwien.ac.at

challenging. Software inspection and testing (more specifically, integration testing) are promising candidates for cross-disciplinary QA activities. Based on the results of a difference analysis (part of the signal change/deletion handling process) domain experts can focus on selected and critical subsets of engineering objects (signals) for conflict resolution and defect detection (focused inspection). Focused inspection enables defect detection in the overlapping areas (where engineers from different disciplines have to collaborate) from different perspectives with respect to detecting defects. Nevertheless, a comprehensive view on the signal (from sensor to variable) is still challenging. Therefore we introduced the End-to-End test to focus on signal traces to identify different classes of defects, e.g., whether all sensors are connected to a switchboard and if all software variables are connected to data values for further operations.

Research Question 3 (RQ3). Support for project and quality managers in collecting and analyzing project data (from heterogeneous sources) more effective and efficient to enable continuous project monitoring and control. Isolated and heterogeneous data source also hinder efficient PM because - similar to synchronization and QA - data from various sources have to be captured and analyzed manually and on request. The VCDM as data source of common data from different sources is the foundation for a comprehensive view on engineering data across disciplines. We introduced the Engineering Cockpit (Moser *et al.*, 2011) providing the 'Window to engineering data' aiming at (a) providing stakeholder related data derived from the engineering project data bases and (b) enabling control of project steps based on the analysis results.

Nevertheless the presented prototype implementations are a starting point for supporting AS development projects in an ongoing research project.

6.2 Future work

Based on the results from the research project and after discussing them with our industry partner we identified a set of future work aspects related to QA and PM.

- *Common concepts.* Signals have been identified as common concepts in the domain of hydro power plant systems integration. Because of generalization opportunities of the ASB approach, we plan to investigate additional application domains to include a wider range of common concepts.
- *Process observation and control.* The signal change/signal deletion handling processes have been considered as very critical processes in the application domain. Nevertheless, future work will include the consideration of more and more complex processes. Next steps will also include tool-supported definition, implementation and evaluation of processes and process results (Sunindyo *et al.*, 2010).
- *Focused inspection.* This work highlighted the applicability of (perspective-based) inspection for inspecting the overlapping areas of related disciplines. Nevertheless, a more detailed inspection approach has to be developed and evaluated to enable the applicability in industry context.
- *Consistency.* Future work also includes checks for consistency during the check-in processes (a) within one signal and (b) across signals based on pre-defined rules for signal data sets. The goal is to identify deviations early in the development process, i.e., during the specification and design phase of the engineering project.

- *End-to-End-Test.* First results of a prototype evaluation showed benefits of the 'integration test' across disciplines. Nevertheless open issues will focus on different defect types and how to address them properly. In addition, empirical studies are necessary to evaluate the effectiveness and efficiency of end-to-end test approaches.

Finally, empirical studies in industry context are necessary to evaluate the presented concepts (a) in academic and (b) industry environments for the validation of the ASB approach and related industry applications.

7. Acknowledgements

This work has been supported by the Christian Doppler Forschungsgesellschaft and the BMWFJ, Austria. We want to thank the domain experts at industry partners for their insight into engineering projects and their feedback on draft versions of this chapter.

8. References

(Baker et al, 2008) Baker, P., Zhen, R. D., Gabrowksi, J., & Oystein, H.. *Model-Driven Testing: Using the UML Testing Profile*, Springer, 2008.

(Basili, 1997) Basili, V. *Evolving and Packaging Reading Techniques*, Journal of Software and Systems, 38(1), pp3-12, July 1997.

(Beck et al, 2004) Beck, K., Andres, C. *Extreme Programming Explained - Embrace Change*, Addison-Wesley, 2004.

(Biffl, 2001) Biffl S. *Software Inspection Techniques to support Project and Quality Management*, Shaker, 2001.

(Biffl et al, 2003) Biffl S., Halling M.: *Investigating the Defect Detection Effectiveness and Cost Benefit of Nominal Inspection Teams*, IEEE Transactions of Software Engineering 29(5), pp 385-397, 2003.

(Biffl et al., 2009a) Biffl, S, Schatten A. & Zoitl A. (2009). *Integration of Heterogeneous Engineering Environments for the Automation Systems Lifecycle* Proceedings of the IEEE Industrial Informatics (INDIN) Conference, Cardiff, UK, June 2009.

(Biffl et al, 2009b) Biffl, S., Sunindyo, W.D., and Moser, T. *Bridging Semantic Gaps Between Stakeholders in the Production Automation Domain with Ontology Areas*, Proceedings of the 21st International Conference on Software Engineering and Knowledge Engineering (SEKE), Boston, USA, 2009.

(Biffl et al., 2011) Biffl, S., Moser, T., Winkler, D.: Risk Assessment in Multi-Disciplinary (Software+) Engineering Projects. International Journal of Software Engineering and Knowledge Engineering (IJSEKE), Special Session on Risk Assessment, Volume 21(2), March 2011.

(Chappell et al., 2004) Chappell D.A.: *Enterprise Service Bus: Theory in Practice*, O'Reilly Media, ISBN: 978-0596006754, 2004.

(Gamp, 2008) GAMP 5. *Good Automated Manufacturing Practice*. International Society for Pharmaceutical Engineering (ISPE), 2008.

(Hametner et al, 2011) Hametner, R., Kormann, B., Vogel-Heuser, B., Winkler D., Zoitl, A., *Test Case Generation Approach for Industrial Automation Systems*, Proceedings of the 5th International Conference on Automation, Robotics and Applications (ICARA), Wellington, New Zealand, December 2011.

(Kaner et al, 1999) Kaner, C., Falk, J., Hguyen, H.Q.: *Testing Computer Software, Wiley, 1999,*

(Kollanus et al, 2007) Kollanus, S., & Koskinen, J., *Survey of Software Inspection Research: 1991-2005*, Working Paper WP 40, University of Jyväskylä, 2007.

(Laitenberger et al, 2000) Laitenberger, O., DeBaud J.M. *An encompassing life cycle centric survey of software inspection*, Journal of Software and Systems (JSS), 50(1), pp5-31, 2000.

(Meyers et al., 2004) Meyers, G.J., Sandler, C., Badgett, T. & Thomas, T.M. *The Art of Software Testing*, 2nd edition, John Wiley & Sons, 2004.

(Moser et al., 2010a) Moser, T., Waltersdorfer, F., Zoitl, A. & Biffl. *Version Management and Conflict Detection Across Heterogeneous Engineering Data Models*. Proceedings of the 8th IEEE International Conference on Industrial Informatics (INDIN), Osaka, Japan, July, 2010.

(Moser et al., 2010b) Moser, T., Biffl, S., Sunindyo, W.D & Winkler, D.: *Integrating Production Automation Expert Knowledge across Engineering Stakeholder Domains*. Proceedings of the 4th Interational Conference on Complex, Intelligent and Software Intensive Systems (CISIS), Krakow, Poland, February 2010.

(Moser et al., 2011) Moser, T., Mordinyi, R., Winkler, D., & Biffl, S. *Engineering Project Management using the Engineering Cockpit: A collaboration platform for project managers and engineers*. Proceedings of the 9th International Conference on Industrial Informatics (INDIN), Lisbon, Portugal, July 2011.

(Sage et al., 2009) Sage, A.P., Rouse, W.B. *Handbook of Systems Engineering and Management*, 2nd Edition; Wiley, 2009.

(Schäfer et al, 2007) Schäfer, W. & Wehrheim, H. *The Challenges of Building Advanced Mechatronic Systems*, Proceedings of the Intenational Conference on Software Engineering (ICSE), Future of Software Engineering, Washington DC, USA, 2007.

(Schatten et al, 2010) Schatten, A., Biffl, S., Demolsky, M., Gostischa-Franta, E., Östreicher, T., & Winkler D., *Best-Practice Software-Engineering : Eine praxisorientierte Zusammenstellung von komponentenorientierten Konzepten, Methoden und Werkzeugen*, Spektrum Akademischer Verlag, 2010.

(Schulmeyer, 2008) Schulmeyer, G.G. *Handbook of Software Quality Assurance*. 4th edition, Artech House Inc, 2008.

(Schwaber, 2004) Schwaber K.: *Agile Project Management with Scrum*, Microsoft Press, ISBN: 978-0735619937, 2004.

(Sommerville, 2007) Sommerville, I. *Software Engineering*, 8th Edition, Addison Wesley, 2007.

(Sunindyo et al., 2010) Sunindyo W.D., Moser T., Winkler D., & Biffl S.: *Foundations for Event-Based Process Analysis in Heterogeneous Software Engineering Environments*, Proceedings of the 36th Euromicro Conference, Software Engineering and Advanced Applications (SEAA), Lille, France, September 2010.

(Winkler, 2008) Winkler. D. *Improvement of Defect Detection with Software Inspection Variants: A Large-Scale Empirical Study on Reading Techniques and Experience*, VDM, May 2008.

(Winkler et al, 2009) Winkler, D., Biffl, S., & Östreicher, T. *Test-Driven Automation – Adopting Test-First Development to Improve Automation Systems Engineering Processes.* Proceedings of the 16th European Software & Systems Process Improvement and Innovation (EuroSPI) Conference, Madrid, Spain, September 2009.

(Winkler et al., 2011) Winkler, D., Moser, T., Mordinyi, R., Sunindyo, W.D. & Biffl, S. *Engineering Object Change Management Process Observation in Distributed Automation Systems Projects*, Proceedings of the 18th European Systems & Software Process Improvement and Innovation Conference (EuroSPI), Roskilde, Denmark, June 2011.

Implementation of CVR / IT Methodology on Outsourced Applied Research to Internship Environment, Case, Information Technology Directorate of Bina Nusantara Foundation

Renan Prasta Jenie
Information Technology Directorate,
Bina Nusantara University,
Indonesia

1. Introduction

1.1 Nature of problem

Bina Nusantara Foundation is a Teaching and Learning Corporation. It consists of several Bodies with distinct responsibilities, and one of them is Information Technology Directorate (ITD). ITD is responsible in developing and maintaining information technology solution in whole corporate.

ITD consists of three main divisions, as in Fig 1. The Information System Development Division (IS Dev) is responsible in developing and maintaining whole foundation Information System Solution, such as web sites and console applications. The Network, Data, and Operational Division (Oper) are responsible in building and maintaining Network Connections, Data Storage and Servers, and daily operational procedure in foundation. Technology Development Division (Tech Dev) is responsible in maintaining applied research researchers and Internship partner students for helping the whole directorate.

ITD is ought to publish 20 National Level Research Paper yearly, but they only have 5 non dedicated researcher, and they called them Research Specialist. ITD need at least about 20 Research Teams to create those paper, or 40 teams for safety. ITD Research is divided into two Semester yearly, 1st year for Feasibility Study, and 2nd year for Product Research. They are currently supported by 54 students / 15 teams powered Associate member Research Internship program. The students are considered as partners, with full turnover per semester.

1.2 Scope of problem

This study is done to analyze how to Internship Student can be used by Tech Dev as a Research Outsourcing Body, what to tweak and modify. This study explains how to maintain students in Research Internship Program, including rewarding and punishment system.

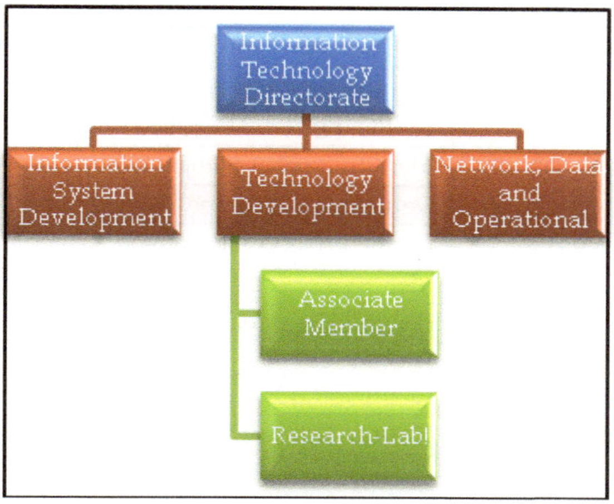

Fig. 1. Information Technology Directorate structure.

This study also describes how the Project Management Process Model is implemented. What changes should be made to modify Project Management Process Model to suite a Student Outsourced Software Development Environment.

1.3 Importance of problem

This study should be done to copes several issues. Internship Researcher should be utilized well and rewarded systematically. Failure to do so can cause serious impact to student trust to ITD, such as lost of faith and even truancy.

Current available Project Management Model normally did not include model for Research Environment that Outsourced to Student Internship. There is model for Professionally Outsourced System, but could not be adapted fully due to different philosophy on how to work with between Students and Professionals.

It is suggested to use CVR / IT Model on Research Project Management System Utilization. What the impact on core system and Social Impact should be documented. How fast the adaption time and what to modify should be found. Failure to do so can stop the whole Research Management System.

1.4 Literary review

We are using (Boehm & Ross, 1989) as definition of software project management, which shall activate several soft skills (Sukhoo, Bernard, Eloff, Poll, & A., 2008). We chose (CVR/IT Consulting LLC, 2002) as in Table 1 over several software management processes such as (Paul, Kunii, Shinagawa, & Khan, 1998), (Goebel III, 2003), (Liu, Kane, & Bambroo, 2003), (Alba & Chicano, 2006), and (Jalote, 2007); subsequently omitting needs of tools like (Quantitative Software Management Associates, 2004), (Reid & Wilson, 2007), and (Callahan & Ramakrishnan, 2007), for internship management as in (True, 2007).

Group		Activities
Project Initiation Activities		Assign an initiating Project Manager
		Identify the Project Sponsor
		Define the Business Need/Opportunity
		Identify Business Objectives and Benefits
		Define Overall Scope
		Define Project Objectives
		Identify Project Constraints and Assumptions
		Ensure Alignment with Strategic Direction and Architecture
		Identify and Engage Key Stakeholders
		Identify Key Potential Risks
		Procurement and Resourcing Requirements
		Determine Cost/Benefit and Schedule Estimates
		Develop a Project Phase Exit Plan
Project Planning Activities	Analysis Phase	Assign a Project Manager
		Refine Project Scope
		Determine Procurement and Sourcing Strategy
		Refine Project Schedule
		Resource Planning
		Identify Other Resource Requirements
		Establish Project Life-Cycle Phase Checkpoints
		Refine Project Cost Estimate and Budget
		Identify Potential Project Risks
		Determine Process for Issue Identification and Resolution
		Develop a Change Management Process
		Develop an Organizational Change Management Approach
		Develop a Quality Management Approach
		Develop A Project Communication Approach
		Develop A Configuration Management (CM) Approach
		Project Performance Commitment
		Consolidate the Project Plan
	Design Phase	Detailed Designs
		Establish a Requirements Traceability Matrix
		Manage scope, cost, schedule and quality (Manage Change)
		Manage issues, change and risk

Group		Activities
		Establish and report project status
		Update Project Performance Commitment document
		Obtain approval from the Governance Body at Design Review
Project Execution and Control Activities		Manage Risk
		Communicate Information
		Manage Schedule
		Document the Work Results
		Manage Organizational Change
		Manage Scope
		Manage Quality
		Manage Costs
		Manage Issues
		Conduct Status Review Meetings
		Review Project Life-Cycle Phases Checkpoints
		Execute the Procurement Plan
		Administer Contract/Vendor
		Update Project Planning Documents
		Build & Test
		Implementation
		Conduct Rollout Acceptance Meeting
Project Closeout Activities		Conduct Final Contract Review
		Conduct Lessons Learned Meeting
		Conduct Knowledge Transfer
		Post Project Review
		Distribution of Resources
		Celebration

Table 1. Project Management Process (CVR/IT Consulting LLC, 2002).

2. Methods

2.1 Whole design

The whole design shall be explained in 3 parts, stakeholders, methods templates, and procedures, as in Table 2. The CVR / IT Standard were mapped as in Table 3. The mapping is necessary to alleviate several problems. In research world, resources often come from grant that has fixed schedule, so grant must be pursued before researcher procurement. Furthermore, to maintain whole University research map, all research planned must go through committee examination. After the permission granted, researcher must procure student for running the research.

Methods Template		This part explains how the activities represented.
Procedures	Research Communication Activities	This part explains how research plan are communicated to every researcher and lecturer.
	Research Monitoring Activities	This part explains how researches are monitored.
	Research Proposal and Budgeting Activities	This part explains how the systems accept researcher proposal and compiling budget on it.
	Research Budget Sounding Activities	This part explains how budget are communicated to management level.
	Research Permission Grant Activities	This part explains how every proposal granted or denied.
	Research Execution and Control Activities	This part explains how to control and execute every research plan granted.
	Student Research Advertisement Activities	This part explains how the researches are communicated to student body.
	Student Research Proposal and Budgeting Activities	This part explains how the systems accept student proposal and compiling budget
	Student Research Execution and Control Activities	This part explains how to control and execute every research plan granted.
	Research Evaluation Activities	This part explains how every researcher and student is evaluated based on their achievement in their research.
	Research Training Schedule Commission Activities	This part explains how research training schedule are controlled.
	Research Schedule Commission Activities	This part explains how research schedule are controlled.
	Monthly Research Specialist Monitoring Activities	This part explains how monthly researcher monitoring done.
	Monthly Research Student Monitoring Activities	This part explains how monthly research student monitoring done.
	Weekly Research Student Monitoring Activities	This part explains how weekly research student monitoring done.
	Researcher Reward Activities	This part explains how weekly researcher and research student rewarded.

Table 2. Whole design.

CVR / IT Activities	Research Communication Activities	Research Monitoring Activities	Research Proposal and Budgeting Activities	Research Budget Sounding Activities	Research Permission Grant Activities	Research Execution and Control Activities	Student Research Advertisement Activities	Student Research Proposal and Budgeting	Student Research Execution and Control	Research Evaluation Activities	Research Training Schedule Commission	Research Schedule Commission Activities	Monthly Research Specialist Monitoring Activities	Monthly Research Student Monitoring Activities	Weekly Research Student Monitoring Activities	Researcher Reward Activities
Assign an initiating Project Manager	O		O	Unique for Research	Unique for Research		Unique for Research									
Identify the Project Sponsor	O		O													
Define the Business Need/Opportunity	O		O													
Identify Business Objectives and Benefits	O		O													
Define Overall Scope	O		O													
Define Project Objectives	O		O													
Identify Project Constraints and Assumptions	O		O													
Ensure Alignment with Strategic Direction and Architecture	O		O													
Identify and Engage Key Stakeholders	O		O													
Identify Key Potential Risks	O		O													
Procurement and Resourcing Requirements	O		O													
Determine Cost/Benefit and Schedule Estimates	O		O													
Develop a Project Phase Exit Plan			O													
Assign a Project Manager			O													
Refine Project Scope		O	O													
Determine Procurement and Sourcing Strategy			O													
Refine Project Schedule			O													
Resource Planning			O													
Identify Other Resource Requirements			O													
Establish Project Life-Cycle Phase Checkpoints			O													
Refine Project Cost Estimate and Budget			O					O								
Identify Potential Project Risks			O					O								
Determine Process for Issue Identification and Resolution			O					O								

CVR / IT Activities	Research Communication Activities	Research Monitoring Activities	Research Proposal and Budgeting Activities	Research Budget Sounding Activities	Research Permission Grant Activities	Research Execution and Control Activities	Student Research Advertisement Activities	Student Research Proposal and Budgeting	Student Research Execution and Control	Research Evaluation Activities	Research Training Schedule Commission	Research Schedule Commission Activities	Monthly Research Specialist Monitoring Activities	Monthly Research Student Monitoring Activities	Weekly Research Student Monitoring Activities	Researcher Reward Activities
Develop a Change Management Process			O					O								
Develop an Organizational Change Management Approach			O					O								
Develop a Quality Management Approach			O					O								
Develop A Project Communication Approach			O					O								
Develop A Configuration Management (CM) Approach			O					O								
Project Performance Commitment			O					O								
Consolidate the Project Plan			O					O								
Detailed Designs			O					O								
Establish a Requirements Traceability Matrix			O					O								
Manage scope, cost, schedule and quality (Manage Change)		O				O			O		O	O	O	O	O	
Manage issues, change and risk		O				O			O		O	O	O	O	O	
Establish and report project status		O				O			O		O	O	O	O	O	
Update Project Performance Commitment document		O				O			O		O	O	O	O	O	
Obtain approval from the Governance Body at Design Review	O															
Manage Risk		O				O			O		O	O	O	O	O	
Communicate Information		O				O			O		O	O	O	O	O	
Manage Schedule		O				O			O		O	O	O	O	O	
Document the Work Results		O				O			O		O	O	O	O	O	
Manage Organizational Change		O				O			O		O	O	O	O	O	
Manage Scope		O				O			O		O	O	O	O	O	
Manage Quality		O				O			O		O	O	O	O	O	
Manage Costs		O				O			O		O	O	O	O	O	
Manage Issues		O				O			O		O	O	O	O	O	
Conduct Status Review Meetings		O				O			O		O	O	O	O	O	

CVR / IT Activities	Research Communication Activities	Research Monitoring Activities	Research Proposal and Budgeting Activities	Research Budget Sounding Activities	Research Permission Grant Activities	Research Execution and Control Activities	Student Research Advertisement Activities	Student Research Proposal and Budgeting	Student Research Execution and Control	Research Evaluation Activities	Research Training Schedule Commission	Research Schedule Commission Activities	Monthly Research Specialist Monitoring Activities	Monthly Research Student Monitoring Activities	Weekly Research Student Monitoring Activities	Researcher Reward Activities
Review Project Life-Cycle Phases Checkpoints		O				O			O		O	O	O	O	O	
Execute the Procurement Plan		O				O			O		O	O	O	O	O	
Administer Contract/Vendor		O				O			O		O	O	O	O	O	
Update Project Planning Documents		O				O			O		O	O	O	O	O	
Build & Test						O			O							
Implementation						O			O							
Conduct Rollout Acceptance Meeting										O						
Conduct Final Contract Review										O						
Conduct Lessons Learned Meeting										O						
Conduct Knowledge Transfer										O						
Post Project Review		O								O						
Distribution of Resources																O
Celebration																O

Table 3. CVR / IT Mapping.

2.2 Stake holders

These system stake holders are as follows.

- Community Development Coordinator. He is in charge of running supporting events like admission, training, and others.
- Technology Development Manager. He is the one who in charge of Technology Development Division where research done. Research Committee. This is the team to judge whether one research is feasible to run.
- Research Coordinator. He is in charge of daily task in research coordination.
- Research Specialist. Researcher for one specific research fields.
- Head of Information Technology Directorate, Head of Information Technology Department and Board of Management. They are in charge of judging research product and its availability for incubation. Thesis Student. Main source of internship program.

2.3 Methods template

Every activity shall be presented in following fashions.

1. Visual representation of procedures. It shall give the reader who shall do certain activities and where they shall get and give the deliverables.
2. What points should be considered so the activities can be called success, and failure to do so can ruin the project.
3. What activities should be done so every success factor can be paid. What deliverables are made in each activity.

2.4 Procedures

2.4.1 Research Communication Activities

Research Communication Procedure is as in Fig 2. Community Development Officer describes how the researches are communicated to every researcher.

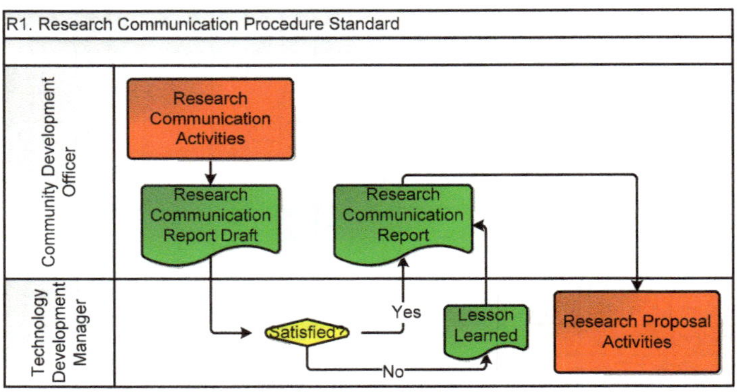

Fig. 2. Research Communication Procedure Standard.

Research Communication Activities critical success factor is:

1. all Lecturers and Research Specialist in University and Information Technology Directorate recognize the program,
2. problems emerged and potential remedies are identified, and
3. Technology Development Manager already approves the research communication report draft.

Research Communication Activities activities is

1. run the Researcher and Lecturers Briefing and report the realization,
2. identify Problems emerged and its potential remedies, and
3. Obtain approval of Research Communication Report Draft from Technology Development Manager.

Research Communication Activities deliverables is Form R1. Research Communication Report.

2.4.2 Research Proposal and Budgeting Activities

Research Proposal and Budgeting Procedure is as in Fig 3. Research Coordinator gathers every research proposal available.

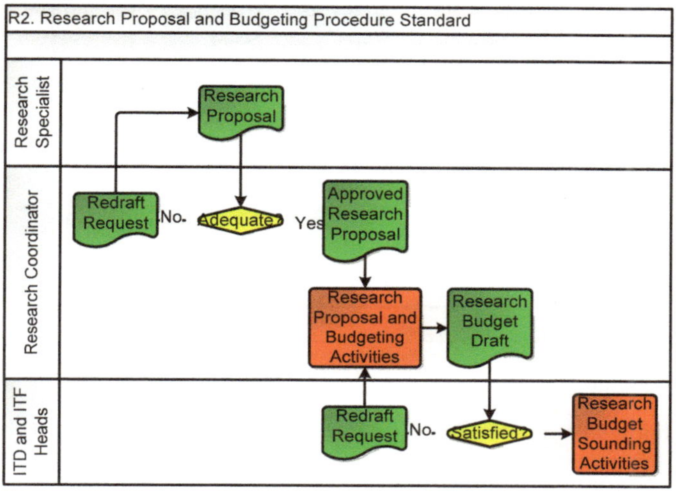

Fig. 3. Research Proposal and Budgeting Procedure Standard.

Research Proposal and Budgeting Activities critical success factor is

1. Every proposal received is reviewed,
2. The following budget is calculated: Man Power demand, Tools / Machine budget, Research object budget, Research operational budget, Literature budget, Seminar / Proceeding / Journal Budget, Awards Budget, Knowledge Sharing Budget, and Student Briefing Budget, and
3. Every member of research committee already approves the research budget plan draft.

Research Proposal and Budgeting Planning Activities activities are

1. Review every research proposal,
2. Calculate: Man Power demand, Tools / Machine budget, Research object budget, Research operational budget, Literature budget, Seminar / Proceeding / Journal Budget, Awards Budget, Knowledge Sharing Budget, Student Briefing Budget, and
3. Obtain approval of Research Budget Plan Draft from Research Committee.

Research Proposal and Budgeting Planning Activities deliverables is Form R2A. Research Budget and Form R2B. Research Proposal.

2.4.3 Research Budget Sounding Activities

Research Budget Sounding Procedure is as in Fig 4. Research Budget is to be communicated with board of management.

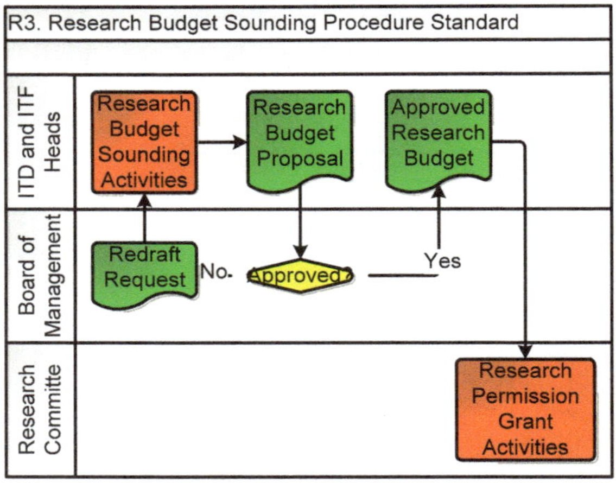

Fig. 4. Research Budget Sounding Procedure Standard.

Research Budget Sounding Activities critical success factor is

1. Research budget plan is presented to board of management, and every related member of board of management already approves the research budget plan draft.

Research Budget Sounding Activities activities is

1. Research budget plan presentation and Obtain approval of Research Budget Plan Draft from Every related member of board of management.

Research Budget Sounding Activities deliverables is as follows. Form R3.Approved Research Budget.

2.4.4 Research Permission Grant Activities

Research Permission Grant Procedure is as in Fig 5. Research Specialist and Research Coordinator are to run every research planned.

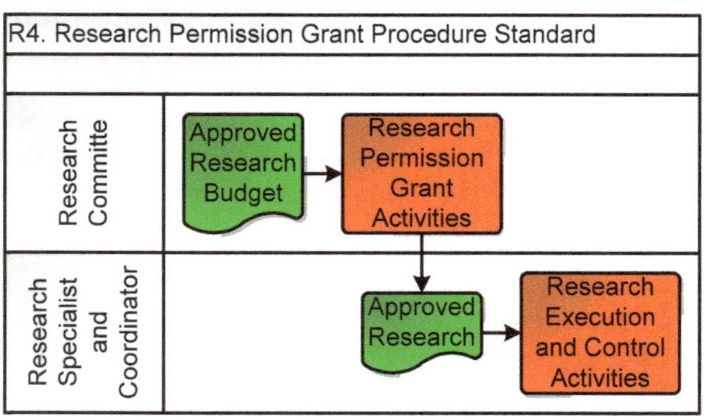

Fig. 5. Research Permission Grant Procedure Standard.

Research Permission Grant Activities critical success factor:

1. every proposal received is reviewed,
2. Research Approved Budget is reviewed,
3. Status of every proposal permission is made, and
4. Every related member of research committee already approves the research permission grant draft.

Research Permission Grant Activities activities is

1. Review every proposal received,
2. Review Research Approved Budget,
3. Grant or deny every proposal permission, where the end choice should be made by all the research committee members.
4. And Obtain approval of Research Permission Grant Draft from Every related member of research committee.

Research Permission Grant Activities deliverables is Form R4. Research Permission Grant.

2.4.5 Research Execution and Control Activities

Research Execution and Control Procedure is as in Fig 6. Research Specialist and Research Coordinator are in charge of execution of every research plan.

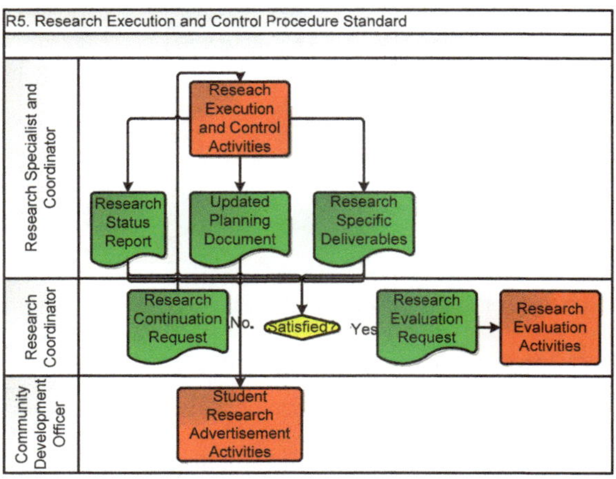

Fig. 6. Research Execution and Control Procedure Standard.

Research Execution and Control critical success factor is:

1. The research proposal is reviewed,
2. Every research prerequisite is satisfied,
3. Every phases described in research methodology is committed,
4. Every milestones / deliverables needed is made, At minimum, it consist of: Research Report, Solutions and it's Manual or Training Materials, Research Journal in English and Indonesian Language, Business Journal in English and Indonesian Language, Script, Thesis, or Dissertation, if available, and

5. Technology Development Manager already approves the research report draft.

Research Execution and Control Activities activities are

1. Review the proposal,
2. Satisfy every research prerequisite,
3. Ensure that every phases described in research methodology is committed,
4. ensure that every milestone / deliverables needed is made, and
5. Obtain approval of Research Report Draft from Technology Development Manager.

Research Execution and Control Activities deliverables is Form R5. Research Execution and Control.

2.4.6 Student Research Advertisement Activities

Student Research Advertisement Procedure is as in Fig 7. Student is drafted based on research plan.

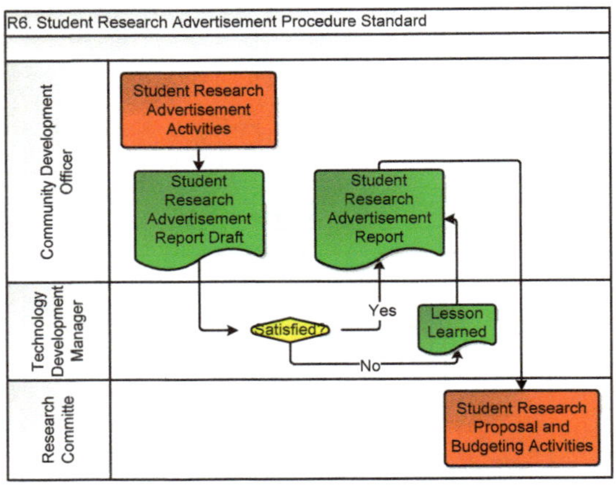

Fig. 7. Student Research Advertisement Procedure Standard.

Student Research Advertisement Activities critical success factor is

6. All Thesis Student in University and Information Technology Directorate recognize the program,
7. Problems emerged and potential remedies are identified, and
8. Technology Development Manager already approves the Student Research Advertisement report draft.

Student Research Advertisement Activities activities is

1. Run the Thesis Student Briefing and report the realization,
2. Identify Problems emerged and its potential remedies, and
3. Obtain approval of Student Research Advertisement Report Draft from Technology Development Manager.

Student Research Advertisement Activities deliverables is Form R6. Student Research Advertisement Report.

2.4.7 Student Research Proposal and Budgeting Activities

Student Research Proposal and Budgeting Procedure are as in Fig 8. Research Coordinator is gathering every student research proposal.

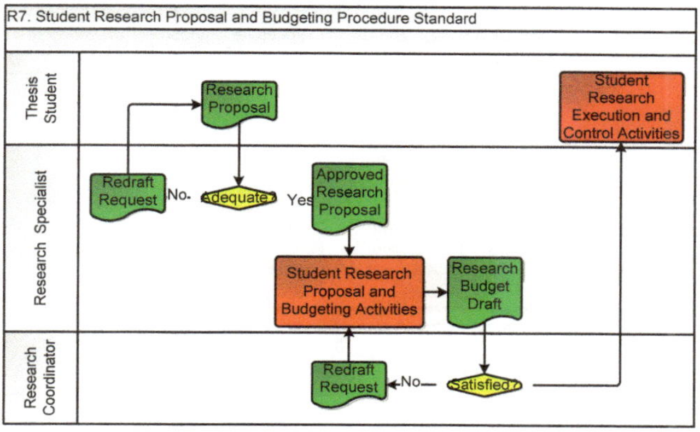

Fig. 8. Student Research Proposal and Budgeting Procedure Standard.

Student Research Proposal and Budgeting Activities critical success factor is

1. Every proposal received is reviewed, The following budget is calculated: Man Power demand, Tools / Machine budget, Research object budget, Research operational budget, and Literature budget, and
2. Every related research specialist and coordinator already approves the student research budget plan draft.

Student Research Proposal and Budgeting Planning Activities activities are

1. Review every research proposal, Calculate: Man Power demand, Tools / Machine budget, Research object budget, Research operational budget, and Literature budget, and
2. Obtain approval of Research Budget Plan Draft from Every related research specialist and coordinator.

Student Research Proposal and Budgeting Planning Activities deliverables is Form R7A. Student Research Budget and Form R7B. Student Research Proposal.

2.4.8 Student Research Execution and Control Activities

Student Research Execution and Control Procedure are as in Fig 9. Research Specialist handle day to day control of Thesis Student.

Student Research Execution and Control critical success factor is

1. The research proposal is reviewed,
2. Every research prerequisite is satisfied,
3. Every phases described in research methodology is committed, Every milestones / deliverables needed is made, At minimum, it consist of: Research Report, Solutions and

it's Manual or Training Materials, Research Journal in English and Indonesian Language, Business Journal in English and Indonesian Language, and Script, Thesis, or Dissertation, if available, and

4. Technology Development Manager already approves the student research report draft.

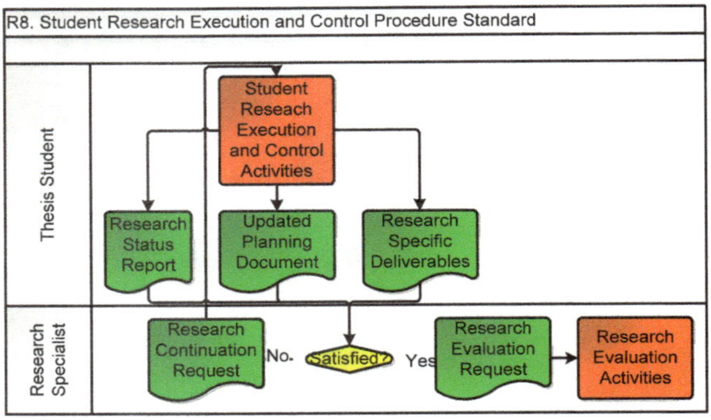

Fig. 9. Student Research Execution and Control Procedure Standard.

Student Research Execution and Control Activities activities are

1. Review the proposal, Satisfy every research prerequisite, Ensure that every phases described in research methodology is committed, ensure that every milestone / deliverables needed is made, and
2. Obtain approval of Student Research Report Draft from Technology Development Manager.

Student Research Execution and Control Activities deliverables is Form R8. Student Research Execution and Control.

2.4.9 Research Evaluation Activities

Research Evaluation Procedure is as in Fig 10. Research Committee is to evaluate deliverables of every research done.

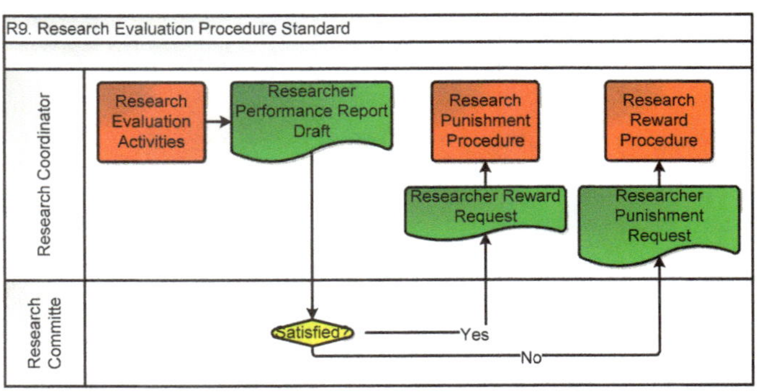

Fig. 10. Research Evaluation Procedure Standard.

Research Evaluation Activities critical success factor is

1. every finished research is reviewed, every research and student research report is reviewed, every research phases described in research methodology is committed, every research milestones / deliverables needed is made, and
2. Every member of research committee already approves the researcher performance report draft.

Research Evaluation Activities activities is

1. Review every finished research, Review every research and student research report, and
2. Obtain approval of Researcher Performance Report Draft from every member of Research Committee.

Research Evaluation Activities deliverables is Form R9. Researcher Performance Report.

2.4.10 Research Training Schedule Commission Activities

Research Training Schedule Commission Procedure is as in Fig 11. Researcher attendance in training must be documented.

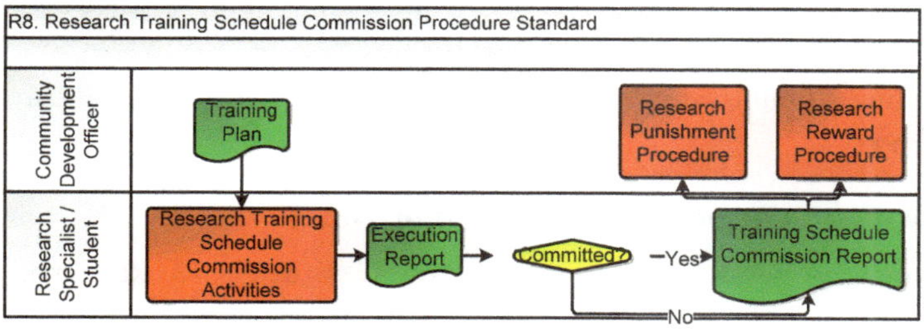

Fig. 11. Research Training Schedule Commission Procedure Standard.

Research Training Schedule Commission Activities critical success factor is

1. Every Research Training Attendance List Report is reviewed, and
2. Technology Development Manager already approves the research training schedule commission activities report draft.

Research Training Schedule Commission Activities activities is

1. Review every Research Attendance List Report, and
2. Obtain approval of Research Training Schedule Commission Report Draft from Technology Development Manager.

Research Training Schedule Commission Activities deliverables is Form R10. Research Training Schedule Commission Report.

2.4.11 Research Schedule Commission Activities

Research Schedule Commission Procedure is as in Fig 12. Researcher attendance in research lab must be documented.

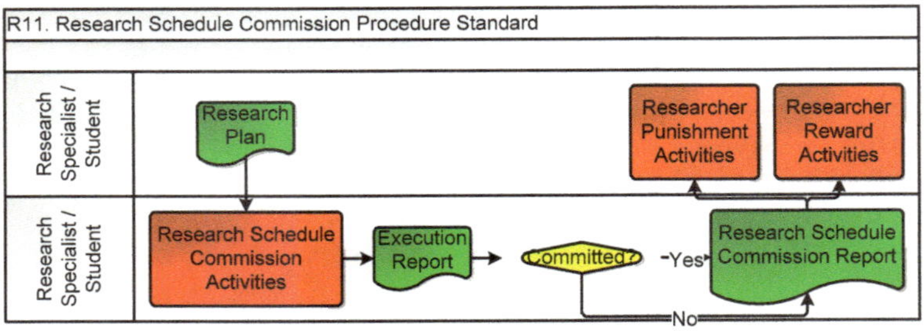

Fig. 12. Research Schedule Commission Procedure Standard.

Research Schedule Commission Activities critical success factor is

1. Every Research Attendance List Report is reviewed, and
2. Technology Development Manager already approves the research schedule commission activities report draft.

Research Schedule Commission Activities activities is

1. Review every Research Attendance List Report, and
2. Obtain approval of Research Schedule Commission Report Draft from Technology Development Manager.

Research Schedule Commission Activities deliverables is Form R11. Research Schedule Commission Report.

2.4.12 Monthly Research Specialist Monitoring Activities

Monthly Research Specialist Monitoring Procedure is as in Fig 13. Research Specialist is ought to report to Research Coordinator Monthly.

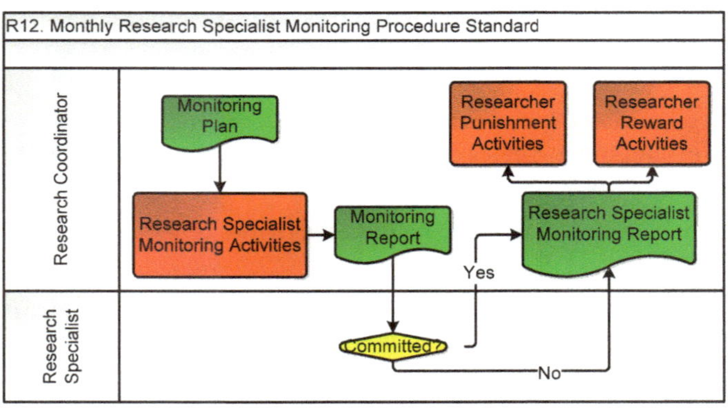

Fig. 13. Monthly Research Specialist Monitoring Procedure Standard.

Monthly Research Specialist Monitoring Activities critical success factor is

1. Every Plan for meeting is scheduled, in every meeting, every intended information from every owner is captured, and

2. Technology Development Manager already approves the Monthly Research Specialist monitoring report draft.

Monthly Research Specialist Monitoring Activities activities are

1. Schedule every Meeting, in every meeting, the intended information from every owner must be captured an documented, and
2. Obtain approval of Monthly Research Specialist Monitoring Report Draft from Technology Development Manager.

Monthly Research Specialist Monitoring Activities deliverables is Form R12. Monthly Research Specialist Monitoring Report.

2.4.13 Monthly Research Student Monitoring Activities

Monthly Research Student Monitoring Procedure is as in Fig 14. Research Student is ought to report to Research Coordinator Monthly.

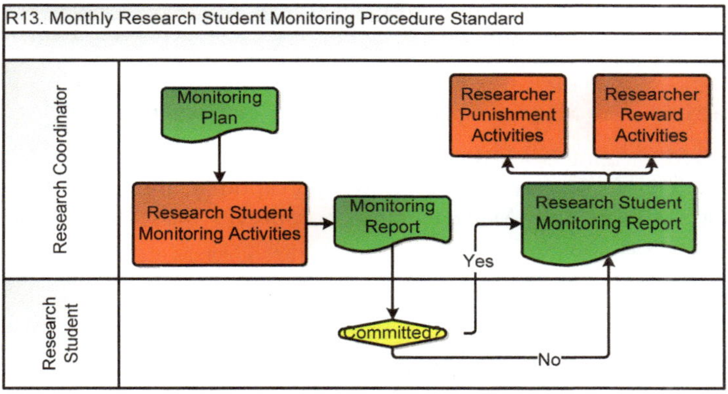

Fig. 14. Monthly Research Student Monitoring Procedure Standard.

Monthly Research Student Monitoring Activities critical success factor is

1. Every Plan for meeting is scheduled, in every meeting, every intended information from every owner is captured, and
2. Technology Development Manager already approves the Monthly Research Student monitoring report draft.

Monthly Research Student Monitoring Activities activities are

1. Schedule every Meeting, in every meeting,
2. the intended information from every owner must be captured an documented, and
3. Obtain approval of Monthly Research Student Monitoring Report Draft from Technology Development Manager.

Monthly Research Student Monitoring Activities deliverables is Form R13. Monthly Research Student Monitoring Report.

2.4.14 Weekly Research Student Monitoring Activities

Weekly Research Student Monitoring Procedure is as in Fig 15. Research Student is ought to report to Research Coordinator Monthly.

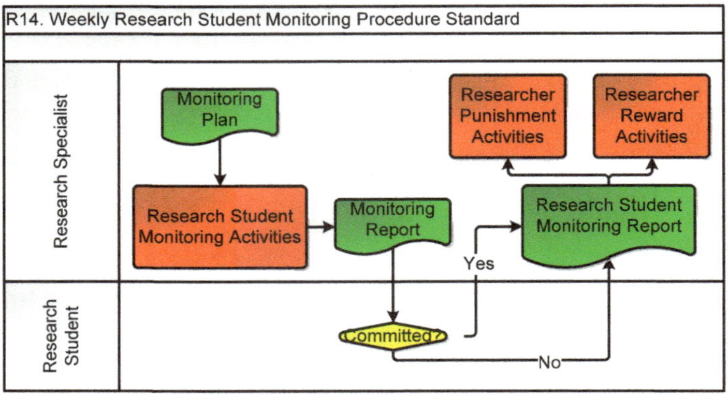

Fig. 15. Weekly Research Student Monitoring Procedure Standard.

Weekly Research Student Monitoring Activities critical success factor is

1. Every Plan for meeting is scheduled, in every meeting,
2. every intended information from every owner is captured, and
3. Technology Development Manager already approves the Weekly Research Student monitoring report draft.

Weekly Research Student Monitoring Activities activities are

1. Schedule every Meeting, in every meeting,
2. the intended information from every owner must be captured an documented, and
3. Obtain approval of Weekly Research Student Monitoring Report Draft from Technology Development Manager.

Weekly Research Student Monitoring Activities deliverables is Form R14. Weekly Research Student Monitoring Report.

2.4.15 Researcher Reward Activities

Researcher Reward Procedure is as in Fig 16. System shall pay every researcher based on their achievement in research.

Researcher Reward Activities critical success factor is

1. Every Researcher Evaluation Report is reviewed,
2. Ensure that all work done is within budget,
3. Technology Development Manager already approves the researcher reward report draft,
4. Reward and Level Advancement is granted to the deserved Researcher, and
5. Budget Balance is updated.

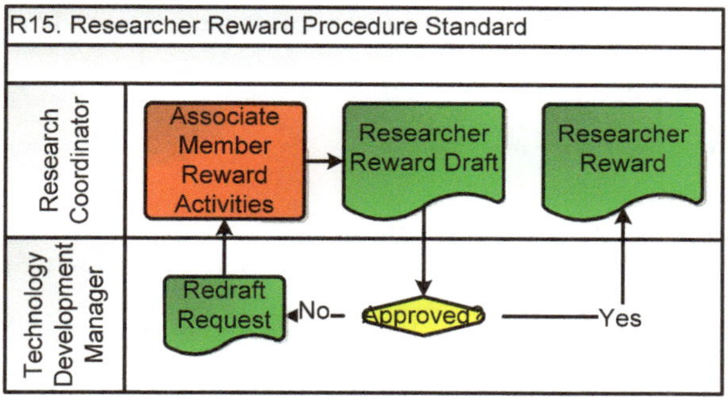

Fig. 16. Researcher Reward Procedure Standard.

Researcher Reward Activities activities is

1. Review every Researcher Evaluation Report,
2. Review available budget,
3. Obtain approval of Researcher Reward Report Draft from Technology Development Manager,
4. Pay every Researcher Reward, and
5. Update Budget Balance.

Researcher Reward Activities deliverables is Form R15. Researcher Reward Report

2.4.16 Researcher Punishment Activities

Researcher Punishment Procedure is as in Fig 17. System shall punish every researcher based on their achievement in research.

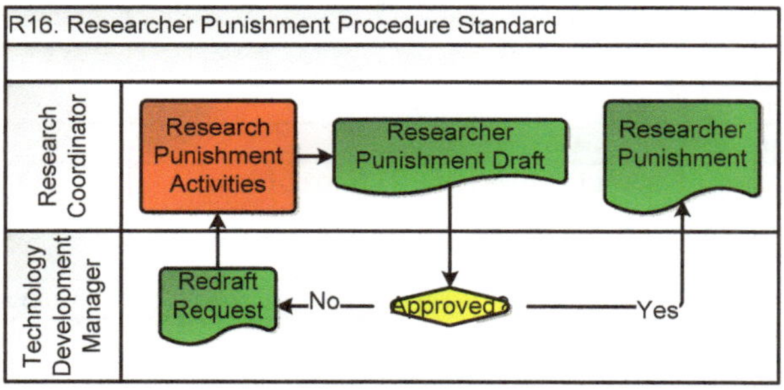

Fig. 17. Researcher Punishment Procedure Standard.

Researcher Punishment Activities critical success factor is

1. Every Researcher status is reviewed,
2. Every Researcher Evaluation Report is reviewed,
3. Verify Researcher misconduct to themselves, related Research Specialist, Research Coordinator, or other related person / group,
4. Ensure that the Researcher is deserved to be punished, If the misconduct is made not by them, they did not deserve to be punished,

5. Technology Development Manager already approves the Researcher punishment report draft, and
6. Punishment is mandated to the deserved Researcher.

Researcher Punishment Activities activities is

1. Review every Researcher status,
2. Review every Researcher Evaluation Report,
3. Verify Researcher misconduct to themselves, related Research Specialist, Research Coordinator, or other related person / group,
4. Obtain approval of Researcher Punishment Report Draft from Technology Development Manager, and
5. Punish the Researcher

Researcher Punishment Activities deliverables is Form R16. Researcher Punishment Report.

2.5 System implementation test

The CVR / IT system adoption rate is calculated using simple equation, how much activities mandated from CVR / IT already implemented on current system.

3. Result

The CVR / IT system adoption rate is as in Table 2. Furthermore, the system adoption status is in Table 3.

Activities	Adoption Status
Assign an initiating Project Manager	O
Identify the Project Sponsor	O
Define the Business Need/Opportunity	O
Identify Business Objectives and Benefits	O
Define Overall Scope	O
Define Project Objectives	O
Identify Project Constraints and Assumptions	O
Ensure Alignment with Strategic Direction and Architecture	O, but using IT Directorate Research Tree
Identify and Engage Key Stakeholders	O
Identify Key Potential Risks	O, using simple test
Procurement and Resourcing Requirements	O, handled by Binus Procurement
Determine Cost/Benefit and Schedule Estimates	O
Develop a Project Phase Exit Plan	O
Assign a Project Manager	O
Refine Project Scope	O
Determine Procurement and Sourcing Strategy	
Refine Project Schedule	
Resource Planning	O
Identify Other Resource Requirements	O
Establish Project Life-Cycle Phase Checkpoints	O, based on project

Activities	Adoption Status
Refine Project Cost Estimate and Budget	O
Identify Potential Project Risks	
Determine Process for Issue Identification and Resolution	
Develop a Change Management Process	
Develop an Organizational Change Management Approach	
Develop a Quality Management Approach	O
Develop A Project Communication Approach	
Develop A Configuration Management (CM) Approach	
Project Performance Commitment	
Consolidate the Project Plan	
Detailed Designs	O
Establish a Requirements Traceability Matrix	
Manage scope, cost, schedule and quality (Manage Change)	
Manage issues, change and risk	
Establish and report project status	O
Update Project Performance Commitment document	
Obtain approval from the Governance Body at Design Review	O
Manage Risk	
Communicate Information	O
Manage Schedule	
Document the Work Results	
Manage Organizational Change	
Manage Scope	O
Manage Quality	O
Manage Costs	O
Manage Issues	
Conduct Status Review Meetings	
Review Project Life-Cycle Phases Checkpoints	
Execute the Procurement Plan	
Administer Contract/Vendor	
Update Project Planning Documents	O
Build & Test	O
Implementation	O
Conduct Rollout Acceptance Meeting	O
Conduct Final Contract Review	O
Conduct Lessons Learned Meeting	O, but problem with documentation
Conduct Knowledge Transfer	
Post Project Review	O, with problem with different POV
Distribution of Resources	O
Celebration	O

Table 4. CVR / IT adoption current status.

Activities	Adoption Status
Research Communication Activities	O
Research Monitoring Activities	O
Research Proposal and Budgeting Activities	O
Research Budget Sounding Activities	O
Research Permission Grant Activities	O
Research Execution and Control Activities	O
Student Research Advertisement Activities	O
Student Research Proposal and Budgeting Activities	O
Student Research Execution and Control Activities	O, with problem with different POV
Research Evaluation Activities	O, with problem with different POV
Research Training Schedule Commission Activities	O
Research Schedule Commission Activities	O
Monthly Research Specialist Monitoring Activities	O
Monthly Research Student Monitoring Activities	O
Weekly Research Student Monitoring Activities	O
Researcher Reward Activities	O

Table 5. System adoption status.

There are about 54 Internship Researcher in IT Directorate until this paper is written, compared to 120 Internship Researcher needed. Their Research Speed is on par with ITD Standard. IT Directorate currently can only handle around 15 researches. All of them yields system adoption rate to around 40 %.

4. Discussion

Low adoption rate is come from several traits, based on post research observation. There some traits in Binus student, that they seldom documenting their problem, but they like to documenting their achievement in their research, so Research Specialist sometimes late in recognizing problems in research.

The IT Directorate research program still unpopular in student body, due to fear to fail in research. Further research is needed to alleviate this problem.

Preservation of Research tree is still uncommon in Binus University by the time this paper is written. IT Directorate must establish its own research tree, based on Research Directorate inputs.

Different Point of View (POV) between IT Directorate and IT Department does ignite problem, mostly in how the project conducted and roll out of project. Some monitoring and documentation activities and rules cannot be done due conflict with IT Department rules.

To alleviate these problems, writer suggested to start from changing student paradigm, start from early semesters. The class should be shaped to expose student to project development culture, where they learn to address real world problem, came up with solution, and essentially, to work together and to track progress, because most project fail in the last two items.

Several alternative solutions can be made. First is to encourage project based activities on class. Although this model can be considered artificial, at least that could train student responsibility on keeping with documentation and schedule.

Second is, to bridge the class to real life industries. The first variant is to get industrial agreement to be the sand box for class project. The second variant is to us real life project as either the complement or substitute of relevant credit.

The third one is, to actually shape the class as a software factory, where student accomplishment measured by task and project based style. The project should come from out school. Another variant of this model is to consider the whole university as a software factory, where the upper classes act as analyst and managers, and lower classes as programmers.

Only when student understand the project environment and the consequences behind it, the CVR / IT adoption rate (or, any project management standard in general) shall rise.

5. Conclusion

Based on observation, the system adoption rate is 40 % at most. Further activity is needed to raise the value to 100 %, addressing how to reduce fear in student, and how to manage research documentation and monitoring.

Solution made should aim to introduce and raise student awareness to project management environment o early semesters.

6. Acknowledgement

The author would like to thank Mr. Gintoro FX, this system is based on his idea to expand current associate member system to accommodate IT Directorate needs. This work will not taken place to begin with if without his idea.

The author would like to Thanks Mr. Eddy Santosa Jaya, Head of Information Technology Directorate, and everyone within the directorate to support the program adoption and let the study takes place.

7. References

CVR/IT Consulting LLC. (2002). *The Project Management Template Library.* Retrieved January 12, 2009, from CVR/IT Consulting: http://www.cvr-it.com/PM_Templates/

Alba, E., & Chicano, J. F. (2006, December 24). *Software project management with GAs.* Retrieved January 3, 2009, from KUEC: http://kuec.sci.ku.ac.th/KM/418571/SW%20Engineer/SWProjMangwGA.pdf

Boehm, B. W., & Ross, R. (1989). *Theory-W Software Project Management: Principles and Examples.* Retrieved February 28, 2009, from CiteSeerX beta: http://citeseerx.ist.psu.edu/viewdoc/download?doi=10.1.1.105.2928&rep=rep1&type=pdf

Callahan, J., & Ramakrishnan, S. (2007). *Software Project Management and Measurement on the World-Wide-Web (WWW).* Retrieved January 26, 2009, from CiteSeerX beta: http://citeseerx.ist.psu.edu/viewdoc/download?doi=10.1.1.30.2191&rep=rep1&type=pdf

Goebel III, J. C. (2003). *Extreme Programming Practices Used to Facilitate Effective Project Management.* Retrieved January 13, 2009, from Menlo Innovations: http://www.menloinnovations.com/freestuff/whitepapers/Project%20Management.pdf

Jalote, P. (2007). *Knowledge Infrastructure for Project Management.* Retrieved January 12, 2009, from CiteSeerX beta: http://citeseerx.ist.psu.edu/viewdoc/download?doi=10.1.1.96.7171&rep=rep1&type=pdf

Liu, X., Kane, G., & Bambroo, M. (2003). *An Intelligent Early Warning System for Software Quality Improvement and Project Management.* Retrieved january 12, 2009, from Scholars' Mine: http://scholarsmine.mst.edu/post_prints/pdf/01250167_09007dcc8030d05e.pdf

Paul, R. A., Kunii, T. L., Shinagawa, Y., & Khan, M. F. (1998, September 21). *Software Metrics Knowledge and Databases for Project Management.* Retrieved January 12, 2009, from College of Engineering: http://cobweb.ecn.purdue.edu/~ee461/Paul-Metrics.pdf

Quantitative Software Management Associates. (2004). *Slim Suite Tools for Software Management.* Retrieved January 23, 2009, from Quantitative Software Management Associates: http://www.qsma.com/docs/SLIMSuite6.0.pdf

Reid, K. L., & Wilson, G. V. (2007). *DrProject: A Software Project Management Portal to Meet Educational Needs.* Retrieved January 21, 2009, from ACM Portal: http://portal.acm.org/citation.cfm?id=1227421&dl=GUIDE&coll=GUIDE&CFID=79879641&CFTOKEN=91755329

Sukhoo, A., Bernard, A., Eloff, M. M., Poll, V. d., & A., J. (2008). *Accommodating Soft Skills in Software Project Management.* Retrieved January 12, 2009, from Informing Science Institute: http://informingscience.org/proceedings/InSITE2005/I55f42Sukh.pdf

True, M. (2007). *Starting and Maintaining a Quality Internship Program.* Retrieved January 21, 2009, from Messiah College: http://home.messiah.edu/~mtrue/documents/Starting_And_Maintaining_A_Quality_Internship_Program.pdf

21

Optimization of Optical Inspections Using Spectral Analysis

K. Ohliger, C. Heinze and C. Kröhnert
Westcoast University, Heide
Institute Machine Vision Technology
Germany

1. Introduction

Machine vision is proving its potential to be an efficient tool for increasing product quality and for decreasing manufacturing costs. Many applications use automated optical inspection systems for measuring mechanical dimensions and for monitoring quality features of manifold products. Optical inspection systems consist mainly of a handling system, a vision system and a user interface. The vision system can be divided into an acquisition and processing part. While there exists much literature about image processing there is less attention spent for designing an image acquisition system. The system accuracy and processing speed correlates directly to image quality defined as the contrast between the different features (e.g. objects) in the image. Caused by the increasing complexity and difficulty of the inspection tasks and the demand for increasing throughputs, image quality gets more important for reliable vision systems. This chapter deals with spectral methods for the selection of hardware components within the acquisition system to optimize image quality. We show that our analytical methods will increase image quality and decrease classification error rates of a vision system.

The outline of this chapter is as follows. The chapter is divided into four sections. Section 2 gives a short introduction into image acquisition and introduces the components involved and the modelling of their spectral characteristics afterwards. This includes the lightsource, the reflection behavior of materials, the optics and the sensor. The section concludes with a model developed to describe the feature of image pixels due to the acquisition chain.

Section 2.2 explains the methods developed how to select the acquisition components for the visible and infrared wavelength range. This optimization is done by using different statistical approaches resulting in two types of image quality measures.

Experimental results for simulated and real data are presented in Section 4. It will be shown that our approach of selecting the hardware components leads to a much lower error rate of the respective inspection task.

This chapter concludes with a summary and discussion of the results.

2. Image acquisition

The image acquisition chain is shown in Fig. 1. The resulting image is influenced by four different components: the lightsource, the observed scene, the optics and the camera. The incident light is emitted by a lightsource which has a specific spectral power distribution. The observed scene includes different objects reflecting the incident light depending on spectral reflection behavior. The reflected light is focussed by optic components transmitting the light according to their optical transmission characteristics. Finally the camera generates an image using a sensor element which has a wavelength dependent degree of efficiency.

In Sections 2.1 to 2.4 the different types of these components and their characteristics with respect to their spectral behavior are shown. Section 2.5 introduces a model for the acquired image with respect to the components included in the image acquisition chain.

2.1 Lightsource

The lightsource within a vision system plays an important role. Primary properties of lightsources are intensity, propagation direction, frequency or wavelength spectrum, and polarization. Lightspeed in vacuum, $\approx 3 \cdot 10^8$ meters per second, is one of the fundamental constants of nature. With respect to machine vision tasks the relevant information of the light emitted by a lightsource is the total radiant intensity I_{total} and the spectral radiant intensity distribution $p_{I_{\text{total}}}$ at a point x, y, z of the observed scene resulting in the wavelength dependent radiant energy I_L of the incident light:

$$I_L\left(\lambda; x, y, z\right) = p_{I_{\text{total}}}\left(\lambda\right) \cdot I_{\text{total}}\left(x, y, z\right). \tag{1}$$

The assumption that $p_{I_{\text{total}}}$ is spatially invariant is valid in most of the machine vision tasks.

If the intensity of light is graphed over the corresponding wavelengths, the spectral behavior of a lightsource can be analyzed. Fig. 2 shows the spectrum of light that is emitted by the sun. The figure shows that the human eye can only perceive a small amount of the radiation emitted by the sun. The visible spectrum for a typical human eye corresponds to a range from about 390nm to 750nm (Wald (1945)). A rendering of this spectrum is shown in Fig. 3. Sunlight is rarely used as a primary lightsource in a vision system. The most common lightsources are incandescent light bulbs, gas-discharge lamps and light emitting diodes. Light emission by incandescence can be designed to have the same type of emission by which sunlight is

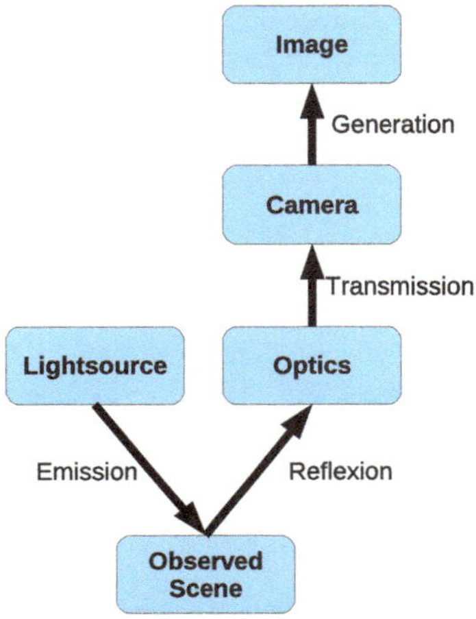

Fig. 1. Image acquisition chain

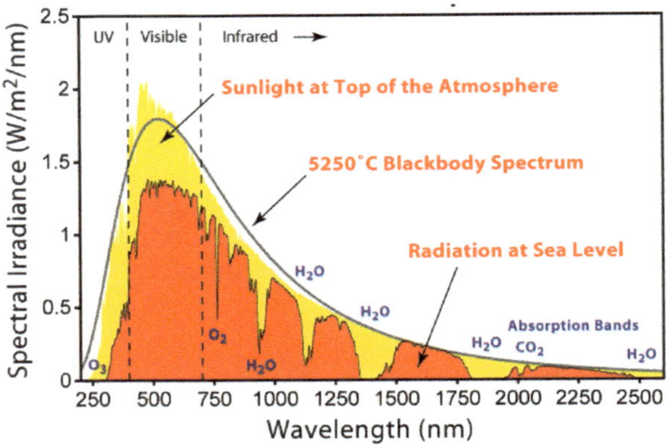

Fig. 2. Spectrum of sunlight (outside atmosphere and at sea level) (Wiki (2007))

created. Both the surface of the sun and the metal filament in an incandescent light bulb glow because of their high temperature (thermal radiation). All materials glow if they are above a temperature of $0K$, but below the Draper Point ($798K$) the intensity is too small to be perceived with the human eye. A model of the emitted thermal radiation can be found by using an idealized material called a black body. A black body is both a perfect absorber and emitter of radiation. The spectrum of the emitted radiation depends on the temperature only and can be expressed by Planck's Law:

$$u(\lambda, T) = \frac{2hc^2}{\lambda^5} \frac{1}{e^{\frac{hc}{\lambda kT}} - 1}. \tag{2}$$

The intensity of the emitted light u at a certain wavelength λ and temperature T can be calculated as shown above with h as Planck's constant, c as the speed of light in a vacuum and k as the Boltzmann constant. Fig. 4 shows the spectra of blackbody radiation at different temperatures. These spectra show the general characteristics for the spectrum emitted by the sun ($T = 5500K$) and that of an incandescent bulb ($T = 3500K$). The spectrum emitted by thermal radiation is always continuous and covers all frequencies of the visible spectrum and beyond. In vision systems, incandescent bulbs are used as a lightsource with a broadband spectrum. This means that the intensity of the light reflected from the object is mostly dependent on the material of the object and not on the spectrum of the lightsource only.

Another type of lightsource often used in vision systems is the gas-discharge lamp. Several different types of these lamps exist, such as neon lamps, sodium-vapor lamps and fluorescent lamps. All of these lamps emit radiation due to the same principle. The atoms of the gas inside the lamp are ionized by electrical energy. This means that some of the electrons are free to move from their atoms. These free electrons are accelerated by an electrical field within the lamp and will collide with other atoms. Some electrons in the atomic orbitals of these atoms are excited to a higher state of energy by these collisions. When the excited atom falls back to a lower state of energy, it emits a photon of a characteristic energy, resulting in infrared,

Fig. 3. Visible spectrum of electromagnetic radiation (Wiki (2010))

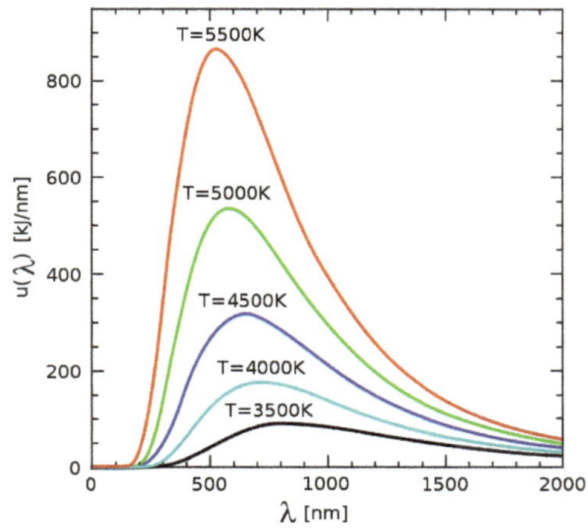

Fig. 4. Spectra of blackbody radiation at different temperatures (Wiki (2009))

visible, or ultraviolet radiation. Since the wavelength λ of a photon is related to it's energy E

$$\lambda = \frac{h}{E}, \tag{3}$$

these lamps can emit light in very specific parts of the spectrum.

The spectrum of a low pressure sodium-vapor lamp is given as an example for this smallband emission of radiation in Fig. 5. The light emitted by this lamp is almost monochromatic and has a characteristic yellow-orange color. The characteristic energy which is responsible for this behavior can be altered by using different gases in the lamps. Fluorescent lamps use a mercury vapor that emits radiation in the ultraviolet part of the electromagnetic spectrum. This radiation will excite a fluorescent coating inside the lamp, which in turn emits light over a broader part of the visible spectrum. The spectrum of such a fluorescent lamp is shown in Fig. 6. The spectrum is characterized by many spectral lines, some of which are due to the mercury vapor, while others are due to the fluorescent coating. In total the light emitted by this lamp is perceived as white.

Light emitting diodes (LED) are used more and more frequently in vision systems. This type of lightsource uses the recombination of electrons and holes in a semiconductor to emit radiation. The wavelength of the emitted radiation depends on the energy gap between the

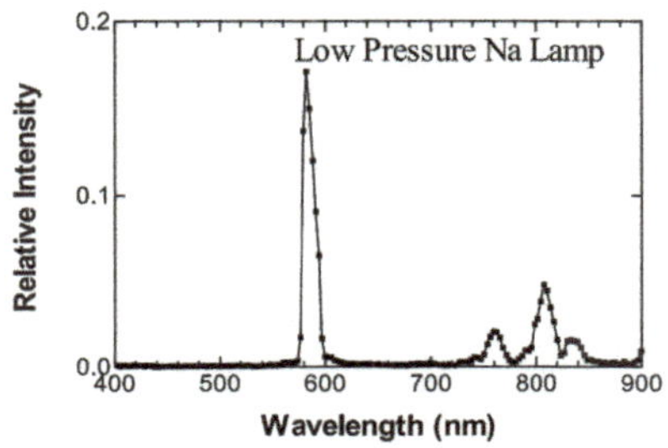

Fig. 5. Spectrum of a low pressure sodium-vapor lamp (Rechtsteiner (1998))

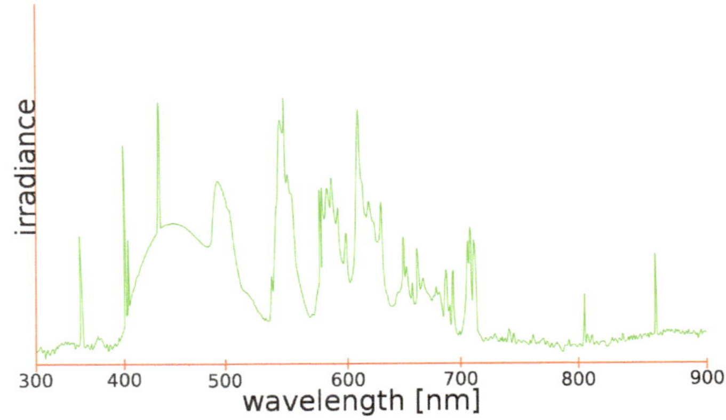

Fig. 6. Spectrum of a fluorescent lamp (Wiki (2011))

electron-energy and the hole-energy. Using different materials, this gap can be designed to specific values. By this, the color of the LED can be influenced. The first LEDs developed in the 1960s were red or infrared types. Modern LEDs are available in several colors across the visible spectrum, ultraviolet and infrared wavelengths. The resulting spectra of these different LEDs usually have small bandwidths. Fig. 7 shows the spectra of several color LEDs.

White LEDs can be designed in two different ways. The combination of three LEDs (blue, green and red) with the spectra shown in Fig. 7 will make the LED appear white. If all three LED types can be electrically contacted individually, the color of the LED can be changed during use, which can be beneficial in certain applications.

The other way to design a white LED is to use a phosphor coating. The phosphor coating shifts a part of the emitted light to a higher wavelength. This is shown in Fig. 8, where the small peak at the shorter wavelength is the original light emitted by the LED, and the broader peak at longer wavelengths is due to the phosphor coating.

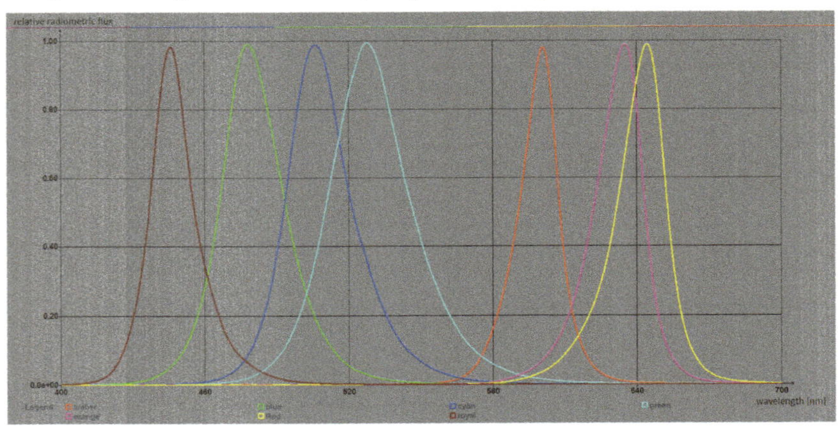

Fig. 7. Spectra of different single colour LEDs

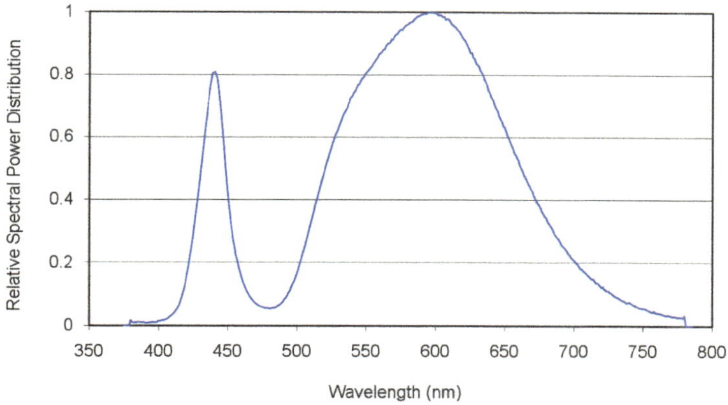

Fig. 8. Spectrum of white LED with original peak at lower wavelengths and broader, shifted peak at higher wavelengths (Philips (2011))

2.2 Observed scene

An observed scene includes one or more objects that have to be inspected by the vision system. The knowledge of the reflection behavior of these objects will give us the possibility to choose optimal components for the vision system. In this Section different models for the reflection of surfaces are introduced.

2.2.1 Lambert's model

Lambert's Model, which was advanced by Johann Heinrich Lambert in 17th century (Lambert (1760)), is the basic of the most reflection models and remains one of the most widely used models in computer graphics (Oren & Nayar (1994)). If a surface reflects the incoming light equally in all direction, the surface is called a Lambertian reflector (a perfect diffuse reflector). This means the luminosity of the surface is independent of angle of the observation. According the Lambert's cosine law, the reflected luminosity I_d depends on the cosine of the angle α between the direction of the incoming light $\vec{L}$ and the normal of the surface $\vec{N}$ (see Fig. 9). With increasing angle the intensity of the reflected light decreases:

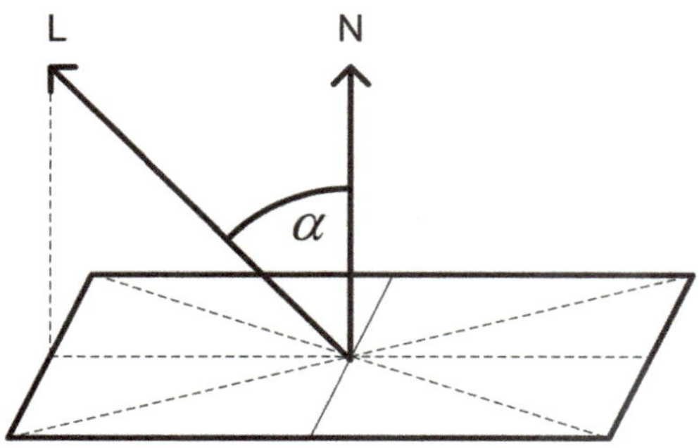

Fig. 9. Lambert law

$$I_d = I_L \cdot k_d \cdot \cos\alpha, \tag{4}$$

with

$$\cos\alpha = \frac{\vec{N} \cdot \vec{L}}{\left|\vec{N}\right| \cdot \left|\vec{L}\right|}. \tag{5}$$

Hence, the luminosity of a Lambertian surface depends on the normal vector $\vec{N}$ and the direction of the normalized vector $\vec{L}$ in the direction between the light and the surface, and also from the diffuse reflection material coefficient k_d.

Using (4) and (5) assuming that $|L| = 1$, $|N| = 1$ leads to

$$I_d = I_L \cdot k_d \cdot \left(\vec{N} \cdot \vec{L}\right). \tag{6}$$

This model can be extended by regarding indirect light with an ambient term. The ambient component is independent from the direction of the incoming light and the direction of the viewer. The term depends on an ambient light I_a and the ambient-material coefficient k_a and is homogeneous over the complete scene

$$I_{Refl,LM} = I_a \cdot k_a + I_L \cdot k_d \cdot \left(\vec{N} \cdot \vec{L}\right). \tag{7}$$

2.2.2 Blinn-Phong model

Since the introduction of computers and in particular 3D graphics, the light models became more and more important. Bui-Tuong Phong extended the Lambert's light model with a component that simulates light that is reflected in a certain direction (Eriksson (2006)). This component is called specular component of light reflection. The model was presented 1975, nearly two hundred years after the publication of Lambert's model. The main aspects of the Phong model are

- light sources are point-like
- the geometry of the surfaces unless the surface normal is ignored
- diffuse and specular reflection are modeled locally
- ambient reflection is modeled globally.

The Blinn-Phong Model is derived from the Lambert's Model by adding a specular term to (7)

$$I_{Refl,BP} = I_a \cdot k_a + I_d + I_s. \tag{8}$$

Where I_a and I_d are the ambient and the diffuse components from the Lambert's model. The effect of the specular term I_s can be seen for example when a billiard ball is illuminated by single non-diffuse light. The light source produces a spot on the ball, which varies with the position of the viewer. Reason for this effect is that surfaces are not perfect mirrors. A perfect mirror would reflect the incoming light only in one direction, so that a viewer would see the reflection only in one direction. In practice the light intensity is a more complicated function of the angle ϕ between the reflected light ray $\vec{R}$ and the view angle (Eriksson (2006)).

Phong expressed an empirical approach to model the amount of reflected light that is described by

$$I_s = I_L \cdot W(\alpha) \cdot C_p \cdot [\cos(\phi)]^n \tag{9}$$

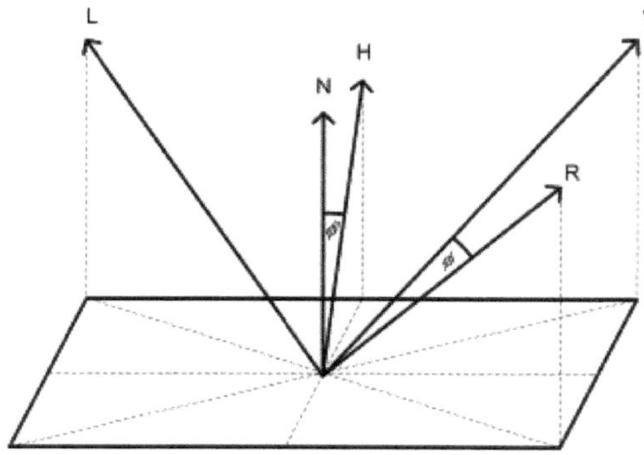

Fig. 10. Halfway vector H

Where $W(\alpha)$ is the ratio of reflected light and incident light as function of the incident angle α. Phong determined the range of $W(\alpha)$ in percent, varying from 10 to 80 percent. ϕ presents the angle between the reflected light $\vec{R}$ and the vector $\vec{V}$ (the direction of the viewer). The parameter n describes the characteristics of the surface, Phong used values between one and ten. For a highly reflecting surface the parameters n and $W(\alpha)$ are large. C_p is the reflection coefficient of an object at Point P for a certain wavelength.

Phong's original equation is also often written (Duff (1979), Lewis (1994), Kajiya (1989)) as:

$$I_s = I_L \cdot k_s \cdot (\cos\phi)^n , \tag{10}$$

with k_s as specular material coefficent and

$$\cos\phi = \frac{\vec{R}\cdot\vec{V}}{\left|\vec{R}\right|\cdot\left|\vec{V}\right|} . \tag{11}$$

This leads to

$$I_s = I_L \cdot k_s \cdot (\vec{R}\cdot\vec{V})^n , \tag{12}$$

under the assumption that $|V|$ and $|R| = 1$.

The Phong light model was extended by James Frederick Blinn 1977. He reformulated the Phong model by a physically based light model. He introduced the half-vector $\vec{H}$, instead of the $\vec{R}$. $\vec{H}$ is defined as the vector halfway between the incoming light- and the viewer vector:

$$H = \frac{\vec{L}\cdot\vec{V}}{|\vec{L}+\vec{V}|} . \tag{13}$$

In addition Blinn defines ϕ' as the angle between the normal vector of the object surface and the halfway vector

$$\cos\phi' = \frac{\vec{N}\cdot\vec{H}}{\left|\vec{N}\right|\cdot\left|\vec{H}\right|} , \tag{14}$$

which is proportional to the angle ϕ. Thus the angle ϕ could be replaced by the angle ϕ' (see Fig. 10):

$$I_s = I_L \cdot k_s \cdot (\cos\phi')^n, \tag{15}$$

with (14) resulting in:

$$I_s = I_L \cdot k_s \cdot (\vec{N} \cdot \vec{H})^n. \tag{16}$$

Reason for this substitution is that the halfway vector can be simpler calculated than the reflection vector.

The overall reflected intensity can hence be determined by:

$$I_{Refl,BP} = I_a \cdot k_a + I_L(k_d \cdot (\vec{N} \cdot \vec{L}) + k_s \cdot (\vec{N} \cdot \vec{H})^n). \tag{17}$$

2.2.3 Cook-Torrance model

The disadvantage of the light models from Blinn and Phong is that the models have no physical basis. Robert L. Cook and Kenneth E. Torrance developed the Cook-Torrance light model in 1981 on a physical model. The model describes the surface by many little facets (mirrors), which allows to consider the refraction ratio, the roughness and shadowing. Torrance describes the reflectivity in He (2001), as follows:

$$I_{Refl,CT} = I_{a,CT} + I_{d,CT} + I_{s,CT}. \tag{18}$$

In the equation $I_{a,CT}$ describes the ambient-, $I_{d,CT}$ the diffuse- and $I_{s,CT}$ the specular reflectivity. The following equations present the content of this three terms:

$$I_{a,CT} = a(\lambda). \tag{19}$$

$a(\lambda)$ is the ambient term, which is like the other light models uniform diffuse over the scene.

The diffuse reflected light is expressed as:

$$\begin{aligned} I_{d,CT} &= I_L \cdot \frac{|F|^2}{\pi} \cdot \frac{G \cdot S \cdot D(\sigma_{rough}, \lambda)}{(\vec{N} \cdot \vec{V})(\vec{N} \cdot \vec{L})} \\ &= I_L \cdot \frac{k_{d,CT}\left(\lambda, \sigma_{rough}\right)}{(\vec{N} \cdot \vec{V})(\vec{N} \cdot \vec{L})}. \end{aligned} \tag{20}$$

Parameter G is the geometry attenuation factor and S the shadowing function, both together describing the selfshadowing due to the microfacets. D is the distribution function of the roughness which is based on a physical model of microfacets distributions. D is dependent on the wavelength. $k_{d,CT}$ contains the influence of the material and its surface attributes. The ratio of refraction of the surface, will be described by the Fresnel function F, which presents the reflection and refraction of light at uniform planar interfaces.

The specular part is

$$I_s = I_L \cdot \frac{|F|^2 \cdot e^{-g(\lambda)} \cdot S}{\cos(\Theta_i) d\omega_i} \cdot \Delta, \tag{21}$$

where g is the surface roughness function, which is also dependent on the wavelength. In (21) the parameter $d\omega_i$ describes the incident solid angle and Θ_i is the polar angle, which is the angle between normal vector and the incident light vector. Δ describes the specular influence and is one if in specular cone otherwise it is zero (He (2001)).

Material	Transmission range	Features
BK7	330nm - 2100 nm	High transmission for visible to near infrared applications, the most common optical glass
UV Grade Fused Silica	185nm - 2500nm	Excellent homogeneity and low thermal expansion, high laser damage resistance
CaF_2	170nm - 8000nm	High transmission for deep UV to infrared applications
MgF_2	150nm - 6500nm	Birefringent material, excellent for use in the deep UV to infrared
ZnSe	600nm - 16000nm	Excellent choice for IR lens due to its broad wavelength range.

Table 1. Transmission range and further features of different glass materials

2.3 Optics

The optics of a vision system have strong influence on many image characteristics e.g. the spatial resolution, depth of focus, image intensity. Additionally the contrast between different objects depends on the spectral behavior of the optics. This Section will focus on the variation of spectral transmission for optical materials and lens types. In general two different aspects influence the spectral transmission of optics: the lens material and the lens alignment and coating. Hence the transmission of the optics t_{opt} can be described as:

$$t_{opt}(\lambda; x, y) = t_{mat}(\lambda; x, y) \cdot t_{alig}(\lambda; x, y) \ . \tag{22}$$

While t_{mat} is the transmission of the lens material, t_{alig} includes the transmission influenced by the coating and alignment. x, y indicates the position of light beam entering the optical system.

2.3.1 Spectral transmission of different lens materials

In optics, transmission is the property of a medium allowing light to pass through the medium with some or none of the incident light being absorbed or reflected during the process. Different glass materials have different transmission ranges, allowing light to pass at different wavelengths. Standard optical glasses offer high transmission throughout the entire visible spectrum and also the near ultraviolet and near infrared range (Newport (2011)). In the following, the spectral transmission of different glass materials for optical lenses is shown. As Table 1 and Fig. 11 show, the spectral transmission of different materials vary especially outside of the visible spectrum. When glass elements are combined to form a photographic lens, the spectral transmission changes.

2.3.2 Spectral transmission of different lens types

The degradation of spectral bandwidth is mostly due to antireflective coating, which is specifically designed to block ultraviolet or infrared radiation. A typical photographic lens is optimized for the visual spectrum of electromagnetic radiation. In the following, the spectral transmittance of different photographic lenses are shown by real-world examples.

First, the spectral transmission of a Schneider Optics Cinegon 1.4/8mm is shown in Fig. 12 (a). The Cinegon is used as an example for all standard purpose photographic lenses. The diagram shows a relatively flat transmittance between 400nm and 700nm. Since this is a prime lens with a fixed focal length, a relatively simple design with six glass elements can be used. The transmittance of a more complex photographic lens will usually show a different characteristic, caused by an increasing number of glass elements.

Fig. 12 (b) shows the spectral transmittance of a Schneider Optics Variogon 1.8/12.5-75 zoom lens. This lens uses an array of 16 glass elements instead of the six elements used in the Cinegon 1.4/8mm shown above. This transmittance diagram is used as an example of a more complex photographic lens, that has not been optimized for a greater transmission bandwidth. In comparison it can be seen, that the peak transmittance is about 15% lower than that of the simpler lens design. Also the transmittance has a smaller bandwith, with lower transmittance values below 450nm or above 600nm.

For optical imaging outside the visual spectrum, special photographic lenses should be used. These are usually optimized on the infrared or the ultraviolet part of the electromagnetic spectrum. Fig. 13 (a) shows the spectral transmittance of a Jenoptik CoastalOpt UV 4.5 105mm lens. The usable transmission range starts at 250nm and declines again at 650nm. This lens mostly covers the ultraviolet spectrum, which can be useful for defect detection (Richards (2006)).

Next, an example for a lens covering the infrared spectrum is given. The lens in this example (Fig. 13 (b)) is a Jenoptik CoastalOpt Hyperspectral 2/25mm SWIR lens. The largest part of usable transmittance is above the usable range of UV lens mentioned before. Usable range with a transmittance above 80% starts at 400nm and ends at 1650nm.

2.4 Camera

Finally the image is generated by a camera using an image sensor. Depending on the wavelength, different types of sensors are used. The appropriate sensor types for ultraviolet to infrared wavelengths are introduced in this Section. Some differences between sensor types

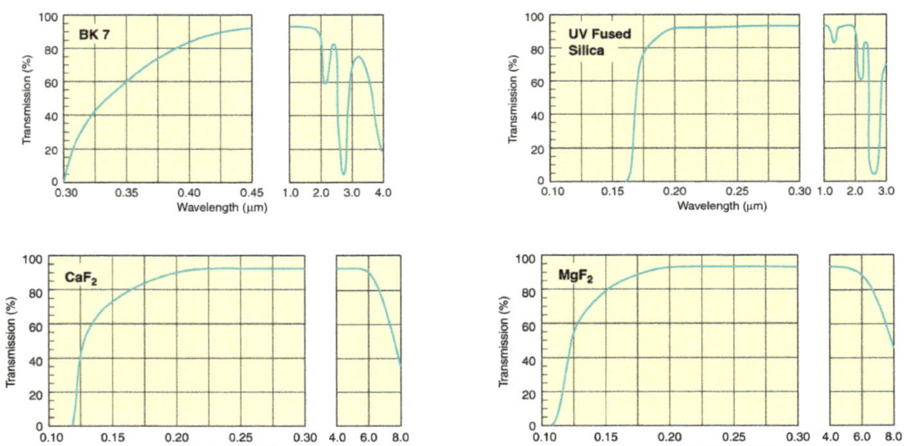

Fig. 11. Spectral transmission of optical materials (Newport (2011))

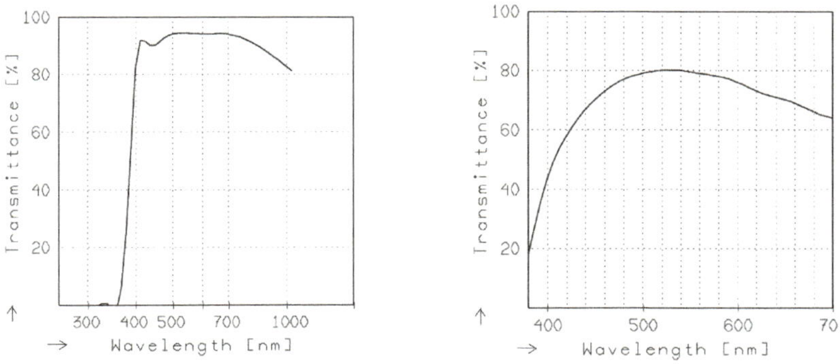

(a) (b)

Fig. 12. Spectral transmittance of Schneider Optics Cinegon 1.4/8mm (a) and Schneider Optics Variogon 1.8/12.5-75mm (b) (Schneider (2011), Schneider (2011b))

as well as the variation for equal sensor types are shown. The key aspect for the spectral behavior of an image sensor is its quantum efficiency (QE). The $QE(\lambda)$ is an attribute which describes how many photons are transformed to electrical charge carriers depending on the wavelength:

$$QE(\lambda) = \frac{n_e}{n_{ph}(\lambda)}. \tag{23}$$

n_e is the count of electrical charge carriers, generated by n_{ph} photons with a wavelength λ.

2.4.1 Quantum efficiency of image sensor materials

Digital image sensors are based on the transformation of electromagnetic waves energy transported by photons to electrical energy. The energy E_{ph} of a photon can be described by

$$E_{ph}(\lambda) = \frac{h \cdot c}{\lambda}, \tag{24}$$

where λ is the corresponding wavelength, h is Planck's constant and c is the speed of light. This means that different wavelengths of light have different energies, with the infrared part of the spectrum being lower in energy than the ultraviolet part. (23) and (24) can be combined

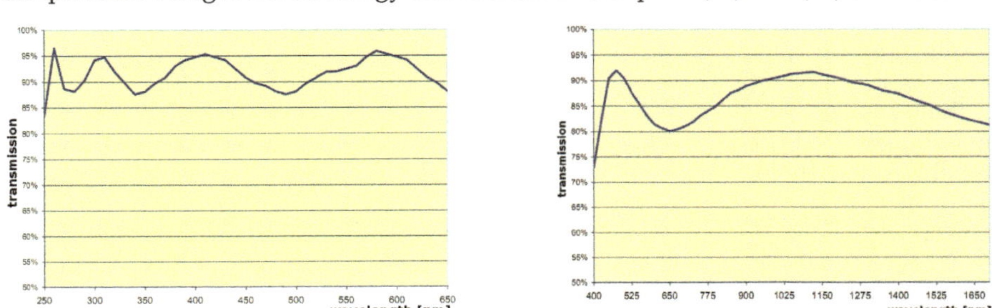

(a) (b)

Fig. 13. Spectral transmittance of Jenoptik CoastalOpt UV 4.5/105mm (a) and Jenoptik CoastalOpt Hyperspectral 2/25mm (b) (Jenoptik (2007), Jenoptik (2011))

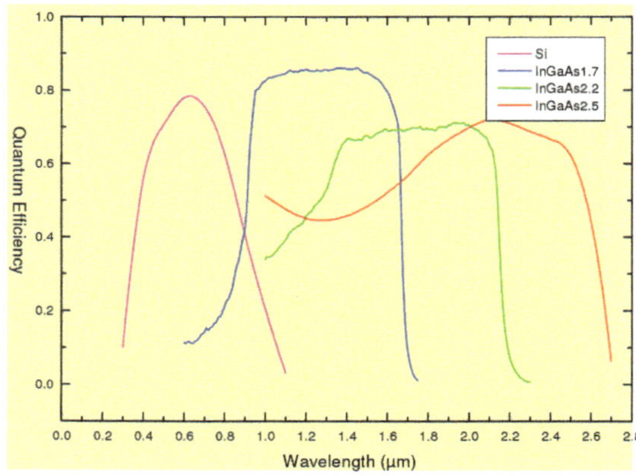

Fig. 14. Spectral sensitivity of different sensor materials: Si (violet), InGaAs1.7 (blue), InGaAs2.2 (green), InGaAs2.5 (red) (Goodrich (2006))

for describing the energy efficiency (EE) of a sensor:

$$EE(\lambda) = \frac{n_e}{n_{ph}(\lambda) \cdot E_{ph}(\lambda)}. \tag{25}$$

Digital image sensor cells use semiconductors to detect energy in different parts of the electromagnetic spectrum and hence different wavelengths of light. The part of the spectrum where a sensor cell is sensitive to light energy is determined by the material (Fig. 14) This diagram shows the quantum efficiency for different materials. A high quantum efficiency over a large frequency range is desirable, so that as much energy as possible can be converted from light to electrical charge. As with lenses, the properties of the sensor materials will determine the maximum range in which a sensor is applicable.

2.4.2 Quantum efficiency of different sensor types

While the quantum efficiency of a single image sensor cell is determined by the factors above, the efficiency for a whole chip is additionally influenced by the fill rate. The fill rate states the percentage of the sensor surface is used to detect light. The fill factor only influences the quantum efficiency as a constant factor.

In the following, the quantum efficiency of different image sensors is shown by real-world examples. First, the quantum efficiency of a Basler A622f camera (Basler (2011)) is shown in Fig. 15 (blue curve). The diagram shows, that the quantum efficiency peaks at 25% between 500nm and 600nm wavelength. This corresponds roughly to the green and orange parts of visible light. While the maximum efficiency of 25% is rather low, this can be countered by increasing the light intensity on the object. In contrast to this, Fig. 15 (yellow curve) shows the quantum efficiency diagram of a Basler A601f camera. This diagram shows a higher quantum efficiency over the whole wavelength range, with a peak of over 30% at 600nm. Not only the peak efficiency is higher, the bandwidth is also higher. Though these two cameras can be considered as relatively common and state of the art, the better quantum efficiency of the latter model has a great impact on image noise in low light situations.

2.5 Image output model

This Section introduces the image output model which will describe the image with respect to the components of the image acquisition chain. Our image output model assumes that the incident light of the observed scene is homogeneous in spectral intensity distribution and total radiant intensity. While the first is nearly common, requires the ladder a light source which illuminates the field of view homogeneously and the distance of the lightsource to the observed scene is nearly constant

$$\forall_{x,y}\ z \approx \bar{z}, \tag{26}$$

with $\bar{z}$ as mean object distance. This simplifies (1) to

$$\widetilde{I_L}(\lambda) = p_{I_{\text{total}}}(\lambda) \cdot I_{\text{total}} \, . \tag{27}$$

Considering only passive objects without any light emission the ambient term of the reflected light has negligible influence. Since the specular reflection is mainly concentrated in a narrow cone small changes in the surface direction will lead to a high variation in the reflected amount of light. This is not appropriate for machine vision solutions. Hence we concentrate on diffuse reflection and (18) using (27) of the Cook Torrance model leads to:

$$I_{Refl}\left(\lambda; \sigma_{rough}; \vec{N}, \vec{V}, \vec{L}; x, y\right) = \widetilde{I_L}(\lambda) \cdot \frac{k_{d,CT}\left(\lambda, \sigma_{rough}; x, y\right)}{(\vec{N} \cdot \vec{V})(\vec{N} \cdot \vec{L})} \, . \tag{28}$$

Further we assume that the roughness of each object is a material constant which does not vary significantly within the object region leading to:

$$\tilde{I}_{Refl}\left(\lambda; \vec{N}, \vec{V}, \vec{L}; x, y\right) = \widetilde{I_L}(\lambda) \cdot \frac{\tilde{k}_{d,CT}(\lambda; x, y)}{(\vec{N} \cdot \vec{V})(\vec{N} \cdot \vec{L})} \, . \tag{29}$$

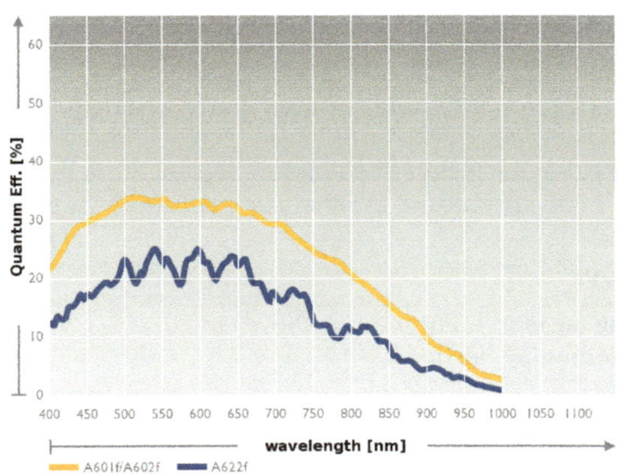

Fig. 15. Quantum efficiency of Basler A601f, A602f (yellow) and Basler A622f (blue) (Basler (2011))

The dependency on x, y of $k_{d,CT}$ is caused by different objects in the observed scene depending on the position. Hence different objects may result in varying $k_{d,CT}$. The reflected intensity I_{Refl} is attenuated by the optics. In case of optics which can be described by small lens model the spatial dependency in (22) has negligible influence. This reduces the intensity transmitted by the optics I_{Opt} combining (29) and (22) to

$$I_{Opt}\left(\lambda;\vec{N},\vec{V},\vec{L};x,y\right) = I_{Refl}\left(\lambda;\vec{N},\vec{V},\vec{L};x,y\right)\cdot\tilde{t}_{opt}\left(\lambda\right). \quad (30)$$

Combining the transmitted intensity to the image sensor (30) with the energy efficiency of the image sensor (25) leads to the sensor output signal (I_{sensor}) and the transmitted intensity:

$$I_{sensor}\left(x,y\right) = C\cdot\int_{\lambda_{min}}^{\lambda_{max}} I_{Opt}\left(\lambda;\vec{N},\vec{V},\vec{L};x,y\right)\cdot EE\left(\lambda\right)\,d\lambda. \quad (31)$$

With C being a constant including specific camera parameter like pixel area, fill factor, gain settings and exposure time. λ_{min} and λ_{min} are the wavelength limits of the regarded wavelength range given by the QE of the camera.

If the alignment of illumination, objects in the observed scene and camera is chosen to assure that the vectors $\vec{N}, \vec{V}, \vec{L}$ are approximately constant, (31) simplifies to

$$I_{sensor}\left(x,y\right) \approx \tilde{I}_{sensor}\left(x,y\right) = C\cdot\int_{\lambda_{min}}^{\lambda_{max}} I_{Opt}\left(\lambda;x,y\right)\cdot EE\left(\lambda\right)\,d\lambda. \quad (32)$$

3. Spectral analysis

The spectral analysis is the determination of the spectral behavior of different objects. As written in Section 2.5 the sensor output depends on the lightsource, the reflection behavior of the observed scene, the transmission of the optics and the quantum or energy efficiency of the camera. Spectral characteristics for the components of an acquisition system are often provided by its manufacturers. On the contrary the spectral reflection behavior of objects in the observed scene are almost always unknown. In Section 3.1 a measurement setup for the spectral reflection behavior is described and in Section 3.2 is the measurement procedure explained. In Section 3.3 the procedure for analyzing spectra is introduced.

3.1 Measurement setup

A block diagram of the measurement setup is shown in Fig. 16. The different components will be introduced in the following Sections.

3.1.1 Lightsource

As written in Section 2.1 there exists a high variation of different lightsources. The main aspects for a lightsource within a spectral measurement setup are the amount of intensity, the spectral range and the flatness of the spectrum. Our setup is equipped with a 150 Watt Xenon lightsource (XE) with a spectrum similar to the spectra shown in Fig. 17. The advantages of the XE lightsource is the wide wavelength range (from 250nm to at least 2000nm) and its high intensity. Its disadvantage are narrow and high peaks between 800nm to 1100nm.

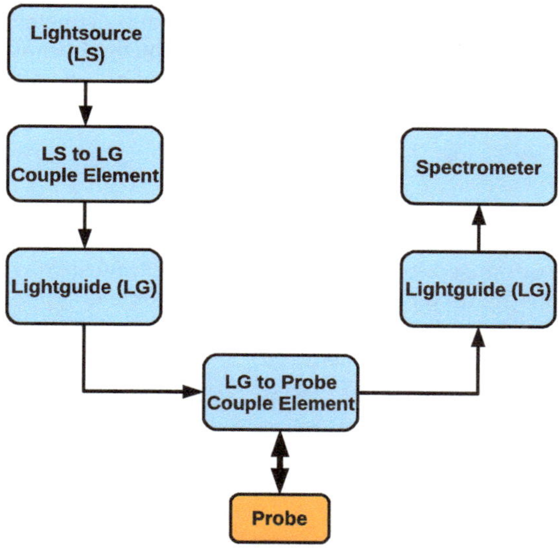

Fig. 16. Blockdiagram of the measurement setup

3.1.2 Lightsource to lightguide coupling element

One important challenge of designing a spectral measurement setup is the coupling of the lightsource to the lightguide without loosing too much light intensity. The output of the XE

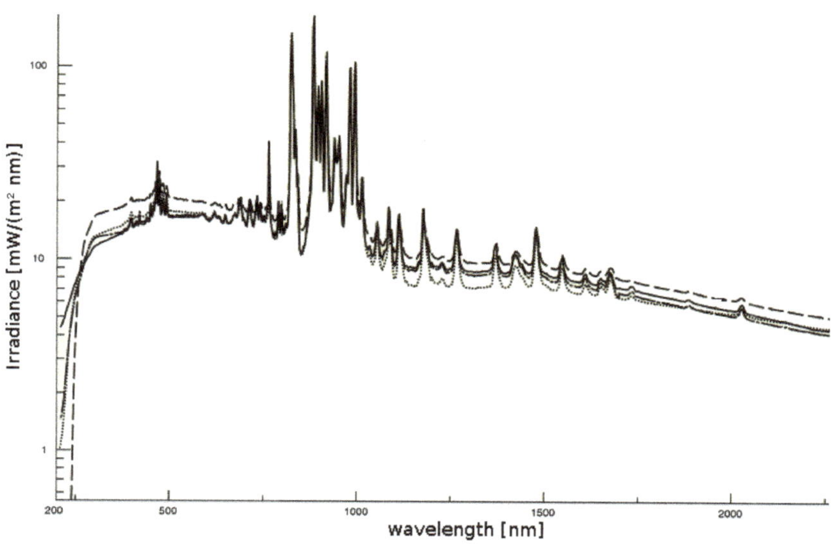

Fig. 17. General spectra of XE lightsources

lightsource is a nearby collimated circular light beam of 8mm diameter and the core diameter of single mode lightguides is only up to 1.5mm. Hence the light beam of the lightsource has to be focussed to the lightguide input. This is accomplished by a parabolic mirror and a three axis adjustment.

3.1.3 Lightguides

Depending on the wavelength range that has to be measured two different types of lightguides are used. One which is optimized for a wavelength range from 200nm to 1100nm and another which is optimized for the wavelength range from 900nm to 2200nm. These two different configurations are used depending on the spectrometer type.

3.1.4 Lightguide to probe coupling element

The light beam transmitted by the lightguide has to illuminate the probe area of interest. In order to measure the diffuse reflection behavior an integrating sphere is used. Advantages of the integrating sphere are the reduction of specular reflection and the angle independent collection of diffuse reflected light. The integrating sphere is coated with a material which has a high reflection (>98%) in the desired wavelength range. While the diffuse reflected light of the probe is reflected several times within the integrating sphere and finally coupled into the output port the specular reflection is absorbed. The probe area is limited to a diameter of 8mm.

3.1.5 Spectrometer

The spectral measurement system can be equipped with two different spectrometers. The first spectrometer includes a charge coupled device sensor which is sensitive from 220nm to 1100nm. It can be used for measuring the spectral reflection behavior in the ultraviolet, visible and near infrared range (NIR). The second spectrometer uses an extended indium gallium arsenide sensor to cover wavelengths from 900nm to 2200nm. This spectrometer can measure NIR and short wave infrared wavelengths.

3.2 Measurement sequence

The spectral measurements are used to calculate the relative reflection coefficients for the different wavelengths. The following procedure has to be accomplished:

1. optimization of the integration time
2. measuring the black and white reference
3. gathering measurement readings of all objects
4. compensation and normalization of the measurement readings.

All of these points are explained in the sections below.

3.2.1 Optimization of the integration time

The integration time of the respective spectrometer has to be chosen with respect to high signal to noise ratio (SNR). In general the SNR increases with higher integration times. The optimal integration time lead to a high reflection signal of the white reference without saturation.

3.2.2 Measuring the black and white reference

The usage of two different reflection standards is required to optimize the resolution and reproduceability of the system. The first standard is a specular black standard which has very less diffuse reflection. The measurement of this standard leads to information of noise influences that are generated by the measurement setup itself. The second standard is the diffuse white reference with a high diffuse reflection. This is used to determine the spectrum of an approximately optimal reflective object including all attenuation given by the measurement system.

3.2.3 Gathering measurement readings

Multiple reflection measurements are gathered for all different objects in the scene. It is important that the measurements are representative for the objects reflection behavior. So the set of measurements has to include all variations given by the object surfaces. Although having objects with low reflection variation the count of total measurements has to be high enough for using the data within a statistical context.

3.2.4 Compensation and normalization of the measurement readings

After taking the measurement readings of the different objects they have to be compensated in terms of degradation caused by the measurement system. This is done by subtraction of all object and white reference measurements R by the spectrum of the black reference R_{black}:

$$R_{comp}(\lambda) = R(\lambda) - R_{black}(\lambda) , \tag{33}$$

R_{comp} is the compensated spectrum. Further the object measurements has to be normalized. They are divided by the spectrum of the compensated white reference $R_{white,comp}$ resulting in the relative reflection spectra for the objects, called normalized spectra R_{norm}:

$$\forall_{\lambda:R_{white,comp}(\lambda)>0} \; R_{norm}(\lambda) = R_{comp}(\lambda) - R_{white,comp}(\lambda) . \tag{34}$$

For all $R_{white,comp}(\lambda) \leq 0$ the normalized spectrum R_{norm} is not valid.

3.3 Image quality with respect to segmentation

As written in Section 1 the main task of spectral optimization is to increase the image quality. This includes the increase of the segmentation ability between different objects. If there exist at least two objects (e.g. foreground and background) the segmentation task is the clustering of regions which contain one single object. In literature there exist several image quality measures driven by the need of compression algorithm benchmarking (see Wang (2002) and Eskicioglu (1995) for details). Due to the fact that we are aiming at the segmentation of an image those measures are not suitable in our case. The following sections introduce two measures for image quality with respect to segmentation.

3.3.1 Signal to noise measure

The first measure is based on standard signal to noise measure SNR. The assumption for the two signals s_1, s_2 to be separated is that their values are outcomes of normal distributions. The signal to noise ratio includes the distance between the mean of two signals μ_1, μ_2 and the variance of both σ_1^2, σ_2^2:

$$SNR(s_1, s_2) = \frac{|\mu_1 - \mu_2|}{\sigma_1^2 + \sigma_2^2} . \tag{35}$$

In our case the signals include intensity information and hence are energy based signals. It is also common to use the decibel notation:

$$SNR_{dB}(s_1, s_2) = 20 \log \left(\frac{|\mu_1 - \mu_2|}{\sigma_1^2 + \sigma_2^2} \right) . \tag{36}$$

The maximum of the SNR will lead to a maximum of image quality.

3.3.2 Classification error estimation

Another measure for the image quality with respect to segmentation is the classification error including information of the rate between pixels assigned to the wrong object and pixels assigned to the correct object. The estimation of this error requires the probability density for the respective feature (e.g. greyvalue) of the different objects. The total error ε_{tot} for two objects ω_1, ω_2 with their respective probability density function p_1, p_2 can be determined by:

$$\varepsilon_{tot}(p_1, p_2) = \frac{1}{2} \left[\int\limits_{\mathcal{R}_{\omega_1}} p_2(x)\, dx + \int\limits_{\mathcal{R}_{\omega_2}} p_1(x)\, dx \right] . \tag{37}$$

The first integration term is in the range $\mathcal{R}_{\omega_1}$ where the classification decides for ω_1 and for ω_2 in the second integration term. Due to the fact that p_1 and p_2 are unknown they have to be estimated. Two different approaches that are implemented in our analysis software are introduced briefly in the following paragraphs. For further information see (Duda (2001), Scott (1992), Wand (1995)).

The classification error is negative correlated to the image quality.

3.3.2.1 Parametric density estimation

The parametric density estimation assumes that the probability density can be described by a certain number of parameters. The most commonly used parametric description of a density is the unimodal Gaussian distribution $\mathcal{N}$ with its parameters mean μ and variance σ^2. Given N samples X_i the expected mean value can be determined by

$$\mu = \frac{\sum\limits_{i=1}^{N} X_i}{N} , \tag{38}$$

and the expected unbiased variance value by

$$\sigma^2 = \frac{1}{N-1} \cdot \sum_{i=1}^{N} (X_i - \mu)^2 . \tag{39}$$

Using (38) and (39) the probability density estimation can be expressed as:

$$\tilde{p}(x) = \frac{1}{\sqrt{2\pi\sigma^2}} \cdot e^{-0.5\left(\frac{x-\mu}{\sigma}\right)^2} . \tag{40}$$

3.3.2.2 Nonparametric density estimation

As written in Section 3.3.2.1 the parametric density estimation is only suitable for unimodal Gaussian distributed signals. To overcome this another approach of density estimation, the nonparametric density estimation, was developed. There exist several types of nonparametric density estimators like histogram, polygonal and kernel density estimators. This section focusses on the kernel density estimators also known as kernel smoothing. The principle of kernel smoothing is a convolution of the sample values with a kernel:

$$\tilde{p}\left(x;h\right)=\frac{1}{N\cdot h}\sum_{i=1}^{N}K\left(\frac{x-X_i}{h}\right). \tag{41}$$

This kernel is configured by the type K and bandwidth h. While the type has limited influence on the estimation the bandwidth plays an important role. Different selection methods are common: normal scale bandwidth, oversmoothed bandwidth, least square cross validation bandwidth and plug-in bandwidth. We focus on the least square cross validation (LSCV) bandwidth h_{LSCV} while the LSCV is defined as (Wand (1995))

$$\text{LSCV}\left(h\right)=\int\tilde{p}\left(x;h\right)^2\,dx-\frac{2}{n}\sum_{i=1}^{N}\tilde{p}_{-i}\left(X_i;h\right), \tag{42}$$

$\tilde{p}_{-i}$ is the density estimate based on the sample set without X_i. The LSCV bandwidth is found by

$$h_{\text{LSCV}}=\arg\min_{h}\text{LSCV}\left(h\right). \tag{43}$$

4. Experimental results

4.1 Simulated data

In this section the spectral analysis is used on simulated data. These contain the simulated behavior of lightsources, objects, optics and cameras. In the next Sections the spectral analysis and the simulated image data using an optimized and a non-optimized configuration are given.

4.1.1 Spectral analysis

The simulated data of the different hardware components and objects are shown in Fig. 18. We simulated two different lightsources (daylight and an LED), two different objects represented by three spectra each, two optics optimized for different wavelength and two cameras with the maximum energy efficiency at 500nm and 670nm respectively. This lead to eight different configurations. The image sensor signals were calculated according to (32). The parameters describing the intensity value (mean and variance) are determined and the classification errors are estimated for each object as shown in Table 2. It is obvious that the selection of the lightsource plays an important role. The classification error is for the LED approximately constant at 0.03%, while for daylight the error varies from 2.72% to 36.3%. This high variation is impressive regarding the small differences between the spectra of different optics and camera types. As a result the 8th configuration can be regarded as optimal setup while the 5th configuration as worst setup.

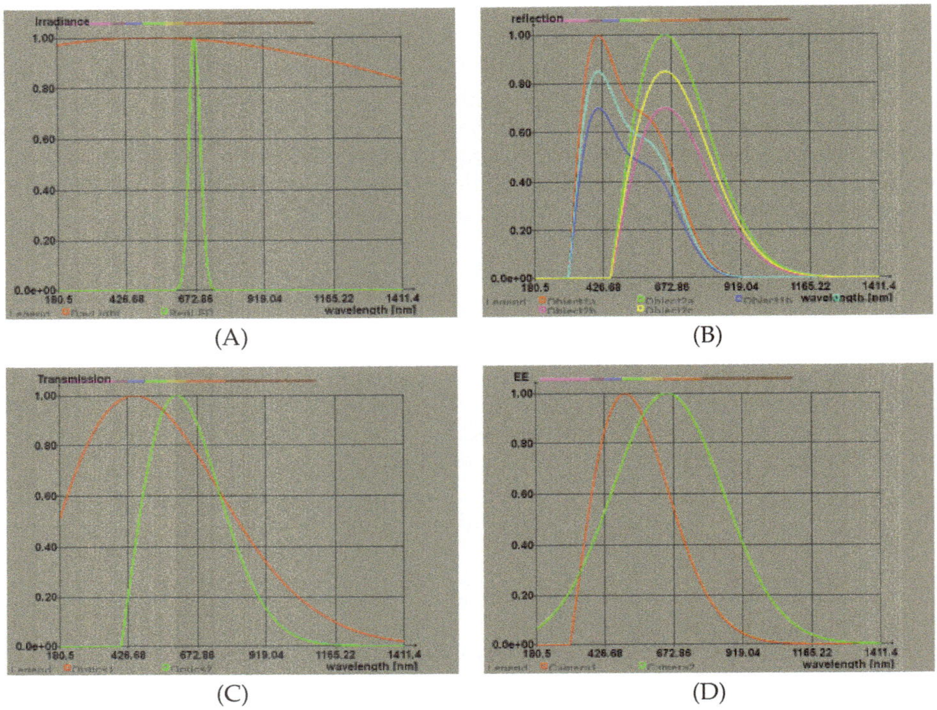

Fig. 18. Merged spectra of two spectra describing different hardware components: lightsources (A), objects (B), optics (C), and cameras (D)

4.1.2 Simulated image data

Using the estimated greyvalue distributions of Section 4.1.1 two images are calculated and shown in Fig. 19. It is obvious that the 8th hardware configuration results in higher contrast image (Fig. 19 (A)) than the 5th hardware configuration (Fig. 19 (B)). The SNR according to (36) is 5dB for the optimized setup and −50dB for the worst setup.

	Configuration			
Id	Lightsource	Optics	Camera	ϵ_{tot}
1	Daylight	1	1	2.86
2	LED	1	1	0.03
3	Daylight	2	1	30.7
4	LED	2	1	0.04
5	Daylight	1	2	36.3
6	LED	1	2	0.03
7	Daylight	2	2	2.72
8	LED	2	2	0.03

Table 2. Classification error for different hardware configurations

(A) (B)

Fig. 19. Simulated images for optimal configuration (A) and non-optimal configuration (B)

4.2 Real data

In this section we will show experimental results using spectral optimization of the lightsource only. The objects used within this experiment are shown in Fig. 20 using daylight illumination in combination with a color camera. First the results of the spectral analysis are provided and then the results are verified by using a real hardware setup optimized by these results and a non-optimized setup are compared in terms of image quality for segmentation as described in Section 3.3.

4.2.1 Spectral analysis

The spectral optimization is aiming at the separation of paperboard A from paperboard B and C. As explained in Section 3.2 we take several spectra (20 for each paperboard) and calculate the normalized spectra for each object. These spectra are shown in merged spectra diagrams in Fig. 21. All diagrams show similar reflection behavior with just small variances within one object as shown in Fig. 21 (A), (B), and (C). The diagram showing the mean spectra in Fig. 21 (D) illustrates only little differences between A and B, C in wavelength ranges from 350nm to 450nm and from 550nm to 600nm. This is confirmed by the classification error estimated parametric (see Fig. 22 (A)) and nonparametric (see Fig. 22 (B)) which show small classification error rates for those wavelength ranges. Low classification error rates indicate that the hardware component selection optimized for this wavelength ranges would be appropriate in order to achieve high image quality for segmentation.

Fig. 20. Image of different paperboards with daylight illumination: paperboard A (right), paperboard B (middle) and paperboard C (left)

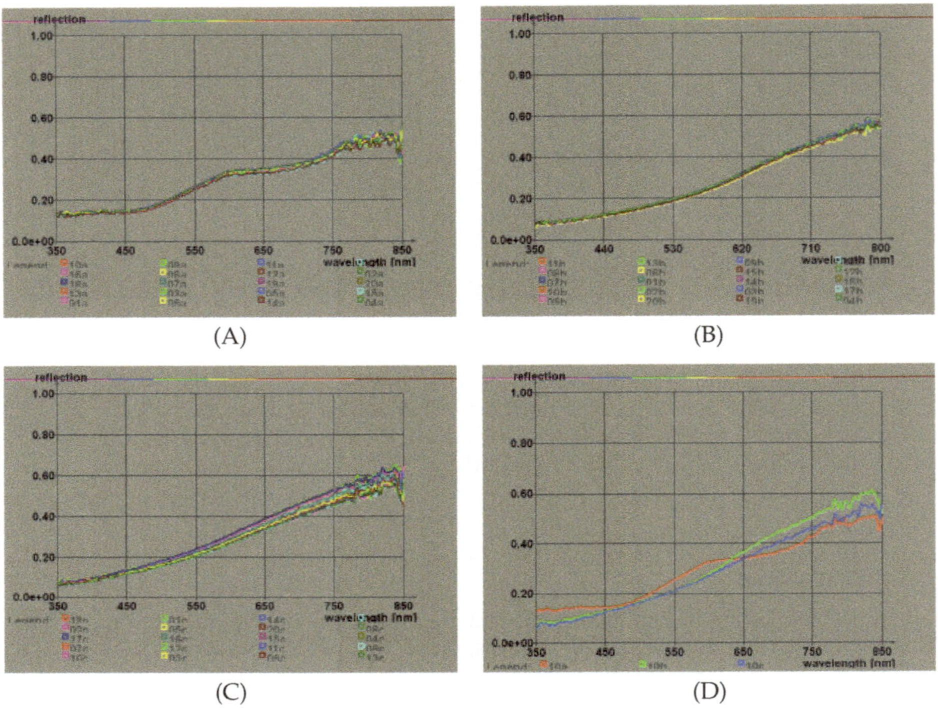

Fig. 21. Merged spectra of paperboard A (A), B (B) and C (C). (D) shows the mean spectra of each paperboard.

4.2.2 Hardware optimization

According to the results of the spectral analysis we choose two different illumination configurations. The first uses the amber LED with a spectrum shown in Fig. 7 (red curve) which is expected to result in low classification error rates and hence to high SNR. The other configuration uses the red LED with a spectrum shown in Fig. 7 (yellow curve) which maximum is located at a high classification error wavelength. Further we use the same camera

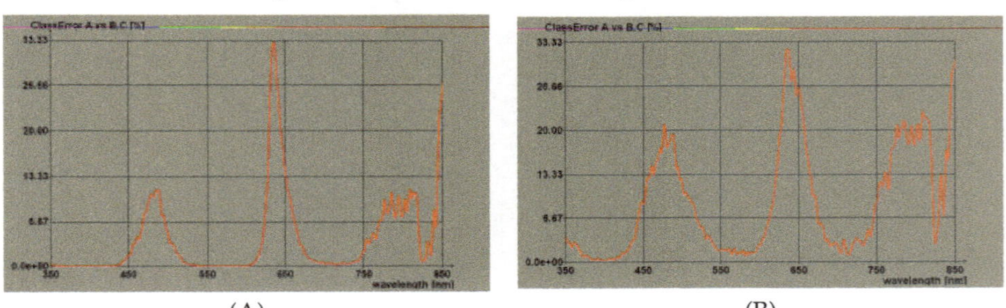

Fig. 22. Classification error of paperboard A versus B and C for parametric density estimation (A) and nonparametric density estimation (b)

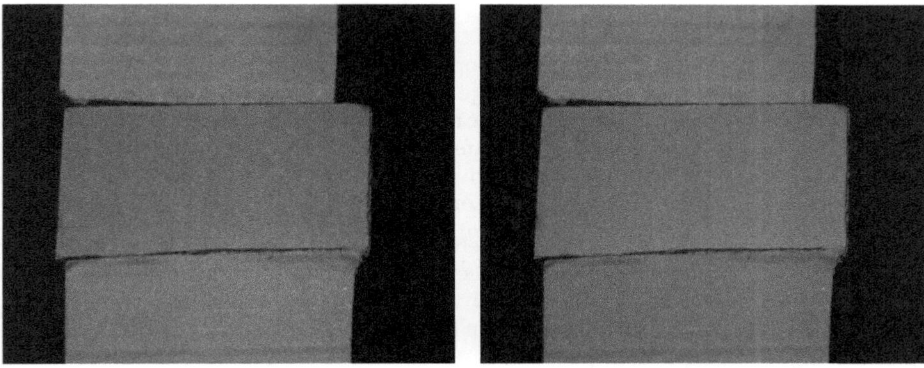

Fig. 23. Images using optimized hardware setup (A) and non-optimized hardware setup (B) showing paperboard A (bottom), B (middle), and C (top)

and optics for both configurations, focussing on optimization of the lightsource only. The resulting monochrome images are shown in Fig. 23 with the upper paperboard is class C, the middle class B and the lower class A. The images look similar but class B appears darker and class A brighter in the optimized setup (Fig. 23 (A)) than in the non-optimized setup (Fig. 23 (B)). In order to measure the image quality we calculate the SNR between paperboard A and the others. The parameters needed are calculated on regions containing the respective object using (38) and (39). The results for parameter estimations are given in Table 4. With these parameters three SNRs are calculated, expressing the separability between paperboard A and B, A and C, and A and B, C together. The values are shown in Table 3. While the SNR between A and C does not increase, the other two SNRs increase significantly using the optimized setup. With the non-optimized setup the SNR between paperboard A and B is very low with -30dB increasing with the optimized setup to -15dB. The overall SNR between A and both other paperboards increases from -26dB to -18dB. It can be seen that the results of the spectral analysis correlate with the image quality of the real images.

Objects	$SNR_{dB,opt}$ [dB]	$SNR_{dB,nonopt}$ [dB]
A vs B	-15	-30
A vs C	-21	-21
A vs B,C	-18	-26

Table 3. SNR of the different configurations

	Optimized		**Non-Optimized**	
Object	μ	σ^2	μ	σ^2
A	127.0	42.3	113.2	37.2
B	111.4	41.0	115.3	34.8
C	119.3	47.6	120.5	44.9

Table 4. Statistical parameters for different hardware configurations

5. Conclusion

In this Chapter we introduced a method for optimization of hardware component selection based on spectral analysis with respect to image quality. The complete image acquisition chain including lightsource, observed scene, optics, camera and the influence on the resulting image was illustrated. The spectral behavior for different types of all components were provided. Models of the spectral characteristics were derived. This lead to an image output model which describes the sensor output signal depending on the components.

Furtheron a spectral analysis measurement setup and its usage was explained focussing on the diffuse reflection behavior of different objects. Two measures for image quality according to segmentation were introduced afterwards: the first is based on signal to noise ratio and the second uses the estimation of classification error by means of analyzing different density estimators.

Finally experimental results for simulated and real data are provided. The first contains eight different setups of simulated components including lightsources, objects, optics and cameras. All combinations of these components were evaluated and resulted in a big variation of image quality as shown by quality measures and simulated images. For the experiments with real data three different types of paperboards with similar appearances were analyzed and an optimal configuration in terms of lightsources was compared with a non-optimal configuration. The spectral optimization increases image quality significantly. Both experiments proved the advantages of hardware component selection due to its spectral characteristics and the spectral behavior of objects in the observed scene.

6. Acknowledgement

This work is part of a project which is supported by the European Union, the European Funds for Regional Development (EFRE), the German Federation and the Federal Land Schleswig-Holstein.

7. References

Basler A601f, A602f, and A622f - Small, Powerful, Easy to Use, http://www.baslerweb.com/downloads/11684/6S_A601f_602f_622f_151208.pdf, August -2011

Duda, R.O.; Hart, P.E.; Stork, D.G (2001). *Pattern classification*, Wiley New York, 2001

Duff, T. (1979). Smoothly shaded renderings of polyhedral objects on raster displays, *Proceedings of the 6th annual conference on Computer graphics and interactive techniques*, pp. 270–275, 1979, ACM

Eriksson, M. (2006). *A light manager for a dynamic light environment*, Sweden

Eskicioglu, A.M. & Fisher, P.S. (1995). Image quality measures and their performance. *Communications, IEEE Transactions on*, 43, 12, 1995 2959–2965

He, X.D.; Torrance, K.E.; Sillion, F.X.; Greenberg, D.P.. A comprehensive physical model for light reflection. *ACM SIGGRAPH Computer Graphics*, 25, 4, 1991 175–186

Goodrich (2006), Application Note: What is InGaAs, http://www.sensorsinc.com/downloads/notes_GaAs.pdf, December -2006

Jenoptik (2007), VIS - NIR color corrected hyperspectral lens introduced for VisGaAs sensor, http://www.jenoptik-inc.com/literature/doc_download/6-3-hyperspectral-lens.html, March -2007

Jenoptik (2011),UV CoastalOpt SLR Lens, http://www.jenoptik.com/cms/jenoptik.nsf/res/CoastalOpt%20UV%20105_V2.pdf, August -2011
Kajiya, J.T. & Kay, T.L. (1998). Rendering fur with three dimensional textures. *ACM Siggraph Computer Graphics*, 23, 3, 1989 271–280
Lambert, J.H. (1760). *Colorum Umbrae*, Eberhard Klett, 1760
Lewis, R.R. (1994). Making shaders more physically plausible. *Computer Graphics Forum*, 13, 2, 1994 109–120
Newport (2011), Optical Materials, http://www.newport.com/Optical-Materials/144943/1033/content.aspx, August -2011
Oren, M. & Nayar, S.K. (1994). Generalization of Lambert's reflectance model, In: *Proceedings of the 21st annual conference on Computer graphics and interactive techniques*, 239–246, ACM
Philips (2011), Luxeon Rebell General Purpose Technical Datasheet DS64, http://www.philipslumileds.com/uploads/28/DS64-pdf, April -2011
Phong, B.T. (1975). Illumination for computer generated pictures. *Communications of the ACM*, 18, 6, 1974 311–317
Rechtsteiner , G.A. & Ganske, J.A. (1998). Using natural and artificial light sources to illustrate quantum mechanical concepts. *The Chemical Educator*, Vol3 , No. 4, 1998, 1–12
Richards, A. (2006). UV imaging opens new applications, http://www.vision-systems.com/articles/print/volume-11/issue-7/features/component- integration/ uv-imaging-opens-new-applications.html ,July -2006
Schneider (2011), CINEGON 1.4/8MM COMPACT - Relative Illumination, Distortion, Transmittance - graphs and data (PDF), http://www.schneiderkreuznach.com/industrieoptik_e/kompakt/pdf/CNG_KOMP_14_8 _ R46045_obj.pdf, August -2011
Schneider (2011), VARIOGON 1.8/12.5-75MM - Relative Illumination, Distortion, Transmittance - graphs and data (PDF), http://www.1stvision.com/ lens/schneider/ dataman/VGN_18_125_75_R34030_1CIE.pdf, August -2011
Scott, D.W. (1992). *Multivariate density estimation*, Wiley Online Library, 1992
Wald, G. (1945). Human Vision and the Spectrum. *Science*, Vol101 , No. 2635, June -1945, 653-658
Wand, M.P. & Jones, M.C. (1995). *Kernel smoothing*, Chapman & Hall/CRC, 1995
Wang, Z. & Bovik, A.C. (2002). A universal image quality index. *Signal Processing Letters, IEEE*, 9, 3, 2002 81–84
Wikipedia (2007), http://upload.wikimedia.org/wikipedia/commons/4/4c/Solar_Spectrum.png, August -2007
Wikipedia (2010), http://upload.wikimedia.org/wikipedia/commons/c/c4/Rendered_Spectrum.png, January -2010
Wikipedia (2009), http://upload.wikimedia.org/wikipedia/commons/thumb/a/a2/Wiens_law.svg/300px- Wiens_law.svg.png, April -2009
Wikipedia (2011), http://upload.wikimedia.org/wikipedia/commons/7/78/Spectra-Philips_32T8_natural_sunshine_fluorescent_light.svg, Mai -2011

Permissions

The contributors of this book come from diverse backgrounds, making this book a truly international effort. This book will bring forth new frontiers with its revolutionizing research information and detailed analysis of the nascent developments around the world.

We would like to thank Prof. Mehmet Savsar, for lending his expertise to make the book truly unique. He has played a crucial role in the development of this book. Without his invaluable contribution this book wouldn't have been possible. He has made vital efforts to compile up to date information on the varied aspects of this subject to make this book a valuable addition to the collection of many professionals and students.

This book was conceptualized with the vision of imparting up-to-date information and advanced data in this field. To ensure the same, a matchless editorial board was set up. Every individual on the board went through rigorous rounds of assessment to prove their worth. After which they invested a large part of their time researching and compiling the most relevant data for our readers. Conferences and sessions were held from time to time between the editorial board and the contributing authors to present the data in the most comprehensible form. The editorial team has worked tirelessly to provide valuable and valid information to help people across the globe.

Every chapter published in this book has been scrutinized by our experts. Their significance has been extensively debated. The topics covered herein carry significant findings which will fuel the growth of the discipline. They may even be implemented as practical applications or may be referred to as a beginning point for another development. Chapters in this book were first published by InTech; hereby published with permission under the Creative Commons Attribution License or equivalent.

The editorial board has been involved in producing this book since its inception. They have spent rigorous hours researching and exploring the diverse topics which have resulted in the successful publishing of this book. They have passed on their knowledge of decades through this book. To expedite this challenging task, the publisher supported the team at every step. A small team of assistant editors was also appointed to further simplify the editing procedure and attain best results for the readers.

Our editorial team has been hand-picked from every corner of the world. Their multi-ethnicity adds dynamic inputs to the discussions which result in innovative outcomes. These outcomes are then further discussed with the researchers and contributors who give their valuable feedback and opinion regarding the same. The feedback is then collaborated with the researches and they are edited in a comprehensive manner to aid the understanding of the subject.

Apart from the editorial board, the designing team has also invested a significant amount of their time in understanding the subject and creating the most relevant covers. They scrutinized every image to scout for the most suitable representation of the subject and create an appropriate cover for the book.

The publishing team has been involved in this book since its early stages. They were actively engaged in every process, be it collecting the data, connecting with the contributors or procuring relevant information. The team has been an ardent support to the editorial, designing and production team. Their endless efforts to recruit the best for this project, has resulted in the accomplishment of this book. They are a veteran in the field of academics and their pool of knowledge is as vast as their experience in printing. Their expertise and guidance has proved useful at every step. Their uncompromising quality standards have made this book an exceptional effort. Their encouragement from time to time has been an inspiration for everyone.

The publisher and the editorial board hope that this book will prove to be a valuable piece of knowledge for researchers, students, practitioners and scholars across the globe.

List of Contributors

Elena Condrea, Anca Cristina Stanciu and Kamer Ainur Aivaz
"Ovidius" University of Constanta, Romania

Cathy Balding
Qualityworks P/L and La Trobe University, Australia

Yasuo Iwaki
Chaos Applied Research Office, Japan

Kozo Koura
Kozo Koura & Associates, Japan

Alberto Amaral
CIPES - Centre for Research in Higher Education Policies, Portugal
A3ES - Agência para a Avaliação e Acreditação do Ensino Superior, Portugal

Maria J. Rosa
CIPES - Centre for Research in Higher Education Policies, Portugal
DEGEI, Universidade de Aveiro, Portugal

Cláudia S. Sarrico
CIPES - Centre for Research in Higher Education Policies, Portugal
ISEG, Universidade Técnica de Lisboa, Portugal

P. Ipsilantis, I. Papadopoulos and D. Kantas
TEI of Larissa, Greece

P. Trivellas
Technological Education Institute (TEI) of Chalkis, Greece

Yuji Okita
Kanazawa Institute of Technology, Japan

Luis Ahumada, Carmen Montecinos and Alvaro González
Pontificia Universidad Católica de Valparaíso, School of Psychology, Chile

Yuan Li
Liaoning University of Traditional Chinese Medicine, China

Houyi Zhu
The University of Manchester, United Kingdom

Juha Kettunen
Turku University of Applied Sciences, Turku, Finland

Beverley Glass
James Cook University, Australia

Alison Haywood
Griffith University, Australia

Cecilia Latrach, Naldy Febré and Ingrid Demandes
Universidad Mayor, Chile

Athanassios Vozikis and Marina Riga
University of Piraeus, Athens, Greece

Víctor Reyes-Alcázar, Antonio Torres-Olivera, Diego Núñez-García and Antonio Almuedo-Paz
Andalusian Agency for Health Care Quality, Spain

Peter Liggesmeyer, Henning Barthel, Achim Ebert, Jens Heidrich, Patric Keller, Yi Yang and Axel Wickenkamp
Fraunhofer IESE / University of Kaiserslautern, Germany

Antonio Almuedo-Paz, Diego Núñez-García, Víctor Reyes-Alcázar and Antonio Torres-Olivera
Andalusian Agency for Health Care Quality, Spain

Vesa Hasu and Heikki Koivo
Aalto University, Department of Automation and Systems Technology, Finland

Dietmar Winkler and Stefan Biffl
Christian Doppler Laboratory, "Software Engineering Integration for Flexible Automation Systems" (CDL-Flex), Institute of Software Technology and Interactive Systems, Vienna University of Technology, Austria

Renan Prasta Jenie
Information Technology Directorate, Bina Nusantara University, Indonesia

K. Ohliger, C. Heinze and C. Kröhnert
Westcoast University, Heide, Germany
Institute Machine Vision Technology, Germany